T0393075

Diagnostics and Gene Therapy for Human Genetic Disorders

Diagnostics and Gene Therapy for Human Genetic Disorders provides an integrative and comprehensive source of information blending classical human genetics with the human genome. It provides a multidisciplinary overview of Mendelian inheritance and multifactorial inheritance, genetic variations, polymorphisms, chromosomal, multifactorial, and mitochondrial disorders.

PCR, electrophoresis, cytogenetics, prenatal, and HPLC based techniques applied for diagnosing genetic disorders are discussed with applications. Symptoms, etiology, diagnosis, treatment of 14 major and 5 minor genetic disorders are discussed in detail. Methods employed for the preparation of kits for the diagnosis of diseases are provided. The role of gene therapy in the amelioration of genetic disorders and the methodology employed are discussed. The success of gene therapy in controlling various disorders such as immune system disorders, neurodegenerative disorders, cardiovascular disorders, eye diseases, and cancer has been described along with type studies.

Features:

1. A blend of classical human genetics with molecular and genome-based applications
2. Techniques applied for the diagnosis of genetic disorders
3. Diagnostics of 19 genetic disorders including symptoms, etiology, diagnosis, and treatment
4. Role of gene therapy in the amelioration of disorders
5. Type studies describing the role of diagnostics in conserving human health

This book attempts to connect the information about classical and modern human genetics, genetic disorders, and gene therapy to all types of diseases in one place. This work provides a comprehensive source of information that can serve as a reference book for scientific investigations and as a textbook for graduate students.

Diagnostics and Gene Therapy for Human Genetic Disorders

K.V. Chaitanya

CRC Press
Taylor & Francis Group
Boca Raton London New York

CRC Press is an imprint of the
Taylor & Francis Group, an **informa** business

First edition published 2023
by CRC Press
6000 Broken Sound Parkway NW, Suite 300, Boca Raton, FL 33487-2742

and by CRC Press
4 Park Square, Milton Park, Abingdon, Oxon, OX14 4RN

CRC Press is an imprint of Taylor & Francis Group, LLC

Library of Congress Cataloging-in-Publication Data
Names: Chaitanya, K. V., author.
Title: Diagnostics and gene therapy for human genetic disorders / K.V. Chaitanya.
Description: First edition. | Boca Raton : CRC Press, 2023. |
Includes bibliographical references and index.
Identifiers: LCCN 2022035247 (print) | LCCN 2022035248 (ebook) |
ISBN 9781032381640 (hardback) | ISBN 9781032381657 (paperback) |
ISBN 9781003343790 (ebook)
Subjects: LCSH: Genetic disorders. | Genetic disorders–Diagnosis. |
Gene therapy. | Medical genetics.
Classification: LCC RB155.5 .C425 2023 (print) |
LCC RB155.5 (ebook) | DDC 616/.042–dc23/eng/20221021
LC record available at https://lccn.loc.gov/2022035247
LC ebook record available at https://lccn.loc.gov/2022035248

ISBN: 9781032381640 (hbk)
ISBN: 9781032381657 (pbk)
ISBN: 9781003343790 (ebk)

DOI: 10.1201/9781003343790

Typeset in Times
by Newgen Publishing UK

Contents

Preface

Genetic material DNA is packed in the nucleus of every living as condensed chromatin. In humans, DNA is further condensed into 46 rod-shaped chromosomes during cell division, with each chromosome having a definite size, shape, and a specific number for its identification. There have been great advances in the field of genetics over the last 25 years. Mendelian genetics, pedigree analysis, linkage analysis, and allele frequency estimations have been identified to predict the chances of inheriting a disease, which has arisen due to mutations or lack of gene expression or differences in the expression of genes within the parents. Epigenetic inheritance and multifactorial inheritance have further explained the significance of the environment in the expression of a gene. Completion of the human genome project has provided us with significant technical advances in DNA analysis. As a result, we can comfortably diagnose a genetic disorder, identify the gene(s) responsible for it, and check for the possibility of eradicating the disease from that human being through gene therapy. With more than 6000 genetic disorders having been identified in humans, 600 disorders are currently being treated.

Diagnostics and Gene Therapy for Human Genetic Disorders provides a blend of classical human genetics and advanced molecular genetics. The latest happenings in monogenic traits, linkage disequilibrium in the human genome, Mendelian inheritance, and multifactorial inheritance offer a complete pedigree chart. Genetic variations, polymorphisms, defects in chromosomes, changes in the genes of nuclear and mitochondrial genomes responsible for chromosomal, multifactorial, and mitochondrial disorders give a comprehensive and detailed understanding of genetics. HPLC, PCR, electrophoresis, cytogenetics, and prenatal-based techniques applied to identify changes in the chromosomes and the DNA will connect the classroom to the laboratory. The significance of diagnostic kits and methods involved in their preparation will also increase the scope of disease diagnosis. Symptoms, etiology, diagnosis, and treatment of 14 major genetic disorders will have a better insight of knowledge into genetics. The role of gene therapy, genome editing, RNAi technology in the amelioration of genetic disorders will help in providing a better understanding of the gene therapy tools and their applications. Applications of gene therapy in controlling various disorders such as immune system disorders, neurodegenerative disorders, cardiovascular disorders, eye diseases, and cancer will further empathize its role in the treatment of human diseases.

Diagnostics and Gene Therapy for Human Genetic Disorders provides the most up to date information, which threads classical human genetics with variations in the genes, polymorphisms, genetic disorders, disease diagnosis and gene therapy, genome editing applications, making human health highly susceptible. Through this work, human classical and modern genetics, genetic disorders, and gene therapy for all types of disorders are provided in one place, which provides a piece of comprehensive information for the reader.

Acknowledgments

I would like to express my deep gratitude to Professor T. Sekhar, Dr. Nageswara Rao Reddy, Dr. Utpal Nath, and Professor P.B. Kavi Kishor for their support. I am thankful to my wife Lalita, my children Abhiram and Aamukta Malyada for their patience. My acknowledgments to genomics and molecular diagnostics students for the discussions that helped me take the topics collectively and comprehensively. Finally, I express my deep sense of gratitude to all authors whose works have been consulted in the writing of this book. I would appreciate any valuable suggestions for further improvement of this work.

K.V. CHAITANYA

Author Bio

K.V. Chaitanya is Professor in the Microbiology and Food Science Technology Department, GITAM University, Visakhapatnam, India. Prof. Chaitanya has a Ph.D. in Life Sciences from Pondicherry University, Pondicherry. Prof. Chaitanya received a postdoctoral fellowship from the Department of Biotechnology, Govt of India, to pursue research at the Indian Institute of Science, Bengaluru, India. He has over 18 years of experience in research and academics in genomics and molecular biology. He has worked in various capacities in academic and research institutes of international reputation and has published 60 research articles. He has authored two books on cell and molecular biology and genomics and has filed five patents to protect the copyright of his experimental products. Prof. Chaitanya has academic awards and fellowships to his credit.

1 Genetic Analysis

1.1 INTRODUCTION

Genetic analysis is the name ascribed to the overall understanding of a process by applying the concepts and thoughts related to classical Mendelian genetics and modern molecular biology. Identifying the genes and genetic disorders inherited by these genes forms the principal basis of genetic analysis. This subject is under study since prehistoric times when early human beings practiced selective breeding to improve the productivity of plants and animals. The scientific basis of genetic analysis has come to light through Mendel's laws. Gregor Mendel was the first to utilize genetic analysis to identify the traits inherited by offspring from parents. These traits are controlled by hereditary units called genes, present in every living cell of an organism (both prokaryotic and eukaryotic). Mendel's work has contributed to the development of hybrid plants, identifying genetically inherited disorders, and differential diagnoses of diseases.

Genetic analysis through modern biology is achieved using cytogenetic analysis such as karyotyping, fluorescence in situ hybridization (FISH), and molecular techniques such as PCR, DNA sequencing technology, and microarrays (Espen et al. 2018). These techniques have helped in the identification of mutations, and changes in the copy numbers. In addition, the development of reverse genetics has helped in the detection of codons and genetic code. Genetic linkage studies have analyzed the spatial arrangement of genes on chromosomes.

1.2 MONOGENIC TRAITS

A trait is a distinct variant of an organism's phenotypic character, which can be inherited or environmentally determined, or a combination of both. For example, eye color is a character or abstraction of an attribute, while blue, brown, and hazel are traits. The human genetic traits are classified into simple monogenic and complex polygenic traits.

The monogenic trait is encoded by a single gene, which means a single gene determines a specific trait. The wild-type allele is normal and healthy whereas the mutated or diseased allele gives a particular disease to the phenotype. Monogenic disorders occur due to mutations in a single gene, which minimizes the function and stability of the corresponding protein by altering its three-dimensional structure (Andreas, 1999). These mutations include point mutations (changes in the single nucleotides that change the amino acid sequence), insertions, deletions in the DNA sequence that encodes the protein, and changes in the non-coding DNA, interfering with the gene splicing. Thus, monogenic traits are strongly influenced by variation within a single gene, recognized by their classic patterns of inheritance within families. As monogenic traits form the primary basis for traditional genetics, it has become clear that conditions whose legacies strictly confirm the Mendelian principles are relatively rare.

DOI: 10.1201/9781003343790-1

Multiple different genes encode a polygenic trait. Each of them might have several possible different alleles, whose combinations can give rise to a range of possible phenotypes. For example, at least three genes control human eye color. Each gene has several different alleles, resulting in many different possible eye colours such as brown, hazel, green, blue, grey, purple, and amber. Complex traits arise from variations within multiple genes and their interactions with behavioral and environmental factors. These traits do not readily follow the predictable patterns of inheritance. There is a significant distinction between a monogenic and polygenic trait, which is overly simplistic. Monogenic traits are also influenced by variation in multiple genes called modifier genes. On the other hand, variations in a single gene can predominantly influence complex traits.

The evolution of single-gene traits as versions of complex traits has been illustrated by: (1) widely different phenotypes accounting for allelic variations in a single gene; (2) the blurring of predicted relationships between genotype and phenotype in several monogenic disorders; and (3) modifier genes and non-genetic factors contributing to the phenotypes of monogenic disorders. The complexity of the relationships between genotype and phenotype, leading to three different monogenic disorders, cystic fibrosis, Hartnup phenotype, and the PiZZ form of α-1 antitrypsin deficiency, has been demonstrated in Figure 1.1. Cystic fibrosis (MIM 219700) occurs due to both allelic variation at the CFTR locus and the expression of a key modifier locus, explaining the difference between the pancreatic-sufficient and insufficient forms of the disease. The Hartnup phenotype (MIM 234500), in which a mutation at the major locus accounts for amino acid transport disorder in the kidneys and intestines. There is also evidence for a multifactorial threshold phenomenon affecting the overall homeostasis of plasma amino acid levels. The latter accounts for the difference between a patient having only the variant Hartnup amino acid transport phenotype (a biochemical trait) and having Hartnup disease, a pellagra-like clinical entity. The PiZZ form of α-1 antitrypsin deficiency, where some of the variations in the clinical phenotype involving the lungs, is explained by environmental factors such as smoking, whereas inter-individual differences involving the liver are related to the handling of mutant protein by the chaperones and proteases.

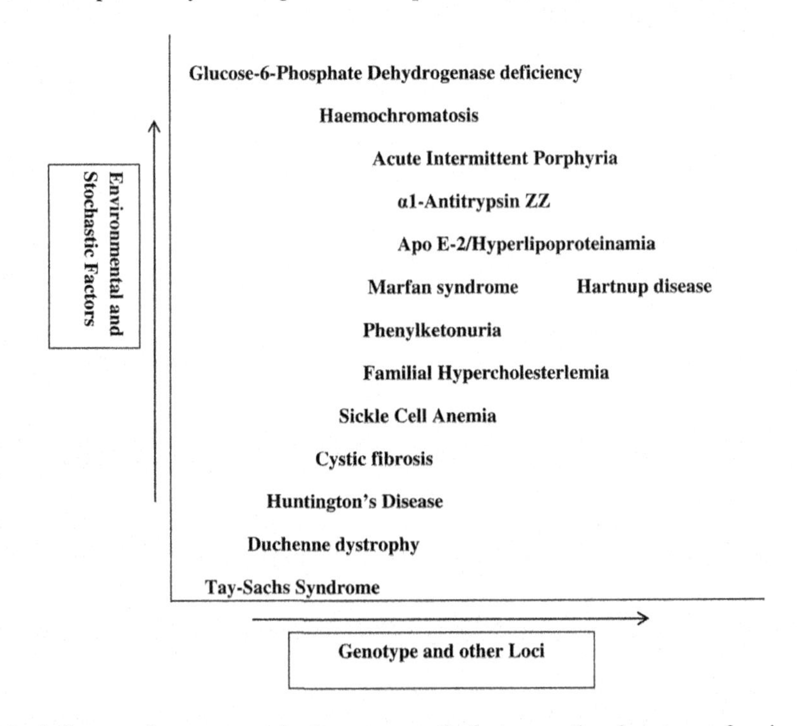

FIGURE 1.1 Influence of genotype and other non-genetic factors on the phenotype of various monogenic disorders.

The above figure indicates the estimated relative importance of background genotype and environment as the major contributors to the phenotype of several monogenic diseases. The equation $VP = V_G + V_E$ implies that variations in genotype and environment contribute to the variation in phenotype. A detailed classification of monogenic traits and their disorders is provided in Chapter 3.

1.3 LINKAGE ANALYSIS

1.3.1 Linkage

Two genetic loci are considered to be in linkage if the alleles at these loci segregate together more often. These two loci locations are so close on the same chromosome that the chances of their separation by a crossover event (also denoted as a recombination) during meiosis are remote. The probability of inheritance of any two alleles at two randomly selected loci together is 0.5. If two loci are closely linked, then the chance of crossover or recombination occurrence is <0.5. The probability of recombination taking place is linked to the distance between any two loci. The recombination fraction [θ] is a measure of the genetic distance between two loci. The distance between two loci is measured in centimorgans. One centimorgan is the genetic distance between two loci with a recombination frequency of 1%. Although the centimorgan is not an exact measure of physical distance, it typically equates to a physical space of one million base pairs. So, two loci close to the *F8* gene with a 5% probability of recombination would be five centimorgans apart: approximately 5 million base pairs. Linkage analysis aims to identify the appropriate markers that co-segregate with the gene of interest, tracking the gene within a family without knowing the mutation (Jurg et al. 2015). By definition, this marker must co-segregate with the gene of interest and so be present in affected family members but absent in unaffected family members.

1.3.2 Linkage Analysis

Linkage analysis is a method applied to establish the carrier status of female 'at-risk' carriers and prenatal diagnoses. In many cases, linkage analysis replaces mutational analysis. In a small number of families in whom the mutation cannot be identified, linkage analysis remains the only method for the genetic diagnosis of carriers. The principle of linkage analysis is straightforward. All human chromosomes come in pairs, with one inherited from the mother and one from the father. Each pair of chromosomes contains the same genes in the same order, but their sequences are not identical. It is easy to determine whether a particular sequence of offspring comes from a mother or father. These sequence variations are called maternal and paternal alleles. In the case of a diseased gene, the alternative alleles will be normal alleles and the diseased allele is distinguished by looking for the occurrence of the disease in a family tree or pedigree. Genetic markers are DNA sequences that show polymorphism (variations in size or sequence) in the population. They are present in every human genome and can be typed (identification of the allele) using techniques such as polymerase chain reaction. This ability to determine the parental origin of a DNA sequence allows us to show whether recombination has occurred or not. Recombination occurs in germ cells that make eggs and sperm. In these cells, the maternal and paternal chromosomes pair up and exchange their fragments. After recombination, the chromosomes contain a mixture of maternal and paternal alleles, placed in the germinal cells and passed to the next generation (Pulst, 1999).

Recombination occurs randomly. During this process, if there is a considerable distance between two DNA sequences on a chromosome, there is a good chance that recombination will occur between them, and the maternal and paternal alleles will be mixed up (A and C in Figure 1.2). In contrast, if two DNA sequences are very close together, they will rarely recombine. As a result, the maternal and paternal alleles will tend to stay together (A and B in Figure 1.2). Diseased genes are mapped by measuring the recombination against a panel of different markers spread over the entire genome. In

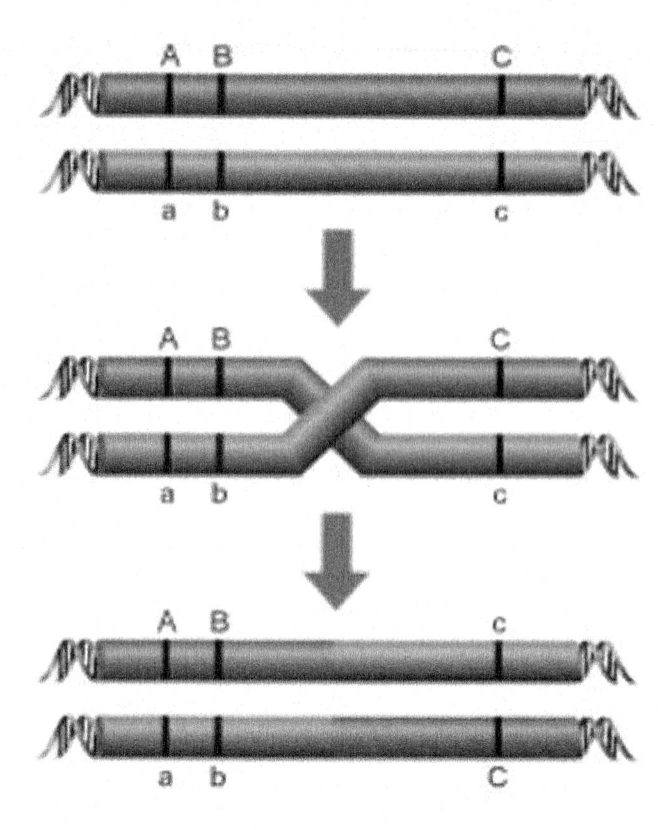

FIGURE 1.2 Linkage analysis: Top panel shows paternal (ABC alleles) and maternal (abc alleles) chromosomes aligned in a germ cell, a cell that gives rise to eggs or sperm. The middle panel shows the physical process of recombination, which involves the crossing over of DNA strands between the paired chromosomes. The bottom panel shows what happens when the crossover is resolved. The maternal and paternal alleles are mixed (recombined), and these mixed chromosomes are passed to the sperm or eggs. If A is the diseased gene and B and C are genetic markers, recombination is likely to occur much more frequently between A and C than between A and B. This allows the disease gene to be mapped relative to markers B and C.

most cases, recombination will occur frequently, indicating that the disease gene and marker are far apart. A few markers will tend not to recombine with the diseased gene due to their proximity. These markers are said to be linked to it. Ideally, close markers have identified that flank, with the diseased gene, and have defined a candidate region of the genome between 1 and 5 million bp in length. The gene responsible for the disease lies somewhere in this region.

1.3.3 Polymorphic DNA Markers

Linkage analysis aims to identify an appropriate DNA marker co-segregating with the gene of interest and can track the gene within a family without knowing the mutation. The markers commonly used to track a gene within a family are polymorphic. There are two major types of polymorphic markers or polymorphisms.

a) Single Nucleotide Polymorphisms (SNPs)
b) Short Tandem Repeat (STRs) or Variable Number Tandem Repeats (VNTRs)

a) Single Nucleotide Polymorphisms [SNPs]: are single nucleotide changes that usually occur in the DNA, which will not change the amino acid sequence of the corresponding protein of interest. The location of SNPs is throughout the human genome as intragenic polymorphisms occurring within a gene, usually in the introns or rarely in the 5' and 3' untranslated regions (UTRs). Extragenic polymorphisms are closely linked to a gene. Historically, SNPs are often designated by a restriction endonuclease or an enzyme that digests the DNA before agarose gel electrophoresis and southern blotting hybridization. For example, within the *F8* gene, the enzyme *Bcl I* identified an intragenic polymorphism located within intron 18 by cutting the DNA into two fragments of 0.8 kb and 1.1 kb. Similarly, the enzyme *Bgl I* identified an SNP located within intron 25 of the same *F8* gene by cutting the DNA into two sequences of 5 kb and 20 kb, The enzyme *Bgl II* identified an SNP located close to, but not a part of the *F8* gene, gives rise to two fragments of 5.8 kb and 2.8 kb. The common feature is that when digested with *Bgl II*, *Bcl I* and *Bgl I*, the DNA fragments generated are of different lengths. The polymorphisms giving rise to these differing fragments are known as restriction fragment length polymorphisms or RFLPs. Southern blotting is rarely performed today. Detection of SNPs is by PCR either with sequence analysis or by the resolution of the DNA fragments on agarose gel electrophoresis.

b) Short Tandem Repeat (STRs) or Variable Number of Tandem Repeats (VNTRs): Short Tandem Repeats are extremely resourceful DNA markers that are highly polymorphic and inherited in a strict Mendelian fashion. Areas of repetitive DNA occur throughout the genome where the repeating unit is tiny, either 1-6 nucleotides (minisatellite) or 2-3 nucleotides (microsatellite). Microsatellite repeats are comprised of dinucleotide repeat sequences such as CA in which the repeated sequence occurs multiple times (CACACACACA) and trinucleotide repeats (ATT)n. Minisatellite repeats contain tetranucleotide repeats [GATA]n or hexanucleotide repeats (CAATAC) n. These are highly polymorphic within a population and can be used for bone marrow transplant engraftment, forensics, identity testing, paternity testing, and so forth. STRs are widely used in genetic linkage studies. The reason for this lies in the greater chance that a particular individual may be heterozygous for a specific marker. Although the number of repeat sequences may change, this change is predicted to occur every 100 generations.

Genetic linkage analysis is a powerful tool for detecting the chromosomal location of diseased genes. It is based on the observation that genes that reside physically close on a chromosome remain linked during meiosis. For most neurological disorders whose underlying biochemical defect was not known, identifying the chromosomal location of the diseased gene was the first step in its eventual isolation. Genes that have been isolated in this way include examples from all types of neurological disorders such as Alzheimer's, Parkinson's, or ataxia, and diseases of ion channels leading to periodic paralysis or hemiplegic migraine, or tumor syndromes such as neurofibromatosis types 1 and 2 (Scriver et al. 1999). With the advent of new genetic markers and automated genotyping, rapid genetic mapping is possible in a minimum amount of time. Generated genetic linkage maps for the human genome provide the basis for constructing physical maps and permit the rapid mapping of disease traits. After establishing the chromosomal location of a diseased phenotype, genetic linkage analysis helps determine the diseased phenotype due to single or multiple gene mutations. Mutations in other genes may also give rise to an identical or similar phenotype. Often it is found that mutations in different genes can cause similar phenotypes. Examples are autosomal dominant spinocerebellar ataxias, caused by mutations in different genes but with very similar phenotypes. In addition to providing novel, genotype-based classifications of neurologic diseases, genetic linkage analysis can also aid in disease diagnosis. However, in contrast to direct mutational analysis such as detecting an expanded CAG repeat in the Huntington gene, diagnosis using flanking markers requires the analysis of several family members.

1.4 LINKAGE DISEQUILIBRIUM

Linkage disequilibrium is a nonrandom association of alleles at two or more loci. Lewontin and Kojima coined the word in 1960. Initially, linkage disequilibrium was not the primary concern of the population geneticists due to the non-availability of data for the study. Its importance to evolutionary biology and human genetics was unrecognized outside population genetics. However, the importance of linkage disequilibrium gained widespread interest in the 1980s after the role of linkage disequilibrium in gene mapping became evident and large-scale surveys of closely linked loci became feasible. It also plays a vital role in evolutionary biology and human genetics, as there are so many factors that affect and are affected by linkage disequilibrium. Linkage disequilibrium provides information about past events and constrains the potential response to natural and artificial selection. Linkage disequilibrium throughout the genome reflects the population history, the breeding system, and the pattern of geographic subdivision (Montgomery, 2008). Linkage disequilibrium in each genomic region demonstrates the history of natural selection, gene conversion, mutation, and other forces that cause gene-frequency evolution. How these factors affect linkage disequilibrium between a particular pair of loci or in a genomic region depends on local recombination rates. The population genetics theory of linkage disequilibrium is well developed and widely used to provide insight into evolutionary history and as the basis for mapping genes in human beings and other species.

1.4.1 LINKAGE DISEQUILIBRIUM AT ONE PAIR OF LOCI

Linkage disequilibrium between alleles at two loci has been defined in many ways, but all definitions depend on the quantity:

$$D_{AB} = pAB - p\,A\,pB \qquad (1)$$

DAB refers to the difference between the frequency of gametes carrying a pair of alleles A and B at two loci (pAB) and the frequency product of those alleles (pA and pB). The initial definition was in terms of gamete frequencies because it allowed the possibility that the loci are on different chromosomes. The second definition is for the loci on the same chromosome. In this case, the allele pair AB is called a haplotype, and pAB is the haplotype frequency. As defined, DAB characterizes a population, and in practice, DAB is estimated from the allele and haplotype frequencies of a sample. Standard sampling theory must be applied to find the confidence intervals of estimated values.

The quantity, D_{AB} is the coefficient of linkage disequilibrium. It is defined for a specific pair of alleles, A and B, and does not depend on how many other alleles are at the two loci. Each pair of alleles has its own D. Constriction of the values for different pairs of alleles is due to the allele frequencies at both loci, and the haplotype frequencies must add up to a value of 1. If both loci are diallelic, as is the case with all SNPs, the constraint is strong enough that only one value of D is needed to characterize linkage disequilibrium between those loci. In fact,

$$D_{AB} = -D_{Ab} = -D_{aB} = D_{ab}$$

Here, a and b are the other alleles. In this case, the D is used without a subscript. Thus, the sign of D is arbitrary and depends on which pair of alleles one starts with.

If either locus contains more than two alleles, no single statistical method can quantify the overall linkage disequilibrium between them. A statistical method is needed when both the loci have numerous alleles, as with loci in the major histocompatibility complex of vertebrates, with more than hundreds of alleles. On the other hand, there might be no single pair of alleles with a particular interest, whether more linkage disequilibrium between one loci pair than another pair or more linkage disequilibrium between a loci pair in one species than in one species another one (John and William, 2018).

1.4.2 Linkage Equilibrium

If D = 0, there is linkage equilibrium (Le), which has similarities with Hardy Weinberg equilibrium in implying statistical independence. When genotypes at a single locus are at H, an allele present on one chromosome is independent of whether it is present on its homolog. Consequently, the frequency of the AA homozygote is the square of the frequency of A ($pAA = pA^2$), and the frequency of the Aa heterozygote is twice the product of pA and pa, the two being necessary to allow for both Aa and aA. Here, the essential feature of Hardy Weinberg equilibrium is that regardless of the initial genotype frequencies, equilibrium is established in one generation of random mating. Any initial deviation from H will disappear immediately. Significant departures from Hardy Weinberg equilibrium indicate that something interesting is going on, as in the case of extensive inbreeding, strong selection, or genotyping error. Le is like Hardy Weinberg equilibrium as it also implies that alleles at different loci are randomly associated. The frequency of the AB haplotype is the product of the allele frequencies (pApB). Linkage equilibrium, however, differs from Hardy Weinberg equilibrium because it is not established in one generation of random mating. Instead, D decreases at a rate that depends on the recombination frequency c, between the two loci:

$$D_{AB}(t + 1) = (1 - c)\, D_{AB}(t) \tag{2}$$

Where t is time in generations. Even for unli nked loci (c = 0.5), D decreases only by a factor of half each generation, as proved by Weinberg in 1909. The general formula was obtained first by Jennings. Although linkage equilibrium will be reached, it will occur slowly in closely linked loci. This is the basis for the use of linkage disequilibrium. Other population genetic forces such as selection, gene flow, genetic drift, and mutation, all affect D. So, substantial linkage disequilibrium will persist under many conditions. When surveying the number of polymorphic loci in a genome is possible, the extent of linkage disequilibrium in a genome can be quantified with great precision, allowing a fine-scale analysis of the forces governing the genomic variation. The co-efficient of linkage disequilibrium and related quantities are descriptive. Their magnitude does not indicate whether there is a statistically significant association between alleles in haplotypes. Standard statistical tests, including the chi-squared and Fisher's exact test, are commonly used to test the significance.

1.4.3 Linkage Disequilibrium at More Than Two Loci

When more than two loci are considered together, a common practice is to distinguish the allelic pairs that show high levels of linkage disequilibrium from those that do not. The result is a graph introduced by Miyashita and Langley (1998) to describe the patterns of linkage disequilibrium in *Drosophila melanogaster*. A 216 kb segment in the class II region of the major histocompatibility complex in humans comprises non-overlapping sets of loci in strong linkage disequilibrium. Each group is called a haplotype block, and boundaries were associated with hotspots of recombination. Similar patterns were found in other regions in the human genome, leading to the hypothesis that most of the human genome had a block-like pattern of linkage disequilibrium. Haplotype blocks in humans vary in size from a few kb to more than 100 kb.

1.4.4 Linkage Disequilibrium Within and Between Populations

When data for more than one population is available, linkage disequilibrium between a pair of loci can be partitioned into contributions within and between populations. Ohta (1982) first suggested this partitioning is like Wright's partitioning of deviations from Hardy Weinberg equilibrium frequencies into F_{IS}, the average deviation within populations, and F_{ST}, the average deviation attributable to differences in allele frequency among populations. Ohta partitioned D_{IT}, the total disequilibrium in a subdivided population, into D_{IS}, the average disequilibrium within subpopulations, and D_{ST} contributing to the overall disequilibrium caused by the differences in allele frequencies within the subpopulations. Computer programs such as Genepop are available to calculate D_{IS} and D_{ST}. These statistics are widely applied in the analysis of data from non-human populations but rarely for human populations. This rarity in application is probably due to the focus on the humans of each population, whereas the focus on other species is often on the overall pattern of linkage disequilibrium. Natural selection favoring adaptations to local conditions will increase D_{ST} whenever alleles at different loci are favored. Partitioning overall linkage disequilibrium is an appropriate first step when determining the differences in linkage disequilibrium resulting only from differences in allele frequency or from other factors that vary among populations.

1.4.5 Population Genetics of Linkage Disequilibrium

1. Natural selection

Initial interest in linkage disequilibrium has been increased because of the questions raised on the operation of natural selection. If alleles at two loci are in linkage disequilibrium and they both affect reproductive fitness, the response to selection on one locus might be accelerated or impeded by selection affecting the other. One line of research in this area concerns the effect of linkage disequilibrium on long-term trends in evolution. Kimura (1965) and Nagylaki (1974) showed that unless interacting loci are very closely linked or selection is very strong, recombination dominates, and as a good approximation, linkage disequilibrium can be ignored. This theory supports Fisher's depiction of natural selection as steadily increasing the average fitness of a population. This theory also shows that when selection is strong, and fitness interactions among loci are complex, average fitness might not increase in every generation because linkage disequilibrium constrains how haplotype frequencies respond to a selection. In this case, linkage must be accounted for explicitly before even qualitative predictions can be made. In some cases, selection alone can increase linkage disequilibrium when fitness values are more than multiplicative, which means that the average fitness of an individual carrying the AB haplotype exceeds the product of the average fitness of individuals carrying A alone or B alone. This pattern is easiest to see with diallelic loci in haploid organisms.

If the relative fitness (w) of the ab, Ab, and aB haplotypes are wAb and waB, then selection will increase linkage disequilibrium only if $wAB > wAbwaB$. If both A and B are maintained by balancing selection, then linkage disequilibrium can persist indefinitely. Further, when more than two loci interact in this way, large blocks of linkage disequilibrium can be maintained by selection, where an individual locus is not an appropriate unit of selection. Interest in this theory decreased during the 1970s, after discovering linkage disequilibrium that was not detected between alleles distinguishable by protein electrophoresis. This theory might become popular again or perhaps be re-invented as studies find increasing evidence of intragenic interactions that can create strong epistasis in fitness.

2. Genetic Drift

Genetic drift alone can generate linkage disequilibrium between the closely linked loci. Even if two loci are in linkage equilibrium, sampling only a few individuals will create some linkage disequilibrium. First results obtained during the late 1960s suggested that a genetic drift balanced by a mutation and recombination would maintain only low levels of linkage disequilibrium. The expectation of $D2$ is small even if there is no recombination because the flux of mutations at both loci tends to eliminate most linkage disequilibrium. To attend to this, genetic drift was largely ignored as a cause of linkage disequilibrium. However, the expectation of $D2$ does not tell the whole story because it includes cases in which one or both loci are monomorphic (when D is necessarily 0). The expectation of $D2$ when both loci are polymorphic cannot be calculated analytically, but simulations show that much more significant values are seen. Genetic drift surprisingly interacts with selection. Selection affecting closely linked loci becomes slightly weakened because drift creates small amounts of linkage disequilibrium that, on average, reduces the response to selection. This effect, known as the Hill–Robertson effect, is relatively weak when only two loci are considered (Felsenstein, 1965). Nevertheless, it is much stronger per locus when many selected loci are closely linked.

3. Inbreeding, Inversions, and Gene Conversion

Other forces that create linkage disequilibrium are inbreeding, inversions, and gene conversion. Inbreeding creates linkage disequilibrium for the same reason as population subdivision. Because of recent common ancestry, inbreeding augments the covariance between alleles at different loci. Genomic inversions significantly reduce the recombination between the inverted and non-inverted segments as the recombination produces aneuploid gametes. Consequently, the inverted and original segments become equivalent to almost completely isolated subpopulations between which linkage disequilibrium accumulates. *Drosophila* geneticists have long appreciated this fact.

Gene conversion affects the linkage disequilibrium at a pair of loci in the same way that reciprocal recombination does. This equivalence can be seen by considering a pair of diallelic loci A/a and B/b. Gene conversion at the B/b locus will result in an individual with haplotype phase AB/ab who will produce Ab or aB gametes depending on whether B converts b or the reverse. However, gene conversion differs from recombination when more than two loci are considered together. Reciprocal crossing over affects linkage disequilibrium between all pairs of loci on opposite sides where the crossing over has taken place. By contrast, gene conversion affects loci only within the conversion track, which is relatively short. Loci that are not within the track are unaffected. For example, three loci A/a, B/b, and C/c are located on a chromosome in order, where B/b is in a conversion track. The linkage disequilibrium between A/a, B/b, and B/b, C/c is affected by the conversion, but the linkage disequilibrium between A/a and C/c is not. Several methods for inferring the relative rates of gene conversion and recombination are based on this idea.

1.4.6 Applications of Linkage Disequilibrium

1. Mutation and Gene Mapping

Mutation has a unique role in creating a linkage disequilibrium. When a mutant allele, M, first appears on a chromosome, it is in low frequency, $pM = 1/(2N)$ (N is the population size) and is in perfect linkage disequilibrium with the alleles at other loci that are on the chromosome carrying the first copy of M. perfect linkage disequilibrium means that $D' = 1$. If $D' = 1$, only three of the four possible haplotypes are present in the population. Perfect linkage disequilibrium will persist until recombination involving an M-bearing chromosome creates a non-ancestral haplotype. Consequently, loci closely linked to M will remain in perfect linkage disequilibrium for a long time and in strong linkage disequilibrium for an even longer time. The persistence of strong linkage disequilibrium between a mutant allele and the loci closely linked to it has many practical implications. Rare marker alleles in strong linkage disequilibrium with a monogenic disease locus must be closely linked to the causative locus.

A relatively simple mathematical theory indicates how close these two are. This method, known as linkage disequilibrium mapping, has been successfully used with several diseases. The same idea underlies the association mapping of complex diseases. Closely linked polymorphic SNPs tend to be in strong linkage disequilibrium with one another. The fine-scale pattern of linkage disequilibrium in humans confirms that the human genome comprises haplotype blocks within which most or all SNPs are in high linkage disequilibrium. These high levels of linkage disequilibrium among SNPs are assumed to be true for alleles that increase the risk of complex inherited diseases. This idea, combined with the development of efficient methods for surveying large numbers of SNPs, has led to the many recent genome-wide association (GWA) studies that have detected SNPs significantly associated with breast cancer, colorectal cancer, and type 2 diabetes, and heart disease. However, one potential problem in genome-wide association studies is creating linkage disequilibrium by unrecognized population subdivision. Several methods have been proposed to account for such linkage disequilibrium. Although genome-wide association studies have successfully found new causative alleles, the overall proportion of risk accounted for is often relatively low. Alleles accounting for a more significant proportion of risk might be found in more extensive studies. It is unclear whether most causative variants will ultimately be found this way. The reason is that GWA studies are more effective in finding causative alleles that are relatively high frequency. Other methods might be needed for low-frequency causative alleles.

2. *Detection of Natural Selection*

Strong positive selection quickly increases the frequency of an advantageous allele, with the result that linked loci remain in unusually strong linkage disequilibrium with that allele. Recently, methods have been developed to detect unusually low heterozygosity regions that indicate past hitch-hiking events. If an advantageous allele has not gone to fixation, variability at linked markers will be lower on chromosomes bearing that allele than on other chromosomes. Several tests of neutrality are based on this idea. One class of methods assumes that a potentially advantageous allele at a locus has been identified and tests whether there is significantly more linkage disequilibrium with that allele than with other alleles at the same locus. The second class of methods assumes only that the potentially selected locus has been identified and tests whether patterns of haplotype variation at that locus are consistent with neutrality.

3. *Estimation of Allele Age*

Strong linkage disequilibrium with an allele in a relatively large region indicates that not much time has passed since the allele arose by a mutation. If the mutant allele has reached a relatively high frequency in a short time, it is likely to have done so under the effect of positive selection. This tendency provides the basis for the selection. In addition to it, linkage disequilibrium can indicate the point in time when an allele arose as a mutation. This approach is straightforward, which takes a reasonable estimate of an allele age, but it does not consider the stochastic nature of recombination and genetic drift and exaggerates the accuracy of the resulting estimates.

1.4.7 Linkage Disequilibrium and Human Genome

Linkage disequilibrium is one of the major processes studied in human population genetics. After resolving the technical problem of efficiently genotyping 500,000 SNPs, which considerably declined the cost of genotyping, there is an increased scope for many very large-scale genome-wide association studies to be carried out. Soon the new technologies will allow large-scale re-sequencing studies, including the 1000 Genomes Project, enabling more data for these studies. The only limiting factor will be the availability of people willing to participate in the genome-wide association studies and the resources needed for the accurate clinical assessment (Erlich et al. 2009).

1.5 MENDELIAN PEDIGREE PATTERNS

The most uncomplicated genetic characters are those whose presence or absence depends on the genotype at a single locus. Sometimes, a particular genotype at a single locus is necessary and sufficient for that character to be expressed, given the organism's typical genetic and environmental background. Such characters are called Mendelian. In humans, more than 10,000 Mendelian characters have been identified. If a family is affected by a disease or a disorder, the pattern of disease transmission from one generation to another can be established by accurate family history. Additionally, family history can also help to exclude a few genetic diseases. Most genes have one or more versions due to mutations or polymorphisms, referred to as alleles. Individuals may carry a normal, diseased, or rare allele, depending on the impact of the mutation or polymorphism and the population frequency of that allele.

1.5.1 Dominance and Recessiveness of Characters

A character is considered dominant if manifested in the heterozygote and is considered recessive if it is not manifested in the heterozygote. Dominance and recessiveness are the properties of characters, not genes. Most of the dominant human syndromes are known only in heterozygote individuals. Sometimes homozygotes have also been described, born from the mating of two heterozygotes, where people are much more severely affected. Examples are achondroplasia (short-limbed dwarfism) and Type 1 Waardenburg syndrome (deafness with pigmentary abnormalities). Nevertheless, achondroplasia and Waardenburg syndrome are described as dominant because these terms also describe phenotypes seen in heterozygotes. When this uncertainty does not exist, the condition is named semi-dominant, where the heterozygote has an intermediate phenotype, reserving the dominant conditions. The homozygote is indistinguishable from the heterozygote, as exemplified in the case of Huntington's disease (an adult-onset progressive neurological deterioration). Males are always hemizygous for the loci on the X and Y chromosomes, where they have only a single copy of each gene. In this case, the question of dominance or recessiveness does not arise in males for both X and Y-linked characters. Autosomal characters in males and females, and X-linked characters in females can be dominant or recessive.

1.5.2 Types of Mendelian Pedigree

No human being has two genetically different Y chromosomes (except in the rare case of XYY males, where the two Y chromosomes are duplicates). There are five archetypal Mendelian pedigrees (i) Autosomal dominant, (ii) autosomal recessive, (iii) X-linked recessive, (iv) X-linked dominant, and (v) Y-linked.

i. Autosomal Dominant

Autosomal dominance is a pattern of inheritance that occurs if the gene of question is located on one of the 1-22 or non-sex chromosomes. "Dominant" means a single copy of the disease-associated mutation is sufficient to cause the disease. Huntington's disease and neurofibromatosis-1 are typical examples of autosomal dominant genetic disorders. Autosomal dominance is one of the several ways that a trait or a disorder can be passed down through families. It describes a trait or disorder in which the phenotype is expressed in those who have inherited only one copy of a particular gene mutation (heterozygotes), explicitly referring to a gene on one of the 22 pairs of autosomes. One of the parents may often have the disease. Inheriting a disease, condition, or trait depends on the type of chromosome affected. It also depends on whether the trait is dominant or recessive. In most of the cases, the abnormal gene dominates. Each child's risk is independent of whether their sibling has the disorder or not. If the first child has the disorder, the next child has the same 50% risk of inheriting the disorder. Children who do not inherit the abnormal gene will not develop or pass on the disease. If someone has an abnormal gene inherited in an autosomal dominant manner, the parents should also be tested for the abnormal gene.

ii. Autosomal Recessive

The autosomal recessive inheritance gene in question is located on one of the 1-22 autosomes or non-sex chromosomes. Autosomes do not affect an offspring's gender. "Recessive" means that two copies of the gene are necessary to have the trait or disorder, with one copy inherited from the mother and the other from the father. If only one recessive gene is present, it is considered a "carrier" for the trait or disease, which will not generate any health problems. Carrier genes do not show any signs of the disease or condition. Most people even do not know that they are recessive gene carriers until they have a child with that disease. Once parents have a child with a recessive trait or disease, there is a 25% chance that another child will be born with the same trait or disorder with each subsequent pregnancy. This means a three out of four or a 75 % chance that another child will not have the trait or disease. The birth of a child with a recessive condition is often a total surprise to a family. In most cases, there is no previous family history of a recessive condition. It is estimated that all people carry about five or more recessive genes that cause genetic diseases or conditions. Examples of autosomal recessive disorders include cystic fibrosis, sickle cell anemia, and Tay-Sachs disease.

iii. X-linked Dominant

Sex-linked dominance is a rare way that a trait or disorder can be passed down through families. A single abnormal gene on the X chromosome can cause a dominant sex-linked disease. Dominant inheritance occurs when an abnormal gene from one parent can cause disease, even though a matching gene from the other parent is normal. The abnormal gene dominates the gene pair. In the case of X-linked dominant disorder, if a father carries the abnormal gene on the X chromosome, all his daughters will inherit the disease, and none of his sons will have the disease because daughters always inherit their father's X chromosome. If the mother carries the abnormal X gene, half of all their children (daughters and sons) will inherit the disease tendency.

If there are four children (two males and two females), and the mother is affected with one abnormal X, and she has the disease, but the father is not, the statistical expectation is

- Two children (one girl and one boy) with the disease and
- Two children (one girl and one boy) without the disease.

If there are four children (two males and two females) and the father is having the affected abnormal gene on the X chromosome, and he has the disease, but the mother is not, the statistical expectation is:

- Two girls with the disease
- Two boys without the disease.

Males are more severely affected than females because, in each affected female, there is one normal allele producing a regular gene product and one mutant allele producing the non-functioning product, while in each affected male, there is only the mutant allele with its non-functioning product and the Y chromosome, no specific gene product at all. Affected females are more prevalent in the general population because the female has two X chromosomes, which could carry the mutant allele, while the male only has one X chromosome as a target for the mutant allele. When the disease is no more deleterious in males than in females, females are about twice as likely to be affected as males.

iv. X-linked Recessive

A mode of inheritance in which a mutation in a gene on the X-chromosome causes the phenotype to be expressed in males who are hemizygous for the gene mutation (they have only one X chromosome) and in females who are homozygous for the gene mutation (they have a copy of the gene mutation on each of their two X chromosomes). Carrier females who have only one copy of the mutation do not usually express the phenotype, although differences in X-chromosome inactivation can lead to varying degrees of clinical expression in carrier females. Examples of X-linked recessive conditions include red-green color blindness and hemophilia A:

Red-green color blindness: Red-green color blindness means that a person cannot distinguish the shades of red and green. Their visual acuity is normal. There are no severe complications. However, affected individuals may not be considered for certain occupations involving transportation or the armed forces where color recognition is required. Males are affected 16 times more often than females because the gene is located on the X chromosome.

Hemophilia: Hemophilia A is a disorder where the blood cannot clot properly due to a clotting factor deficiency known as Factor VIII. This results in abnormally heavy bleeding that will not stop, even from a small cut. People with hemophilia A bruise easily and can have internal bleeding into their joints and muscles. The occurrence of hemophilia A (Factor VIII deficiency) and hemophilia B (Factor IX deficiency) combined is one in 10,000 live male births, with hemophilia A accounting for 80 percent of all cases. Treatment is available by infusion of Factor VIII (blood transfusion). Female carriers of the gene may show some mild signs of Factor VIII deficiency, such as bruising easily or taking longer than usual to stop bleeding when cut. However, not all female carriers present these symptoms. One-third of all hemophilia cases are due to new mutations in the family (not inherited from the mother).

V. Y-linked

Although a few Y-linked characters have been described, no Y-linked diseases are known, apart from disorders of male sexual function. Conceivably such a disease may exist undiscovered, but this is unlikely for two reasons. (1) The pedigree pattern would be strikingly noticeable, especially in societies that trace the family through the male line, yet they have not been noted. (2) The Y-chromosome cannot carry any genes whose function is essential for health because females are perfectly normal without any Y-linked genes. Thus, any Y-linked genes must code either for non-essential characters or male-specific functions, and defects are unlikely to cause diseases apart from defects of male sexual function.

1.6 MENDELIAN PEDIGREE ANALYSIS

1.6.1 Pedigree Analysis

A pedigree is a family relationship chart that uses symbols to represent people and lines to represent genetic relationships. These diagrams make it easier to visualize the relationships within families, particularly extensive families. Pedigrees are often used to determine the dominance and recessiveness of genetic diseases. An example of a pedigree chart showing the four generations of a family with four individuals affected by color blindness is displayed in Figure 1.3. Controlled crosses cannot be made in human beings. Geneticists must emphasize the family records hoping that available informative matings can be used to deduce the dominance and distinguish the autosomal form of X-linked inheritance. The investigator traces the history of some variant phenotypes back through the family's history and draws up a family chart or pedigree using the standard symbols. Standard symbols used for the pedigree analysis are listed in Table 1.1. Clues in the pedigree must be interpreted differently. The interpretation depends on one contrasting phenotype, such as a rare disorder, or where both of the phenotypes of a pair are common morphs of a polymorphism.

In this pedigree chart, the unaffected founding mother, I-1, and affected founding father, I-2 are the parents of two affected daughters, II-1 and II-2. The affected founding daughter II-2 and the unaffected male II-3 who "marries into the family" have two offspring, an unaffected daughter

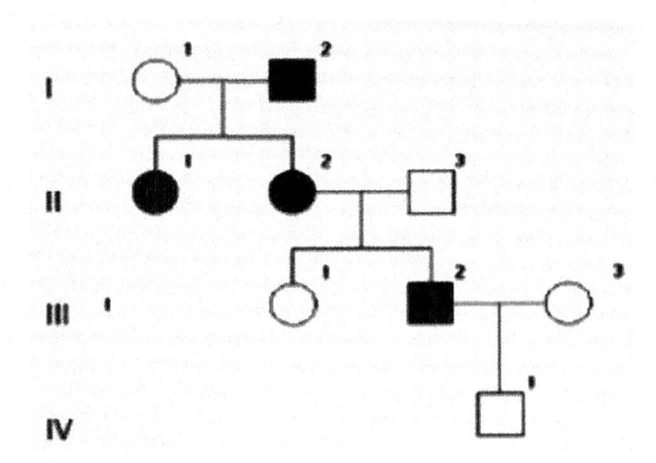

FIGURE 1.3 Pedigree chart of a family affected with color blindness. Circles represent females and squares represent males. A dark circle or a dark square represents an individual affected by the trait. Individuals are represented by the Roman numeral, which stands for the family's generation. The female at the upper left is an individual I-1. Founder parents in this family are the female I-1 and the male I-2 in the first generation at the top. A male and female in this pedigree are directly connected by a horizontal line, have mated, and had children. These three pairs have mated in this tree: I-1 & I-2, II-2 & II-3, III-2 & III-3. Vertical lines connect parents to their children. For instance, the females II-1 and II-2 are daughters of I-1 and II-2. The founding family consists of the two founding parents and their children, II-1 and II-2.

TABLE 1.1
Symbols Used for the Pedigree Analysis

S. No	Symbol	Representation in Pedigree Analysis
1.		Male
2.		Female
3.		Mating
4.		Parents and Children in order of birth. 1 Boy and 1 Girl.

TABLE 1.1 (Continued)
Symbols Used for the Pedigree Analysis

S. No	Symbol	Representation in Pedigree Analysis
5.	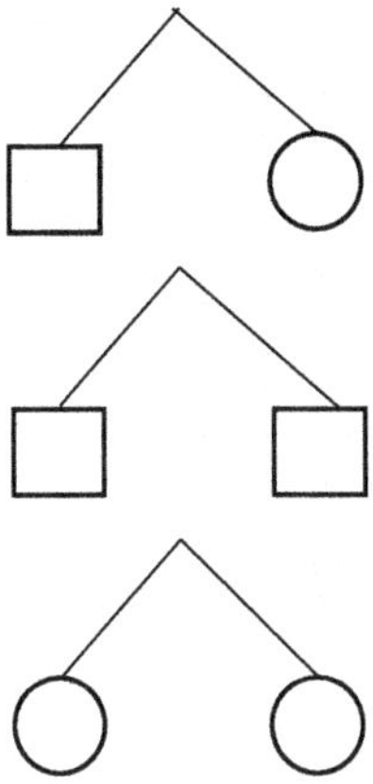	Dizygotic or non-identical twins
6.	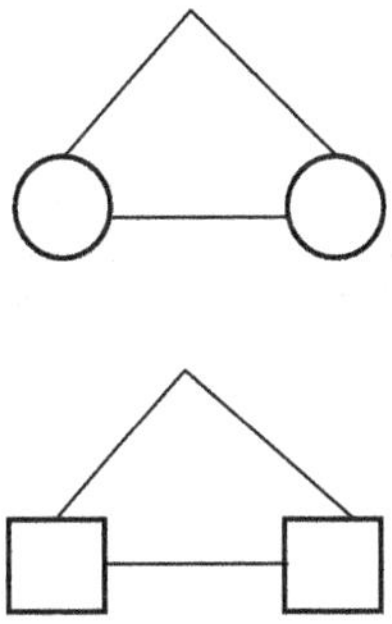	Monozygotic twins
7.		Unspecified Sex
8.		Abortion or Still Birth (Sex unspecified)
9.		Propositus
10.		Carrier of sex-linked recessive
11.		Death
12.		Affected Individuals
13.		Heterozygotes
14.	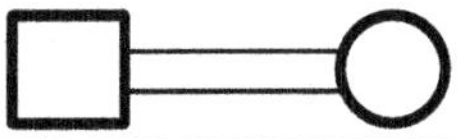	Consanguineous Marriage

III-1 and an affected son III-2. Finally, this affected male III-2 and the unaffected female III-3 who "marries in" have an unaffected son, IV-1. The genetic disorders of human beings can be either autosomal or X-linked.

1.6.2 Patterns Indicating a Recessive Trait

If any affected founding daughter has two unaffected parents, the disease must be autosomal recessive. The disease must be recessive if any affected individual has two unaffected parents. At least one parent must have an allele for the disease (Figure 1.4).

1.6.2.1 Autosomal Recessive

An affected individual must inherit a recessive allele from both parents, so both parents must have that allele. If the father had a recessive X-linked allele, he must have been affected by only one X-linked allele. If an affected founding son has two unaffected parents, we cannot determine if the disease is autosomal, or X-linked (Figure 1.5). If the trait is autosomal, both parents can be unaffected carriers of the disease. If the trait is X-linked, the son must have inherited his allele from his mother only, and his father can be unaffected.

1.6.3 Patterns Indicating a Dominant Trait

The disease must be dominant if every affected child of the non-founding parent has an affected parent. The unaffected mother, who is marrying, does not carry an allele for the disease. So, the affected child inherits an allele only from the affected father. No child could be affected by a single autosomal recessive allele or X-linked recessive allele, so the trait is dominant (Figure 1.6).

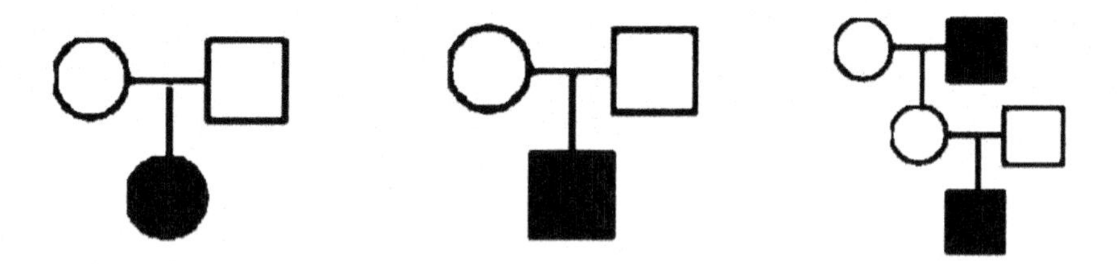

FIGURE 1.4 Pedigree patterns that indicate a recessive Trait.

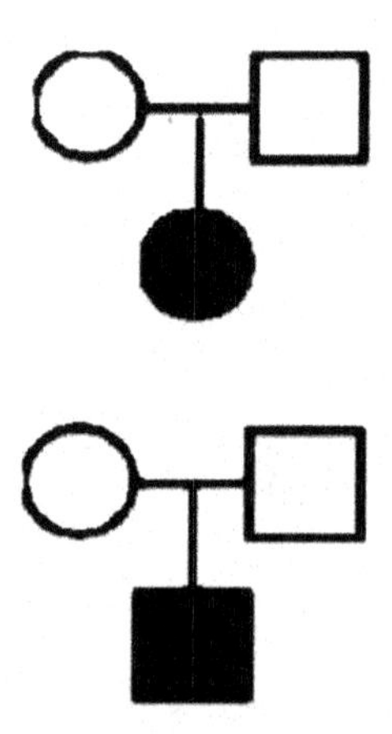

FIGURE 1.5 Pedigree of an autosomal recessive disorder.

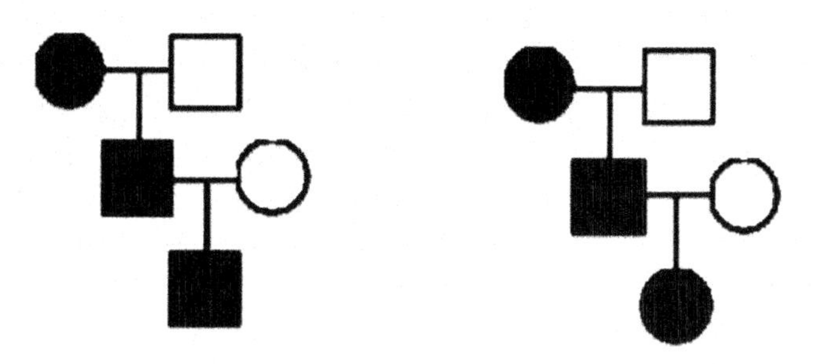

FIGURE 1.6 Pedigree patterns that indicate a dominant trait.

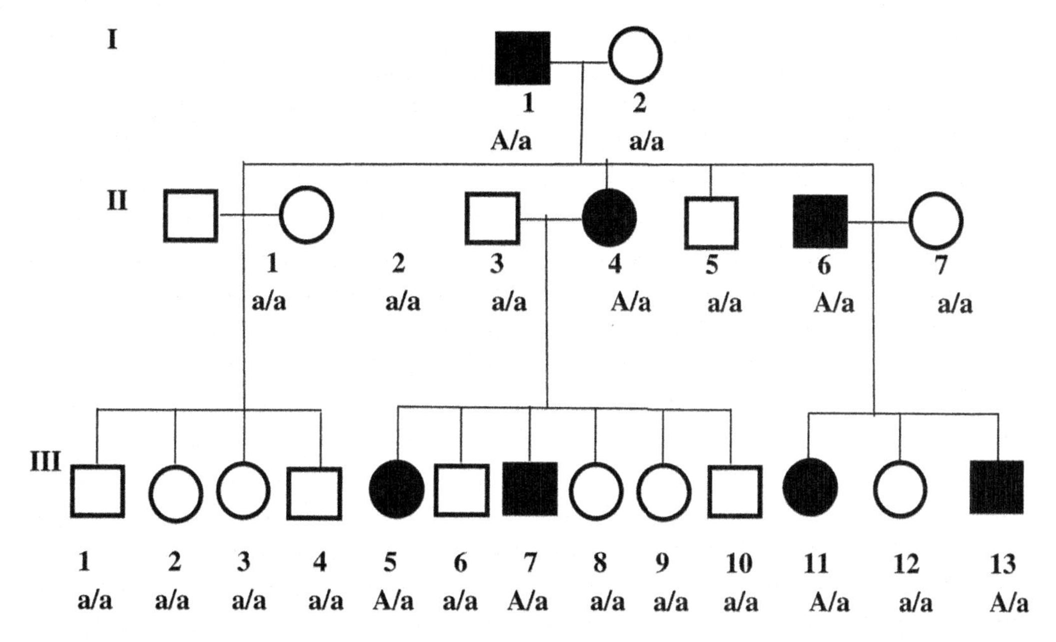

FIGURE 1.7 Pedigree of a dominant phenotype determined by a dominant allele, A with all genotypes deduced.

1.6.3.1 Autosomal Dominant Trait

In autosomal dominant disorders, the normal allele is a recessive one, and the abnormal allele or the diseased allele is dominant. It might seem paradoxical that a rare disorder can be dominant. It can also be considered that dominance and recessiveness reflect how different alleles act, which is not defined in terms of *pre*dominance in the population. During the pedigree analysis, the main clue for identifying an autosomal dominant disorder is the phenotype that tends to appear in every generation of the pedigree and affected parents (both father and mother) transmit the phenotype to both sons and daughters. Here, the phenotype appears in every generation as an abnormal allele, carried by an individual who might have inherited it from his parent in the previous generation. Mendelian ratios need not be necessarily observed in all families. In recessive disorders, individuals bearing one copy of the rare allele (*A/a*) are much more common than those bearing two copies (*A/A*). So, most of the affected people are heterozygotes. Virtually all matings involving dominant disorders are *A/a* × *a/a*. Therefore, when the progeny of such matings is totaled, a 1:1 ratio is expected of unaffected (*a/a*) to affected individuals (*A/a*). In relatively rare cases, there is a possibility that abnormal alleles can also arise due to mutations (Figure 1.7).

Huntington's disease is an example of an autosomal dominant disorder resulting in neural degeneration, leading to convulsions and premature death. It is a late-onset disease, where the symptoms do not appear till the person has begun to have children. Each child, a carrier of the abnormal allele, has a 50% chance of inheriting the allele associated with the disease. This tragic pattern has led to finding ways of identifying people carrying this abnormal allele before they experience the onset of the disorder. The discovery of the molecular nature of a mutant allele and neutral mutations on the DNA close to the affected allele can be considered markers on the specific chromosome and have revolutionized the diagnosis of this disease.

1.6.4 X-linked Recessive

If an affected non-founding son has two unaffected parents, then the disease must be X-linked recessive. The father, who is marrying, does not have any disease alleles since he is marrying into the family. So, the affected son inherits an allele only from his unaffected mother. A male cannot be affected by a single autosomal recessive allele but can be affected by a single X-linked recessive allele (Figure 1.8). The alleles determine a few phenotypes on the X-chromosome's differential regions, related to the sex determination. Phenotypes inheriting the X-linked recessiveness display the following patterns in the pedigree analysis.

1. Male phenotypes are more X-linked recessive than female phenotypes because a female phenotype is the result of mating by a father and mother, with the father bearing the allele ($X^A/X^a \times X^a/Y$), whereas a male with the X-linked phenotype can be produced when only the mother carries the allele. Almost all individuals showing the recessive phenotype are males.
2. None of the offspring of an affected male are affected, but all his daughters will be heterozygous carriers as the females receive one of their X chromosomes from their fathers. As a result, half of the sons born to these carrier daughters will be affected.

The best-known example of X-linked recessive is hemophilia, a malady in which a person's blood fails to clot. Most famous cases of hemophilia have been reported from the pedigree of the European royal families. The original hemophilia allele in the pedigree arose from a spontaneous mutation in the reproductive cells of either Queen Victoria's parents or Queen Victoria herself. Alexis, the son of the last czar of Russia, inherited the allele from Queen Victoria, the grandmother of his mother, Alexandra. Duchenne dystrophy is another lethal X-linked recessive disorder, where the phenotype is the atrophy of muscles. The onset of the disease generally occurs before six years of age, with confinement to a wheelchair by 12 and death by 20 years. The gene responsible for the disorder has been isolated and shown to encode a muscular protein known as dystrophin. A sporadic X-linked recessive phenotype testicular feminization syndrome occurs at a frequency of about 1 in 65,000 males. People afflicted with this syndrome are chromosomally males (44A+XY) but develop as females. They have female external genitalia, a blind vagina without a uterus. Testes may

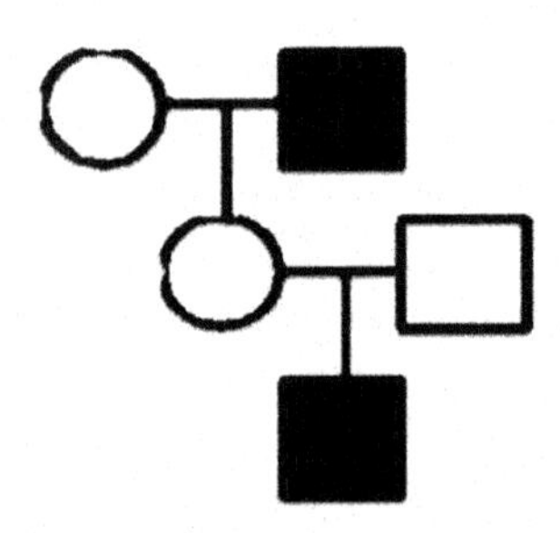

FIGURE 1.8 Pedigree of an X-linked Recessive inheritance.

be present either in the labia or in the abdomen. These people are sexually sterile, and the condition cannot be reversed even by injecting the androgen (male hormone). Hence, it is also called androgen insensitivity syndrome. One of the main reasons for the sensitivity is the allele that codes for a malfunctioning androgen receptor protein. As a result, androgen will not affect the target organs involved in the maleness. In humans, femaleness results when the male determining system is not functional.

1.6.5 X-Linked Dominant

The phenomenon of X-chromosome inactivation complicates the mechanisms of X-linked dominance in human beings. When an affected daughter of non-founding parents has an affected father, it is difficult to determine whether the dominant disease is autosomal or X-linked (Figure 1.9). The affected father can transmit either an autosomal dominant allele or an X-linked dominant allele to his daughter. Hypophosphatemia, a type of vitamin D-resistant rickets, is an example of X-linked dominant inheritance. The pedigree of rare X-linked dominant males and females shows the following characteristics:

1. Affected males transmit the condition to all their daughters but none of their sons.
2. Females married to unaffected males pass the condition on to half their sons and daughters.

1.6.6 Calculations for the Risk Allele by Pedigree Analysis

After confirming the diseased allele in a family, the probability of the prospective parents having a child with the respective disorder is calculated by using simple gene patterns. For example, if a married couple finds out that each had an uncle with a severe autosomal recessive Tay-Sachs disease (a disorder that leads to the destruction of the brain and spinal cord), the pedigree will be as mentioned in Figure 10.

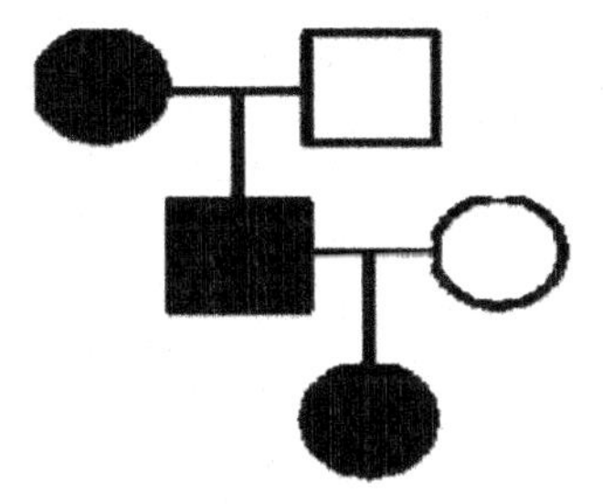

FIGURE 1.9 Pedigree of an X-linked dominant inheritance.

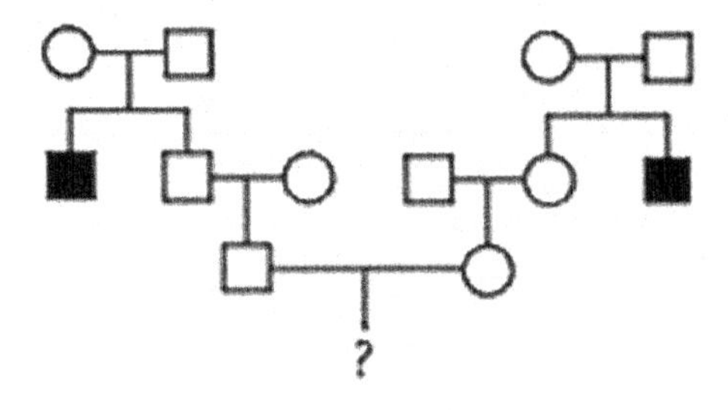

FIGURE 1.10 Pedigree of a married couple, each having an uncle with Tay-Sachs disease.

The probability of having a child with the Tay-Sachs disease by the couple can be calculated in the following way. It is very clear that both are not getting the disease. However, it is not clear whether both are heterozygotes. If, at all, they are heterozygotes, then they stand a chance of having an affected child.

1. Both of the man's grandparents must be heterozygotes *T/t* as they produced a *t/t* child (the uncle) with the disease. Therefore, they effectively constitute a monohybrid cross. The man's father could be either *T/T* or *T/t*. We know that the relative probabilities of these genotypes must be 1/4 T/T and 1/2 T/t and 1/4 t/t, respectively. Therefore, there is a 2/3 probability that the father is a heterozygote [calculated as 1/2 divided by (+1/4 + 1/2)].
2. The man's mother can be assumed to be *T/T* since she married into a family where the diseased alleles are generally rare. Therefore, if the father is *T/t*, the mating to the mother will be *T/t* × *T/T*, and the expected progeny proportions are 1/2 *T/T* and 1/2 *T/t*.
3. The overall probability of a man being a heterozygote can be calculated by a statistical product rule, which states that *"the probability of two independent events that occur is the product of their probabilities."* The probability of a man being a heterozygote is the probability of his father being a heterozygote *multiplied by* the probability of the father having a heterozygous son, which is 2/3 × 1/2 = 1/3.
4. Similarly, the probability of the man's wife being heterozygous is also 1/3.
5. If both man and his wife are heterozygous (*T/t*), the probability of having a *t/t* child is 1/4.
6. Finally, the overall probability of the couple having an affected child is 1/3 × 1/3 × 1/4 = 1/36. In other words, there is a 1 in 36 chance of having an affected child by these parents.

1.7 ALLELE FREQUENCY ESTIMATION

Alleles are variant forms of a gene located at the same position, or a genetic locus, on a chromosome. Allele frequency measures the relative frequency of an allele on a genetic locus in a population. It also represents the incidence of a gene variant in a population. It is the relative proportion of all alleles of a gene. Allele frequency, also known as gene frequency, is defined as a percentage of all alleles at a particular locus in a population belonging to the same gene pool, represented by a particular allele. Allele frequency is the number of copies of a specific allele divided by the total number of all alleles at that locus in each population. Allele frequency is expressed as a percentage (%), decimal, or even as a fraction in the studies involving population genetics for depicting the amount of genetic diversity at the individual, species, generic, and population levels. It is calculated by dividing the number of times an allele of interest is observed in a population by the total number of copies of all alleles at that particular genetic locus in that population (Su et al. 2011). Allele frequencies are the reflection of genetic diversity in a population. Changes in allele frequencies over time can indicate that genetic drift occurs or that new mutations have been introduced into the population.

1.7.1 Determination of Allele Frequency

Determination of an allele frequency in a population has become the most fundamental and salient quantification of human genetics, which forms the basis for many populations and medical genetic studies. Allele frequencies are subjected to changes by many evolutionary forces. Hence, these allele frequencies can be used for inferring past evolutionary events. Allele frequencies at single nucleotide polymorphisms (SNPs) are used as tools to infer the demographic history. Patterns of changes in the allele frequency provide information about the possible effects of natural selection. After a complete selection sweep, a series of low-frequency and high-frequency derived SNPs can be identified around the selected site.

Conversely, SNPs under the direct influence of adverse selection are expected to be at a lower frequency than predicted by demography alone. Many commonly used summary statistics in population genetics are the direct functions of allele frequencies. Allele frequencies also form the primary basis of association studies between SNPs and common diseases. In their simplest form, case-control association studies seek to quantify the difference in allele frequency between cases (individuals with the disease) and controls (individuals without the disease). There is a rapidly growing interest in performing association studies between rare variants and common diseases using next-generation sequencing approaches. Given the importance of allele frequencies in genetic studies, it is critically important to estimate them reliably. Traditionally, allele frequencies were estimated by counting the number of times each allele was noticed in a population sample. This approach has been used quite successfully on SNP genotype data and Sanger's sequencing data as the genotypes for unambiguous determination of each individual (Adrianto and Montgomery, 2012). However, this approach may not be so successful when applied to the data generated by the next-generation sequencing technology, mainly due to the following specific reasons.

1. Data generated by next-generation sequencing has a higher error rate than the traditional Sanger sequencing or SNP genotyping assays.
2. To sequence more samples, researchers often sequence everyone with a shallow coverage. As a result, every base will only be covered by a few reads, making it more difficult to infer an individual's genotype at a particular site accurately.
3. As the reads from next-generation sequencing technologies are short, additional errors can occur while trying to align the short reads back to the reference genome. For these reasons, estimation of allele frequencies remains challenging.

Several different approaches have been proposed for making accurate inferences of allele frequency by using next-generation sequencing technologies. The first set of approaches uses the traditional paradigm of estimating allele frequencies by first inferring individual genotypes and then tabulating their frequencies. Over the last decade, large-scale genome-wide association studies (GWAS) based on genotyping arrays have helped researchers identify hundreds of loci harboring common variants associated with complex traits. One of the significant disadvantages of genotyping arrays is the limited power for detecting a rare disease variance. Rare variants with Minor Allele Frequency (MAF) less than 1% cannot be captured by GWAS. Such low MAF variants may have substantial effect sizes without showing Mendelian segregation. Another most popular approach is the genotyping of the Affymetrix 6.0 array chip that contains nearly 1 million SNPs, with one-third of them residing in the coding regions. Despite identifying statistically significant SNPs through GWAS and locating them either in the intron or in the intergenic regions, their biological function is yet to be identified. A major limitation of genotyping arrays is their incapability of finding novel SNPs, as the SNPs are pre-determined on the array.

Most of the limitations in genotyping arrays can be overwhelmed by using high throughput next generation sequencing technologies, which can target a specific region of interest, such as the whole exome. The functions of variants identified in the coding regions are often easier to explain than variants identified in the intron or intergenic regions. Also, by targeting the exome, ~30 million base pairs in the coding region can be effectively examined rather than just 0.3 million SNPs on the Affymetrix 6.0 array. Sequencing technology has also been used for detecting rare variants with a frequency between 1% and 5%. However, due to the requirement for a large sample size for detecting low frequency variants, the detection of rare variants with a frequency of < 1% can still pose a significant challenge for NGS technology.

1.7.1.1 DNA Pooling

DNA pooling is a strategy that is often applied to reduce the financial burden in a large sample size. Erlich and his associates in 2009 introduced the concept of DNA pooling. The main idea behind this approach is to pool the DNA from multiple individuals into a single large DNA mixture, prepare it as a single genomic DNA library, and subject it to sequencing. In this approach, library preparation costs are considerably reduced as only a single library is prepared for the entire DNA pool instead of one library per every DNA sample. DNA Pooling is divided into two categories. Firstly, the non-overlapping pool method considers each individual in only one pool, and each pool consists of a fixed number of individuals. Secondly, the overlapping pool method enrolls the DNA of every individual into multiple pools and utilizes this information for recovering the genotype of every individual. DNA pooling has been extensively applied for linkage studies in plants, allele frequency measurements of microsatellite markers and single nucleotide polymorphisms (SNPs), homozygosity mapping of recessive diseases in inbred populations, and mutation detection. Data generated from the DNA pooling studies is accurate and highly reliable. Pooling is also associated with the detection of rare alleles and mutations. On the other hand, a few drawbacks reduce the effectiveness of the DNA pooling strategy. Pooled sequencing can generate variances with high false-positive rates when compared with individual sequencing. The ability of pooled sequencing to accurately determine the allele frequency is often limited. There are two different kinds of pooling paradigms.

1. DNA Barcoding: Paul Hebert's group first published the methodology involved in DNA barcoding at the University of Guelph in 2003. They proposed a novel mechanism of species identification and new species discovery using a short fragment of DNA from a standardized region of the genome. A region in the genome that identifies almost all major groups of the animal kingdom is 648 bp in the mitochondrial cytochrome c oxidase 1 gene (COI), proving to be highly effective. This fragment is short enough for quick and economical sequencing, with sufficient length to identify variations among species. Unfortunately, the COI barcode is not applicable for identifying plants as it evolved too slowly in the plant system. During multiplex barcoding on an Illumina HiSeq 2000 sequencer, one lane on average can generate 100 - 150 million reads per run. For exome sequencing, a minimum of 30 - 40 million reads per sample is needed to generate reliable coverage in the exome for variant detection. Thus, the common practice is to multiplex between 3 and 4 samples per lane to minimize costs. By multiplexing the barcode technology, it is possible to identify each read's origination. The only disadvantage of multiplex barcoding is the extra cost of barcoding and labor.
2. Pooling without multiplex: This is a cheaper alternative, which prevents identifying the origin of each read. By applying the simulations, the effectiveness of estimating allele frequency from pooled sequencing data can be determined. These approaches have the advantage that they can directly estimate the quantity of interest without inferring the other uncertain information (individual genotypes). The utility of this approach is yet to be fully explored for different types of population and genetic studies.

1.8 THE HARDY-WEINBERG LAW

The *Hardy-Weinberg law* is a fundamental law, considered to be the fundamental basis for population genetics. This law is denoted as the unifying concept of population genetics and provides theoretical support for the description of evolutionary phenomena. Two eminent personalities independently postulated this theorem: Godfrey Harold Hardy, a mathematician, and professor at the University of Cambridge, and Wilhelm Weinberg, a doctor from Germany in 1908. This law predicts how the gene frequencies are transmitted from one generation to another, given a specific set of assumptions. Specifically, the law is, ***"If an infinitely large, random mating population is free from outside evolutionary forces such as mutation, migration, and natural selection, then the gene frequencies will not change over time, and the frequencies in the next generation will be p^2 for the AA genotype, 2pq for the Aa genotype and q^2 for the aa genotype."*** The assumptions and conclusions made by this law are discussed here.

1.8.1 Infinitely Large Population

A population must be sufficiently large so that the chance occurrences cannot significantly change allelic frequencies. This point can be better understood by the random flipping of a coin, which is just as likely to land on heads as it is on tails. If a coin is flipped 1000 times, it will likely land on heads 50% of the time. If the same coin is flipped only ten times, it is much less likely to land on heads five times. The same holds for allele distributions in populations. Large populations are unlikely to be affected by chance changes in the allele frequencies as these chance changes are minimal compared with the total number of allele copies. In small populations with few allele copies, chance changes can considerably alter the allele frequencies. In small populations, a change in the allelic frequencies and phenotypes based on random occurrences is known as genetic drift.

1.8.2 Random Mating

For all alleles to have an equal chance of being passed down to the next generation, mating within the population must be random. Non-random mating can provide an advantage to specific alleles, allowing them to be passed down to offspring more than other alleles, which increases their relative frequency in the population. According to the theory and processes of natural selection, individuals with great fitness in a specific environment will work against random mating, and the fittest organisms are likely to mate.

Random mating refers to matings in a population that occur in proportion to their genotypic frequencies. For example, if the genotypic frequencies in a population are MM= 0.83, MN = 0.16 and NN = 0.01 then, 68.9% (0.83 x 0.83 x 100) of the matings would occur between MM individuals. If any significant deviation from this expected value occurs, it can be considered that random mating did not occur in this population. If significant deviations such as MM x MN or MN x NN occurred in the other matings, the assumption of random mating would be violated again. In humans, random mating will occur for many traits, such as blood group, where individuals do not select their mate based on their blood group. However, for the traits, such as intelligence or physical stature, the population is not randomly mating. These traits will not preclude the analysis of the population for those traits at which random mating is occurring.

1.8.3 No Mutation

For allelic frequencies to remain constant, any change should occur in the number of copies of an allele due to mutation. This condition can be met in two possible ways. Either, by the population experiencing little or no mutation, or by the population experiencing balanced mutation. Balanced mutation occurs when the rate at which copies of an allele are lost to a mutation equals the rate at which new copies are created by mutation.

1.8.4 No Immigration

Allele frequencies will remain constant in a population only if the individuals do not move in and out of that population. Whenever an individual enters or exits a population, it makes copies of alleles with it, changing the overall frequency of those alleles in that population.

1.8.5 No Evolutionary Forces Affecting the Population

Evolutionary forces may or may not show their effect on a population. If mating in that population is random, some loci may be more affected by these forces. For the loci affected by the forces, this assumption will be violated. For those loci that are not affected by these forces, this assumption will not be violated.

1.8.6 Conclusion

Two conclusions can be mathematically demonstrated for the Hardy-Weinberg Law from the following table. If p equals the frequency of allele A and q is the frequency of allele a, the union of gametes would occur as mentioned below

	p	q
p	p^2	pq
q	pq	q^2

1. In the above table, the genotypic frequency for *AA, Aa,* and *aa* are p^2, $2pq$, and q^2 respectively. These are the values predicted by the law concerning the genotypic frequencies after one generation of random mating.
2. Frequencies of the two alleles will be similar in two generations. This can be mathematically predicted as follows:

$p = f(AA) + ½f(Aa)$ (substitute from the above table)
$p = p^2 + ½(2pq)$ (factor out p and divide)
$p = p(p + q)$ ($p + q = 1$; therefore $q = 1 - p$; make this substitution)
$p = p\,[p + (1 - p)]$ (subtract and multiply) $p = p$

From the above equation, it can be concluded that the gene frequencies do not change in one generation. Analogous calculations would also show that the (q) allele frequency would not change in one generation. In the absence of any factors that change these allelic frequencies, the genotypic and allelic frequencies will remain the same from generation to generation. These two conclusions have been experimentally demonstrated to be valid and form the basis for population and evolutionary genetics research.

1.8.7 Circumstances in which the Hardy-Weinberg Law Cannot Apply

Forces lead to an evolutionary change and provide the conditions where the Hardy-Weinberg law will not be applicable. These are the following forces

1. **Mutations**
   The frequency of gene B and its allele b will not remain in Hardy-Weinberg equilibrium if the mutation rate of B > b or b > B changes. Critically, this type of mutation plays only a minor role in the evolution, and the rates are too low. However, both gene and whole genome duplication is a form of mutation, which might have played a significant role in the evolution of new alleles. After being shuffled in various combinations with the rest of the gene pool, they provide the raw material for natural selection.
2. **Gene Flow**
   Most of the species are made up of local populations, whose members tend to breed within the same population group. Breeding among the local populations might lead to developing a gene pool that is distinct from other local populations. Breeding between the members of one population group with occasional immigrants from the adjacent population can introduce new genes or even alter the existing gene frequencies among the residents. In many plants and some animals, gene flow can also occur between different species, known as hybridization. Breeding these hybrids with either of the parents will lead to the entry of new genes into the gene pool of the parent population. This process is known as introgression. In either case, gene flow increases the variability of the gene pool.

3. **Genetic Drift**
   Most of the interbreeding is limited to the members of local population groups. Chance alone can eliminate certain members out of proportion to their numbers in a population. In such cases, the frequency of an allele may begin to drift toward higher or lower values, ultimately representing 100% of the gene pool, or just disappear from it. Drift produces evolutionary change, but there is no guarantee that the new population will fit better than the original one. Evolution by drift is aimless and non-adaptive.
4. **Non-random Mating**
   One of the cornerstones of the Hardy-Weinberg equilibrium is that mating in the population must be random. If individuals (usually females) are choosy in their selection of mates, the gene frequencies may become altered. Darwin called this phenomenon sexual selection. Non-random mating in a population is more common, with breeding territories, courtship displays, and pecking orders being responsible for it. In each case, specific individuals do not get to make their proportionate contribution to the next generation.
5. **Assortative Mating**
   Humans rarely mate with random preferred phenotypes known as assortative mating. Marriage between close relatives is considered a particular case of assortative mating. The closer the kinship, the more alleles are shared and the greater the degree of inbreeding. Inbreeding can alter the gene pool as it might prejudice homozygosity. Potentially harmful recessive alleles invisible in the parental generation become exposed to the forces of natural selection in the children due to assortative mating. Many species, including plants and animals, avoid assertive mating through unique mechanisms. Male mice use olfactory cues to discriminate against close relatives while selecting mates.
6. **Natural Selection**
   Individuals with specific genes could produce better offspring than those without them. The frequency of these genes will increase in a population. This mechanism is known as Darwin's natural selection, which results from either differential mortality selection or differential fecundity selection. Mortality selection is another way of describing Darwin's survival of the fittest. Specific genotypes are less successful in surviving till the end of their reproductive period than others. The evolutionary impact of mortality selection can be felt at any time between forming a new zygote to the end of the organism's fertility period. Fecundity selection is another way of describing Darwin's criterion of fitness. Specific phenotypes make a disproportionate contribution to the gene pool by producing a disproportionate number of young. As a result, some phenotypes of the next generation will be better than others in their ability. By Darwin's standards, they are considered fit. The outcome of this fitness is a gradual change in the gene frequencies of that population

1.9 MULTIFACTORIAL INHERITANCE

Multifactorial inheritance is the condition in which many factors or multiple factors are involved in inheriting a trait. They comprise genetic and environmental factors, where a combination of genes from both the parents and unknown environmental factors produces a trait or a condition. Multifactorial inheritance was first studied by Galton, a close relative of Darwin and a contemporary of Mendel. Galton (1886) established the principle and termed it "Regression to mediocrity." Galton has studied the inheritance of a continuous flow of characteristics in humans such as height, and intelligence, in contrast to Mendel's discontinuous flow of characteristics. Galton noticed that extremely tall fathers tend to have shorter sons, and extremely short fathers have their sons taller than them. Here tallness or shortness did not breed true as in Mendel's experiments on pea plants. Here, the offspring has regressed to mediocrity.

TABLE 1.2
Degree of Relationship and Genes That are Shared Between the Family Members in Common

Degree of relationship	Relationship	Common genes (%)
First degree relative	Parents, Children, siblings	50
Second degree relative	Aunt, Uncle, Nephew, Niece	25
Third degree relative	First cousin	12.5

Multifactorial traits recur in families as they are partially genetic. The chance of a multifactorial trait depends on how close the family member is to the trait. The risk will be higher if a brother or a sister has the trait than the first cousin with the trait or disease. Family members share a certain amount of common genes in their pool, depending on their relationship (Table 1.2).

One specific gender, either male or female, is often affected more frequently by multifactorial traits, generating a threshold of expression. As a result, one gender is more likely to show a problem over the other gender. For example, congenital hip dysplasia is a multifactorial disorder that is nine times more common in females than males. Multifactorial inheritance is responsible for a significant number of disorders resulting in hospitalization and special care. More than 10% of newborns will express at least one multifactorial disease at some time in their life. Diseases such as atopic reactions (development of allergies such as asthma, rhinitis, dermatitis), diabetes, cancer, spina bifida/anencephaly, pyloric stenosis, cleft lip, cleft palate, congenital hip dysplasia, and club foot are the most widespread disorders resulting from multifactorial inheritance. A few disorders resulting from multifactorial inheritance are discussed below.

1.9.1 Open Neural Tube Defects (ONTDs)

Spina bifida (Open spinal cord), Anencephaly (Open skull), and Encephaloceles (protrusion of the brain or its coverings through the skull) are the general open neural tube defects (ONTDs) seen in every 2 of the 1,000 births. During pregnancy, every human and spinal cord begins their formation as a flat plate of cells, which further rolls into a neural tube. An entire or a part of the neural tube fails to close, leaving an opening known as ONTD. This opening is covered by the skin or a bone in 20% of the cases and is left open in 80%. Anencephaly occurs when a neural tube fails to close at the base of the skull. Spina bifida occurs when the entire or a part of the spinal cord fails to close. Babies with anencephaly are either stillborn or alive only for a few days after their birth. Babies born with spina bifida have either temporary or permanent physical problems such as paralysis, lack of bowel and bladder control, club feet, hydrocephaly (accumulation of spinal fluid in the head), and mental retardation, usually forcing one or two surgeries. Recent technology has developed fetus surgery for closing the defect before delivery.

The chances of couples producing a baby without any ONTD with no family history is 90%. ONTDs have resulted from genes inherited from both the parents and environmental factors, such as uncontrolled diabetes in the mother and usage of medicine not prescribed by the medical practitioner. ONTDs are noticed seven times more in females than that in males. The birth of a child with ONTD in a family will increase its re-occurrence by 3 to 5%. The type of ONTD might also change. Generally, the neural tube closes after 32 days of conception. Consumption of folic acid during the early days of pregnancy will reduce neural tube defects, especially in couples with ONTD. The chance of a multifactorial condition on ONTD in a future pregnancy would be 1.3 - 5% if the parents already had a child with ONTD, 2% if the aunt or uncle has an ONTD, and 0.5% if the cousin has an ONTD. ONTDs can be diagnosed before birth by measuring the alpha feto protein (AFP) concentration in the amniotic fluid around the baby. Fetal ultrasound during the pregnancy can also detect the possibility of ONTD occurrence but not with 100% accuracy. Minor or closed defects without any leakage of cerebrospinal fluid will not be identified in these tests.

1.9.2 Hip Dysplasia

Hip dysplasia is nine times more prevalent in females than in males. One of the foremost environmental factors contributing to hip dysplasia is the baby's response to the mother's hormones during pregnancy. The birth of a child with hip dysplasia will increase the chance of its re-occurrence in another child by ~ 6%. The specific chance of its re-occurrence is still less if the second child is a male. The difference is the threshold for the trait between a male and a female.

1.9.3 Height

Height is the factor determined both by genetic and environmental factors. Few people (both male and female) are either exceptionally tall or exceptionally short due to an effect of its gene. The population of a specific ethnic group is short, and a few ethnic groups are tall, influenced both by the genome and geographical conditions.

1.9.4 Model for Multifactorial Inheritance

For many years, the applicability of the Mendelian inheritance or Galtonian inheritance for humans has debated. Mendelian inheritance applies to a few genetic diseases, but these are rare, affecting a small portion of the human population, and considered trivial. In contrast, the inheritance of quantitative traits could not be used to predict the outcomes, with only average estimates can be measured in the extensive population studies. Fisher (1918) achieved the balance between these two mechanisms of inheritance. He has demonstrated how the inheritance of quantitative traits can be reduced to Mendelian inheritance at many loci. Fischer demonstrated a model, which consists of one locus for height with three alleles. Allele H2 adds 2 inches to the average 68-inch height. Allele h0 neither adds nor subtracts from the average height of 68 inches, and allele h- subtracts 2 inches from the average height. If the frequency of h0 is twice the frequency of h2 or h-, the Punnett square for the population would be as displayed in Table 1.3.

Each locus consists of three different alleles enough to produce population frequencies indistinguishable from a standard curve. As more loci are included in this Punnett square, this binomial distribution quickly approaches this Gaussian distribution, or the standard bell-shaped curve observed with human quantitative traits. The following conclusions can be drawn from the above multifactorial model.

TABLE 1.3
Punnett Square Demonstrating the Multifactorial Inheritance

		Father's Gametes		
		h2	**2h0**	**h-**
Mother's Gametes	**h2**	h2h2 72"	2(h2, h0) 70"	h2h- 68"
	2h0	2(h2h0) 70"	4(h0h0) 68"	2(h-h0) 66"
	h-	H2h- 68"	2(h-h0) 66"	h-h- 64"

1. Several but not unlimited or innumerable loci are involved in the expression of a trait.
2. There is no clear dominance or recessiveness at each of these loci
3. The loci act in an additive fashion, with each either adding or detracting a small amount from the phenotype.
4. The environment interacts with the genotype to produce a final phenotype. For example, women are three inches shorter than men, with the same genome, with environmental factors such as hormones affecting the final phenotype.
5. Not all human traits that show a continuous distribution in the population are multifactorial.
6. No bimodal distribution can be controlled by multifactorial expression. It is likely to be controlled by a single dominant or recessive gene with modifying environmental factors.
7. All multifactorial traits show a unimodal bell-shaped distribution.

1.10 HERITABILITY

Heritability is the concept that plays a central role in understanding the phenotypic and genotypic differences of an individual and the role of the environment. Heritability is defined as the proportion of a phenotypic variance attributed to its genetic variance and the extent to which individual genetic differences contribute to the individual phenotypic differences. The contribution of a genotype to the phenotypic variation for a specific trait in a population is expressed as the ratio of genetic variance to the phenotypic variance. Thus, heritability denotes the proportion of phenotypic variance due to the genotype. As the heritability is a proportion, its numerical value ranges from 0.0, where the genes do not contribute to the making of a phenotype, to 1.0, with genes being the only reason for individual differences.

Heritability is represented in the following equation:

$$\text{Heritability} = \frac{VG}{VP} \text{ or } \frac{VG}{VG + VE}$$

VG is the genotypic variance, VP is the phenotypic variance, and VE is the environmental component of variance. Suppose every individual in a population has the same allele for a specific trait that shows very little or no variation (differences) on that trait. In that case, heritability for that specific trait is zero because the trait has no genetic variation in the case of hair color among the Eskimo population. The whole Eskimo population has identical alleles for hair color. Hence, the heritability for their hair color is 0.00, even though the hair color is under substantial genetic control (Croston et al. 2015). The concept of heritability was developed for experiments on plant and animal breeding, which are also applied to human genetic processes. Heritability reflects the variations (differences) among the individuals of any given species. These variations among individuals within a species are both genetic and environment specific. This statement can be best attributed to the case of height in human beings. Taller parents will have their children taller than shorter parents, and identical twins are equally tall. These are the genetic differences. Japanese Americans are taller and heavier than their cousins who grow up in Japan, indicating the role of the environment on the phenotypes.

1.10.1 TYPES OF HERITABILITY

Heritability is of two types: broad sense heritability and narrow sense heritability.

1. Broad sense heritability is the ratio of genotypic variance VG to the total phenotypic variance (VP).

$$h2\ (bs) = VG/VP \text{ or } VG/VG + VE$$

Broad sense heritability estimates are valid specifically for homozygous populations. For a segregating population, the genetic variance consists of an additive and a dominance component. This dominant component will not have any contribution to the phenotypes of homozygous populations. Hence, the broad sense heritability is less critical to the segregating generation, but a narrow sense heritability is more critical as it cannot entirely rely on a phenotype.

2. Narrow sense heritability is the ratio of additive genetic variance VA to the total phenotypic variance VP.

$$h2\ (ns) = VA/VP = VA/VG + VE$$

As the narrow sense heritability is based on the individual value, the measurement is highly reliable. Nevertheless, the magnitude of narrow sense heritability is always lesser than that of broad sense heritability (Vischer et al. 2006).

1.10.2 Methods of Estimating the Heritability

Traditionally, heritability is estimated by three different methods:

1. From the ANOVA table of a trial consisting of a large number of genotypes.
2. Estimation of VG and VE from the variance of P1, P2, P3, and P4 generation of a cross.
3. By the regression of parent-offspring, followed by doubling. This method provides the estimates of heritability. Thus, H = 2b, where b is the regression of progeny means on parent value.

Recently, novel methods that exploit the genetic marker data have been applied for estimating heritability. These methods are based on the correlation between phenotypic and genetic similarity within the families. They exploit the fact that there is variation in identity between pairs of individuals with the same expected value, and this variation can be measured with genetic markers. Variation in identity is the result of the random segregation of chromosomes during meiosis. Comparing the percentage of genetic similarity between siblings with phenotypic variance is the latest method to estimate heritability. The genetic similarity was derived from the genome-wide expression data of genetic markers. Methods that exploit the genetic marker data, such as DNA fingerprint markers and maximum likelihood calculation without any prior information on the subject's relatedness, were implied. These methods are based on the correlation between phenotypic and genetic similarity within the families. They exploit the fact that there is variation in identity between pairs of individuals with the same expected value and that this variation can be measured with genetic markers. Variation in identity is the result of the random segregation of chromosomes during meiosis. Heritability might be subject to changes with time and age, caused by the genetic variations in a population. Genetic variations are also known to change in time and age in response to new environmental conditions, altering allelic frequencies, or inducing new mutations in the genome.

1.10.3 Heritability and the Environmentability of a Trait

A trait in an organism is equally influenced and controlled by the environment and genetic components. Every gene will express in an environment, and all environments will influence a genotype. However, all traits are genetically controlled and are significantly less influenced by the environment. For example, the human embryo is genetically influenced to develop hands and legs but might be abnormal if exposed to thalidomide during the early stages of its development. Most factors, such as a person's language, depend on the environment in which he or she grows. All human beings learn some language even under language deprived conditions, implying that the traits in the human brain are genetically influenced to develop a language or languages. Linguists claim that all languages have an essential common feature universally. The environmentability of a trait has an analogous interpretation to heritability, denoted as the proportion of the phenotypic variance attributable to the environmental variance. The extent to which the differences in the environment contribute to the individual differences. For example, if the heritability of human behavior is in the range of 0.3 - 0.6, then the environmentability of human behavior will be in the range of 0.4 - 0.7. Five attributes have been identified regarding estimates of heritability and environmentability:

1. *Heritability and environmentability are abstract concepts.* Whatever the number might be, estimates of heritability tell nothing about the specific genes contributing to a trait. Similarly, numerical estimates of environmentability can provide no information about the environmental variables that influence behavior.
2. *Heritability and environmentability are population concepts.* Both of these concepts tell nothing about an individual. If the heritability of an individual is 0.4, it implies that 40% of the individual differences such as shyness is attributable to individual genetic difference. It does not mean that 40% of a person's shyness is due to genes, and 60% is due to their environment.
3. *Heritability depends on the range of typical environments in the population that is studied.* The heritability will be high if the environment of a population is uniform and will be low if the range of environmental differences is extensive. If every individual in a population is exposed to the same environment, the observed differences are due to the genes. In this case, heritability will be high. If individuals are treated in different environments, then the heritability will be low.
4. *Environmentability depends mainly on the range of genotypes in the population studied.* This is converse to the above made point, which might be strongly applied to animal breeding, specially bred animals, and human populations genetically homogeneous as breeds of dogs, sheep, and the like.
5. *Heritability is no cause for therapeutic nihilism.* As heritability largely depends on the range of typical environments in the studied population, there is very little information about the extreme environmental interventions utilized in some therapies.

1.10.4 Misconceptions or Misunderstandings on Heritability

Heritability is generally misunderstood as an inheritance of a trait. Heritability is a statistical estimate, which does not give any information about the genetic structure of an individual. Moreover, it indicates that genes may be attributed to a certain fraction of a variation in a trait at the population level. A heritability estimate is unique to a particular population, and is non-transferable. However, it has been almost conserved across populations and species. Another common misconception is that heritability is not the proportion of the phenotype, but the proportion of phenotypic variations explained by genetic factors.

1.10.5 Heritability of Human Height

Human height is a highly heritable, polygenic, classic anthropometric quantitative trait. The study of human height has been a tradition in genetics for a long time, and studies on human height eventually resulted in quantitative genetics at the beginning of the 20th century. Galton published his observations on the relationship between parent and offspring height in 1886. Pearson and Lee have calculated correlations of height between relatives, providing a shred of evidence for the inheritance of height in humans. The first heritability estimate of human height was calculated by Fisher (1918) and explained the proportion of total variation by genetic variation. He also demonstrated that continuous characters result from multiple loci combinations with minor effects, replacing the blending inheritance hypothesis proposed by Galton (1886). After this, considerable evidence has emerged from twins, adoption, and family studies, estimating the role of genetic factors in determining height, revealing that it is one of the most heritable human quantitative phenotypes.

Further, genetic linkage studies have established the influence of the human genome on height and genome wide association studies (GWAS) have identified loci consistently associated with height in populations of different ancestry. Besides these genetic factors, a multitude of environmental factors also affect human height. These factors have their influence during the whole growth period, with infancy as the most sensitive phase. The growth of children, and adults, is affected by adverse environmental conditions, such as nutrition, lack of dietary protein, diseases, and infections. These biological determinants are further associated with socio-economic differences in height both within and between populations.

The heritable nature of human height is the subject of study over the past one hundred years, including the genetic variations during childhood and adolescence. Twin studies have estimated that the heritability of height is lowest during infancy (0.2 - 0.5), which rapidly increases during childhood with varying values, and reaches from 0.70 - 0.90 during adolescence and adulthood. However, these studies did not involve the role of environmental factors, essential during infancy and childhood, persist in adolescence, or after the cessation of growth. A study conducted in four countries on 12,000 twin pairs from birth to 19 years of age has shown that the effect of the environment will remain up to 12 years and will influence again at 16 years. Height in human beings is also considered a classic example of a sexually dimorphic trait. On average, men are taller than women among all human populations. However, the genetic and environmental contributions responsible for the gender differences in the height variation are unknown. Further, an excellent mean difference in height has been observed among Western populations compared with East-Asian populations. A multinational study on the adolescent twins representing eight countries has shown that the total height variation was higher in Western populations, and the heritability estimates were mostly the same between Western and East-Asian populations.

The contribution of DNA sequence variations leading to the differences in adult height has been evaluated by genome-wide association studies (GWAS). Through this approach, 697 independent variants located within 423 loci have been identified, explaining about 27.4% of the heritability regarding human height. Most of the alleles that affect the height have been identified in the genome with a minor allele frequency (MAF) > 5% and are mainly located outside coding regions. A study performed on 200,000 coding variants in 7,11,428 individuals has identified 32 rare and 51 low frequency coding variants associated with the height. Exome chip assay and deep DNA sequencing have identified 89 genes that are likely to modulate human growth. This study has also identified 24 alleles segregating in the general population, affecting height by more than 1 cm. Rare and low-frequency variants are responsible for 1.7% of the heritable variation in height.

1.10.6 Heritability and Human Behavior

Determination of heritability is an essential measure in the current quantitative and behavioral genetics. The contribution of a gene, and of the environment on its expression in the phenotype, are studied for a long time. Psychologist Eric Turkheimer (2000) has proposed three laws of heritability for behavioral studies:

1. All human behavioral traits are heritable.
2. The effect of a shared family environment is less than genes.
3. A substantial proportion of variation in complex behavioral traits cannot be explained either by environment or genes.

These laws contrast with the early statements regarding the biological basis for human behavior, emphasizing the environment's role. Conceptually, heritability does not ensure that any percentage of a behavioral trait will either solely be decided by genes or by the environment. For the individuals who share a common genotype, any variance may be attributed to the environment. For the individuals who share a common environment, genes may have considerable control in determining the observed variance in a trait. The environment and environmental conditioning can further amplify genetic influence on the behavior. The individuals who do not share environmental conditions are unique, and this will have specific impacts on their behavior.

1.10.7 Designs for Estimating the Heritability of Behavior

1.10.7.1 Twin Studies

Classical twin studies are amongst of the most frequently used methodologies applied for estimating heritability. Comparing monozygotic twins with dizygotic twins, the monozygotic twins share 100% of genes, whereas dizygotic twins share only 50% of genes when exposed to similar environments. Deviation in their behavior is also studied, which can be ascribed to the genes.

1.10.7.2 Adoption Studies

This study provides data on the role of the environment. When siblings with similar genetic backgrounds (monozygotic twins) grow up in a different environment, deviations in the behavioral trait can be attributed to the environment due to similarity in the genes. The adoption study design is also applied to siblings with a certain level of genetic similarity, evaluated by DNA markers when they grow in a different environment. The heritability of a behavioral trait is estimated from a graph of phenotype variation against shared genes. Children of the monozygotic twins grown under a rearing-up environment continue the deviation in their behavior which is due to the genes, and the appearance of novel behavior, which parents did not originally show is due to the environment (Ashutosh Kumar et al. 2017). Molecular tools such as GWAS and SNP microarrays are used to identify the variance in the behavior. This study is commonly applied to estimate heritability. Overutilization of statistics complicates the whole issue, and even a slight variation in the statistical parameters will have changed results, especially in the GWAS analysis. Hence, these studies are to be performed with more skepticism.

1.10.7.3 Developmental Models

Developmental models are based on the concept that heritability changes with the age of individuals as they grow. If a child is followed throughout life for continuity of a particular behavioral trait, a change may be assigned to the child's changing environment or gene expression while growing up. However, a gene-environment interaction influences the genes in selecting or rejecting the environmental conditions, and individual adjustment to the environment may not be denied.

1.10.8 Candidate Genes for Studying Heritability

Several candidate genes in the human genome are tagged with various behavioral aspects such as the MAOA gene with antisocial behavior, dopamine receptors DRD2 and DRD4 with aggression and antisocial behavior, serotonin transporter 5HTTLPR for depression, SLC6-a noradrenalin polymorph for attention deficit hyperactivity disorder (ADHD), and GABRA2-a GABA polymorph to childhood conduct disorder and substance abuse in adults. To date, even a single significant causal relationship has not been proved for any such gene.

1.11 POLYGENIC INHERITANCE

A polygenic inheritance or multigenic inheritance is a type of inheritance in which a single trait is the cumulative effects of many genes. Polygenic inheritance contrasts with monogenic inheritance.

A single trait results from a single gene expression. A few examples of polygenic inheritance in human beings are skin color, eye color, height, weight, and intelligence. These traits do not follow the Mendelian pattern of inheritance.

1.11.1 Human Skin Colour

Human Skin color is one of the classic examples of polygenic inheritance. Assume that three genes A, B, and C, control dark pigmentation producing more melanin. The recessive alleles of these three genes a, b, and c, control light pigmentation as less melanin is produced. A genotype with all dominant genes AABBCC produces the maximum amount of melanin, resulting in dark skin, whereas a genotype with all recessive genes aabbcc produces the lowest amount of melanin, responsible for very light skin. Among these three genes, each dominant gene controls one unit of color, producing a wide range of intermediate colours. Depending on the number of dominant and recessive gene combinations such as AaBbCc (three dominant and three recessive gene combinations), a medium amount of melanin is produced, resulting in an intermediate skin color, a characteristic feature of **mulatto.** As mentioned in Table 1.4, the cross between two mulatto genotypes (AaBbCc X AaBbCc), each parent produces eight different types of gametes, which combine in 64 different ways in a total of seven skin colours. These skin colours can be represented by the number of capital letters ranging from 0 (no capital letters) – 6 (all capital letters).

The cross in the above table can also be mentioned with a binomial expansion $(a + b)^6$, where the letter a stands for the number of capital case letters and the letter b represents the number of small case letters. Each term in the expression represents the number of offspring with a specific skin color phenotype based on the number of capital letters in the genotype. For example, 20 offspring have three capital letters in their genotype. Hence, they have a skin color intermediate between very dark with all caps (AABBCC) and very light with no caps (aabbcc).

$(a + b)^6$	$= a^6$	$+ 6 a^5b$	$+ 15 a^4b^2$	$+ 20 a^3b^3$	$+ 15 a^2b^4$	$+ 6 ab^5$	$+ b^6$
	6 Caps	5 Caps	4 Caps	3 Caps	2 Caps	1 Cap	0 Caps

Polygenic inheritance explains many traits with a wide variation between extreme phenotypes and most individuals with intermediate phenotypes. In polygenic inheritance, the dominant genes are additive, with each dominant gene adding one unit of color to the genotype. With increasing dominant genes, the phenotype gets darker. The pea traits studied by Gregor Mendel involved pairs of alleles with only three possible genotypes and two phenotypes per trait. The round seed shape (R) gene is dominant over the wrinkled seed shape (r). Here, only three possible genotypes RR, Rr, and rr produce two phenotypes round (RR Rr) and wrinkled (rr), with no intermediate traits between

TABLE 1.4
Combinations of Skin Colours Due to a Cross Between Two Different Mulatto Types

Gametes	ABC	ABc	AbC	Abc	aBC	aBc	abC	abc
ABC	6	5	5	4	5	4	4	3
ABc	5	4	4	3	4	3	3	2
AbC	5	4	4	3	4	3	3	2
Abc	4	3	3	2	3	2	2	1
aBC	5	4	4	3	4	3	3	2
aBc	4	3	3	2	3	2	2	1
abC	4	3	3	2	3	2	2	1
Abc	3	2	2	1	2	1	1	0

them. If all human characters are controlled by the dominant and recessive alleles like the ones which Mendel studied, humans would have been either tall or short without any intermediates. The polygenic model is considered another exception to Mendel's ratios.

1.12 HAPLOTYPING

A haplotype is a set of genes of an organism inherited from a single parent, namely, mother or father. The word haplotype is derived from haploid, describing the cells having a single set of chromosomes, and from the genotype, which provides the genetic makeup of an organism. A haplotype can also describe a pair of inherited genes from both the parents on a single chromosome due to genetic linkage or describe all genes on a single chromosome inherited together from a single parent (Stacey et al. 2002). Haplotype also refers to the inheritance of a group or a cluster of single nucleotide polymorphisms (SNPs), which are the variations in single positions of DNA among different individuals. Haplotype also consists of alleles at multiple linked loci with one allele per locus on a single chromosome.

1.12.1 The Importance of Haplotype Analysis

Haplotypes help in the identification of disease patterns or genetic variations that are concerned with human health. For instance, if a haplotype is associated with a particular disease, then stretches of DNA near an SNP cluster can be examined to identify the gene or genes responsible for causing the disease. Haplotype information is helpful in the analysis of complex traits, population histories, evolutionary genetics, and linkage analysis using low-density markers on long chromosomal regions. Haplotype analysis can predominantly enhance the information content, which can be attributed to any single marker. Haplotypes for a pedigree can estimate the identity by descent probabilities among pedigree members, which provides the primary basis for many linkage analysis methods. The use of haplotypes on a selected short chromosomal region will be more beneficial than using individual high-density markers to analyze complex traits, association studies, and linkage disequilibrium analysis. Haplotyping can also be used to identify errors in genotyping, identify double recombination in short chromosomal regions, and infer missing genotypes in pedigrees.

1.12.2 Methods of Haplotyping

Haplotypes in diploid species are determined by using molecular techniques, which are more time consuming and expensive. Hence, they are not suitable for large-scale applications. The silico haplotyping method infers haplotypes from observed genotype data by statistical and computational methods, which is more valuable because the estimation is accurate. Haplotyping can be performed by using pedigree data as well as DNA samples.

1.12.2.1 Haplotyping for Pedigree

Haplotyping for a pedigree refers to the re-construction of a true and unknown haplotype based on the observed genotype data and available pedigree structure. The space for all consistent haplotype configurations is often monumental, particularly with the missing data. Statistics-based and genetic rule-based methods are developed for estimating a true haplotype by identifying either a single or a set of consistent haplotype configurations. A consistent haplotype configuration is a set or a cluster of ordered genotypes of all members in the pedigree at all loci, consistent with observed data and Mendelian segregation. A set of ordered genotypes can specify a haplotype configuration

$$\mathbf{G} = (\boldsymbol{G}_{i,j};\ \boldsymbol{i} = \mathbf{1}, \ldots, \boldsymbol{n}, \boldsymbol{j} = \mathbf{1}, \ldots, \boldsymbol{L})$$

$G_{i,j}$ is the ordered genotype of pedigree member i at locus j, n is the pedigree size, and L is the number of markers.

Likelihood-based methods are ideal for pedigree-based haplotyping for a long chromosomal region with relatively low-density markers. These methods re-construct configurations by maximizing their likelihoods or conditional probabilities. They assume Hardy-Weinberg equilibrium at individual loci and linkage equilibrium among markers because most of these methods were developed for an extended chromosomal region with relatively low-density markers such as 2 - 10 cM microsatellite maps. To utilize the densely clustered SNP markers for haplotyping, Abecasis and Wigginton have developed a likelihood-based method accounting for marker-marker linkage disequilibrium within the clusters of tightly linked markers and model recombination between clusters while assuming no recombination within the clusters. Recently available high-throughput technologies such as SNP arrays for genotyping many tightly linked SNP markers pose new challenges for haplotyping (Andrew, 2004). Traditional haplotyping methods developed for pedigrees with low-density markers can produce inaccurate results when applied to high-density markers with moderate to large amounts of linkage disequilibrium. Genetic rule-based methods are highly appropriate for tightly linked markers for a tiny chromosomal region. These methods re-construct the configurations by minimizing the total number of recombinants in pedigree data. A few rule-based genetic algorithms also account for a marker-marker linkage disequilibrium by estimating the haplotype frequencies of founders. Thus, population-based haplotyping approaches can be used for genome-wide association studies and the study of population and evolutionary genetics.

1.12.2.2 DNA Sample Based Haplotyping Methods

1.12.2.2.1 Microfluidic Whole Genome Haplotyping

Microfluidic whole genome haplotyping is applied for the physical separation of individual chromosomes from the spindle plate of a metaphase cell, followed by the immediate resolution of the haplotype for each allele. Whole genome haplotyping is used for resolving the personal haplotypes on a whole genome basis. Haplotyping invariably contributes to identifying polymorphisms existing in the same or allelic strand of DNA, which cannot be done by applying the next generation sequencing technologies (Jung-Ying et al. 2007). The inference frequently resolves haplotypes by comparing parental genotypes or population samples using statistics or computational methods. However, direct haplotyping is possible only through the isolation of full-length chromosomes. Most of the current haplotyping is performed through a limited genome by isolating chromosome segments. Whole genome haplotyping resolves the haplotype at the whole genome level, usually by isolating individual chromosomes.

1.12.2.2.2 Micro Dissection of a Chromosome

Chromosome microdissection is another technique applied for the isolation of single chromosomes for genetic analysis. In this process, the cells are arrested at the metaphase of mitosis. The nucleus is lysed on a glass slide, and the genetic material is partitioned into two under the microscope. The dissection of the whole chromosome is performed by careful use of a fine needle. Nowadays, the needle is replaced by a computer-directed laser. The genome area isolated can range from a single chromosome to several. The micro-dissected section of the genome is either genotyped or sequenced after the amplification of the DNA sample using a specialized platform to accomplish the whole genome haplotyping.

1.13 CONCLUSION

Genetic analysis is the process involved in understanding the phenomenon of heredity and its complexities by parsing it into unit entities, measuring the relationships, and empowering the role of the environment. Modern genetic analysis involves support through cloning, sequencing, and studying the expression of a gene, understanding its regulation, and applying these studies for the diagnosis of single gene disorders such as cystic fibrosis (CF), phenylketonuria, hemophilia, or multi-gene disorders such as diabetes or perhaps cancers with a clear heritable component (breast cancer). Even though genetic disorders are rare, they are responsible for 80% of rare disorders without proper treatment. The genetics of the immune system and its enormous variations across the population will determine our response to an infection. Further, most cancers result from an accumulation of genetic changes that occur throughout an individual's lifetime, under the influence of environmental factors. Finally, a clear understanding of genetics and the genome variations in the human population is vital for applying processes, which provide the foundation for quick diagnosis, curative therapies, beneficial treatments, and preventative measures.

REFERENCES

Adrianto, I., Montgomery, C. (2012). Estimating allele frequencies. *Methods Mol Biol.* 850:59–76.

Adrianto, I., Montgomery C. (2012) Estimating Allele Frequencies. In: Elston R., Satagopan J., Sun S. (eds) Statistical Human Genetics. Methods in Molecular Biology (Methods and Protocols), vol 850. Humana Press.

Andreas, Z. (1999). Basic mechanisms of monogenic inheritance. *Epilepsia*, 40(Suppl. 3):4–8,

Andrew, G. C. (2004). The role of haplotypes in candidate gene studies. *Genetic Epidemiology* 27: 321–333.

Ashutosh, K., Muneeb A. F., Vikas P., Maheswari, K. (2017). Heritability of behavior. In: J. Vonk, T.K. Shackelford (eds.), Encyclopedia of Animal Cognition and Behavior, Springer International Publishing.

Charles, R. S., Paula J. W. (1999). Monogenic traits are not simple lessons from phenylketonuria. *Trend in Genetics* 15 (7): 267–272.

Clayton, E. W., Halverson, C. M., Sathe, N. A., Malin, B. A. (2018). A systematic literature review of individuals' perspectives on privacy and genetic information in the United States. *PLoS ONE* 13(10): e0204417.

Croston, R., Ranch, C., Kozlovsky, D., Dukas R., Pravosudov, V. (2015). Heritability and the evolution of cognitive traits. *Behavioral Ecology* 26: 1447–1459.

Eirini, M. et al. 2017. Rare and low-frequency coding variants alter human adult height. *Nature* 542: 186–190.

Erlich, Y., Chang K., Gordon, A., Ronen, R., Navon, O., Rooks, M., Hannon, G. J. (2009). DNA Sudoku--harnessing high-throughput sequencing for multiplexed specimen analysis. *Genome Res.*19(7):1243–53.

Espen, R., Ragnhild B. Nes., Nikolaio, C., Olav, V. (2018). Genetics, personality and wellbeing. A twin study of traits, facets and life satisfaction. *Scientific Reports* 8: 12298.

Felsenstein, J. (1965). The effect of linkage on directional selection. *Genetics* 52: 349–363.

Fisher, R. A. (1918). The correlation between relatives on the supposition of Mendelian inheritance. *Transactions of the Royal Society of Edinburgh* 52: 399–433 (1918).

Galton, F. (1886). Regression towards mediocrity in hereditary stature. *Journal of the Anthropological Institute* 15: 246–262.

Getabalew, M., Alemneh, T., Akeberegn, D. (2019). Heritability and its use in animal breeding. *Int. J. Vet. Sci. Technol.* 4(1): 001–005.

Hebert, P. D., Cywinska, A., Ball, S. L., deWaard, J. R. (2003). Biological identifications through DNA barcodes. *Proc. Biol. Sci.* 270 (1512):313–321.

Jelenkovic, A. et al. (2016). Genetic and environmental influences on height from infancy to early adulthood: An individual-based pooled analysis of 45 twin cohorts. *Sci. Rep.* 6: 28496.

John, A. S., William, G. H. (2018). One hundred years of linkage disequilibrium. *Genetics* 209 (3): 629–636.

Jung-Ying, T., Daowen, Z. (2007). Haplotype-based association analysis via variance-components score test. *Am. J. Hum. Genet.* 81:927–938

Jurg, Ott., Jing, W., Suzanne M. L. (2015). Genetic linkage analysis in the age of whole-genome sequencing. *Nature Reviews* 16: 275–284.

Kimura, M. (1965). Attainment of quasi linkage equilibrium when gene frequencies are changing by natural selection. *Genetics* 52:875–890.

Montgomery, S. (2008). Linkage disequilibrium-understanding the evolutionary past and mapping the medical future. *Nat. Rev. Genet.* 9(6): 477–485.

Miyashita, N., Langley, C. H. (1988). Molecular and phenotypic variation of the white locus region in *Drosophila melanogaster*. *Genetics* 120:199–212.

Nagylaki, T. (1974). Quasi linkage equilibrium and the evolution of two-locus systems. *Proc. Natl. Acad. Sci. USA*. 71:526–530.

Ohta, T. (1982). Linkage disequilibrium due to random genetic drift in finite subdivided populations. *Proc. Natl. Acad. Sci. USA*. 79:1940–1944.

Pearson, K., Lee, A. (1903). On the laws on inheritance in man. *Biometrika* 2: 356–462.

Pulst, S. M. (1999). Genetic linkage analysis. *Arch. Neurol.* 56(6):667–672.

Schwabe, I., Janss, L., Van Den Berg S.M (2017). Can we validate the results of twin studies? a census-based study on the heritability of educational achievement. *Front. Genet.* 8:160.

Scriver, C. R., Waters, P. J. (1999). Monogenic traits are not simple: lessons from phenylketonuria. *Trends Genet.* 15(7):267–72.

Stacey B. G., Stephen F. S., Huy, N., Jamie, M. M., Jessica R., Brendan B., John H., Matthew D. F., Amy, L., et al. (2002). The structure of haplotype blocks in the human genome. *Science*.1069424.

Su, Y. K., Kirk, E. L., Anders A., Yingrui L., Thorfinn K., Geng, T., Niels, G., Tao J., et al. (2011). Estimation of allele frequency and association mapping using next-generation sequencing data. *BMC Bioinformatics* 12:231–246.

Tsutomu, O. (2006). Genetics of human growth. *Clin. Pediatr. Endocrinol.* 15(2): 45–53.

Turkheimer, E. (2000). Three laws of behavior genetics and what they mean. Current Directions in Psychological Science 9(5): 160–164.

Visscher, P. M., Medland, S. E., Ferreira, M. A. R., Morley, K. I., Zhu, G., Cornes, B. K., Martin, N. G. (2006). Assumption-free estimation of heritability from genome-wide identity-by-descent sharing between full siblings. *PLoS Genetics* 2(3): e41.

2 Genetic Variations in Human Beings

2.1 INTRODUCTION

Genetic variation forms the primary basis of the origin and evolution of new lifeforms on this planet. Charles Darwin (1859) was one of the pioneers to recognize genetic variation and mentioned its importance in his Origin of Species, long before modern genetics. The contributions of many early investigations have made significant observations regarding the differences in skin color, hair color, form, blood group, body size, and other characteristics. With the development of molecular tools, genetic variations are studied by using genomic DNA, organellar DNA, antigens in the blood cells, and so forth.

The discovery of the genetic basis for human genetic diseases and their impact on the human population needs an overall understanding of human genetic variations. Different patterns of its variation, including the origin and genetic diversity, discovery of the variants, accurate genotyping, elucidation of the haplotype structure, and mutations, have immensely contributed to shaping the present-day human genome. Genetic variations are applied to explore the origin and history of human beings, from studying their evolutionary relationships with existing primates to estimating the importance of selection and genetic drift in shaping the genetic structure of populations and determining the time of origin and spread of specific mutations. Information on the genetic variations will also increase our understanding of the patterns of genetic diversity among different individuals, different ethnic groups, and sub-populations. This chapter will discuss the contribution of genetic variations in origin, similarities, differences, and evolution of humans.

2.2 THE ORIGIN OF HUMAN BEINGS

After analyzing the patterns of variation in 1,327 genetic markers belonging to 121 African populations, 4 African American populations, and 60 non-African populations, it has been identified that Africa consists of more human genetic diversity than any region on the planet. The genetic structure of Africans was traced back to 14 ancestral population clusters. The ancestral origin of humans was presumed to be in southern Africa near the border of Namibia and South Africa (Masatoshi, 1995). Human genetic diversity decreases with increasing the migratory distance from Africa, which might be due to the reason responsible for the migration of human beings. The variations in the skull measurements were decreased with increasing distance from Africa, correlating with the decrease in the genetic diversity.

DOI: 10.1201/9781003343790-2

2.2.1 Out of Africa Theory and Supporting Evidence

Multiregional theory of human evolution regarding the origin of modern human beings (*Homo sapiens*) mentions that *Homo erectus* was the first to move out of Africa ~ 1 million years ago and occupied various parts of the world. *Homo sapiens* have evolved gradually from *Homo erectus* through gene flow and natural selection. This hypothesis also mentions that certain regional characteristics such as the shovel-shaped incisors in East Asians and the prominent brow ridge in Australian aborigines have remained unchanged for more than 1 million years since the time of their ancestral species, *H. erectus*.

Out-of-Africa theory proposes that *H. sapiens* probably originated in Africa 100,000 - 200,000 years ago, and all the existing human populations outside sub-Saharan Africa are the primary descendants of a population that has migrated out of Africa ~100,000 years ago. The basis of the Out-of-Africa theory was initially on the phylogenetic analysis of RFLP data on mitochondrial DNAs (mtDNA), sampled from different parts of the world supported by the paleontology data. However, the mtDNA study was criticized due to its statistical instability. The probable time of the split has a high standard error, which takes time back to 80,000 years. The inheritance of the mtDNA through a single gene makes it extremely difficult to infer the phylogenetic tree of human populations from mtDNA variation. One of the approaches to resolve this controversy is using microsatellite DNA polymorphisms to estimate the time of the most profound split of human populations. Microsatellite DNA is the segments of tandemly repeated DNA of length, 2- 5 nucleotides. These repeats display an extensive amount of polymorphism concerning the number of repeats. Hence, they help in understanding the phylogenetic relationships of populations. Experimental evidence suggests that polymorphisms are generated either by the gain or loss of a repeat unit, with some exceptions. A pattern of mutation can be described approximately by the stepwise mutation model in population genetics. The genetic distance between two populations is the average square distance (ASD), which increases linearly with evolutionary time when the mutation-drift balance is maintained.

$$\text{The expected value of ASD} = Vx + Vy + 2\,\beta t$$

Vx and Vy are the variances of allele size in populations X and Y, respectively, β is the mutation rate per locus and T is the number of generations since the two populations X and Y, have diverged.

However, the presence of the terms Vx and Vy makes the ASD variance very large. To rectify this problem, a new distance measure exists, designated as

$$(\delta\mu)^2 = (\mu x - \mu y)^2$$

where μx and μy are the mean allele sizes in populations X and Y, respectively. The expectation of $(\delta\mu)^2$ is $2\beta t$, thus $(\delta\mu)^2$ has a smaller variance than that of ASD, used for estimating the time of population divergence without the knowledge of Vx and Vy. If we know β from other sources, the divergence time between two populations can be estimated by

$$t = (\delta\mu)^2/(2\beta) \text{ generations or } T = gt \text{ years}$$

where g is the generation time in years.

However, $(\delta\mu)^2$ is still a significant variance compared with other distance measures, which need a large number of microsatellite loci for obtaining a reliable estimate of evolutionary time. The $(\delta\mu)^2$ is applied to 30 microsatellite repeats to construct a phylogenetic tree for 14 different human populations worldwide. This tree consists of the deepest root that has separated from the non-Africans and the Africans. Using the mutation rate for dinucleotide microsatellite repeats 5.6×10^{-4}

per locus per generation, with a generation time of g = 27, the estimated time of divergence between the African and the non-African populations is 156,000 years with a 95% confidence interval of 75,000-287,000 years (Nei and Roychoudury,1993).

Out of Africa theory originated with Charles Darwin's "Descent of Man." This theory also identified the location for the origin of human migration as Southwestern Africa, near the coastal border of Namibia and Angola, and the exit point out of Africa is East Africa. According to the available genetic and fossil evidence, after the evolution of Homo sapiens in Africa between 100,000 to 200,000 years, members of one branch migrated from Africa 60,000 years ago. Out of Africa theory also mentions that existing African sub-populations have undergone speciation during the first migration, inhibiting the gene flow between African and other human sub-populations.

2.2.2 Evidence from Population Genetics

The distribution of neutral polymorphisms among contemporary human sub-populations reflects a demographic history. It is believed that the human population in Africa might have passed through a population bottleneck before their population expansion and migration out of Africa, leading to an African-Eurasian divergence ~100,000 years ago (5,000 generations), followed by a Euro-Asian divergence ~40,000 years ago (2,000 generations). Richard Klein, Nicholas Wade, and Spencer Wells have postulated that human beings successfully migrated from Africa 60,000 - 50,000 years ago and colonized the rest of the world. The rapid expansion of the subpopulations has two significant effects on the distribution of genetic variation.

1. Founder effect occurs when founder populations bring only a subset of the genetic variations from their ancestral population.
2. When the founders are geographically separated, the probability of mating between two individuals belonging to two different founder populations will become low. This assortative mating will reduce the gene flow between geographical groups and increase the genetic distance between them.

The expansion of humans after their migration also affected the distribution of genetic variation in two possible ways.

1. Founder populations experience more significant genetic drift because of increased fluctuations in neutral polymorphisms.
2. New polymorphisms arising in one founder population are very much less likely to be transmitted to other groups due to the restriction of gene flow.

Speciation also has a certain amount of impact on the genetics of subpopulations. Populations in Africa tend to have a lower amount of linkage disequilibrium than the populations outside Africa. Low linkage disequilibrium could be due to the large size of the human population in Africa throughout human origin and evolution. The number of human populations that have migrated from Africa could be relatively low. In contrast, populations that have undergone a drastic reduction in their size or the populations that have experienced a rapid expansion have high linkage disequilibrium. The size of the population is very well supported by geographic, climatic, and historical factors in human genetic variation seen in today's existing population. Population processes associated with colonization, periods of geographic isolation, socially reinforced endogamy, and natural selection have shown an immense effect on the allele frequencies of specific populations. However, the recency of common ancestry and continual gene flow among human groups has limited the genetic differentiation in the population.

2.3 GENETIC DIVERSITY AMONG HUMAN BEINGS

Knowledge of genetic diversity distribution in humans is essential for understanding human evolution and genetic diseases among populations. The genetic diversity among human populations is relatively lower than that of many other species. This low genetic diversity might be due to their recent origin and the small size of the ancestral population. The proportion of the diversity existing between human sub-populations is also relatively low. Based on the studies performed on protein polymorphisms, blood groups, and craniometrics, the estimate of genetic diversity was 15%, and a genetic variation of 85% exists within the local populations. Around 7% of genetic diversity exists between local populations within the same continent, and 8% variation occurs between the large groups living in different continents. The genetic diversity of Africa was predicted to be 100%, higher than any other place on this planet. The genetic diversity decreases with increasing distance from Africa, such as New Guinea, which consists of 70% genetic diversity.

2.3.1 The Salient Features of Human Genetic Diversity

People from different population groups can be identified based on their consistent physical distinctions which are readily observed (skin color, height, hair color), but also those which cannot be observed (blood type, Rh factor). Genetic variations among those groups cause these distinctions. Cultures divide people into races based on their physical and genetic differences. Genetic differences between the races are essential for the identification of a specific individual. Below are a few findings from the genetic diversity of humans.

2.3.1.1 From Genetic Research

1. Human beings have a very low genetic variability, which might be due to descendance probability (entire species descended from a single family around 200,000 years ago).
2. 85% of genetic variation occurs within the human populations but not amongst them. Examination of their DNA and protein sequences confirms the same. Statistical divisions of humanity based on different kinds of genetic data do not group people consistently into races.
3. Genetic variations among different races are based on the commonness versus the rarity of specific alleles. Identical alleles are present in all human populations. A few alleles, including the one that causes Tay-Sachs disease, are found only in one race, and these alleles are relatively rare within that race.
4. Genes (MC1R) that vary consistently between populations will significantly affect skin, hair, and eye color.
5. There is an incredible genetic diversity among humans living in Africa today.
6. Genetic differences that separate the races are new and specific to a geological location. Adaptations to changing environments such as skin color, and differences in blood group are the outcomes of harmless mutations.
7. Traits such as intelligence and self-control are considered to be necessary by people for the selection of a mate. Hence, the alleles that enhance these traits will tend to spread throughout the human gene pool of all habitats and increase in frequency to a greater extent than alleles that do not. Consequently, alleles that favor beneficiary traits will be widespread in all populations.
8. Mutations that occur in a population have produced more significant genetic differences among individuals when compared with the mutations that occur among populations.
9. Restriction mapping analysis of the mitochondrial DNAs belonging to 147 individuals stemmed from a single woman hypothesized to have lived about 200,000 years ago in Africa. All of the populations, except the Africans, have multiple origins, implying that each geographical land area was a subject of repeated colonization.

2.3.1.2 Findings from Fossil Remains

1. The remains of modern humans (*Homo sapiens*) are found in Africa tens of thousands of years before they appear in the rest of the world.
2. The mitochondrial DNA extracted from the bones of Neanderthal man indicate that they are not related to any other living humans. The *Homo sapiens* of Europe have probably immigrated either from Asia or Africa.

2.3.1.3 Techniques for the Measurement of Genetic Diversity

1. Single nucleotide polymorphisms: Genetic variations in a single nucleotide of DNA are called, single nucleotide polymorphisms (SNPs, also called snips). These are the most common type of genetic variations amongst people. The predicted nucleotide diversity between humans is 0.1% as there is a difference of 1 base per 1,000 bases between two human beings chosen randomly. As the human genome length is 3 billion nucleotides, the number of nucleotide differences goes to 3 million. Most of the SNPs are neutrally occurring between gene loci, but some of them are functional, generating phenotypic differences between humans through alleles. An estimated 100 million SNPs exist in the entire human population, with at least 1% of them functional. SNPs are used as biological markers to identify the diseased loci with which they are linked. A detailed discussion of SNPs is provided in the next section.
2. Copy number variation: Copy number variations (CNVs) are responsible for more genetic variations than single nucleotide diversity. These variations are the genetic differences in the population due to deletions, duplications, inversions, and insertions of a stretch of DNA ranging from 1 Kb to 1 Mb. To date, more than 6.2 million different CNVs mapping 500,000 genomic regions have been identified in the human genome. CNVs contribute significantly by generating a difference of 0.8% of the human genome length between two human individuals, much more than a single nucleotide content per genome. CNVs also exhibit a higher mutation rate per locus than SNPs. Since CNVs reside in the coding regions, they can alter the gene dosage, disrupt coding sequences, or even modify the level and timing of gene expression responsible for the diseases.
3. Clines: Evolutionary biologist Julian Huxley coined the word, cline to describe the biological variation among some species. A cline is a gradation in measurable characteristics. As the traits change gradually and continuously across geographic space, a cline cannot be categorized, and any line that separates a cline will be arbitrary. In biology terms, a cline is a continuum of populations, races, species, varieties, or different forms of organisms exhibiting gradual phenotypic or genetic differences over a specific geographical region due to environmental heterogeneity. Frank Livingstone (1962) hypothesized that the genetic continuum of humans occurs in the form of clinal variation. According to Livingstone, genetic variation in human beings results from constant migrations across large geographic areas, and there were no significant reproductive barriers among them. As a result, each characteristic varied non-concordantly for each individual, which was the main reason why humans weren't placed into distinct classes or subspecies or races. Plenty of evidence is available to describe the human species variation, with both gene frequencies and phenotypic traits such as skin color or cranial morphology varying clinally.

   Genetic and phenotypic traits vary non-concordantly, with different traits exhibiting distinct patterns of variation across geographic space. For example, skin color indicates a continuous latitudinal gradient, with individuals living at lower latitudes near the equator being exposed to high intensity of UV radiation, exhibiting darker skin color than the individuals living at high latitudes. Other traits also express a distinctive clinal gradient across the world. Thus, skin color, hair texture, and facial features will not co-vary but will vary independently across space, irrespective of a racial group. Non-concordant clinal variation did not subdivide human beings into discrete, genetically distinct, and biologically homogeneous racial or continental groups (Joan et al. 2014). Human populations appear to be genetically more similar despite continental boundaries, and there is an inverse correlation in the degree of the genetic similarity between populations and the geographic distance.

4. Haplogroups: A group of similar haplotypes that share a common ancestor with a single nucleotide polymorphism (SNP) is a haplogroup. These pertain to deep ancestral origins back for thousands of years. The most studied haplogroups in humans are the Y-chromosomes haplogroup and the mt DNA haplogroup. The Y-chromosome provides a patrilineal line, and mt DNA provides a matrilineal line of genetic diversity. These DNA fragments escape from the processes such as re-shuffling or recombination that commonly occur in the rest of the genome, giving a concise history based on the mutational events without any recombination or re-assortment. These two haplogroups provide a picture of sex-specific processes in the past and present, which can be illuminated by comparing the patterns of DNA diversity in these two molecules. As recombinations are absent in these two haplogroups, it is easy to construct phylogenetic trees by using sequence variants and superimposing them on maps for making certain inferences on evolution and migration in the field of phylogeography. Based on haplogroup analysis, it is also possible to estimate the age of a specific lineage if its molecular clock (the rate of mutations in DNA markers) is available. However, the data obtained has been controversial, particularly in the Y-chromosome haplogroup regarding the appropriate mutation rate and molecular clock settings. The two main disadvantages of using the uni-parentally inherited systems are firstly that they provide biased information regarding their evolutionary history, and secondly, they are highly susceptible to the influence of genetic drift and natural selection.

2.4 GENETIC POLYMORPHISMS

Inherited differences in the DNA sequence among individuals greatly influence their anthropometric characteristics, risk of inheriting a disease, and response to the environment, significantly contributing to the phenotypic variation. One of the central goals of genetics is to identify the genetic variants, which significantly contribute to the population variation of each trait. Genome-wide linkage analysis and positional cloning have so far identified hundreds of genes that are responsible for human diseases with a mutation in a single gene being necessary and sufficient for the generation of disease. Most of these mutations are rare. Genome-wide linkage studies had limited access to the common diseases, necessitating intense methods for their identification. A promising approach for identifying common diseases is the systematic exploration of common gene variants for their association with the disease. These variants are rare in the human population, but a small number of polymorphisms generated by these variants explain the heterozygosity.

Variations in the human phenotype are determined both by genetic and environmental factors. Many forms of these genetic variations, such as deletions, duplications, and repeated sequences in the form of VNTRs resulted when comparing the DNA sequences on identical chromosomes of any two random individuals. These variations are also known as mutations and polymorphisms (Mehdi et al. 2015). A mutation changes the DNA sequence from normal, which implies a normal allele prevalent in the population, and mutation in it changes it to a rare and abnormal variant. Polymorphism is a DNA sequence variation that is more common in a population, considering no allele as the standard sequence but as equally acceptable alternatives. The common point between a mutation and a polymorphism is around 1%. It means a variation should have a frequency of greater than or equal to 1% in a given population to be considered a polymorphism. If an allele frequency is lower than 1%, then it will be regarded as a mutation. Even though 99% of DNA sequences are similar across the human population, 1% of minor variations in their DNA are responsible for their disease resistance, response to changes in the environment, and so forth.

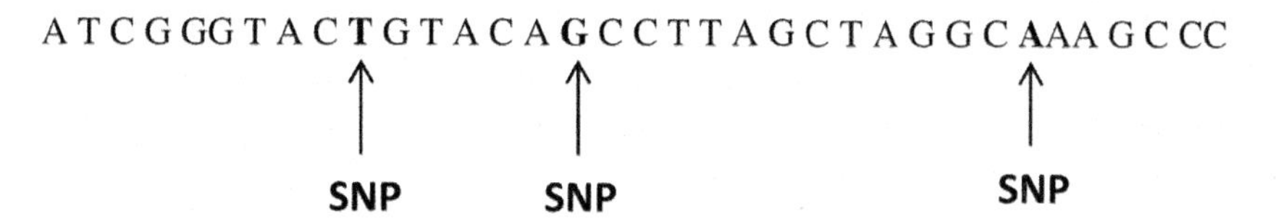

FIGURE 2.1 DNA sequences showing SNPs.

2.4.1 Single Nucleotide Polymorphisms (SNPs)

One of the most common sources of variations in the human genome is the single base mutation substitutions of one nucleotide by another, known as single nucleotide polymorphisms (SNPs) (Figure 2.1). Two types of nucleotide substitutions are responsible for SNPs.

1. Transition: This substitution occurs between either purine (A, G) or between pyrimidines (C, T). Transition makes 2/3 of all SNPs.
2. Transversion: A substitution occurs between a purine and pyrimidine.

Most of the SNPs are bi-allelic. The nucleotide substitution at these positions will be one or two possibilities. These are the most common types accounting for 90% of all genetic variations in humans. The estimated density of SNPs is 1 per 1,000 bases of human DNA, leading to 3 million SNPs. Single nucleotide polymorphisms may occur within the coding and non-coding regions of genes, intergenic regions between genes, and even in their regulatory regions. As only 3 - 5 % of a human DNA sequence codes for protein synthesis, changes in non-coding sequences are more prevalent than the changes in the coding sequences. SNPs generated within the coding region will not necessarily change the amino acid sequence of the produced protein due to the degeneracy of the genetic code. An SNP that leads to the same polypeptide sequence in both forms is synonymous (also known as a silent mutation) and non-synonymous if it produces different polypeptide sequences. SNPs generated in the coding region are of particular interest due to their capability to alter the biological function of a protein. A few SNPs in the coding regions are also responsible for the generation of diseases. Hence, they can be used as markers to identify the disease-causing gene. In the intergenic regions and non-coding regions, SNPs are responsible for gene splicing, transcription factor binding, or the sequence of non-coding RNA. 99% of the SNPs are reported to occur in the non-coding regions. SNPs generated in the gene regulatory regions are difficult to determine. The numbers of SNPs that occur in each chromosome have been listed in Table 2.1. Variations in the DNA sequences are responsible for humans developing diseases, and responding to pathogen infections, chemicals, drugs, and therapy. However, it is of utmost importance to compare and match the regions in the genome of a diseased cohort with a regular cohort (Richard et al. 2006).

2.4.1.1 Identification of SNPs

It is essential to identify the number of SNPs present in a genome before considering them as markers or tools for performing genetic studies. Currently, four methods are commonly used for the detection of SNPs.

2.4.1.1.1 The Identification of Single Stranded Conformation Polymorphisms (SSCPs)

This is one of the widely used screening techniques for the identification of different genomic variants across a large number of organisms. This analytic method detects the sequence variations of

TABLE 2.1
Chromosome Wise SNPs Number in the Human Genome

Chromosome	Length	SNPs	Kb per SNP
1	214,066,000	129,931	1.65
2	222,889,000	103,664	2.15
3	186,938,000	93,140	2.01
4	169,035,000	84,426	2.00
5	170,954,000	117,882	1.45
6	165,022,000	96,317	1.71
7	149,414,000	71,752	2.08
8	125,148,000	57,834	2.16
9	107,440,000	62,013	1.73
10	127,894,000	61,298	2.09
11	129,193,000	84,663	1.53
12	125,198,000	59,245	2.11
13	93,711,000	53,093	1.77
14	89,344,000	44,112	2.03
15	73,467,000	37,814	1.94
16	74,037,000	38,735	1.91
17	73,367,000	34,621	2.12
18	73,078,000	45,135	1.62
19	56,044,000	25,676	2.18
20	63,317,000	29,478	2.15
21	33,824,000	20,916	1.62
22	33,786,000	28,410	1.19
X	131,245,000	34,842	3.77
Y	21,753,000	4,193	5.19
TOTAL	2,710,164,000	1,419,190	1.91

even a single base change (SNPs) through the difference in electrophoretic mobility when subjected to non-denaturing (or) partially denaturing conditions. The DNA fragment is amplified on a PCR, denatured, and separated on a non-denaturing polyacrylamide gel to detect SNPs. During separation in the gel, single-stranded fragments adopt a secondary structure according to their sequence. The fragments bearing SNPs are identified by their aberrant migration pattern and with confirmation by sequencing. SSCP is a widely used technique, which gives a variable success rate of 70 - 95% for SNP detection.

2.4.1.1.2 Heteroduplex Analysis

This technique relies on detecting a heteroduplex formed during re-annealing of the denatured strands of a PCR product derived from a heterozygous individual for an SNP. Further, this heteroduplex can be detected as a band shift on a gel or by differential retention on an HPLC column. The application of HPLC to detect SNPs through heteroduplex is high due to its simplicity, low cost, high efficiency rate (95 - 100%), and average time of 10 minutes per sample.

2.4.1.1.3 Direct DNA sequencing

This is one of the highly preferred high-throughput methods for SNP detection, based on a next generation sequencing technology such as Applied Biosystems 3700 capillary system, capable of generating sequences of more than 1500 DNA fragments of size ~500 bp in 48 hours with minimal human intervention. In contrast, dye termination sequencing can detect 95% heterozygotes, but is more expensive and labor-intensive. Furthermore, insilico detection of SNPs in the DNA sequence can enhance the wealth of redundant sequence data deposited in public databases.

2.4.1.1.4 SNP Arrays

DNA microarray technology allows the identification of an SNP, based on hybridization with a series of oligonucleotides arrayed on glass or a silicon chip. Two platforms are available for SNP microarrays.

1. Affymetrix Platform: Affymetrix was the first to commercialize SNP arrays, in 1999. The initially prototyped HuSNP assay was designed for the genotyping of 1,494 SNPs on a single chip. Subsequent versions have increased SNP genotyping from 10,000 to nearly one million in the current release. A set of probes of 25 nucleotides in length will interrogate every SNP. A probe is designed complementary to the DNA sequence harboring the SNP site. In its first few versions, each SNP was interrogated by 24 - 40 distinct probe sequences, forming a probe set. Within a set, each probe is associated either with allele A or B. Each of them is either a perfect match (PM), entirely complimentary for one of the target alleles, or a mismatch (MM). A mismatch is identical to a perfect match probe with an alteration at the center base perfectly complementary to neither of the alleles.

   For the 10K version, a point accepted mutation (PAM) based algorithm is used. This algorithm relies on examining the probe intensities across multiple arrays. At each SNP, this algorithm uses the maximum difference between PM and MM probes as a proxy for allele A abundance and the analogous measure for allele B abundance. The 100K version of the array is based on a dynamic model algorithm. Unlike the PAM-based algorithm, the dynamic model approach operates without any need for training data. Consider the representation of three genotypes by three different models relating each genotype to a signal intensity value of each probe quartet. The AA model stipulates that the PMA intensity predominates, while the intensities of the other three probes have smaller and relatively equal means.

   Similarly, the BB model stipulates a PMB foreground and approximately equal background for the other three. The AB model assumes equal PMA and PMB means in the foreground and equal MMA and MMB means in the background. This algorithm also adds a null model for equal mean values to all types of probes, with the score for each model being the difference between model likelihood and the highest likelihood among the other models. The same dynamic model algorithm was initially applied for the 500K array, which was later replaced with a Bayesian Robust Linear Model. In the latest version of the Affymetrix array system Human SNP Array 6.0, each SNP is interrogated by 6 - 8 perfect match probes, with 3 - 4 replicates of the same probe for each of the two alleles, making two sets of repeated measurements for each SNP. Further, these SNP probes are augmented with almost 1 million copy number probes to interrogate the regions of the genome that do not harbor SNPs but are polymorphic for their copy number.
2. Illumina Platform: Similar to the Affymetrix platform, the Illumina based array has gradually increased in its capacity from 100,000 of the Human-1 to the current one million Human Hap1M. However, the data output format of the array has remained relatively consistent, with one measurement for the A allele and one for the allele B at each SNP. Due to the stability of the data output, there is very little change in the algorithmic evolution of this platform.

   The raw file from a single array consists of one million data points, with half a million pairs undergoing normalization before being analyzed. Instead of normalizing across the arrays, Illumina software performs internal normalization of each sample individually using six transformation parameters for all allele pairs. These parameters capture relevant factors for shifting, scaling, and rotating the X- and Y-coordinates and are inferred using these pairs. Each of these three genotypes AA, AB, and BB, represents a cluster in one-dimensional space. Proximity to a cluster determines a test sample's genotype, and cluster separation determines the SNP quality score. Similar to the Human SNP Array 6.0 of Affymetrix, the HumanHap1M also includes copy number probes meant for interrogating the non-SNP human genetic variations (Wang et al.1998).

Recent advances in molecular cytogenetics have improved the testing sensitivity by applying a microarray-based technique on comparative genomic hybridization technology (CGH). This technology involves the immobilization of numerous probes on glass slides as targets for the comparative hybridization of two different DNA samples. Tagging of each DNA source with different fluorochromes allows the identification of changes in the copy number in the targeted region by measuring the intensity of the fluorescence signal. The addition of CGH technology to SNPs has increased sensitivity far beyond the possibility allowed by standard hybridization techniques. Specifically, it highlights homozygosity regions for the identification of genetic change. SNP chromosomal microarray technology is also being used for assessing the individual genomes for determining the variations in their copy number (CNVs) by identifying the regions of genetic deletions and duplications, which are too small to be detected by prior methods. They also help uncover the regions of homozygosity reflecting the consanguinity that occurs due to mating between related persons or by partners from an inbred population. Although these consanguineous relationships are accepted in some societies, they may not be in others. In some cases, consanguinity may act as evidence indicating the commitment of a crime, as in the case of an incestuous relationship.

In a case study, multiple congenital anomalies were diagnosed in an infant born to a non-consanguineous married couple with no family history of congenital disabilities or miscarriages. Diagnosis of the specific genetic syndrome was not possible even with a careful medical evaluation. SNP microarray has demonstrated several large regions of homozygosity in the infant genome, which indicates that these areas were identical by descent (both copies inherited from a common ancestor). These homozygous regions were most consistent with a second or third-degree relative mating, thus revealing unsuspected consanguinity, specifically incest. Sharing this information with the infant's mother, she unwillingly admitted to being raped by her half-brother around the time of conception, and she was not willing to share this information with her husband. He believed that he is the father of the infant. She also declined the treatment of the infant from the same hospital authorities and finally got separated from that family after receiving the SNP microarray results.

2.4.1.2 SNP Genotyping

SNP genotyping is the measurement of single nucleotide variations in the genome of members the same species. The most common methodologies applied for the SNP genotyping are hybridization, primer extension, and cleavage methods. Hybridization involves annealing one strand of a PCR product bearing an SNP to the oligonucleotides complementary to that SNP. SNP genotyping based on the primer extension depends on the ability of the DNA polymerase to extend an oligonucleotide across a polymorphic site in the presence of nucleotides that will only allow the extension of one of the two variants. Cleavage technology-based genotyping measures the SNP status based on the ability of the cleavage enzyme to cut a matched or a mismatched DNA strand. Despite these traditional methods, biotechnology companies have developed a variety of high throughput technologies for SNP detection and genotyping, including fluorescent micro- array-based systems (Affymetrix), fluorescent bead-based technologies (Luminex, Illumina, Q-dot), automated enzyme-linked immunosorbent (ELISA) assays (Orchid Biocomputer), fluorescent detection of pyrophosphate release (Pyrosequencing), fluorescence resonance energy transfer (FRET)-based cleavage assays (Third Wave Technologies) and mass spectroscopy detection techniques (Rapigene, Sequenom).

2.4.1.3 SNP Frequency in Human Genome

After the elucidation of the HapMap project, a good flow of information regarding SNP frequency is available, very well supported by the recent analysis of nearly 300 genes. SNP content in recent years has undergone a massive change. Even though different methods were employed to determine the SNP content in different populations, many common observations were made. These are outlined below.

1. Synonymous changes in the non-coding and coding regions of the genome are more common than the non-synonymous changes, reflecting a very high selection pressure responsible for a reduction in diversity.
2. Transitional changes are more common than transversions, with CpG dinucleotides displaying a high mutation rate, probably due to the deamination.
3. Significant genetic diversity in the SNP frequency has been observed between the genes, owing to different selection pressures on each gene combined with different mutation and recombination rates across the genomes. Genome Wide Association Studies (GWAS) have identified many genetic variants associated with various traits responsible for diseases. The 1,000 Genomes Project Consortium has identified 15 million SNPs, 1 million short insertions and deletions (Indels), and over 20,000 structural variants through low coverage whole genome sequencing and exon targeted sequencing (Shen et al. 2013). Phase 1 of this project also has identified a large number of previously unidentified variants.

2.4.1.4 SNP Based Genome Wide Association Studies (SNP-GWAS)

An association study attempts to establish a relationship between a diseased phenotype or a phenotype with a specific trait and one or more genome regions. The achievement of this association is measuring the SNP frequencies of two populations differing in a phenotype and detecting the SNPs contributing to a significant difference in their frequency. Suppose a specific factor is contributing to an increased risk of a phenotype occurrence. In this case, it is sure that the event of the factor is at a high frequency in that phenotype or individuals compared with the control phenotype, also known as linkage disequilibrium (LD). The polymorphism may have direct functional consequences bearing on the phenotype state. Alternatively, mapping the position of those SNPs in the genome might help in identifying the gene involved in the phenotype under consideration, as the groups of closely linked alleles near SNP sites tend to be inherited together as haplotypes. Several association studies have been performed by focusing on the candidate genes and regions of their linkage to disease (Ozaki and Tanaka, 2005). Many large-scale studies have provided the path for identification of genetic risk factors for disorders such as Alzheimer's disease (APOE), type 1 diabetes (human leukocyte antigen (HLA), type 2 diabetes (PPARG), deep vein thrombosis (factor V), myocardial infarction (LTA), stroke (PDE4D), and asthma (ADAM33).

2.4.2 Short-Tandem-Repeat Polymorphisms (STRPs)

Short Tandem Repeats (STRs), also known as variable numbers of tandem repeats (VNTRs), is the most comprehensive term for a DNA sequence motif, repeated several times in the genome, continuously inherited in a Mendelian fashion. Earlier, these tandem repeats were considered noncoding DNA or even junk DNA without any specific function. However, rapid advancement in genome sequencing technology has enabled identifying their role and the repetition number of these sequences differs between individuals making them an entity for a polymorphism.

Repetitive sequences were identified during the mid-1970s while performing studies on the behavior of DNA in centrifugal fields, which yielded a single broad centered band corresponding to its density along with one or more minor bands, hence named satellite DNA. The behavior of satellite DNA is abnormal from its predicted buoyant density, which could be due to its methylation.

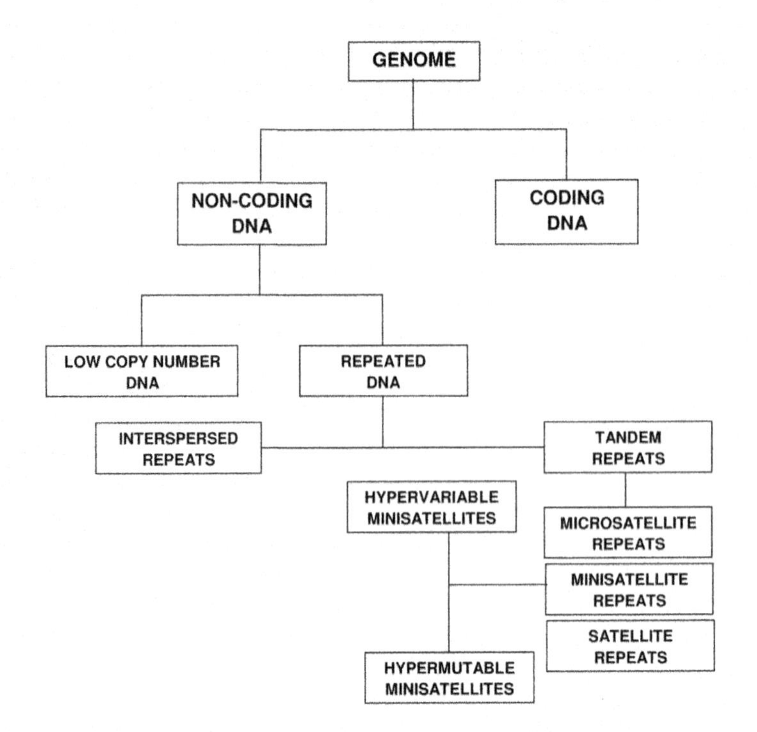

FIGURE 2.2 Organization of repeated sequences in a eukaryotic genome.

In situ hybridization studies reveal that satellite DNA is located in the heterochromatin, around the centromere and telomere of chromosomes. Most of the satellite DNA is non-coding. When subjected to restriction digestion, satellite DNA has yielded one or a few low, molecular weight bands, upon separation on agarose gel by electrophoresis, a sign of tandem repeated sequences, confirmed by the sequencing of a band after eluting the gel. Three possible reasons for the origin of the VNTRs are, slippages during DNA replication, the outcome of unequal crossing-over, and the repetitive nature of the sequence, have allowed further increase or decrease in the number of repeats, generating polymorphism.

Tandem repeats contribute up to 3% of the human genome. They are also found in the promoters of the genes and in potential functional locations, impacting phenotypes responsible for the genetic disorders. VNTRs are classified into microsatellite, minisatellite, and satellite repeat sequences, according to the length of the repeated sequence. Microsatellite repeats are less than 10 bp per repeat, minisatellites are 10 - 100 bp per repeat, and satellites are >100 bp per repeat. Satellite repeats are further categorized into macrosatellite repeats, up to several Kb per repeat (Figure 2.2).

2.4.2.1 Minisatellite Repeat Sequences and Their Functions

Minisatellites are tandem repeat sequences of 14 - 500 bp in length. They are frequently associated with length polymorphism, resulting in variations in the number of internal copies, making them valuable genomic markers. Minisatellites played an active role in the earlier versions of DNA profiling and the early stages of genome mapping. They were the first highly polymorphic and multiallelic markers used in linkage studies. Minisatellite repeats are in the telomeric regions of human chromosomes. Due to their location, highly polymorphic minisatellite repeats are considered ideal for detecting macro and microdeletions in the ends of chromosomes associated with mental retardation disorders. Polymorphic minisatellite repeats are highly abundant in the human genome.

These repeats are amongst the fastest evolving sequences. An abundance of minisatellite repeats in the human genome contributes to the genome function in three ways:

1. Some minisatellite repeats are a part of the ORF, which may or may not display polymorphism.
2. Minisatellite repeats located in 50 regions of different genes are involved in the regulation of transcription, and intron interference with splicing.
3. Minisatellite repeats constitute chromosome fragile sites found in the vicinity of several recurrent translocation breakpoints and the switch recombination site of immunoglobulin heavy chain genes. Minisatellites at imprinted loci are presumed to play a significant role in imprint regulation. They act as intermediates in the initiation of chromosome pairing in eukaryotic genomes.

The distribution of minisatellite sequences in the human genome is uneven. They are localized in the ends of the chromosomes, especially in the sub-telomeric regions, implying a limitation in using these repeated sequences in linkage analysis. Sub telomeric localization of the minisatellite repeats can be correlated with a high-density chiasmata formation during meiosis, indicating their role in the meiotic crossing over. The number of minisatellite repeat loci in the haploid human genome was estimated to be 1500, containing more than 6,000 minisatellite repeats (Gilles and France, 2000). Among them ~2000 are highly polymorphic forms.

2.4.2.1.1 Techniques for Determining Minisatellite Repeats

Variations in minisatellite repeats are studied by restriction digestion analysis followed by hybridization with probes, capable of hybridizing with many minisatellite repeat loci. The application of PCR further strengthened this analysis. A minisatellite variant repeat (MVR) PCR system amplifies the samples with long repeats and subtle variations in the internal repeats with increased sensitivity. DNA amplification by MVR PCR uses the primers specific for repeat variants, enabling PCR amplification of long stretches of repeated DNA sequence with rare variants, standard for most of the minisatellite repeats (Matthieu et al. 2007). The amplified repeats are sequenced by using multiplex sequencing machines.

2.4.2.1.2 Types of Minisatellite Repeats

Minisatellite repeats are subdivided into three types, hypermutable minisatellite repeats, hypervariable minisatellite repeats, and telomeric DNA repeats.

1. Hypermutable Minisatellite Repeats and Genotoxicity: New alleles exhibiting changes in minisatellite repeat number, only in a specific locus, are known as hypermutable minisatellite repeats. These repeats may serve as a potent source of information for understanding the mechanism of minisatellite repeat instability in humans. Repeat instability arises through gene conversion events, either during or immediately after meiosis, involving intra and interallelic information transfers. Hypermutable minisatellites serve as markers against exposure to genotoxic agents such as ionizing radiation. A study was conducted on humans exposed to chronic doses of radiation contaminated by the release of the radioactive material after the explosion at Chernobyl Power Station in 1986. The frequency of hypermutable minisatellite loci of control genotypes and exposure to radiation was studied through southern blotting (Wayne et al.1990). The data indicated that the frequency of the mutant alleles in the exposed genotypes was twice the frequency observed in the control. Several chemicals released into the environment are also suspected of inducing mutations in the minisatellite repeats of germinal cells undergoing meiosis.
2. Hypervariable Minisatellite Repeats and DNA Fingerprinting: Hypervariable minisatellite DNA sequences are the short tandemly repeated sequences of 6 - 50 bp. These sequences are present

throughout the human genome, implicating an enhancement in the recombination rate. These repeat sequences were first identified in 1985 by Alec Jeffreys and his colleagues at Leicester University who observed a core sequence GGAGGTGGGCAGGA, repeated hundreds of times, making a DNA fragment size of 20 Kb. The number of repeats at any given locus is highly polymorphic between individuals. Hypervariable minisatellite repeats are beneficial resources for DNA fingerprinting, which involves the hybridization of short synthetic probes consisting of sequences such as GGGCAGGANG, to an individual's genomic DNA, digested with restriction enzymes, separated on an agarose gel, and transferred onto a nitrocellulose membrane. The probe will bind to several DNA fragments, whose sizes will be determined by the number of repeats they contain. As a result of the probe binding, a pattern of DNA fingerprinting will be produced, which will be unique to an individual's DNA. Further, every offspring's DNA would display a fingerprinting pattern generated by a similar combination of bands present in their parent's fingerprints. This technique is the primary basis for paternity testing. Hypervariable minisatellite DNA sequences provide a powerful means of discrimination between individuals with a very high order magnitude compared to traditional methods. One of the essential forensic applications of DNA fingerprinting is the identification of rapists. Here, the protein-based techniques will not yield a good result as both vaginal fluid and semen have enzyme and blood group activity, whose mixtures are difficult to interpret. It is easy to separate the sperm DNA from other cellular material, effectively separating the male from a female component. Figure 2.3 displays a typical example of a rape case, in which a mixture of male and female DNA profiles has been obtained. With the discovery of PCR in the 1980s by Kary Mullis, the new possibilities of forensic analysis, such as petty crime that strains quickly and economically, were invented. Quick forensic analysis using DNA has led to the development of the United Kingdom National DNA Database. In 1995, The Forensic Science Service Department, under the initiative of the Home Secretary, had set up the national DNA criminal intelligence database to generate the DNA profiling of more than 200,000 samples per year covering England and Wales and their analysis with a target of 5 million profiles. DNA will be isolated from any individual's buccal scrapes or hair roots, either suspected or convicted with a recordable offense and sequenced using a second-generation multiplex sequencing technology. The results in the database will be a digital code based on the number of tandem repeats at each locus. During casework, forensic laboratories analyze the crime material and compare the existing profiles in the database. If any profile matches the database, police will investigate further, and if suspects are innocent, their profiles are subsequently removed from the database. To date, over 700,000 samples have been stored in the database, and more than 65,000 matches between suspects and samples have been obtained. Minisatellite repeats of degraded samples are also considered active, as demonstrated in the analysis of samples from the Waco, Texas disaster of 1993, in which tissues and bones that were damaged during fire and decomposition were collected from the remains of 61 people. Among them, samples of 50 bodies gave a complete result, displaying the minisatellite repeats.

Even though the DNA profiling technique has revolutionized the forensic sciences, it has certain disadvantages:

1. There is no practical way of determining the pairs of bands in the fingerprint and their corresponding alleles at a specific minisatellite locus.
2. It is not possible to calculate the allele frequencies.
3. Technical difficulties are more with high possibilities of human error.

3. Telomeric Minisatellite DNA: Minisatellites are composed of hexanucleotide repeats TTAGGG, 15 Kb long, and are added to the telomere of the chromosomes by the enzyme telomere terminal transferase also, known as telomerase. These repeat sequences are involved

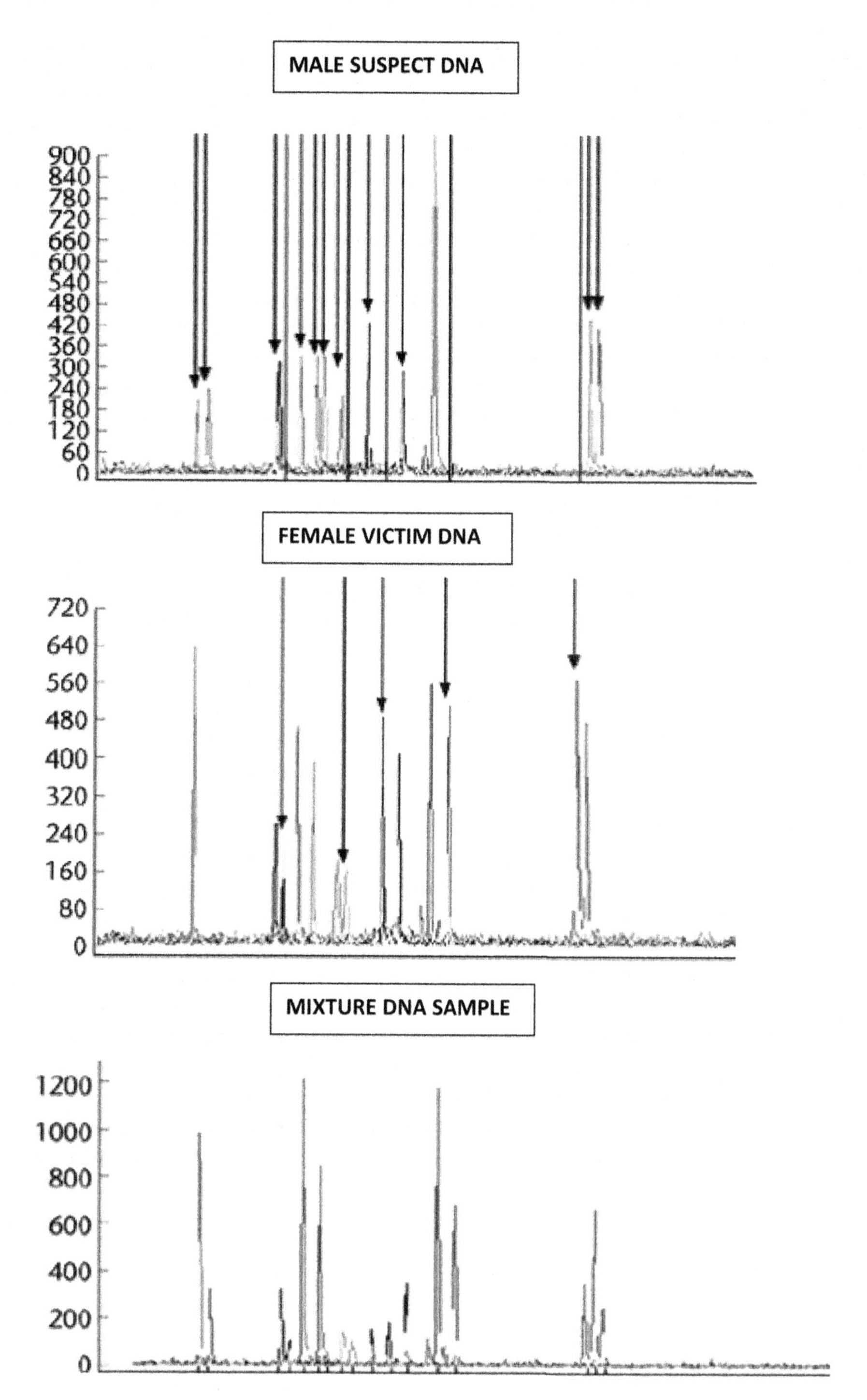

FIGURE 2.3 Crime sample comprising the DNA of mixture, male suspect, and female victim. Peaks comprising males and females are displayed in arrows.

in protecting the chromosomal ends from degradation and provide a means for complete replication. These telomeric minisatellite repeats are also known to play a vital role in the pairing and orientation of chromosomes during cell division.

2.4.2.1.3 Mutations in Minisatellite DNA

The frequency of mutations in the minisatellite repeats is dependent on their length. Short repeats are stable over millions of years, whereas long repeats have an extremely high frequency of mutations ranging up to 15%. This high instability of the human minisatellites is due to its repetition as demonstrated in human minisatellite MS1, which retained instability after its insertion into the yeast genome. In five highly unstable minisatellite loci, the length change mutations were correlated with their heterozygosity, demonstrating that these changes were selectively neutral. The distribution of mutations in the minisatellite repeats is non-random in a genome. These mutations also occur predominantly at one end of the locus, similar to germline mutations. Four different mechanisms are responsible for the change of a minisatellite repeat sequence, replication slippage, intramolecular recombination, unequal sister chromatid exchange, and unequal inter-allelic recombination. A few mutations involve gene conversion through which an allele in one chromosome replaces an allele in the other chromosome. Some minisatellite mutations are intra-allelic, and some are anomalous repeats, not corresponding to either of the alleles. A mismatch repair might have also generated some mutations.

2.4.2.1.4 Minisatellite Repeats and Human Diseases

A few minisatellites are responsible for diseases in humans. One of the best studied is the minisatellite repeat connected with the Ha-*ras* protooncogene locus *HRASI* VNTR located 1000 bp downstream of the polyadenylation signal, consisting of 28 bp repeating units, forming up to 30 alleles. Among these, four alleles are present in 94% of the population. However, the rare alleles are three times more common in patients suffering from multiple cancer types. The possible reason behind the association between HRASI minisatellites and cancer is the exhibition of linkage with the diseased locus.

Mutations in the minisatellite repeats of the insulin gene (INS), located 600 bp upstream of the transcription initiation site, are also linked. The minisatellite comprises of 14 bp repeated units arranged in three allelic classes with lengths of 600, 1200, and 3,200 nucleotides for class I, Class II, and Class III, respectively. The Class I minisatellite allele is associated with the doubling of risk for Type I Diabetes mellitus. The INS minisatellite binds to a specific transcription factor, Pur-1. High-risk alleles experience a weaker binding than the low-risk alleles, governing the transcriptional response. A minisatellite upstream of a human immunoglobulin heavy-chain gene IGH enhancer will have a suppression effect on the immunoglobulin gene expression in a similar fashion.

A wide occurrence of minisatellite repeat sequences provides an indispensable tool in genetic linkage analysis, forensic medicine, and paternity testing. It is also clear that the utilization of these repeats will play an essential role in the research and practical applications. Even though the utilization of these repeat sequences in forensic science has been a subject of controversy and debate, the application of novel techniques such as PCR and novel laboratory procedures can remove these problems in the near future.

2.4.2.2 Microsatellite Repeats

Microsatellite repeats, also known as short tandem repeats, are among the most variable loci in the human genome. These repetitive sequences are usually di, tri, or tetranucleotide repeats with rare penta and hexanucleotide repeats. In the human genome, the frequency of dinucleotide repeat occurrence is every 30,000 bp. These form 3% of the human genome, constituting a significant part of the DNA (Table 2).

Microsatellite repeats are formed as errors during DNA replication, further amplified through strand slippage during successive DNA replications, giving rise to long microsatellite repeats.

TABLE 2.2
Most Common Microsatellite Repeat Sequences of the Human Genome

S. No.	Sequence	Count	Frequency (%)
1.	A/T	104,373	19.4
2.	AC/GT	91,786	17.0
3.	AT/TA	37,219	6.91
4.	AAAT/ATTT	30,771	5.71
5.	AAT/ATT	26,782	4.97
6.	AG/CT	23,680	4.39
7.	AAAC/GTTT	21,156	3.92
8.	AAC/GTT	17,974	3.33
9.	AATG/CATT	15,045	2.79
10.	AAAG/CTTT	14,865	2.75
11.	AAAAC/GTTTT	12,610	2.33
12.	AAGG/CCTT	10,681	1.98
13.	AGG/CCT	10,438	1.93
14.	AGGG/CTTT	10,314	1.91
15.	AGC/GCT	6,169	1.14
16.	CCG/CGG	3,820	0.70
17.	CCCG/CGGG	1,098	0.20

Accumulation of dispersed simple nucleotide repeat sequences will increase the risk of homologous recombination between chromosomal segments leading to inversions, deletions, and translocations. To counteract the rapid accumulation of the microsatellite repeats, the genome is taking a risk of developing high mutations, which can be attributed to their high levels of variation. Mutations occur in the human genome at 10^3 - 10^5 per generation, which is an order of magnitude that is much higher than that of SNPs (10^8 - 10^9).

Although there are functional differences in the mini and microsatellite repeats, the primary basis for the difference is the number of base lengths and repeats. There are little pieces of evidence, which suggest that minisatellites evolved from microsatellites. The transposition model proposed by Alec Jeffreys et al. (1985) indicates that the core sequences between the mini and microsatellites are the result of transpositions mediated by the sequences flanking the minisatellite sequences. The association of minisatellite repeats with dispersed repetitive elements such as Alu SINES and transposable elements is the observation supporting this hypothesis. However, most of the minisatellite repeats with core sequences are not exhibiting any association with dispersed repeats, making it unlikely that they emanate from this type of transposition. Another model, called the expansion hypothesis, is based on the motifs containing core sequences capable of enhancing the expansion of tandem repeats independently at different loci. This model predicts the expansion of microsatellites into minisatellites by the microsatellite fossils. Several examples signifying the association of micro and mini satellite sequences has been recorded, indicating their generation.

2.4.2.2.1 Instability of Microsatellite Repeats

Most of the microsatellite repeat sequences are exceedingly unstable, especially CG-rich trinucleotide and CA-dinucleotide repeats, exhibiting a high instability and more variability when compared to other satellite repeat sequences (Meltzer et al. 1994). The specific reason for the instability is unknown, but its instability influences the length of the microsatellite repeat, showing an increased instability with increasing length. CG-rich trinucleotides and CA dinucleotides form four groups according to their stability (Table 3). Microsatellite short length repeat sequences are

TABLE 2.3
Stability of the CG-Rich Trinucleotide and CA Dinucleotide Microsatellite Repeats

S. No.	Repeats	Stability
1.	Short length repeats	stable
2.	Middle length repeats	polymorphic, stable
3.	Long Alleles	instable
4.	Long Alleles	extremely instable

highly stable, whereas the repeats of moderate length are polymorphic but are stable between the generations.

With an increase in their length, their instability increases by several orders of magnitude, as in the case of long alleles, which constitute pre-mutational alleles, without stability between the generations. Extremely unstable long alleles also exhibit mitotic instability as identified in fragile X syndrome, where the likelihood of instability is as follows; 50 repeats, no risk. 60 repeats, low risk. 70 - 86 repeats, high risk. Above 86 repeats, the absolute likelihood of the disease.

Strand slippage during DNA replication is one of the main reasons for the repeat length mutations, leading to increased DNA stability. Other meDNA replication slippages facilitate an increase in the length of the trinucleotide repeats. A 1000-fold increase in the human dihydrofolate reductase gene expression involving an episomal mechanism for increased expression was observed. This gene is presumed to be excised and copied by a rolling circle mechanism and is re-integrated into non-homologous chromosomal sites. However, this mechanism fails to explain the amplification of trinucleotide repeats. An in-situ model explains the mechanism of trinucleotide repeat amplification and replication of CG-rich sequences by polymerases, giving rise to multiple incomplete strands containing pre-mature stop and start codons. There can be an extensive increase in the length of the strand switching between the incomplete strands. This model predicts an increase in rate and expansion with an increasing initial length of the trinucleotide repeat, which was also observed experimentally.

The stability of the microsatellite repeats is also dependent on the intact mismatch DNA repair. The loss of repair function in human DNA has drastically increased the instability of microsatellites, leading to diseases such as hereditary nonpolyposis colon cancer (HNPCC). The gene involved in this cancer is located on chromosome 2, linked with a locus of dinucleotide microsatellite AC repeats. Mutations in this gene have caused many changes in the array of the AC repeats, followed by extensive instability. Other than HNPCC, several cancer forms such as gastric, pancreatic endometrial, Barrett's esophageal, and lung cancers are associated with instability in the micro-satellite repeat sequences. Deficiency in the enzyme exonuclease can also generate instability in microsatellites.

2.4.2.2.2 Mutations in Microsatellite Repeats

The mechanisms responsible for the development of variations in microsatellite repeat sequences have not been completely understood. One possibility for these variations is the rate of mutations in these sequences. Mutations in microsatellites are capable of changing one or more repeat units with a very high mutation rate of 10^{-2} - 10^{-6} per microsatellite locus per generation, compared to other mutation types such as point mutations (10^{-9} nucleotides per generation for the whole genome). One of the most common sources of mutations in microsatellites is replication slippage, involved in either the gain or contraction of repeat sequences. Other sources of mutations are unequal crossing over, nucleotide substitutions, and duplication events, along with factors such as allele, motif size and GC content, gender, the effectiveness of the mismatch repair enzymes, and stabilization patterns. The base position of the microsatellite, its location in the genome, type and number of the repeats, its flanking sequence, the speed at which the replication and transcription occur, and heterozygosity of the microsatellite alleles also influence the rate of mutations in microsatellite repeats. Microsatellites in the non-coding regions tend to mutate more frequently than those located in the coding regions. Dinucleotide repeats are subject to mutations more frequently than tri and tetranucleotide repeats. Long, pure repeats undergo mutation at a higher rate than short repeats with low purity. The rate at which the mutations occur in microsatellite repeats is also species and gender-specific in humans, with males showing a high mutation rate of $0\text{-}7 \times 10^{-3}$ per locus per gamete per generation compared to females. Repeat expansion mutations are faster in humans than in chimpanzees. Evolutionarily, long repeat microsatellite sequences are of recent origin with a high biasedness towards repeat expansion mutations compared with shorter ones (Jarl et al. 2008). Two models predict the role of mutations in the evolution of microsatellite repeats.

1. The Infinite Allele Model (IAM) assumes that mutations in the microsatellite repeat sequences might have created an infinite number of repeated units and alleles that exist in the population.
2. The Stepwise Mutation Model (SMM) shows that the mutations in the microsatellite repeat units will have the same probability of gaining or losing one repeat unit. This model also takes back mutations into account. The generalized stepwise model (GSM) also known as the two-phase model (TPM) is an extension of SMM, which considers the probability of a mutation in a microsatellite sequence involving more than one repeat unit.

Studies on the mutation process in microsatellite repeat sequences of the human genome are based on three approaches:

1. Pedigree Analysis: A pedigree study in humans will directly view and tabulate the mutations for each generation. The limitation of this approach is that the number of generations that can be studied and analyzed will decrease.
2. Comparative genomics: Comparing the human genome with closely related species will provide information about microsatellites and their evolution. Comparative genomics of humans and chimpanzees have identified specific factors involved in the evolution of microsatellite repeats, subjected to multiple mutations per locus, due to the combination of long divergence and a very high mutation rate.
3. Studies of polymorphisms in a population: Studies of polymorphisms in a population will provide the necessary descriptions of the variations in a population. The comparison of trinucleotide microsatellite repeat sequences belonging to the exons and introns of the human reference genome sequence, Craig Venter genome sequence, Watson sequence, and chimpanzee genome sequence have provided a positive correlation between polymorphism within humans and divergence with chimpanzee microsatellite repeat numbers, suggesting that the long loci are subjected to mutation faster than the shorter ones.

A comparison of divergence with polymorphism has not yielded any evidence on the variations in the natural selection of trinucleotide repeats. The main reason for this pattern of variation in most microsatellite repeat sequences falling outside the genes is also not known.

2.4.2.2.3 Analytical Methods for Microsatellite Repeats

An efficient system for the isolation of microsatellite repeats was developed by Armour et al. (1994), based on the enrichment of DNA fragments containing tandem repeats by hybridization with long tandemly repeated targets. Initially, a library of restriction fragments with appropriate linkers for

PCR amplification is constructed. PCR amplification yields fragments of size 400 - 1000 bp. The selection of fragments containing tandem repeats by hybridization to the long arrays of trimeric or tetrameric repeat sequences enables the rapid isolation of many microsatellite clones.

2.4.2.2.4 *Functions of Microsatellite Repeats*

Traditionally, non-coding DNA was considered to be junk DNA without any specific function. As the microsatellite repeats form the non-coding DNA, they were also considered non-functional neutral elements without any biological function. However, recent developments have identified multiple functions of microsatellite repeat sequences. The occurrence of microsatellites in coding, regulatory, introns, and intergene regions supports the processes such as gene expression, gene silencing, alternative splicing, mRNA transport, chromatin organization, and cell cycle regulation. An increase or decrease in the microsatellite repeat number is associated with at least 40 neurological disorders in humans (discussed in the next section). Polymorphism in the microsatellite repeats of the vasopressin 1a receptor gene is responsible for autism, monogamy, social behavior, and socialization skills. Reductions and expansions in the intron repeat numbers of genes such as Asparagine synthase, NOS3, and EGRF genes might cause lymphoblastic leukemia, hypertension, and osteosarcoma, respectively. Microsatellite repeats of the human genome are conserved throughout the Vertebrata. Highly conserved mammalian microsatellites are over-represented in the promoter regions of many human genes involved in the regulation of growth and development. Changes in the microsatellite repeat sequence length in the promoters can alter the phenotypes. This alteration can occur in the phenotypes when these repeats are not transcribed. Untranslated microsatellite repeats located proximally to the transcription initiation site have a profound effect on the phenotypes. Microsatellite repeats are also involved in determining evolutionary fate, survival, plasticity, and adaptation to constantly changing and potentially harmful environments.

Variation in humans is also linked to regulatory microsatellites consisting of the AC/GT motif. Neural development genes such as PAX6, and NOS1 consisting of these regulatory microsatellites, are expressed in the neuronal cells of the human eye, brain, and other organs. Promoters of genes involved in neural development consist of numerous conserved microsatellite repeats, forming various DNA secondary structures. Microsatellite repeats with AC/GT motifs can form ZDNA, which is a double helix turned towards the left hand. Repeats with AC/GT motifs can also form H-DNA, a triplex. Repeats with (TGGG)4z can also form G-quadruplex, inducing the single strandedness in the complement C-rich strand, forming a I motif. G-repeats show a strong preference for promoter regions regulating transcription by altering the RNA polymerase activity. They can also affect DNA folding when located in the UTR regions.

2.4.2.2.5 *Microsatellite Markers*

Microsatellites are potential genetic markers for solving numerous tasks. For the past three decades, they have been considered the best markers due to their abundance and distribution throughout the genome. These repeat markers are highly mutagenic and polymorphic in humans. They are co-dominant, hence suitable for the detection of heterozygotes. These satellite repeat markers are multi-allelic experimentally reproducible and can be transferred within the related taxa. Economically, they are cost-effective and easy to detect. Microsatellite markers can be amplified even from low-quality and low quantity DNA and are highly neutral markers. Microsatellite markers are handy during the construction of a genetic map of large genomes when no reference map is available. They are best suited to small scale genetic studies. They can detect ample genetic information and physiological parameters of the genome with a limited budget. They are involved in the construction of genetic linkage and integrated maps (Maria et al. 2016). They are used for correlating phenotypic and genotypic variations using linkage disequilibrium associated approaches, for the analysis of parentage and ancestry, for understanding migration, and demographic processes, for

TABLE 2.4
Neurological Diseases Associated with Trinucleotide Microsatellite Repeat Reiteration in Humans

Repeat	Normal range	Disease range	Disease
CAG	11–33	40–62	Spinal and bulbar muscular dystrophy
CAG	11–33	40–62	Kennedy's disease
CAG	11–34	42–100	Huntington's disease
CAG	29–36	43–60	Spinocerebellar Ataxia Type 1
CGG	6–54	250–4000	Fragile X syndrome
CTG	5–30	>50	Myotonic dystrophy

population differentiation and kinship, for the assessment of mutagenic contaminants, in forensics, and in the diagnosis of diseases.

2.4.2.2.6 Microsatellite Repeats and Human Diseases

Microsatellite repeats have taken a central role in disease diagnosis since their identification as the primary sources of many human diseases. Trinucleotide microsatellite repeats are predominantly involved in the generation of several human neurological disorders (Table 2.4).

Microsatellite-dependent diseases are represented in two classes. Trinucleotide repeats located in the non-coding regions of a gene form Class I, represented by diseases such as Fragile X syndrome and Myotonic dystrophy. Microsatellite trinucleotide linked to a gene's coding region gets affected by their unusual expansion and forms Class II, leading to Huntington's disease, spinal and bulbar dystrophy, Kennedy's disease, and so forth. An initial increase in the number of trinucleotide repeats in the sequence is pre-mutational. After exceeding the critical number of repeats, the sequence will become unstable, resulting in symptoms. Another significant feature of the diseases oriented with the repeats is, that the symptoms will tend to be more severe during subsequent generations due to their amplification during gametogenesis or in the zygote. This process is known as genetic anticipation, a sex-linked process that occurs through the mother in Fragile X-syndrome and the father in Huntington's disease. Genetic anticipation is largely connected with the methylation of cytosine bases and genetic imprinting. The probable mechanism for expanding trinucleotide repeat sequences is the slippage of leading and lagging DNA strands during DNA replication, which involves the spanning of 150 - 200 bp Okazaki fragments comprising of ~ 70 trinucleotide repeats as identified during Fragile X syndrome. Unusual expansion of trinucleotide repeats in the UTR regions eliminates the transcription of a gene leading to no mRNA formation. Increased nucleosome binding of such repeats leads to transcriptional repression, as observed in the case of myotonic dystrophy. Spinal and bulbar muscular atrophy is an X-linked neurological disorder that occurs due to the expansion of microsatellite repeats in the ORF of the androgen receptor gene leading to its repression. Microsatellite repeats are one of the highly studied DNA sequences in the non-coding and coding regions of the genome. They form one of the most reliable genetic markers, successfully used in science and research. These repeats are responsible for numerous diseases, that are being diagnosed. They also represent a few concerns and caveats that require attention and correction for their proper utilization.

2.4.2.3 Alu Insertion Polymorphisms

Rapid progress in human genome analysis has identified novel genetic markers based on retrotransposable elements, whose insertions generate polymorphisms. Recent studies have identified numerous retroelement insertions in the human genome (Ilgar et al. 2010). Retro elements are deemed as genetic markers for the following reasons:

1. Retro transposable elements are highly stable in the human genome, and there is no specific mechanism for the withdrawal of this element from its insertion point.
2. The probability of an independent insertion of a retroelement is negligibly less.
3. The presence of a retroelement in the genome indicates its identity by descent.
4. Retro element polymorphism can be typed by using a locus-specific PCR and agarose gel electrophoresis.
5. There are several retroelement families in the human genome. Among them, the most abundant retroelement type is *Alu* elements.

Alu family retroelements are a fraction of renatured repetitive DNA elements of ~ 300 bp long SINEs, which can be distinctly cleaved by a restriction enzyme *AluI*. *Alu* elements share 90% of the homology with the 7SL RNA gene, whose mRNA is an essential constituent of the endoplasmic reticulum signal recognition particle, which indicates that these elements are the derivatives of 7SL RNA. It is predicted that over one million *Alu*s are present in human DNA, representing 11% of the haploid genome. Among them, 0.5% are polymorphic. Direct repeats flank *Alu* repeats in the human genome. *Alu* repeats are mobilized through the retroposition of RNA polymerase III located in the left half of the element. Chromosomal distribution of *Alu* elements displays a preference for the R bands or AT rich regions of the genome.

It was hypothesized that the *Alu* elements might have originated 65 million years ago. A few source genes named Master *Alu* genes are still undergoing amplification at a rate of 8 x 10^{-3} *de novo Alu* insertions per year. Depending on the changes that occurred to the Master genes, the *Alu* family is classified into Old (Jo and Jb subfamilies), intermediate, and Young (Y). Y family of *Alu* elements consists of ~one million members in the human genome. Among them, 500 - 2000 belong to several closely related Y subfamilies Yc1, Yc2, Ya5, Ya8, Yb8, and Yb9, consisting of recently inserted human-specific *Alu* members not found in the genomes of other closely related genomes. Among these sub-families, Ya5 and Ya8 are the most recently formed. As these two subfamilies share an equal number of diagnostic mutations (5 regular and 3 additional), this lineage is collectively referred to as Ya5/8. This subfamily lineage consists of 500 - 2000 elements that are exclusive to the human genome. Another Alu sub-family, which is an independent derivative of the Y lineage, is Yb8. This subfamily consists of more than 500 copies that are specific to the human genome. These two Ya5/8 and Yb8 elements are recently retroposed and are highly polymorphic for their presence or absence at a specific location in the human genome. The youngest Alu sub-family is Ya5a2, reported in 2002, with a majority of its members being *Alu* insertion polymorphisms.

Insertion polymorphisms of Y sub-families exhibit a biallelic and co-dominant pattern of insertion without inheritance, reflecting a common ancestry with the absence of insertion, which is the ancestral state. Hence, the *Alu* insertions shared by two different individuals are identical by descent and not by the state, which means that if two individuals who share an insertion are likely to share a common ancestor in whom this insertion was generated. As there is no known mechanism for the complete removal of the inserted element, the lack of insertion represents the ancestry. As the insertion rate of two *Alu* elements is 100 - 200 per one million years, the probability of two independent *Alu* element insertions at a similar genomic location is nil.

2.4.2.4 Hypervariable Region Polymorphism of Mitochondrial DNA

Human mitochondrial DNA (mtDNA) is a 16,569 bp circular molecule comprising up to 37 introns fewer genes. It is the most used DNA for ancestral determination in archeological specimens, evolutionary studies, and human migration. As the mitochondria are maternally inherited (sperm DNA is eradicated upon fertilization), they exist in a high copy number of 1000 - 10,000 copies in each cell. The outcome of the maternal inheritance is that the offspring will receive the same mtDNA from the mother without any recombination. The mitochondrial genome consists of coding and non-coding regions. The non-coding region is also known as the hypervariable region (HV) or displacement loop (D loop) region. In human mt DNA, the HV region is 11.1 Kb long, divided into hypervariable region I (HVRI 16024-16365 nucleotides), hypervariable region II (HVRII, 73-340 nucleotides), and hypervariable region III (HVRIII, 438-576 nucleotides) which are regions of 400 bp each. These regions are the focal sources of polymorphisms in the control region including the D-loop. Sequence variations among individuals are largely found in these two HV1 and HV2 regions. HVIII with additional polymorphic positions is used to resolve the un-distinguishable HV1 and HV2 individuals (Swarkar et al. 2005).

The small size and high inter-person variability of the HV regions are extremely useful in the forensic sciences. A single mitochondrion unit of a somatic cell consists of 2 - 10 copies of mtDNA and each cell contains up to 1000 mitochondria. mtDNA is the only source for the forensic analysis of the degraded samples with low-quality DNA such as hair shafts, due to its abundance. As the mtDNA is exclusively maternally inherited, its sequence in all siblings and maternal relatives is identical. This unique feature can be helpful in the analysis of a missing person among the maternal lineage where the maternal relatives of several generations can be used as reference samples. The haploid nature of mtDNA in most individuals will ease DNA sequencing analysis and the interpretation of the result.

2.4.2.5 Y Chromosome Polymorphisms

Chromosomal polymorphism is the term used for defining the variations in size, shape, and morphology of a chromosome. The human Y chromosome is unique and is the male counterpart of mtDNA, exhibiting strict paternal transmission. Unlike the other 22 sets of human autosomes, the Y chromosome exists in the non-homologous form, which does not recombine with a partner chromosome during meiosis. 95% of the human Y chromosome is inherited from father to son as a single block. Only short regions in the tip of the Y chromosome, called pseudoautosomal regions (PAR) undergo recombination with the X chromosome. Comprehensive sequencing of the Y chromosome was achieved in 2003 by Skaletsky and his group. It is an acrocentric chromosome consisting of a large heterochromatin region. 95% of the Y chromosome comprises of the Male Specific Y Region (MSY), consisting of 568 genes. Among them, only 71 are capable of coding for proteins. However, only 27 genes code for distinct MSY proteins, and other genes encode for the proteins belonging to the same protein families. The expression of 9 genes is ubiquitous, and 14 are testis-specific, also expressed in the brain (PCDH11Y), thyroid (TBL1Y), and the like. No data is currently available for the expression location of the remaining 4 genes. 109 genes produce non-coding RNAs with unknown functions. They might be involved in the regulation of gene expression. All other 388 genes are pseudogenes.

2.4.2.5.1 Y Chromosome Polymorphisms and Population Studies

The Y chromosome contains various repeat elements subjected to rapid and frequent mutations. Y chromosome polymorphism refers to an increase or decrease in the repeats, variations in the chromosome short arm, the fluorescent intensity of the centromere region, lengthening, and shortening of the constriction area. One of the vital Y chromosome polymorphisms utilized in population studies is the DYS1 locus, detected through southern blot hybridization. Two probes 49a and 49f have recognized 18 fragments in human DNA digested with Taq1. Among them, 8 fragments exhibit polymorphism by either present or absent or even exhibit different sizes. In addition to this, three polymorphic loci were detected on the Y chromosome, which can be screened using PCR- based assays (Timo and Jukka, 2015):

1. DYS287 polymorphism generated by the insertion of an Alu element (Y Alu element). The frequency of this insertion is found to be high in the African Negroids (78%), and African Khoisans (47%) and is low in European populations (0 - 11%).

2. The DYS271 YAP insert polymorphism, generated by an A→G transition, leads to a *Nla*III site loss. This polymorphism, known as YAP+ chromosomes, is found only in the African populations.
3. A C→T transition at the DYS199 locus, was found to be present in the Y chromosomes of Native Americans.

2.4.2.5.2 Y Chromosome Polymorphisms and Human Diseases

Variation in the SRY locus of the Y chromosome is responsible for a disposition to hypertension. A hypertense father is responsible for determining blood pressure in male offspring, not a hypertense mother. Further, a variant of the Y chromosome defined by the Hind III polymorphism in its centromeric region is associated with hypertension in the male population of Australia, the USA, Scotland, and Poland. Hind III variant is also involved in the cholesterol concentration of the Polish population and myocardial infarction of the Spanish population These studies indicate a broad role of human Y chromosome polymorphism in determining cardiovascular risk (Kostrzewa et al. 2013). Hind III Y chromosomal variants with cardiovascular risk are also identified in Caucasians located in the UK, Belgium, Italy, and South Asians. Y chromosome polymorphisms are harnessed in male-specific diseases such as spermatogenic failure, testis and prostate cancer, and autism. The human Y chromosomal DNA variations play a vital role in investigations related to human evolution, paternity testing, and forensic sciences.

2.5 IMMUNOGENIC POLYMORPHISMS

Genetic polymorphism is a significant hallmark of human biology. The immune system is essential for an organism to confer protection against pathogens such as bacteria and viruses. The immune system is spread throughout the human body, involving many cells, tissues, and organs for distinguishing the host and foreign tissue, host cells from the foreign cells, living and dead cells, and normal and abnormal functioned (faulty) cells. The human immune system comprises two subsystems, closely linked and working together to generate an immune response. Firstly, an innate immune system or non-specific immune system involved in generating an overall immune response against germs and substances. The Innate immune system generates immune response through cells such as phagocytes and killer cells and prevents harmful foreign substances and germs from entering the human body, either through the skin or through the digestive system. Secondly, an adaptive or acquired or specific immune system, which prepares antibodies and uses them specifically for combating the germs and other foreign bodies in contact with the body (Jana and Lucie, 2016). The adaptive immune system constantly adapts itself to fight a variety of foreign cells or foreign bodies. For that, the host immune system should show much polymorphism.

2.5.1 Polymorphism in Human Major Histocompatibility Complex (MHC)

The human immune system is most profoundly affected by its genetic variations. Jean Dausset is one of the pioneers in the identification of immunogenic polymorphisms in humans. His observation of the individuals who received several blood transfusions from strangers has developed antibodies against the donor's leukocytes, which eventually led to identifying the human leukocyte antigen (HLA) system, a nomenclature that refers to the human major histocompatibility complex (MHC). MHC molecules in human blood bind to the peptides of pathogens and display them on the cell surface for their recognition by appropriate T cells. MHC has two exclusive and highly distinguishable properties, making it difficult for pathogens to evade the immune system. Firstly, MHC is polygenic, comprising MHC class I and MHC class II genes, making every individual with a set of MHC molecules have a different range of peptide binding specificity. Because of this polygeny, every individual expresses a minimum of three different MHC class I molecules and 3 - 4 class II molecules in their cells. Secondly, MHC is highly polymorphic, with multiple variants for each gene in the whole population. MHC genes are the most polymorphic genes known to date. Due to this extreme property, the number of MHC molecules expressed in human cells is more significant and is co-dominant. More than 200 alleles of MHC class I and II genes were identified in humans, with each allele present at a high frequency in the population. A high frequency of alleles in the population generates a minimal chance of the same MHC locus on both homologous chromosomes of an individual. Hence, all individuals are heterozygous regarding the MHC loci. Specific combinations of MHC alleles found on a single chromosome are known as MHC haplotypes. The expression of MHC alleles in humans is co-dominant, as the resulting genes of both the alleles of a locus are expressed together in the cell giving their product antigens to the T cells. The extensive MHC polymorphism in humans can double the number of MHC molecules expressed and further expand their diversity through polygeny. An individual can express 6 different MHC class I molecules and 8 different class II molecules with 3 MHC class I and 4 MHC class II genes through this mechanism, respectively. Further, the number of MHC class II molecules may be increased by the combination of α and β chains encoded by different chromosomes (Fehling et al. 1994). All proteins generated through the expression of the MHC genes are also highly polymorphic, except the DRα chain. This protein chain does not show any variation in its sequence between different individuals and is monomorphic, indicating a specific function, preventing its polymorphism. However, this mechanism is yet to be identified. Polymorphism in the MHC complex genes of the human genome and its implications are mainly related to organ transplantation, immune response, and conservation of an autoimmune disease. Genes of the MHC complex also undergo extensive duplication with redundant functions for increasing their antigens.

2.5.2 Immune Polymorphism

Other polymorphic molecules related to the immune system were also extensively studied along with the HLA molecules. Non-classical MHC loci have been demonstrated to contain various degrees of polymorphism. Besides class I HLA molecules, humans also encode for three non-classical, selectively expressed leucocyte antigens E, F, and G. Unique transcription patterns, protein structure, and immunological function are the characteristic features of these antigens.

Chain genes MIC-A and MIC-B related to class I MHC, located in the MHC region, exhibit a very high polymorphism rate, with more than 50 alleles identified so far. Molecules encoded by these genes do not bind to peptides and even do not associate with β2-microglobulin. These polymorphic variants do not concentrate around the peptide binding region. Still, they seem to have functional significance as most of the mutations are non-synonymous, suggesting the role of selection pressure on these polymorphisms. The expression of these two genes is restricted to epithelial, endothelial, and fibroblasts. The exact role of MIC genes is still ambiguous, but it is predicted to be involved in the functional modulation of NK and CD8+ T cells by binding to the stimulating receptor NKG2D. The role of MIC has also been implicated in graft rejection. The detection of alloantibodies in the recipients of the transplants might exert the complement-mediated cytotoxicity against the endothelial cells of the transplanted organ. Another set of unusual MHC molecules in the human immune system is lipid antigens such as CD1, found to be involved in the transport of immunoglobulins and the regulation of iron metabolism.

Killer cell immunoglobulin-like receptors (KIR) are another major group of genes involved in the human immune system with increased polymorphism. Products of these genes are expressed on the surface of Natural Killer (NK) and CD8+ T cells and are known to possess a solid regulatory hold on their function. HLA class I are the ligand molecules involved in the inhibition of KIR ubiquitously present on the surface of normal cells. KIR expression is encoded by individual genes, which can either inhibit or activate NK cell function. The human genome has evolved with many similar immune-related genes in their genome, known as the leukocyte receptor cluster, located on chromosome 19. This gene collection is in a firm linkage equilibrium that results in several haplotypes incorporating inhibitory and stimulatory KIR sequences. Inhibitory KIR is known to interact with HLA class I molecules. As humans contain several inhibitory KIR genes, it is very much possible that each individual carries at least one KIR with the capability of recognizing one autologous HLA class I allele.

2.5.3 Cytokine Polymorphism

Cytokines are glycoproteins with immunomodulatory functions in humans. They regulate the activities of target cells involved in the hematopoietic system by binding to their specific cytokine receptor ligands and initiating the signaling pathways and second messenger pathways, resulting in gene activation, cell division (mitosis), cell growth and differentiation, and apoptosis. A wide range of cell types in the human body produce these proteins, classified into either monokines produced by the monocytes or their lineage, or the lymphocytes producing lymphokines. Cytokines co-ordinate a complex network, which either induces or represses their synthesis along with other cytokines and their receptors. Many cytokines are pleiotropic, making a highly flexible network due to the overlap between the function and redundancy of individual cytokines. Cytokines are produced both by antigen and non-antigen-specific stimuli. Genes that are coded for cytokine and cytokine receptors are highly conserved in humans. A slight variation has been identified in the amino acid sequence of the IL-4 receptor, LT α, TGFβ and GM-CSF receptor β in healthy individuals and the IL-2 receptor gene of people suffering from severe combined immune deficiency (SCID). Even though conservative silent mutations do not alter their amino acid sequence, they might influence protein synthesis by altering the mRNA splicing, mRNA stability, and different gene transcription levels. Polymorphisms in the regulatory regions or the introns of a gene will impact its transcription, as these polymorphisms will alter the structure of transcription factor binding sites within the gene promoter sequence or the structure of enhancers or silencers. They may also alter the binding sites of transcription factors involved in the regulation of cell architecture, modulating their promoter geometry.

Many polymorphisms occur in the non-coding and regulatory regions of cytokine genes. These polymorphisms are attributed to the understanding of several disorders in humans. Being a part of the cell signaling the expression of cytokines, is tightly regulated and most of the cytokines are not constitutively expressed. The perturbations in cytokine balance have severe implications on human health, which is an outcome of polymorphisms leading to many immune system related diseases, and even problems during organ transplantation. Regarding the role of cytokines and their receptor polymorphisms in human disease, to date, no concrete information is available supporting their significance and practical effects. There is the possibility that balanced polymorphism of immune regulatory genes could have been evolutionarily selected for conferring a selective advantage during outbreaks of disease. A list of cytokine polymorphisms in humans is displayed in Table 2.5.

2.5.3.1 Polymorphisms and cytokine signaling

Cytokines are not constitutively synthesized and stored in the cellular compartments for their ready release in response to specific stimulation. The synthesis and expression of cytokines are triggered by stimulation. The secretion of cytokines is due to their protein synthesis. Stimulation-based cytokine synthesis provides regulation of their availability in the cellular space. As a result, increased synthesis of cytokines in response to an inflammatory stimulus is regulated by the transcription rates of their genes, controlled by transcription factors. Several transcription factors are polymorphic in the regions that regulate cytokine transcription. The nuclear factor (NF-KB) is a transcription factor involved in modulating cytokines. NF-KB is the fundamental requisite for the maximum transcription of cytokines such as tumor necrosis factor-α (TNF α), interleukins IL-1, IL-6, and IL-8,

TABLE 2.5
Cytokine Polymorphisms in Humans

S. No.	Gene	Polymorphism
1.	IL-1α	Intron 6, 46 bp VNTR
2.	IL-1α	Dinucleotide repeat
3.	IL-1α	Trinucleotide (TTA) repeat
4.	IL-1β	nt5810 A→T (*Bso*FI)
5.	IL-1β	nt5887 C→T (*Taq*I)
6.	IL-1Ra	Intron 2 86 bp VNTR
7.	IL-1Ra	nt8006 T→C (*Msp*I)
8.	IL-1Ra	nt8061 C→T(*Mwo*I)
9.	IL-1RI	2 *Pst*I RFLPs
10.	IL-2	Dinucleotide repeat
11.	IL-2Rα	*Taq*I RFLP
12.	IL-2Rβ	Dinucleotide repeat
13.	IL-3	*Bgl*II RFLP
14.	IL-3	Enhancer nt232, nt236 and nt283
15.	IL-4	Intron 3, GT repeat
16.	IL-4	Intron 2, 70 bp VNTR
17.	IL-4R	nt148 A→G
18.	IL-4R	nt426 C→T
19.	IL-4R	nt747 C→G
20.	IL-4R	nt864 T→C
21.	IL-4R	nt1124A→C
22.	IL-6R	(CA)n repeat
23.	IL-8	*Hind*III RFLP
24.	IL-9	Dinucleotide repeat
25.	IL-10	5′ proximal (CA) repeat (*IL10*.G)
26.	IL-10	5′ distal (CA) repeat (*IL10*.R)
27.	IL-11	5′ dinucleotide repeat
28.	TNFα	−862 (*−863)
29.	TNFα	−856 (*−857)
30.	TNFα	−376 G→A

vital for generating acute inflammatory responses. The TNF-receptor-associated factor (TRAF) is a member of a highly conserved scaffold protein family, which links IL-1 and TNF receptors to signaling cascades to activate NF-KB and mitogen-activated protein kinases (map kinases). TRAF proteins also serve as docking platforms for a variety of regulators involved in the signaling. These proteins are also regulated at posttranscriptional and posttranslational levels. Most of these genes vary individually through SNP, which plays a crucial role in determining individual susceptibility to a specific disease.

2.6 TRANS-SPECIES POLYMORPHISM

Trans-species polymorphism (TSP) is the mechanism of identical allele occurrence in related species. Similarity developed due to either convergence or introgression is excluded from trans-species polymorphism. These polymorphisms are the ancient genetic variants whose origin predates speciation, resulting in shared alleles between evolutionarily related species. TSP alleles of related species show more sequence similarity when compared to alleles of individual species. Trans-species polymorphism arises while passing the alleles from ancestral species to descendent species by incomplete lineage sorting. TSPs are of two types. Firstly, neutral TSPs that are more prevalent in closely related newly diverged species and disappear gradually. These types of TSP can spread across loci only in a short period after speciation. Secondly, balanced TSPs that arise from balancing selection (selection for the maintenance of variability), which are long lasting, and are maintained in the genes related to the immune system for millions of years. Identification of balanced TSP variants is one of the novel and robust approaches to shortlist various naturally occurring resistance alleles with an active role in nature conservation. Balancing selection is the primary evolutionary force responsible for the long-lasting trans-species polymorphisms in immune response genes. The maintenance of genetic variation by the balanced selection in human populations over a significant period is probably due to three possible mechanisms.

1. Heterozygote advantage also known as overdominance, arises when heterozygous individuals for a particular gene can resist a pathogen infection better than both the homozygotes. In this case, polymorphism is maintained by selection for heterozygosity.
2. Negative frequency-dependent selection is the mechanism by which only a particular genotype provides a resistance advantage over a period, which is negatively linked to the allele frequencies of a population. Pathogens tend to repeatedly infect and adapt to the most common genotype of the host in a population, leaving rare genotypes. Here, rare alleles are favored until they increase and reach a specific equilibrium beyond which they are not considered rare alleles.
3. Spatio-temporally fluctuating selection is based on variations in the selective pressures of space and time. Most of the pathogens exist in a limited area, and their abundance changes from time to time, generating a definite pressure on their host populations, leading to, or linking to migration. Even though there is no concrete evidence supporting fluctuating selection, theoretical approaches such the maintenance of MHC polymorphism make this mechanism possible.

2.6.1 Trans-Species Polymorphism in Major Histo Compatibility (MHC) Genes

MHC comprises of many highly polymorphic and dynamically evolving members of the immunoglobulin superfamily that plays a pivotal role in the adaptive immune defense of humans against pathogens. Even though the strength of association between MHC polymorphism and the resistance of humans to infectious diseases varies among individuals, the type of selection involved in the resistance is partially natural and partially sexual. Since these two mechanisms are not mutually exclusive, both can contribute to a balanced parasite-mediated polymorphism. In MHC genes, trans-species polymorphism mainly covers the variable exons of peptide binding regions encompassing exons 2 and 3 of class I, exon2 of class II MHCIIA, and MHCIIB genes, respectively. These two genes have been the targets of strong diversifying positive selection and balancing selection. In contrast, the domains of non-protein binding regions in the MHC are evolutionarily conserved, often under selection without any trans-species polymorphism. Even though the trans-species polymorphism is considered a general phenomenon in MHC genes, most studies indicate a polymorphism in most variable exon 2 of MHC IIB genes.

2.6.2 Trans-Species Polymorphism in Other Genes Related to the Immune System

Trans-species polymorphism was studied in the heavy immunoglobulin chain genes (IgVH). The genetic variability in these genes might reduce the binding specificity of the antigen-binding sites of antibodies. TSP is also reported in exon 2 of the Cα constant region of the igG-A heavy chain. Exon 2 encodes the hinge region lying at the base of immunoglobulin heavy chain regions. Variability in this region will influence the antigen-binding site of the IgA antibody and the spectrum of antigens recognized.

Proteosome subunit β-type 8 gene (PSMB8) encodes the immunoproteosomal catalytic subunit of β-ring involved in the cleavage of peptides processed for presentation on MHC class I molecules is located on the MHC class 1 cluster along with another proteosomal gene PSMB9, whose expression is induced by interferons. Polymorphism at the amino acid position 31 of this gene will affect its catalytic function. A-isoform of tripartite motif protein5 (TRIM5 α) is one of the longest isoforms of a viral restriction factor, interacting with the cytosolic viral capsid proteins of retrovirus during its infection, and prevents reverse transcription. TRIM genes are highly polymorphic, and their evolution is driven by factors such as gene loss, pseudogenization, and duplication. Host defense peptides, also known as antimicrobial proteins, are small, diverse evolutionarily conserved, effector molecules mainly involved in pathogen killing, immunomodulation, wound healing, and cell development. Many host defense peptides are present in organisms of all categories, with a balancing section maintaining their intra-specific sequence variability. In host defense peptides, trans-species polymorphism has been identified only in the β-defensin, and AvBD12 genes, rich in cysteine peptides. In humans, polymorphism in the β-defensin gene increases their susceptibility to pathogens, including HIV.

2.7 GENETIC VARIATIONS IN DRUG METABOLIZING ENZYMES

Drug metabolizing enzymes were initially considered to be liver detoxification systems responsible for the degradation of drugs and other environmental pollutants for their excretion from the body. These enzymes are involved in the synthesis and degradation of every known non-peptide involved in the ligand modulated transcription processes responsible for growth, differentiation, apoptosis, homeostasis, and other functions. Genetic polymorphisms have been well characterized in several human drug metabolizing enzymes. Some of these polymorphisms were correlated with an enhanced risk of toxicity or even cancer. The drug metabolizing enzymes are classified into two phases. Phase I drug metabolizing enzymes prepare a drug for its metabolism by activating them into electrophilic intermediates. Phase I enzymes are represented mainly by Cytochromes P450. Phase II drug metabolizing enzymes such as glutathione transferases, N-acetyltransferases, and UDP glucuronosyltransferases are proper detoxification enzymes, which conjugate the intermediates into water soluble derivatives and complete the detoxification cycle.

In humans, the liver is the main organ responsible for the phase I and phase II reactions. Phase I enzymes and UDP glucuronosyltransferases are localized in the endoplasmic reticulum, whereas the phase II enzymes are localized in the cytoplasm. Genetic changes in the drug metabolizing enzymes are responsible for polymorphisms in human beings (Einosuke, 2001). The details of drug metabolizing enzymes are displayed in Table 2.6. These polymorphisms represent one of the

TABLE 2.6
Details of the Polymorphisms in the Drug Metabolizing Enzymes

Isoenzyme	Mutation Type	Consequence	No. of variant Alleles
Cytochrome P450 drug metabolizing enzymes			
CYP2C9	Point mutation	reduced affinity for substrates	4
CYP2C19	Point mutation	inactive enzyme	4
CYP2D6	Point mutation and Deletion	inactive, unstable enzyme	22
CYP2E1	Point mutation	inactive enzyme	6
Non-cytochrome P450 drug metabolizing enzymes			
N-acetyltransferase-2	Point mutation	inactive enzyme	14
Thioprine methyltransferase	Point mutation	inactive enzyme	8
dihydropyrimidine dehydrogenase	Point mutation and Deletion	inactive enzyme	3

earliest and the most documented examples of genetic variation in human beings. Genetically variable enzymes are discussed here.

2.7.1 Cytochrome P-450

Cytochrome P-450 (CYPs) are essential enzymes involved in the biosynthesis and degradation of endogenous compounds such as steroids, lipids, and vitamins. They metabolize many compounds that form the main diet, as well as medications. They also metabolize many toxins that enter the human body from the environment. These enzymes reduce or sometimes even alter the pharmacologic activity of many drugs and facilitate their excretion. To date, over 57 different CYP isoenzymes are present in humans. Among them, 42 enzymes participate in the metabolism of exogenous xenobiotics, endogenous steroids, and prostaglandins. 15 isoenzymes are involved in drug metabolism, with a significant variation in the inter individual enzyme activities mainly due to induction, inhibition, and inheritance. Many functional genetic polymorphisms were reported in the CYP isoenzymes *CYP2A6*, *CYP2C9*, *CYP2C19*, and *CYP2D6*. Around 80% of drug metabolism in humans is by 7 CYP isoenzymes CYP1A2, CYP 2A6, CYP2C9, CYP2C19, CYP2D6, CYP2E1, and CYP 3A4/5 (Li et al. 2017). Table 2.7 displays the level of drug metabolism by these isoenzymes.

Several CYP isoenzymes are functional in humans, whose nomenclature employs a three-tiered classification. Individual cytochrome P-450 enzymes are classified based on the similarities in their amino acids, which are designated by a family number, a subfamily letter, a number for an individual enzyme within the subfamily, and an asterisk followed by a number and a letter for each genetic (allelic) variant. Here, members of the same family are 40% homologous in respect of their amino acid sequences, with 55% homology within the sub-family, and 55% amongst genes. Polymorphisms in CYP genes result from changes in the nucleotides, which consequently change

TABLE 2.7
The Level of Drug Metabolism Carried Out by 7 Major CYP Isoenzymes and Their Substrates

CYP isoenzyme	Substrate	Level of drug metabolism (%)
CYP1A2	Phenacetin, Caffeine, Aflatoxins, Clozapine, Theophylline, Acetaminofen, Haloperidol	13
CYP2A6	Nicotine, Coumarin, Dimethylnitrosamine	4
CYP2C9	Ibuprofen, Phenytoin, Hexobarbital, S-warfarin, Tolbutamide	18
CYP2C19	S-Mephenytoin, Amitriptyline, Imipramine, Diazepam, Clomipramine, Omeprazole, Lansoprazole, Chloroproguanil, Moclobemide, Proguanil	1
CYP2D6	Sparteine, Debrisoquine, Antiarrhythmics, Antihypertensives, Tricyclic antidepressants, Selective serotonin reuptake inhibitors (SSRIs), Beta-adrenergic blockers, Morphine derivatives, Bufuralol	2
CYP2E1	Acetaminofen, Ethanol, Organic solvents, Dimethylnitrosamine, Chlorzoxazone, Anaesthetics	7
CYP 3A4/5	Calcium antagonists, Antiarrhythmics, Antiinfectives, Immunosuppressants, Endogenous substrates, Chemotherapeutics, Anaestheticanalgesics, Antimicrobials, Benzodiazepines, Antihistamines, Anti-emetics, Anti-HIV agent	35

the amino acid chains of the proteins. CYP2C9, CYP2C19, CYP2D6, and CYP2E1 are the main CYP450 isoenzymes studied for genetic polymorphism in humans.

CYP2C9

CYP2C9 is one of the vital cytochrome P450 isoenzymes that metabolize narrow therapeutic index drugs such as phenytoin, hexobarbital, tolbutamide, anticoagulant warfarin. *CYP2C9*2* and *CYP2C9*3* are the two most common CYP2C9 variants found in Caucasians, exhibiting a single amino acid substitution at positions critical for the enzyme activity. In the *CYP2C9*2* variant, Cys residue at position 144 substitutes Arg, and Leu replaces Ile residue at 359 positions in *CYP2C9*3*. This substitution is responsible for the catalytic efficacy of the enzyme. In Caucasians, allele frequencies of 12.5% and 8.5% have been reported for CYP2C9*2 and CYP2C9*3, respectively. Another variant *CYP2C9**8 allele occurs in 10% of African Americans, having certain implications for CYP2C9 substrate metabolism in this population. Variations in *CYP2C9*2* and *CYP2C9*3* alleles have clinically significant consequences in warfarin-treated patients amongst the Caucasian population. *S*-warfarin has reduced up to 90% in *CYP2C9*3* homozygotes over the homozygotes of wild type**1* variant. Similarly, *CYP2C9* variant alleles in 81% of the Caucasian patients require low quantities of warfarin (≤1.5 mg/day). This group of patients has reported more difficulty with warfarin induction, requiring an extended hospital stay to stabilize the warfarin regimen. In addition, a profound therapeutic response to regular doses of warfarin was observed in a patient homozygous for *CYP2C9*3*, necessitating dose reduction to 0.5 mg/day.

CYP2C19

CYP2C19 is the least abundant CYP isoenzyme among the four. The two significant alleles responsible for genetic polymorphism in CYP2C19 are *CYP2C19*2* due to an aberrant splice site and *CYP2C19*3* with a premature stop codon, resulting in an inactive CYP2C19 enzyme in a polymorphism phenotype. Seven CYP2C19 genes CYP2C19*2A, CYP2C19*2B, CYP2C19*3, CYP2C19*4, CYP2C19*5A, CYP2C19*5B, and CYP2C19*6 are present in the human genome. The most common mutation of CYP2C19 is CYP2C19*2, which occurs both in Caucasians and Japanese. 2 - 5% of Caucasians, 13 - 23% of Asians, 17% of Chinese, 13% of Koreans, and 23% of Japanese are deficient in the substrate Smephenytoin 4'-hydroxylase. 60-70% of the defective alleles are accounted for by CYP2C19*2, followed by CYP2C19*3. These alleles are detected only in the poor Japanese metabolizers. Another poor metabolizing allele CYP2C19*4, consisting of A→G mutation in its start codon, was found in Caucasians. The clinical implications of *CYP2C19* genetic polymorphisms were not examined extensively. The poor metabolizers for CYP2C19 polymorphism have shown a ten-fold rise in the area under the curve of CYP2C19 substrate omeprazole when compared with the enhanced metabolizers. Asia's population metabolizes CYP2C19 substrates more slowly than Caucasians.

CYP2D6

CYP2D6 involves debrisoquin panel metabolism, comprising over 30 drugs for antiarrhythmics, antihypertensives, P-blockers, monoamine oxidase inhibitors, morphine derivatives, antipsychotics, and tricyclic antidepressants along with environmental agents. While studying the polymorphism of debrisoquin, a hypertensive agent, Robert L. Smith has identified a polymorphism in its oxidation. He noticed a difference in the behavior of the drug in the UK and the USA populations. The drug in the United States has generated a high incidence of side effects (idiosyncratic reactions), probably due to the genetic variations in the patients. A large population was screened and found that the poor metabolizers of the drug debrisoquin were present in 6 - 10% of Caucasian populations compared with the extensive metabolizers. The capability to handle debrisoquin in extensive metabolizers has an efficiency of 10,200 times. Frequencies of poor metabolizers are -5% in Blacks and <1% in Asians. This drug metabolizing enzyme was identified to be a cytochrome P450 and was named CYP2D6. The gene of CYP2D6 is localized on chromosome 22, locus 13.1, located near the SIS oncogene. Poor metabolizing alleles of CYP2D6 code for a defective protein or incorrect splicing of the gene (sometimes both), resulting in decreased or utterly absent enzyme activity. CYP2D6 Polymorphisms are the best characterized among CYP 450 enzymes, with more than 75 gene variants and 120 alleles identified so far in this gene. Among these alleles, poor metabolizers and extensive metabolizers can be identified with 99% accuracy. *CYP2D6*1* is the wild-type variant with regular activity. *CYP2D6*1* and *CYP2D6*2* are the extensive metabolizing alleles with the same enzyme activity. *CYP2D6*1* is capable of duplication or amplification. Polymorphism in the CYP2D6 is vital for 20% of metabolism among all commonly prescribed drugs. Even though human beings have more than 60 unique P450 genes, only six CYP1A2, CYP2C17, CYP2D6, CYP2E1, CYP3A4, and CYP4A11 are responsible for the metabolism of a vast majority of prescribed and over-the-counter drugs.

Numerous epidemiological studies have linked the allelic differences of CYP2D6 with toxicity and cancer. Such studies indicate that the CYP2D6 extensive metabolizer phenotype is associated with increased malignancies in the urinary bladder, liver, pharynx, stomach, and cigarette smoking-induced lung cancer. The CYP2D6 mediated metabolism of one or more unknown dietary and environmental reagents forms a reactive intermediate involved in urinary bladder, liver, pharynx, stomach, and lung cancer initiation. Individuals with deficient metabolizing phenotypes develop a 2-5-fold risk of Parkinson's disease. A positive correlation between the poor metabolizing phenotype and a decreased tolerance to chronic pain reflects the differences in the endogenous synthesis of morphine by CYP2D6 in the human brain. Recent studies have implicated the relationship between CYP2D6 and dopamine neurotransmission in the human brain. An association between the CYP2D6 phenotype and the tendency to opiate addiction is underway.

CYP2E1

CYP2E1 is responsible for the metabolic activation of many low molecular weight compounds such as ethanol, benzene, vinyl chloride, and N-nitrosamines. This isoenzyme has a high redox potential compared with other cytochrome P450 members and can induce the peroxidation of lipids and other reactive oxygen species (ROS) capable of causing oxidative stress. This enzyme is considered to be an essential source of ROS in alcohol-induced liver injury. Additionally, exo and endogenous substrates are associated with human susceptibility to the toxicity. Industrial and environmental chemicals associated with the human carcinogenicity are also metabolized by CYP2E1. The human CYP2E1 gene is located on the 10q24.3-qter locus of chromosome 10. CYP2E1 enzyme polymorphisms in humans are correlated with cancer and other diseases.

CYP2A6

This enzyme is a significant metabolizer of nicotine, and the clinical relevance of this enzyme polymorphism lies in tobacco abuse. Several variants of CYP2A6 polymorphism for the wild type CYP2A6*1, such as CYP2A6*2 with single amino acid substitution, *CYP2A6*4* with gene deletion, CYP2A6*5 with gene conversion, and CYP2A6*20 with frameshift mutation were identified in the humans. All these genetic variants have abolished their enzyme activity. The *CYP2A6*4* gene deletion variant is the most common among Asian populations, responsible for a dramatic difference in the frequency of poor metabolizers in Asian (20%) against Caucasian (1%) populations. Non-smokers are more likely to carry a defective CYP2A6 allele than smokers. Smokers with defective alleles are very likely to quit smoking after a few trials. A consequence of the defective CYP2A6 allele is its inability to metabolize nicotine, enhance nicotine tolerance, and decrease its adverse effects. These observations show that CYP2A6 inhibition will have a definite role in tobacco dependency.

2.7.2 Glucose-6-Phosphate Dehydrogenase

This enzyme is expressed in all tissues of the human body, controls the carbon flow through the pentose phosphate pathway, produces NADPH for reductive biosynthesis, and maintains oxidation-reduction in the cell to keep glutathione in a reduced state. Maintenance of reduced glutathione levels is vital in phase II detoxification pathways. G6PD polymorphism was reported for the first time during the Second World War. After being transported by aircraft at 5,000 - 10,000 ft altitude soldiers being treated for malaria developed hemolytic anemia. Most of the Afro-American soldiers with only 10% G6PD deficiency exhibited the symptoms. Here, a combination of the antimalarial drug, primaquine and the low oxygen levels in their RBC led to a total block of the G6PD pathway. G6PDH deficiency reduces glutathione levels, allowing oxidative drugs to oxidize sulfa hydroxyl groups of hemoglobin, leading to hemolysis. Over 25 drugs, including primaquine, sulfones, sulfonamides, nitrofurans, vitamin K analogs, cefotetan, and chloramphenicol, cause hemolytic anemia in G6PDdeficient humans. The location of the G6PD gene is on the X chromosome. Over 400 G6PD variants exist in humans. At least 10% of the world's population (400 million) comprises of G6PD variants. Most of the variants are asymptomatic. However, only 30 different mutations have been reported in the gene functional region, coding for a protein. 29 of these functional mutations are point mutations, with around 15 being nucleotide conversions from cytosine to guanine.

Ethnic differences in G6PDH deficiency within the population groups were >100 fold between Ashkenazic and Sephardic Jews. Males of African and Mediterranean descent express this trait more frequently. Two mutations have been identified in the African population, namely G6PD A, where the protein produces regular red cell activity, and G6PD A (–) where it produces only 10% of the usual red cell activity, which is also unstable. Among G6PD A population, an A→G substitution at 376 nucleotide of the gene causes the replacement of asparagine residue with aspartic acid. There are three different variants in the G6PD A (–) allele.

1. A→G mutation at 376 nucleotide.
2. G→A mutation at 202 nucleotide leading to the replacement of valine by methionine.
3. Substitution at the 68th codon.

2.7.3 N- Acetyltransferase polymorphism

Also known as N-acetylation polymorphism or isoniazid acetylation polymorphism, this illustrates the polymorphism of a drug metabolizing enzyme N- Acetyltransferase, a phase-II conjugating liver enzyme, that catalyzes the *N*-acetylation (deactivation) and *O*-acetylation (activation) of arylamine carcinogens and heterocyclic amines. Polymorphism in N- Acetyltransferase was identified in the 1940s during the treatment of tuberculosis patients with isoniazid. There was a high incidence of peripheral neuropathy representing an unusually slow clearance of isoniazid. The patients were phenotyped into slow and rapid acetylators. The phenotype corresponding to the slow-acetylator was inherited as an autosomal recessive trait, often experiencing toxicity from drugs such as isoniazid, sulfonamides, procainamide, and hydralazine. Phenotypes corresponding to the rapid acetylator may not respond to isoniazid and hydralazine during treatment for tuberculosis and hypertension, respectively. During the development of isoniazid, plasma concentrations were observed in a distinct bimodal population after a standard dose. Patients with the highest plasma isoniazid levels were slow acetylators suffering from peripheral nerve damage, while rapid acetylators were not affected. Two human N-acetyltransferase functional genes NAT1, NAT2, and one pseudogene NATP are located on the pter-qll locus of chromosome 8. NAT2 gene expression controls the phenotype corresponding to rapid and slow acetylations. This gene encodes for a NAT2 enzyme, having a 10-times-lower Km for aromatic amines than that of NAT1. Significant variations in the slow acetylator phenotypes among different ethnic groups. 40 - 70% of Caucasians, 60% of African Americans, 10 - 20% of Japanese and Canadian Eskimo, 82% of Egyptians, and 2 - 5% of Jewish populations are slow acetylators. There are three major slow acetylator alleles, two in Caucasians and one in Asians, and several minor, rare NAT2 alleles.

Studies correlating the epidemiological associations between the acetylation phenotype and cancer, or toxicity have identified that slow acetylators exhibit a high risk of bladder cancer and a low risk of colorectal carcinoma. Cigarette smoking and occupational exposure to arylamines are vital factors that work in conjunction with the slow-acetylator phenotype to develop bladder cancer. There is no relationship between acetylator phenotype and smoking-related bladder cancer in the absence of arylamines. It was also found that cigarette-smoking postmenopausal women with the slow acetylator phenotype are more likely to develop breast cancer than those with the rapid-acetylator allele. The development of breast cancer is due to toxic nitrosamines in tobacco smoke being degraded by NAT2 much slowly, making the individual more susceptible to breast cancer. Slow acetylator phenotypes are even more susceptible to hydralazine or lupus syndrome and suffer from hemolytic anemia caused by certain sulfonamides.

2.8 METHODS OF GENOTYPING

Genotyping is the process of determining the genetic makeup or genetic constitution of an organism by examining its DNA sequence. It is the technology used for detecting minor differences in the genomes that could lead to a significant change in phenotypes. Genotyping determines the differences or variations in the genetic complement of an organism by comparing its DNA sequence with that of a reference sequence. It also identifies variations in the genetic makeup within the populations. Approaches for the identification of the genetic variations are listed in Table 2.8.

2.8.1 Human HapMap Project

The human genome sequence has enabled the positional cloning of the genes responsible for monogenic disorders. Over 1500 genes have been identified through this project. Most of the genes are responsible for the rare disorders caused due to mutations. The identification of complex, polygenic disorders requires a whole-genome association approach for analyzing thousands of samples using 100,000 genetic markers. For validation of the markers, the human HapMap project was launched on 27th October 2005 to capture genetic variations among human beings through SNPs and to provide the complete haplotype sequence information of European, African, and Asian Populations. Phase III of the project was completed in 2009, in which global ancestral groups from Africa, Utah, Han Chinese, Gujarati Indians, Japanese, Luhya, Mexicans, Maasai, Tuscans, and Yoruba were assembled. Phase III of this project has identified and validated over 6 million SNPs (www.hapmap.org/).

2.8.2 Genome Wide Association Study

HapMap and SNP databases have provided valuable information for SNP assay designs. The information on genomic variations should generate high throughput, cost-efficient genotyping products, suitable for the linkage and genome-wide association (GWA) studies. A genome-wide association study is an approach that uses rapidly scanning markers belonging to the complete genomes of many

TABLE 2.8
Approaches for Studying the Variations Among Genomes

S. No.	Approach	Technique
1.	Enzymatic Approach AFLP CFLP Ligation based Assays	RFLP
2.	PCR based AP-PCR DT-PCR	RAPD
3.	Electrophoretic Discrimination Heteroduplex Analysis Fragment Analysis	SSCP
4.	DNA sequencing Doublex sequencing Single Sequencing Reaction (SSR) Clipper Sequencer	High performance DNA sequencing
5.	Solid Phase Determination High density Oligonucleotide Arrays Fiberoptic DNA sensor array	Oligonucleotide Arrays
6.	Chromatography based Approach	DHPLC
7.	Physical Methods Fluorescence exchange-based methods	Differential sequencing with MS

people to determine genetic variations associated with a specific disease. Researchers can use the information to develop better strategies to detect, treat and prevent that particular disease. GWAS is particularly useful in bringing out genetic variations that significantly contribute to common, complex diseases, using the reference human genome sequence, and the human genetic variation map. A set of new technologies provide a quick and accurate analysis of whole-genome samples for genetic variations, contributing to the onset of a disease.

GWAS studies have successfully identified genetic variations for many common diseases, such as asthma, diabetes, prostate cancer, bipolar disorder, coronary heart disease, Crohn's disease, type I and type II diabetes, rheumatoid arthritis, and genetic variations that influence responses to antidepressant medications. In 2005, three independent studies found a common form of blindness associated with variations in the gene responsible for complement factor H, whose protein regulates inflammation. Earlier, it was thought that inflammation might contribute to this type of blindness, known as age-related macular degeneration. Structural variation is also known to be a significant contributor to human genetic variation and sequence variants. Copy number variants (CNVs) are the fragments of the genome larger than 1 kb and vary in copy number between individuals. The role of copy number variants in common diseases such as autism is under study. The next challenge is to cover the regions of the genome deemed "unSNPable" owing to recombination hotspots and copy number variations.

For the human genome-wide association studies, at least 1,00,000 SNPs assays are needed. Four technologies, Invader assays, the Perlegen genotyping platform, Affymetrix GeneChips, and Illumina's Infinium Beadchips have been invented for achieving this. The Affymetrix and Illumina chips offer the highest marker densities with 1.8 and 1.2 Mi assays, respectively.

2.8.3 Next-Generation Sequencing Technologies for Genotyping

The rapid development of high-throughput next-generation sequencing technology, such as massively parallel sequencing, offers a great potential to revolutionize genotyping. The next-generation sequencing technologies can sequence millions of DNA or cDNA fragments in a single instrument run in a massively parallel fashion. Such high-throughput sequencing technology is used for *de novo* sequencing, genome re-sequencing to detect SNPs and other variants, transcriptome sequencing, immunoprecipitation-based protein-DNA or protein-RNA interaction mapping, and DNA methylation using bisulfite-mediated cytosine conversion. When the cost of sequencing a human genome has become feasible at a cost of $1000, this current technology has every chance of replacing genotyping for GWAS. Three major platforms are commercially available for the massive parallel sequencing, the GS FLX 454 sequencer of Roche, the Illumina Genome Analyzer from Solexa, and the Applied Biosystems SOLiD sequencer. The Roche 454 Sequencer can produce reads of length 250 bp per fragment, in contrast to the 35 - 50 bp reads of the other two platforms (Solexa and SOLiD). The short-sequence instruments produce many more reads for each instrument run. Capacities of these instruments are increasing rapidly with new and rapid developments in technology.

2.9 SINGLE STRANDED CONFORMATIONAL POLYMORPHISM (SSCP)

The single stranded conformation polymorphism (SSCP) is a technique for recognizing the sequence variations in a 150 - 250 bases single strand of DNA. It is one of the simplest, and also instrumental methods for mutation detection (Steve et al. 1997). Under non-denaturing conditions, a DNA single strand will have a specific conformation and folding, depending on its base composition. This conformation will vary with a change of every base in its sequence. Most of the conformational changes will sufficiently alter the physical configuration of the sequence. Even though the original sequence and variant sequence has the same charge, their specific size-to-charge ratio will differ. This slight mobility difference between two sequences is enough for its detection on a non-denaturing acrylamide gel upon electrophoresis. Mutation-induced changes in the tertiary structure of the DNA strand are also responsible for the differences in mobility, detected as the appearance of new bands on the gels stained with ethidium bromide or through autoradiograms in the case of radioactive based detection. The tertiary structure of the single stranded DNA will change under varying physical conditions such as temperature, ionic environment, and the likes. The sensitivity of SSCP also depends on these conditions. PCR is used for amplifying the specific regions in the DNA of interest. The mutation detection capability of a PCR-SSCP is >80% in a single run for the fragments of sizes shorter than 300 bp, which decreases with an increase in its length. For mutation detection in long DNA fragments, overlapping short primer sets or long PCR products digested with appropriate restriction enzymes can be used.

2.9.1 SSCP Procedure

SSCP of a DNA fragment involves four steps

1. PCR amplification and labeling of DNA: 10 μl reaction mixture consists of 1 μg DNA, 5 pM of forward and reverse primers, 0.5 units of Taq polymerase, each dDNTP at 0.2 mM, 10 mM Tris-HCl (pH 8.3), 50 mM KC1, 1.5 mM $MgC1_2$, and 0.01% gelatin. DNA is labeled by adding 0.5 μCi of ^{32}p dCTP (3000 Ci/mM) to the PCR mixture. PCR conditions are 30 cycles of 94°C for 30 sec, 55°C for 30 sec, and 72°C for 30 sec.
2. Denaturation of the PCR product: The amplified DNA fragments are mixed with an equal volume of sample buffer consisting of 95% formamide, 20 mM EDTA (pH 8.0), 0.05% xylene cyanol, and 0.05% bromophenol blue, denatured at 85°C for 5 min and immediately cooled on ice.

3. Gel conditions: 100 ml of a 5% non-denaturing gel containing 17 ml of 30% polyacrylamide in lx TBE is prepared freshly and cast into standard sequencing gel plates using 0.4 mm thick spacers and a shark's tooth comb. The cast gel is cooled in a cold room for at least 30 minutes before sample loading. Denatured DNA is loaded (2 µl/well) and electrophoresed at 4°C at a constant voltage of 500 V for 12 - 15 hrs. in lx TBE electrophoresis buffer. After electrophoresis, the gel is transferred onto a sheet of Whatman 3MM paper and vacuum-dried before autoradiography.
4. The detection of single stranded DNA fragments: Detection methods for single strands involve autoradiography of ^{32}P incorporated during PCR or ethidium bromide staining of the gel, intercalating with 5 to 10-fold high affinity and silver staining with 1 - 10 pg/mm^2 silver nitrate. Further, DNA can be blotted onto the membrane and is detected conveniently by direct blotting hybridization. Automated DNA sequencers perform electrophoresis and detection of fragments on the gel using fluorescently labeled oligonucleotides.

2.9.2 High Throughput Methods for SSCP

Several significant improvements in the techniques used for sample handling allow sample analysis with mutations in a rapid manner. A reduction in the tube handling and labeling time by storing DNA from individuals in a 96 well microtiter plate. Alteration of the DNA concentration to 16 ng^{-1} mL water. 20 replicas from one master array can be prepared with an eight-channel pipette in 2 hours with this concentration. Storage of pre-PCR dried plates at room temperature also makes it possible to prepare many identical replicas in advance. Setting up PCRs is extremely simple. There is the possibility of minimizing the imprecision introduced by pipetting. Application of microtiter array diagonal gel electrophoresis (MADGE) for the 96-well array, with a modified gel system, allows a 5 to 10-fold higher throughput than the normal ones.

2.9.3 Applications

1. SSCP has been used extensively to screen for inherited mutations or to detect somatic mutations in cancer cells.
2. Because of its sensitivity to single-base changes, SSCP is applied to search for polymorphisms in cloned or amplified DNA for its utilization as genetic markers.

2.9.4 Limitations

1. SSCP screening only hints at the existence of a mutation. Subsequent DNA sequencing will determine the nature of the mutation responsible for a shift in the electrophoretic mobility of the given sample.
2. Not all point mutations in the given sequence will cause a detectable change in electrophoretic mobility. However, sensitivity can be increased by optimizing PCR reactions and run conditions before attempting a large-scale screening.

2.9.5 Modifications of SSCP

RNA-SSCP: Single stranded RNA should have a larger repertoire of secondary structure since RNA base pairing is more stable than RNA-DNA base pairing. High RNA base stability might adopt a more conformational structure and hence be more sensitive to sequence changes. The efficiency of detection is 70% for rSSCP in comparison to 35% for dSSCP (for 2.6 kb fragment). However, it is more inconvenient to make RNA strands involving RNA polymerase promoters compared to DNA strands.

Restriction endonuclease fingerprinting-SSCP: A modification of SSCP, where restriction digestion of a 1 Kb segment with 5 restriction enzymes produces fragments of ~150 bp. After digestion, the products are mixed, end-labeled with ^{32}P, denatured, and electrophoresed under non-denaturing conditions. Two components will be evident, the gain or loss of the restriction site (informative restriction component) and abnormal mobility of the 5 restriction fragments (10 strands) known as the 'SSCP component'. The efficiency of detection, in this case, is 96%.

2.10 CONCLUSION

Genetic polymorphism is the existence of more than one genetic variant of a gene sequence, chromosome structure, or phenotype. The human genome comprises of 6 billion nucleotides packed into 23 sets of chromosomes, with one set inheriting from each parent. The probability of polymorphism is very high in humans due to the relatively large genome size, varying from a single base change to many base pairs and repetitive sequences. Single nucleotide polymorphisms are amongst the most common and highly abundant polymorphisms in humans, which are very important for mapping diseases and performing population and evolutionary studies. 30% of the human genome comprises of repetitive sequences, such as variable numbers of tandem repeats (VNTRs), which are also highly polymorphic, used in forensic investigations, DNA fingerprinting technologies, paternity testing, and the like. Recent work on VNTRs confirms their role in pathological studies. Variations in these tandem repeats also led to the development of DNA markers for studying genetic variations in human health and diseases.

Genes are inherited strictly in two copies in a genome. The rapid advancement in the technology facilitating genome sequencing has revealed that the segments of DNA ranging from thousands to millions of bases in size vary in copy number. Such copy number variations (CNVs) are important sources for the high genetic diversity in humans. There is the development of a global CNV map to detect the genes connected with the diseases, study familial genetic conditions, and identify the variations among individuals, facilitating the complete understanding and utilization of the human genome. Other critical applications of genetic polymorphism focus on improving human health through gene therapy, discovering novel drugs, and their targets, and upgrading of discovery processes with advanced technologies to generate more accurate information. The development of high throughput DNA sequence technology and microarrays has facilitated the generation and analysis of data. Further identification of the variations in the human genome before geographical location and personalized medicine will be of need to human society.

REFERENCES

Altshuler, D., Hirschhorn, J. N., Klannemark, M., Lindgren, C. M., Vohl, M. C., Nemesh, J., et al. (2000). The common PPARgamma Pro12Ala polymorphism is associated with decreased risk of type 2 diabetes. *Nat. Genet.*26: 76–80.

Amarger, V., Gauguier, D., Yerle, M., Apiou, F., Pinton, P., Giraudeau, F., Monfouilloux, S., Lathrop, M., Dutrillaux, B., Buard, J., et al. (1998). Analysis of the human, pig, and rat genomes supports a universal telomeric origin of minisatellite sequences. *Genomics* 52: 62–71.

Armour, J. A. L., Neumann, R., Gobert, S., Jeffreys, A. J. (1994). Isolation of human simple repeat loci by hybridization selection. *Hum. Mol. Genet.* 3:599–605.

Astrid, M., Roy-Engel., Marion, L. C., Erika, V., Randall, K., Garber, S. V. N., et al. (2001), Alu Insertion Polymorphisms for the Study of Human Genomic Diversity. *Genetics* 159: 279–290.

Azevedo, L., Serrano, C., Amorim, A. et al. (2015). Trans-species polymorphism in humans and the great apes is generally maintained by balancing selection that modulates the host immune response. *Hum Genomics* 9: 21.

Bennett, P. (2000). Demystified . . . Microsatellites. *J. Clin. Pathol. Mol. Pathol.* 53:177–183.

Beth, A. T., Laura, K., Aaron, J. G., Edward, B. G., Shawn, E. Mc. (2013). The perils of SNP microarray testing: uncovering unexpected consanguinity. *Pediatr. Neurol.* 49(1): 50–53.

Bidwell, J., Keen, L., Gallagher, G., Kimberly, R., Huizinga, T., McDermott, M. F., Oksenberg, J., Nicholl, Mc., Pociot, F., Hardt. C., D'Alfonso, S. (1999). Cytokine gene polymorphism in human disease: on-line databases. *Genes and Immunity* 1: 3–19.

Bret, A. P., Peicheng, J., Ryan, J. H. (2011). A genomic portrait of human microsatellite variation. *Mol. Biol. Evol.* 28(1):303–312.

Buido B., Arianna, M., Eric, M., Luca, C-S. (1997). An apportionment of human DNA diversity (genetic variation microsatellite loci restriction polymorphisms racial classification) *Proc. Natl. Acad. Sci.* USA. 94: 4516–4519.

Cann, R. L., Stoneking, M., Wilson, A. C. (1987). Mitochondrial DNA and human evolution. *Nature* 325:31–36.

Choongwon, J., Benjamin, M. P., Peter, B. B., Maniraj, N, Cynthia, M. B., Geoff, C., Sienna, R. C., John, N., Anna, D. R. (2017). A longitudinal cline characterizes the genetic structure of human populations in the Tibetan plateau. *PLoS One* 12(4): e0175885.

Choy, Y. S., Dabora, S. L., Hall, F., Ramesh, V., Niida, Y., Franz, D., Kasprzyk-Obara, J., Reeve, M. P. and Kwiatkowski, D. J. (1999). Superiority of denaturing high performance liquid chromatography over single-stranded conformation and conformation-sensitive gel electrophoresis for mutation detection in TSC2. *Ann. Hum. Genet.* 63: 383–391.

Csilla K., Lluis, Q-M., Gianni, F. (2004). Y chromosome polymorphisms in medicine. *Annals of Medicine* 36 (8): 573–583.

Dahlback, B., (1997). Resistance to activated protein C caused by the factor VR506Q mutation is a common risk factor for venous thrombosis. *Thromb Haemost* 78: 483–488.

Daly, A. K. (2010). Pharmacogenetics and human genetic polymorphisms. *The Biochemical Journal.* 429(3):435–449.

Danie, M. B., Lucio, P-V., Angel, O-P. (2005). The relationship between environmental stability and avian population changes in Amazonia. *Ornitologia neotropical* 16: 289–296.

Darwin, C. (1859). On the origin of species by means of natural selection, or preservation of favored races in the struggle for life. John Murray, London.

Einosuke, T. (2001). Polymorphism of drug metabolizing enzymes in humans. *Sepsis* 4:247–254.

Nebert, D. W. (1997). Polymorphisms in drug-metabolizing enzymes: what is their clinical relevance and why do they exist? *Am. J. Hum. Genet.* 60:265–271.

Daniel, D., Christian S. (2003). Two distinct modes of microsatellite mutation processes: evidence from the complete genomic sequences of nine species. *Genome Reseracch* 13(10): 2242–2251.

David, S., Svante, P. (2204). Evidence for gradients of human genetic diversity within and among continents. *Genome Research* 14:1679–1685.

Dobson-Stone, C., Cox, R. D., Lonie, L., Southam, L., Fraser, M., Wise, C., Bernier, F., Hodgson, S., Porter, D. E., Simpson, A. H., Monaco, A. P. (2000) Comparison of fluorescent single-strand conformation polymorphism analysis and denaturing high-performance liquid chromatography for detection of EXT1 and EXT2 mutations in hereditary multiple exostoses. *Eur. J. Hum. Genet.* 8: 24–32.

Dong, Y., Zhu, H. (2005). Single-strand conformational polymorphism analysis: basic principles and routine practice. *Methods Mol. Med.* 108:149–57.

Dorman, J. S., LaPorte, R. E., Stone, R. A., Trucco, M. (1990). Worldwide differences in the incidence of type I diabetes are associated with amino acid variation at position 57 of the HLA-DQ beta chain. *Proc. Natl. Acad. Sci.* U S A. 87: 7370–7374.

Fehling, H. J., Swat, W., Laplaceeta, C. (1994). MHC class I expression in mice lacking the proteasome subunit LMP-7. *Science* 265 (517): 1234–1237.

France, D., Gilles, V., Gary, B. (2003). Predicting Human Minisatellite Polymorphism. *Genome Res.* 13(5): 856–867.

Gilles, V., France, D. (2000). Minisatellites: mutability and genome architecture. *Genome Research* 10:899–907.

Giraudeau, F., Taine, L., Biancalana, V., Delobel, B., Journel, H., Moncla, A., Bonneau, D., Lacombe, D., Moraine, C., Croquette, M. F., et al. (2001). Use of a set of highly polymorphic minisatellite probes for the identification of cryptic 1p36.3 deletions in a large collection of patients with idiopathic mental retardation: Three new cases. *J. Med. Genet.* 38: 121–125.

Goldstein, D. B., Ruiz-Linares, A., Cavalli-Sforza, L. L. and Feldman, M. W. (1995). An evaluation of genetic distances for use with microsatellite loci. *Genetics* 139: 463–471.

Goldstein, D. B., Ruiz-Linares, A., Cavalli-Sforza, L. L., Feldman, M. W. (1995). Genetic absolute dating based on microsatellites and the origin of modern humans. *Proc. Natl. Acad. Sci.* USA. 92: 6723–6727.

Gross, E., Arnold, N., Goette, J., Schwarz-Boeger, U., Kiechle, M. (1999). A comparison of BRCA1 mutation analysis by direct sequencing, SSCP and DHPLC. *Hum. Genet.* 105: 72–78.

Hellgren, O., Sheldon, B. C. (2011). Locus-specific protocol for nine different innate immune genes (antimicrobial peptides: *β*defensins) across passerine bird species reveals within-species coding variation and a case of trans-species polymorphisms. *Molecular Ecology Resources* 11 (4): 686–692.

Horing, W. J. Papagiannes, E., Dray, S., Rodkey, L. S. (1980). Expression of cross-reacting determinants of the immunoglobulin heavy chain variable region a3 allotype in *Oryctolagus* and *Lepus*, *Molecular Immunology* 17: 111–117.

Huddleston, J., Chaisson, M. J. P., Steinberg, K. M., Warren, W., Hoekzema, K., Gordon, D., Graves-Lindsay, T. A., Munson, K. M., Kronenberg, Z. N., Vives, L., Peluso, P., Boitano, M., Chin, C. S., Korlach, J., Wilson, R. K., Eichler, E. E. (2017). Discovery and genotyping of structural variation from long-read haploid genome sequencing data. *Genome Research.* 27(5):677–685.

Ian, C. G., David, A. C., Nigel, K. S. (2000). Single nucleotide polymorphisms as tools in human genetics. *Human Molecular Genetics* 9(16): 2403–2408.

Ilgar, Z. M., Irina, A. S., Marya, A. K., Sergey, N. N., Dmitry, A. S., Yury, B. L. (2010). A new set of markers for human identification based on 32 polymorphic Alu insertions. European *Journal of Human Genetics* 18: 808–814.

Jana, S., Lucie, P., Nicolas, P. (2016). Genetic polymorphisms in the immune response: A focus on kidney transplantation. *Clinical Biochemistry* 49 (4–5): 363–376.

Janina, E. L., Adrian, E. (2016). Impact of host genetic polymorphisms on vaccine induced antibody response. *Human Vaccines & Immunotherapeutics* 12 (4): 907–915.

Jarl, A. A., Oddmund, K., Lutz, B., Jan, T. L. (2008). Microsatellite evolution: Mutations, sequence variation, and homoplasy in the hypervariable avian microsatellite locus *HrU10*. *BMC Evolutionary Biology* 8:138.

Jeffrey, T. L., Michael, D. B., Theodore, G. S., Rem, I. S., Yelena, B. S., Antonio, T. L. G . Moore, G. M. T., Douglas, C. W (1997). Y chromosome polymorphisms in Native American and Siberian populations: identification of Native American Y chromosome haplotypes. *Hum Genet.*100:536–543.

Jeffreys, A. J., Wilson, V., Thein, S. L. (1985). Hypervariable minisatellite region in human DNA. *Nature* 314:67–73.

Jiannis, R. (2009). Genotyping Technologies for Genetic Research. Ann. Rev. *Genomics Hum. Genet.* 10:117–133.

Jin, P, Panelli, M. C., Marincola, F. M., Wang, E. (2004) Cytokine polymorphism and its possible impact on cancer. *Immunol Res.* 30(2):181–90.

Joan, H. F., Deborah A. B., Ramya R., Jay S. K., Richard, C. L., Troy, D., Pilar, O., Jonathan, M. 2014. Clines without classes: how to make sense of human variation. *Sociological Theory* 32(3): 208–227.

Johnson, R. E., Kowali, G. K., Prakash, L., Prakash, S. (1995). Requirement of the yeast RTH1 5' to 3' exonuclease for the stability of simple repetitive DNA. *Science* 269:238–240.

Jorde, L. B., Watkins, W. S., Bamshad, M. J., Dixon, M. E., Ricker, C. E., Seielstad, M. T., Batze, M. A. (2000). The distribution of human genetic diversity: a comparison of mitochondrial, autosomal, and Y-chromosome data. *Am. J. Hum. Genet.* 66:979–988, 2000.

Kostrzewa, G., Broda, G. K. M., Krajewki, P. P. R. (2013). Genetic Polymorphism of Human Y Chromosome and Risk Factors for Cardiovascular Diseases: A Study in WOBASZ Cohort. *PLoS ONE* 8(7): e68155.

Li, W., Guoxia, R., Jingjie, Li., Linhao, Z., Fanglin, N., Mengdan, Y., Jing, L., Dongya, Y., Tianbo, J. (2017). Genetic polymorphism analysis of cytochrome P4502E1 (CYP2E1) in a Chinese Tibetan population. *Medicine* 96:47 e8855.

Livingstone, F. B. (1962). On the non-existence of human races. Current Anthropology 3: 279–281.

Luigi, L. C-S., Paolo, M., Alberto, P. (1995). The history and geography of human gene. *The Journal of Asian Studies* 54(2):154–201.

Mansour, S. C., Pena, O. M., Hancock, R. E. (2014). Host defense peptides: front-line immunomodulators. *Trends in Immunology* 35 (9): 443–450.

Maria, C. T., Miguel, A. A-S., Gabriel, E. N., Javier, R. L., Harlette, L., Robert K. L., Maria, R., Rene, J. H. (2009). Insights on human evolution: an analysis of Alu insertion polymorphisms. *Journal of Human Genetics* 54: 603–611.

Maria, L. C. V., Luciane, S., Augusto L. D., Carla de Freitas, M. (2016). Microsatellite markers: what they mean and why they are so useful. *Genetics and Molecular Biology*, 39 (3): 312–328.

Mark, S., Jennifer, J. F., Stephanie, L. C., Himla, S., Santosh, S. A., Nilmani, S., Trefor, J., Mohammad, A. T., Prescott, L. D., Mark, A. B. (1997). Alu insertion polymorphisms and human evolution: evidence for a larger population size in Africa. *Genome Research.* 7:1061–1071.

Masatoshi, N (1995). Genetic support for the out-of-Africa theory of human evolution. *Proc. Natl. Acad. Sci.* USA. 92: 6720–6722.

Matthias, K., Anne, S., Ralf, W., Schmitz, H. K., Mark, S., Svante, P. (1997). Neandertal DNA sequences and the origin of modern humans. Cell 90: 19–30.

Matthieu, L., Nathalie, P., Theodore, P., Kevin, J. V. (2007). Sequence-based estimation of minisatellite and microsatellite repeat variability. *Genome Research* 17:1787–1796.

Mehdi, Z., Jeffrey, R. M. D., Daniele, M., Stephen, W. S. (2015). A copy number variation map of the human genome. *Nature Reviews Genetics* 16:172–183.

Meltzer, S. J., Yin, J., Manin, B., Rhyu, M. G., Cottrell, J., Hudson, E., Redd, J. L., Krasna, M.J., Abraham, J. M., Reid, B. J. (1994). Microsatellite instability occurs frequently and in both diploid and aneuploid cell populations of Barrett's- associated esophageal adenocarcinoma. *Cancer Res.* 54:3379–3382.

Mourant, A. A. K., Domaniewska, K. (1976). The distribution of the human blood groups and other polymorphisms. 2nd Ed., Oxford University Press, UK.

Nei, M., Roychoudhury, A. K. (1993). Evolutionary relationships of human populations on a global scale. *Mol. Biol. Evol.* 10(5):927–43.

O'Donovan, M. C., Oefner, P. J., Roberts, S. C., Austin, J., Hoogendoorn, B., Guy, C., Speight, G., Upadhyaya, M., Sommer, S. S., McGuffin, P. (1998). Blind analysis of denaturing high-performance liquid chromatography as a tool for mutation detection. *Genomics* 52: 44–49.

Orapan, S., Suthat, F. (2007). Genetic polymorphisms and implications for human diseases. *J. Med. Assoc. Thai.* 90 (2): 394–398.

Ovchinnikov, I. V., Götherström, A., Romanova, G. P., Kharitonov, V. M., Lidén, K., Goodwin, W.(2000). Molecular analysis of Neanderthal DNA from the northern Caucasus. *Nature.* 404(6777):490–493.

Ozaki, K., Tanaka, T. (2005). Genome-wide association study to identify SNPs conferring risk of myocardial infarction and their functional analyses. *Cell Mol. Life Sci.* 62: 1804–13.

Peter, G., Rebecca, S., Gillian, T. (2001). DNA Profiling in Forensic Science. Encyclopedia of Life Sciences. John Wiley & Sons, Ltd.

Ping, J., Ena, W. (2003). Immunological Polymorphisms in the Humans. Polymorphism in clinical immunology – From HLA typing to immunogenetic profiling. *Journal of Translational Medicine* 1:8

Pinheiro, A., Lanning, D., Alves, P. C. et al. (2011). Molecular bases of genetic diversity and evolution of the immunoglobulin heavy chain variable region (IGHV) gene locus in leporids. *Immunogenetics* 63(7):397–408.

Ram, B. R., Chris G., Bhuwon, S., Devra, J. (2007). Influence of socio-economic and cultural factors in rice varietal diversity management on-farm in Nepal. *Agriculture and Human Values* 24:461–472.

Rebecca, L., Cann, M. S., Allan, C. W. (1987). Mitochondrial DNA and human evolution. *Nature* 32: 31–36.

Relethford, J. H., Harpending, H. C. (1994). Craniometric variation, genetic theory, and modern human origins. *Am. J. Phys. Anthropol.* 95(3):249–270.

Richard, R. Shumpei, I., Karen, R. F., Lars, F., George, H. P., Daniel, A. Heike, F., et al. (2006). Global variation in copy number in the human genome. *Nature* 444(7118): 444–454.

Roger, B. (2013). DNA microarrays: Types, Applications and their future. Curr. Protoc. *Mol. Biol.* 22: Unit–22.1.

Romualdi, C., Balding, D., Nasidze, I. S., Risch, G., Robichaux, M., Sherry, S. T., Stoneking, M., Batzer, M. A., Barbujani, G. (2002). Patterns of human diversity, within and among continents, inferred from biallelic DNA polymorphisms. *Genome Res.* 12(4):602–612.

Sahidan, S., Shahrul, H., Zainal, A., Rus, D. R. D., Rohaya, M. A. W., Intan, Z. Z. A., Zaidah, Z. A. (2014). Haplogroup determination using hypervariable regions 1 and 2 of human mitochondrial DNA. *Journal of Applied Sciences* 14: 197–200.

Salwa, H. T. (2018). DNA polymorphisms: DNA-based molecular markers and their application in medicine. In: Yamin Liu, Genetic Diversity and Disease Susceptibility. Intech Open Publishers, Croatia.

Schasch, H., Aitman, T. J., Vyse, T. J. (2009). Copy number variation in the human genome and its implication in autoimmunity. *Clinical and Experimental Immunology* 156: 12–16.

Shen, H., Li, J., Zhang, J., Xu, C., Jiang, Y., et al. (2013) Comprehensive characterization of human genome variation by high coverage whole-genome sequencing of forty-four Caucasians. *PLoS One* 8(4): e59494.

Shi, L., Guo, Y., Dong, C., Huddleston, J., Yang, H., Han, X., Fu, A., Li, Q., Li, N., Gong, S., Lintner, K. E., Ding, Q., Wang, Z., Hu, J., Wang, D., Wang, F., Wang, L., Lyon, G. J., Guan, Y., Shen, Y. et al. (2016). Long-read sequencing and de novo assembly of a Chinese genome. *Nature Communications*. 7:12065.

Skaletsky, H., Kuroda-Kawaguchi, T., Minx, P. J. et al. (2003). The male-specific region of the human Y chromosome is a mosaic of discrete sequence classes. *Nature* 423: 825–837.

Slatkin, M. (1995). A measure of population subdivision based on microsatellite allele frequencies. *Genetics* 139: 457–462.

Steve, E. H., Vilmundur, G., Ros, W., Ian, N. M. D. (1997). Single-strand conformation polymorphism analysis with high throughput modifications, and its use in mutation detection in familial hypercholesterolemia. *Clinical Chemistry* 43: 427–435.

Stringer, C. B., Andrews, P. (1988). Genetic and fossil evidence for the origin of modern humans. *Science* 239: 1263–1268.

Strittmatter, W. J., Roses, A. D. (1996). Apolipoprotein E and Alzheimer's disease. Ann. Rev. *Neurosci.* 19: 53–77.

Swarkar, S., Anjana, S., Ekta, R. A. B., Ramesh, B. (2005). Human mtDNA hypervariable regions, HVR I and II, hint at deep common maternal founder and subsequent maternal gene flow in Indian population groups. *J. Hum. Genet.* 50: 497–506.

Tanaka, E. (1999). Update: genetic polymorphism of drug metabolizing enzymes in humans. Journal of Clinical Pharmacy and Therapeutics 24: 323–329.

Tanaka, E. (2001). Polymorphism of drug metabolizing enzymes in humans. *Sepsis* 4:247–254.

Timo, T., Jukka, S. M. (2015). Human Chromosome Y and Haplogroups; introducing YDHS Database. *Clinical and Translational Medicine* 4:20–29.

Underhill, P. A., Jin, L., Lin, A. A., Mehdi, S. Q., Jenkins, T., Vollrath, D., Davis, R. W., Cavalli-Sforza, L. L., Oefner, P. J. (1997) Detection of numerous Y chromosome biallelic polymorphisms by denaturing high-performance liquid chromatography. *Genome Res.*7: 996–1005.

Vessela, N. K., Dimitris, K., Tom, K., Anne Lise, B-D. (2001). High-throughput methods for detection of genetic variation. *BioTechniques* 30:318–332.

Wahls W. P., Wallace, L. J., Moore, P. D. (1990). Hypervariable minisatellite DNA is a hotspot for homologous recombination in human cells. *Cell* 60 (1): 95–103.

Wang, D. G., Fan, J-B., Siao, C-J., Berno, A., Young, P., Sapolsky, R., Ghandour, G., Perkins, N., Winchester, E., Spencer, J. et al. (1998). Large-scale identification, mapping, and genotyping of single-nucleotide poly- morphisms in the human genome. *Science* 280: 1077–1082.

Watkins, W. S., Ricker, C. E., Bamshad, M. J., Carroll, M. L., Nguyen, S. V., Batzer, M. A., Harpending, H. C., Rogers, A. R., Jorde L. B. (2001). Patterns of ancestral human diversity: an analysis of Alu-insertion and restriction-site polymorphisms. *Am. J. Hum. Genet.* 68:738–752.

Wayne, P., Wahls L. J., Wallace P, D. M. (1990). Hypervariable minisatellite DNA is a hotspot for homologous recombination in human cells. *Cell* 60:95–103.

Wenz, H. M., Baumhueter, S., Ramachandra, S. and Worwood, M. (1999). A rapid automated SSCP multiplex capillary electrophoresis protocol that detects the two common mutations implicated in hereditary hemochromatosis (HH). *Hum. Genet.* 104: 29–35.

Yu, J., Jingbo, W., Maulana, B., Samuel, S. C., Caroline, G. L. L (2018). Architecture of polymorphisms in the human genome reveals functionally important and positively selected variants in immune response and drug transporter genes. *Human Genomics* 12:43.

3 Genetic Disorders

3.1 INTRODUCTION

Human genome sequencing and its annotation have rapidly increased the knowledge of genetic disease. We can mark the traditional category of genetic diseases with a specific genetic contribution through the human genome, facilitating a list of diseases contributed by genes and the environment. Future genomics knowledge makes a significant contribution to the rapid diagnosis of genetic diseases. All humans are at risk of acquiring these genetic diseases due to exposure to mutagens and radiation, which lead to a high rate of genetic mutations. The high prevalence of these genetic diseases, particularly in specific communities is also due to social and cultural factors, such as consanguineous marriage, resulting in a high rate of autosomal recessive conditions including congenital malformations, stillbirths, and mental retardation. Further, maternal age above 35 years is associated with a high frequency of chromosomal abnormalities in the offspring.

A genetic disease can vary in severity from being lethal right before birth to requiring continuous monitoring and support during all stages of human life. Genetic disease at birth is a severe burden leading to early death or lifelong chronic morbidity. Around 8 million children are born globally per year with severe genetic malformations. Among them, 90% of the infants are born in mid or low-income countries, and the data in the developing countries is uncertain primarily because of the genetic diversity and non-diagnosis of genetic diseases. In developed countries, genetic and congenital disorders are the second most common reason for infant death, occurring at the rate of 25 - 60 per 1000 births. The current chapter focuses on the main reasons for genetic disorders and their classification.

3.2 CLASSIFICATION OF GENETIC DISORDERS

Control of human health and the occurrence of diseases is by the interaction of genes with each other and with environmental factors. Genetic diseases form three categories, chromosomal disorders, single gene disorders, and multifactorial disorders. Chromosomal and single gene disorders are the result of a defect or a mutation in one gene or chromosomes leading to a pathological condition. Multifactorial disorders occur due to genetic and environmental factors, which are also traditionally considered to be genetic diseases due to their association with genetics. Multifactorial disorders are further sub-grouped into congenital disorders (neural tube disorders, cleft lip) and genetic predisposition diseases (chronic and communicable diseases). Congenital disorders are frequently associated with genetic diseases due to their occurrence during pregnancy, childbirth, or early childhood.

One of the main factors necessary for the generation of chromosomal imbalances is the disease-causing copy number variations (CNV). These copy number variations can affect either a single

DOI: 10.1201/9781003343790-3

gene or several genes, with random genomic localization, producing extreme phenotypic heterogeneity. Further, monogenic disorders and chromosomal disorders exhibit very high variability in phenotypic manifestations and molecular mechanisms. In total, the classification and designation of a genetic disease as monogenic, polygenic, or chromosomal, indicate a genetic cause without reflecting its etiology (Ivan et al. 2019). The genetic cause of disease provides a multilateral evaluation, which can be further classified. According to recent classification, genetic diseases are divided into two sectors, gene centric and factor centric. This classification divides a genetic disorder into:

1. A monogenic monogryphic disorder, caused by a mutation in a single gene involved in a metabolic pathway associated with the disease. *Example:* phenylketonuria, a disorder caused due to mutations in the PAH gene, encoding phenylalanine hydroxylase catalyzing the hydroxylation of phenylalanine to tyrosine
2. A monogenic-polygryphic disorder, caused by a mutation in a single gene, leading to a pathogenetic cascade, altering numerous metabolic pathways and giving rise to chromosome instability syndromes and chromatin remodeling diseases. An example of this is Rett syndrome, a genetic disorder, occurs due to defects in the MECP2 gene involved in several pathways regulating genome activity
3. Chromosomal oligogryphic diseases are chromosomal abnormalities involved in the alteration of several biochemical pathways. An example of this is ataxia-telangiectasia, a chromosome instability syndrome caused by defects in the ATM gene involved in a multitude of pathways regulating genome stability maintenance, cell cycle, programmed cell death, and so forth.
4. Chromosomal digryphic disorders are chromosomal syndromes, whose occurrence led to the rearrangement of chromosomal loci containing two genes. Each gene alters a single specific pathway, leading to syndrome manifestation. One example of this is familial Alzheimer's disease, considered to be a multifactorial disorder, caused by mutations in a single gene implicated in multiple pathways. Another example is chromosomal microdeletion at 3p22.1p21.31, which is a unique case, responsible for altering two biochemical pathways.
5. Monogenic homeogryphic disorders are a set of comorbid disease caused by mutations in different genes involved in a single pathway. An example is sporadic Alzheimer's disease, a multifactorial disorder involved in various genetic defects, leads to alterations in multiple pathways.
6. Polygenic monogryphic disorders occurs due to a coordinated effort of single gene mutations, chromosomal abnormalities, and interactions of genetic and environmental factors, leading to an alteration of a specific pathway resulting in complex diseases. An example of this is Williams syndrome, a chromosomal disorder caused due to microdeletions at 7q11.23, leading to an imbalance of 20 - 30 genes, adversely affecting several pathways.

This classification leads to a dogma that genes and environmental interactions are the only two main factors described for the occurrence of a disease with a genetic background. However, the recent evidence from human genome research suggests that genetic diseases cannot be wholly designated using genes and other regulatory elements. Further, genetic disorders occurring due to a single gene and multifactorial factors cannot be limited to a specific gene or a single gene-gene interaction but require extensive information about the gene-specific ontological properties and processes occurring at the levels above their interaction. Genetic diseases can also vary in their severity from being fatal before birth to requiring continuous care and support. The onset of a genetic disorder can change from infancy to old age. These disorders, present at birth, are vigorously burdensome, leading to lifelong morbidity. They may also cause early death. Genetic and congenital disorders are the second most common disorders responsible for the death of an infant or a child, occurring at a rate of 25 - 60 per head.

3.3 CHROMOSOMAL DISORDERS

3.3.1 Human Chromosomes

Human nuclear DNA is packed in chromatin. During cell division, the nuclear DNA will condense into 46 chromosomes, which include 22 pairs of homologous autosomes and two sex chromosomes, X and Y. Females have two X chromosomes (XX) and males have an X, and a Y chromosome (XY). Each chromosome has a characteristic size and shape, which permits its identification. Morphologically, chromosomes consist of two sister chromatids joined at a central constriction known as the centromere, consisting of a kinetochore and a microtubule organizing center, responsible for the attachment of the chromosome to the spindle plate during mitosis. Two sister chromatids are held together by paracentric heterochromatin at the opposite ends of the centromere. A typical centromere divides the chromosome into a long arm, designated as the q arm, and a short arm, designated as the p arm (Figure 3.1). The convention of a chromosome takes place at the p arm and terminates at the telomere, a highly conserved DNA sequence repeat region that is responsible for the inhibition of end-to-end fusion for the attachment of chromosome ends to the nuclear envelope during meiosis. The diminution of the telomere is associated with cell aging. Human chromosomes are of three types: acrocentric, metacentric, and telocentric. If the p arm is extremely short and is virtually invisible, then the chromosome is known as acrocentric. Human chromosomes 13, 14, 15,

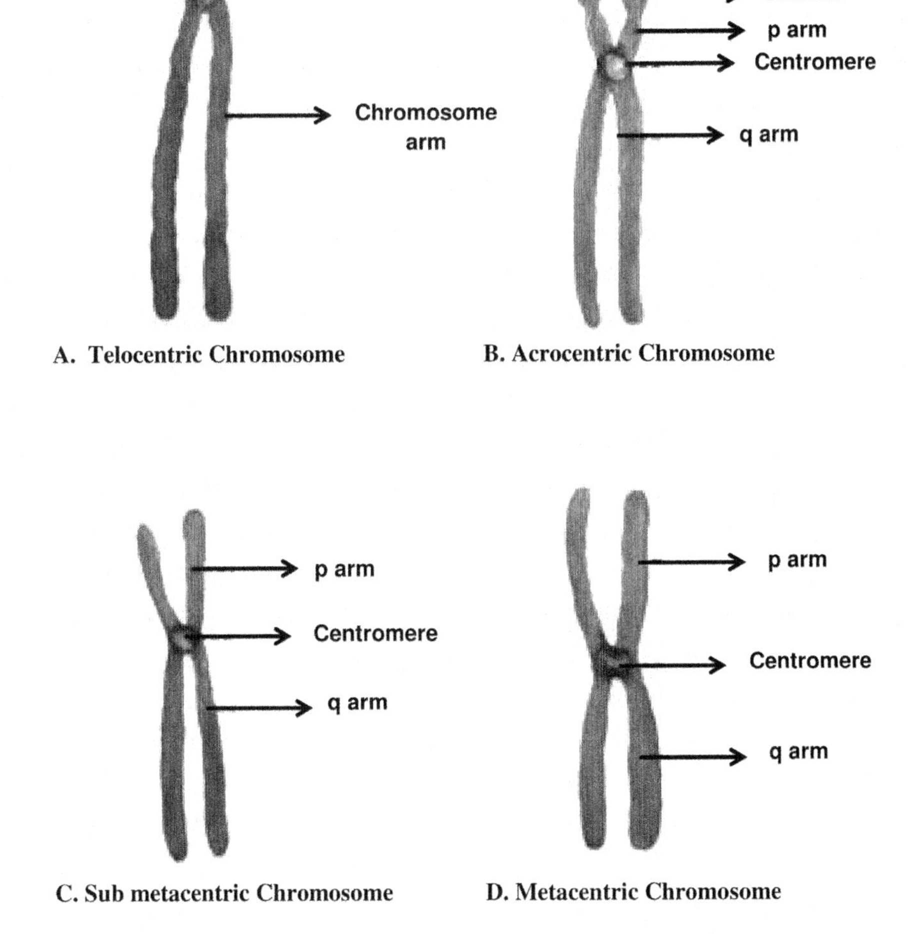

FIGURE 3.1 Types of Human Chromosomes.

21, 22 are acrocentric. The p arm of human acrocentric chromosomes carries nucleolar organizing regions consisting of the genes that code for ribosomal RNA (rRNA). If the p and q arms are of almost equal length, then the chromosome is metacentric and is considered sub-metacentric if the p arm is slightly shorter. A chromosome is telocentric and sub-telocentric if the p arm is very short but is still visible (Sarnataro et al. 2017). The human Y chromosome is sub-telocentric (Figure 3.1).

3.3.1.1 Human Chromosome 1

Chromosome 1 of human beings is one of the immense physical units of the human genome comprising 285 million base pairs and contributes up to 9% of the entire human genome. Genomic synteny of chromosome 1 is highly conserved in apes, some old monkeys, and cetaceans. In most of the placental mammals (40 from 11 orders), human chromosome 1 exists as two unordered conserved segments. In organisms with a slow genome evolution rate, such as spider monkeys and dogs, human chromosome 1 is arranged in six unordered conserved segments. Two possible hypotheses were deduced on the evolution of human chromosome 1. Firstly, the two-segment hypothesis, suggests that human chromosome 1 comprises of two segments. Segment 1 corresponds to 1q21.3-pter, and segment 2 comprises of rest of the long arm up to 1q21.3qter. This hypothesis is based on the reciprocal comparison of the non-human chromosome homologs, suggesting a common evolutionary origin of human chromosome 1 segments in different orders, whose boundaries have been mapped in the 1q21-1q23 regions of pigs, cats, primates, and rabbits. These segments were independently fused in these lineages, leading to the development of a single conserved segment, homologous to primates and cetaceans. Secondly, the alternate hypothesis suggests that human chromosome 1 was very much intact in the genomes of ancestral placental mammals. Extensive work performed by William Murphy (2011) on the evolution of chromosome 1 suggests that the ancestral placental mammals consist of a single intact chromosome 1, which is homologous to the present human chromosome 1. During evolution, chromosome 1 underwent frequent and independent separation into fragments. The p arm of chromosome 1 has conserved its gene order, whereas the q arm has been subjected to inter and intrachromosomal disruption at a higher rate.

3.3.2 Chromosome Staining and Karyotype

Studying chromosomal morphology by their staining is termed karyotype (Figure 3.2). Chromosomes of a cell during its division are analyzed by arresting their division at the metaphase or prometaphase and staining with Giemsa, resulting in a pattern of light and dark bands that is unique to every

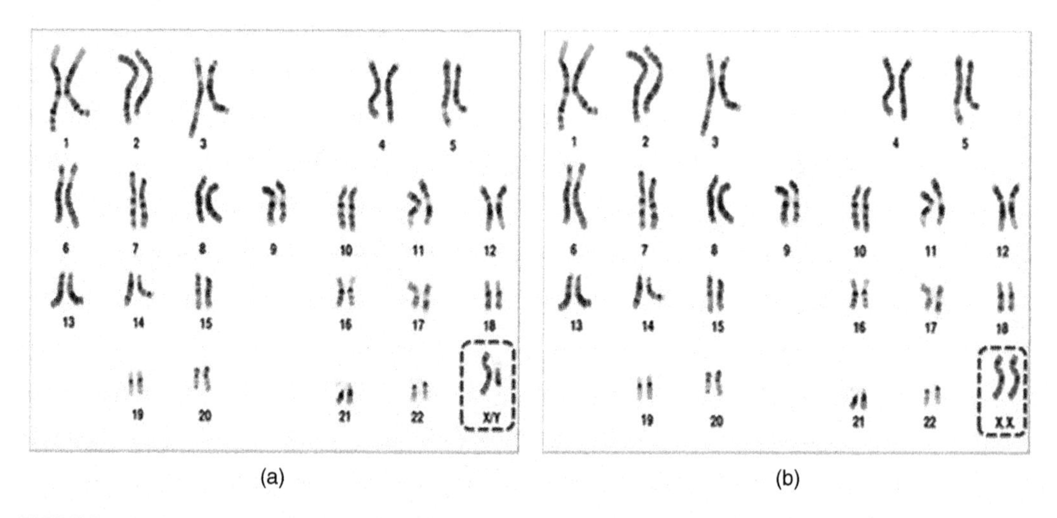

FIGURE 3.2 Karyotype of Human male (A) and female (B) stained with Giemsa.

chromosome, and which is called Giemsa banding or G-banding. Around 450 - 500 bands can be identified on the chromosomes using G-banding. Banding studies have identified that 17 - 20% of the human chromosomes consist of C-bands corresponding to the highly polymorphic heterochromatin region, full of satellite DNA and other repetitive sequences. 80% of the human genome is stained into G, R, and T bands respectively, due to differences in their nucleotide composition and gene density. T-Bands show high G+C content and are also gene-rich, whereas G bands are low in G+C. The euchromatin region of the human genome is divided into four isochores, L, H1, H2, and H3. The L isochore is gene-poor, and the H isochore is gene-rich. Gene density is high in the regions with high G+C contents. Also, a high proportion of genes are present in the G+C poor regions of the human genome. Chromosomes 17,19, and 22 have a high amount of H3 containing bands and the highest gene density in the nucleus of human cells, whereas chromosomes 4, 18, 13, X, and Y have low gene density and low H3 bands (Carole et al. 2013).

3.3.3 Chromosomal Abnormalities

Chromosomal abnormalities are mainly caused either due to a change in their number or their structure. These changes will influence their arrangement or the amount, leading to adverse effects on the human body's growth and development or functioning. Most of the chromosomal abnormalities will occur during spermatogenesis or oogenesis. Sometimes chromosomal changes might also occur spontaneously during the early days after conception. They are also inherited from the parents. Chromosomal abnormalities form a significant category of human disease, represented in 1% of live births and 2% of pregnancies among women older than 35 years. More than 100 human syndromes and single-gene disorders are responsible for 50% of all spontaneous first-trimester abortions (He et al. 2014).

3.3.3.1 Factors Responsible for Chromosomal Abnormalities

The causes of chromosomal abnormalities in humans are pretty much variable. One of the most common factors is the changes that occur to the chromosomes during cell division, and their effect on the development of sperm or ovum. There are two types of cell divisions. 1. Mitosis and 2. Meiosis. During mitosis, duplication of chromosomes occurs initially, followed by splitting of the cell into two halves. Mitosis starts immediately after fertilization and continues until the end of life. If one of the cells responsible for the division has abnormal chromosomes, then the same will be continued throughout the body leading to a mosaic of problems. Meiosis is the process of germinal cell division (sperm and egg cells) with a haploid number of chromosomes (23). Fertilization of a sperm cell with an egg cell leads to a zygote with 46 chromosomes. Abnormalities during meiosis of either sperm cell or egg cell led to an addition (trisomy) or deletion of chromosomes (monosomy). Sometimes, they lead to pregnancy loss or develop health problems for the baby. Chromosomal abnormalities also occur due to a strong influence from a mother's conceiving age. Women aged 35 years or above are at a high risk of a baby with chromosomal abnormalities as their eggs in the ovaries are also 35 years old. Paternal age is less important when compared to maternal age as men keep making new sperms. With age, egg cells in the mother will increase their abnormalities, compared with the sperm cells. External environmental factors also have their effect on the chromosomes. When the parents are exposed to radiation comprising X-rays or consume the food exposed to the X-rays, they have a cumulative character. Teratogens also play a significant role in chromosomal abnormalities. These are the substances that generate or increase the risk of child defects. Alcohol, tobacco, and a few drugs are teratogenic, along with street medicines, chemicals, a few bacteria, and viruses.

3.3.3.2 Types of Chromosomal Abnormalities

Chromosomal abnormalities or aberrations occur due to mutations during mitosis and meiosis, leading to disorders. There are two major types of chromosomal disorders, changes in the chromosome structure, and changes in the chromosome number.

3.3.3.2.1 *Structural Changes*

Changes in the chromosomal structure occur due to a break in the chromosomes. Following the break, the chromosomal segments undergo rearrangements, leading to a displacement of the regions. Most structural rearrangements occur in one or two chromosomes, which can retain a complete complement of the genetic material or lead to either loss (deletion) or gain (duplication) of the chromosome. Duplication of a human chromosome is relatively less harmful than deletion. The structural abnormalities in the chromosomes occur in the ways outlined below.

3.3.3.2.1.1 Translocation Translocations are among the most common types of structural abnormalities in the human population and occur at a frequency of one out of every 1000 live births. Chromosomal translocations occur due to the exchange of chromosomal segments between chromosomes. After translocation, the chromosomes are balanced (exchange of chromosomal segments without any change in the number or DNA loss) or unbalanced (exchange of the segments leading to a change in their number or in losing some of the segments). The only clinical symptom found in the gametes with balanced translocations is impotence. Gametes with unbalanced translocations develop abnormalities in the offspring due to the receipt of altered chromosomes lacking essential genes. The chromosomal translocations are of two types, Robertsonian translocations, and reciprocal translocations. In Robertsonian translocations, breakpoints occur in the short arms of two acrocentric chromosomes (homologous and non-homologous), followed by their fusion to form dicentric chromosomes. The dicentric chromosomes will attain stability by inactivating one of the centromeres. Some Robertsonian translocations also occur due to the breakage and fusion of centromeres (known as centric fusion), and the break and fusion of one long arm and one short arm of acrocentric chromosomes leading to monocentric rearrangements. Loss of acrocentric chromosome short arms during Robertsonian translocations does not produce any phenotypic effects. Most individuals with Robertsonian translocations remain undetected until they reproduce (Chamara et al. 2019).

Reciprocal translocation is the outcome of breakage in two non-homologous chromosomes, with one of them being non-acrocentric. It also occurs due to the interchange of chromosomal fragments leading to the formation of two derivative chromosomes with no deletion or duplication of the genetic material. Reciprocal translocations are highly balanced chromosomal rearrangements that would not significantly affect the phenotype unless one or both breakpoints are involved in the function of a vital gene. Many cases describe the fetuses or children with malformations such as congenital, dysmorphic features, and impaired growth. Most of the phenotypes in these cases are with partial monosomy or partial trisomy of different chromosomal segments. These segments frequently occur in a parent if he or she is the carrier of balanced reciprocal translocations due to unbalanced chromosomal segregation during meiosis.

Carriers of these translocations most commonly display problems with their reproductive systems. These problems are generated due to the unbalanced chromosomal segregation in their gametes during meiosis, leading to infertility, miscarriages, or offspring with congenital anomalies. Even though all human chromosomes are susceptible to chromosomal translocations, the frequency of translocations displays a non-random distribution. Translocations between the acrocentric chromosomes 13, 14, rob (13q14q), and rob (14q21q) constitute the most balanced Robertsonian translocations, whereas Robertsonian translocations involve chromosome 21, common unbalanced ones, which might lead to Down's syndrome. The occurrence of other homologous or heterologous Robertsonian translocations is quite rare.

Translocations are one of the most inherited chromosomal aberrations responsible for pregnancy loss in human populations. In females with translocations, oogenesis progresses without meiosis, resulting in the generation of abnormal oocytes with an unbalanced chromosomal constitution and subsequent partial aneuploidies in the developing fetus. Translocations are also responsible for male infertility. Cytogenetic studies conducted on infertile men show a high range of chromosomal translocations compared to the general population. Chromosomal translocations cause partial or complete spermatogenic arrest with consequential oligospermia and azoospermia.

3.3.3.2.1.2 Deletions Chromosomal abnormalities such as deletions occur due to mutations during meiosis. Deletion is a structural change in a chromosome when a portion or a fragment is lost or deleted. All chromosomal deletions in humans might not lead to disease. For instance, deletion of a fragment (1q21.1) from the q arm of chromosome 1 comprising the genes ACP6, BCL9, CHD1L, FMO5, GJA5, GJA8, GPR89B, and HYDIN delays the development of the fetus, decreases intellectual ability, and might develop physical, neurologic, and psychiatric abnormalities. Deletion of the 1p36 fragment from chromosome 1 leads to intellectual disability, distinctive facial features, and structural abnormalities. However, the genes responsible for these disorders are yet to be identified. Chromosomal deletion in the p arm of chromosome 4 causes Wolf-Hirschhorn syndrome, and a terminal deletion in the q arm of chromosome 11 leads to Jacobsen syndrome (also known as 11q deletion disorder).

3.3.3.2.1.3 Microdeletions and Syndromes As indicated by their name, microdeletions are below the resolution level of a light microscope. A general cytogenetic analysis of a haploid karyotype can resolve up to 400 - 500 chromosome bands on a light microscope, which can visualize deletions to the range of 5 - 10 Mb. This resolution can be further increased to 2 - 5 Mb (650 - 850 bands) using high-resolution methods. Deletions below this size are visualized using molecular techniques such as fluorescent in situ hybridization (FISH).

Microdeletions in the chromosomes lead to abnormal phenotypes as they generate dominant haploinsufficiency and absolute recessive deficiencies. Absolute deficiency occurs when the only functional copy of a specific chromosome is deleted, leading to Duchenne muscular dystrophy and Angelman syndrome. Microdeletions also lead to haploinsufficiency (reduced protein synthesis or reduced gene expression), leading to Digeorge, velocardiofacial, and Willaims syndrome (Ohta et al. 1999). Syndromes that occur due to microdeletions in human chromosomes are listed in Table 1. One of the probable reasons for a high frequency in the occurrence of microdeletion syndromes in the human genome is its duplication during the evolution of mammalian genomes, resulting in the duplication of genes, segments, and repeat clusters. This genome provides a platform for homologous recombination between the non-synthetic regions of chromosomes and unequal crossovers,

TABLE 3.1
List of Syndromes in Humans Occurring Due to Microdeletions in Their Chromosomes

S. No.	Syndrome/disorder	Microdeletion region	Chromosome No.
1	Prader-Willi syndrome	q11 - q13	15
2	Angelman syndrome	q11 - q13	15
3	Duchenne muscular dystrophy	p21	X
4	Azoospermia	q11.21	Y
5	Digeorge syndrome 1	q11.1	22
6	Velocardiofacial syndrome	q11	22
7	Willaims syndrome	q11	7
8	Hypomegakaryocytic Thrombocytopenia	q21.1	1
9	Cat eye syndrome	q11.1	22
10	CHARGE syndrome	q12.2	8
11	Cri du chat syndrome	p15.2	5
12	Digeorge syndrome 2	p14	10
13	Sotos	q35.3	5

resulting in microdeletions. They also arise due to the de novo mutations resulting from genome-wide predisposition to the deletion events.

Significant features of the Prader-Willi syndrome phenotype are developmental delay, neonatal hypotonia, and postnatal development of hyperphagia with obesity, short stature, hypogonadism, and mild to moderate mental retardation. Angelman syndrome, also known as Happy Puppet syndrome, is associated with extrapyramidal symptoms and signs such as unsteady gait with stiff upper arms, hyperactivity, seizures, severe mental retardation with the absence of mental speech, paroxysms of inappropriate laughter. Both syndromes occur at a frequency of 1 - 2 per 15,000 births. More than 75% of the patients with Prader-Willi and Angelman syndromes have a large 4 Mb deletion, either paternal or maternal q11-q13 of chromosome 15 respectively, subjected to parental imprinting. These diseases are associated with microdeletions (75%), uniparental disomy (25% Prader-Willi syndrome with maternal origin and 2% Angelman syndrome with paternal origin), and imprinting mutations in the human genome (2% Prader-Willi syndrome and 5% Angelman syndrome). Imprinting is the process of gene expression in the selected areas of either maternally or paternally inherited alleles. Genes undergoing genome imprinting are known as imprinting genes and are marked differently in the female and male germlines, leading to differences in somatic expression and function. Most imprinted genes in the human genome are in clusters characterized by differential DNA methylation and asynchronous DNA replication of maternal and paternal alleles. Despite the location of the imprinted genes in male and female human genomes and studying their expression patterns, the mechanisms of the imprint establishment are not entirely understood. One of the best studied mechanisms for imprinting is differential DNA methylation of the paternal and maternal alleles. Methylation analysis is the best molecular diagnostic tool for these two syndromes. A gene coding for the small nucleolar RNA (snRNA) named SNRPN is ideal because its maternal allele is completely methylated, and its paternal allele is completely un-methylated. A differential monoallelic expression in the maternal and paternally imprinted genes located on chromosome 15 is attributed to this region's microdeletions. A recent study has identified mutations in the genes of Angelman syndrome that code for a ubiquitin-protein ligase (*UBE3A*) involved in the protein degradation pathway, which undergoes brain-specific imprinting.

3.3.3.2.1.4 Duplications Chromosomal duplications arise due to errors during the homologous recombination and are responsible for abnormal phenotypic effects. Chromosomal duplications are vital for understanding clinical and evolutionary genetics. The presence of an extra gene copy makes it free from selection pressure and contributes to the diversification of protein functions resulting in the formation of protein families such as globin genes, with common functions, that differ in their specialized tasks. Expression of different globin proteins occurs during different development times with each protein specialized in oxygen transport during the specific conditions.

Charcot-Marie-Tooth disorder (CMT) is a non-life-threatening neurological disorder that occurs in humans due to chromosomal duplications, named after three physicians Jean-Martin Charcot, Pierre Marie, and Howard Henry Tooth, who were the first to describe this syndrome in 1988. CMT disorder damages the peripheral nerves and the tracts of nerve fibers that connect the brain and spinal cord to muscles, and sensory organs, which carry feelings, sensation, pain, and temperature from the hands and feet to the spinal cord. They also help in controlling the body balance and aid with locomotion. CMT disorder causes nerve damage or neuropathy, muscle weakness and wasting, and loss of sensation in the body's extremities. This disorder is the result of duplication in the region p12 of chromosome 17. The chromosomal basis of this order varies in severity, symptoms, and the affected region. In CMT type 1A, duplication up-regulates protein synthesis from the genes of the target location, causing abnormal structural and functional changes in the myelin sheath around the nerve fibers, leading to multiple clinical manifestations. CMT disorder occurs upon overexposure to certain chemicals and immune disorders. This disorder is hereditary and transmitted to the next generations. Due to these factors, CMT disorder is also known as hereditary and motor sensory neuropathy (HMSN). CMT disorder exists in multiple forms, such as Dejerine-Sottas disease, expressed during infancy. To date, there is no complete cure for CMT, but treatment is available for reducing the symptoms.

3.3.3.2.1.5 Inversions An inversion is a change in the orientation of a DNA segment within a chromosome. Generally, the chromosomal repeats mediate these inversions, complicating their analysis for a long time. Elucidation of the human genome has focused on the structural chromosomal changes, linking it with natural selection. Most of the inversions are small, less than 1 Kb. However, a few inversions range up to several megabases in size (3RP inversion of Drosophila). These inversions are divided into two classes. Firstly, pericentric inversions include the centromere of a chromosome. During pericentric inversions, a single crossover event between the breakpoints of a heterozygote will produce an unbalanced gamete, which eventually generates deletions and insertions. With 0 - 2 centromeres, unbalanced gametes reduce fertility and make the inversions less dominant (reduced heterozygote fitness). A few pericentric inversions will escape fitness and suppress recombination. Secondly, paracentric inversions are the most common types of inversions that do not include the centromere. These inversions are not dominant due to their high frequency of occurrence and their polymorphisms (Kirkpatrick, 2010).

Polymorphic inversions are known to exist in the human genome, but not much information is available regarding their association with changes in gene expression, adaptation, and disease. Inversions in human chromosomes are studied using G-banding karyotyping, which permits the identification of variations of the fragments up to a few megabases. The variations can be further enhanced by applying FISH, southern blot hybridization, and pulse-field gel electrophoresis. Application of novel genome sequence technology, such as whole genome sequence comparison, helps to identify the inverted segments. Paired-end sequencing and mapping (PEM) provides an insight into the predicted inversions in the sequence. The only drawback with PEM is its inability to detect inversions flanking the long-inverted repeats such as segmental duplications. One of the most well studied human inversion polymorphisms is the inversion at q21.31 on chromosome 17, known as HsInv0573, which inverts a 970 Kb fragment of a single copy sequence and maintains two separated haplotypes, H1 (normal) and H2 (inverted). These two haplotypes have an estimated divergence of 0.476% and consist of different combinations of duplicated sequences at different breakpoints. The H1 haplotype is associated with neurodegenerative disorders such as progressive supranuclear palsy, corticobasal degeneration, Alzheimer's disease, and Parkinson's disease due to the aggregation of a hyperphosphorylated Tau protein, encoded by the MAPT gene in nerve cells. The H2 haplotype is associated with an increased rate of recombination in females, leading to more children. These haplotypes were detected in the Icelandic population. Another polymorphic inversion p23.1 found in chromosome 8, also known as HsInv0501 is the longest inversion polymorphism, of 4.5 Mb, described in human beings. Large and complex repeat sequences mediate this inversion. Two haplotypes are responsible for this inversion. Haplotype 1 (a standard haplotype) is responsible for the risk of lupus erythematosus and rheumatoid arthritis. Haplotype 2 (an inverted haplotype) protects against these diseases.

Along with polymorphic inversions, traditional inversions such as pericentric inversions in chromosomes 1,2,3,5,9,10, and 16 are also present in humans without any clinical significance. Inverted heterochromatin sequences are frequently observed during cytogenetic analysis without any effect on the phenotype. The pericentric inversion in chromosome 9 displays variable frequencies in different human populations, such as 0.26% in Asians and 3.57% in Africans. However, inversions related to several diseases have been found occasionally by disrupting the coding regions or altering their expression levels. These inversions appear either as *de novo* inpatients or as inherited mutations restricted to a specific family. Hence, they can be used as markers for the detection of certain underlying genetic disorders. A few inversions are also responsible for the generation of diseases in a recurrent manner. Diseases such as hemophilia A, and Hunter syndrome arise due to chromosomal inversions. Hemophilia A results in an inversion due to recombination between a 9.5 Kb repeat in intron 22 of the F8 gene with one of the two copies located 565 Kb away on the X chromosome. These inversions disrupt the F8 gene, encoding a coagulation factor VIII in 45% of patients suffering from severe hemophilia A. Two types of inversions lead to hemophilia A. Males inherit an inversion mutation from their unaffected mothers, and the inversion is *de novo.* Truncating the IDS gene by an inversion on the X chromosome, encoding the iduronate 2-sulfatase enzyme leads to Hunter syndrome. In 13% of the patients, this syndrome occurs by recombining this inversion with an adjacent partial pseudogene IDS2. Inversions create fusion genes and disrupt open reading frames, leading to cancer.

Inversions are also involved in pre-disposing alternate re-arrangements. Some chromosomal inversions do not affect phenotype but carrying specific alleles may pre-dispose other re-arrangements in the same region of the genome, leading to disorder. In these cases, parents of the disordered individuals will show an increased frequency of the allele with inversion, compared to the general one as studied in Williams Beuren syndrome. This syndrome occurs due to deletion at the 1.5 Mb region of chromosome 7 and an inversion in the same region with a 12.4% transmitting frequency in the patient's parents and a 2.9% transmitting frequency in normal individuals. Similarly, the deletions responsible for Angelman syndrome have an inversion frequency of 33% in inpatient mothers, whereas it is only 4.5% in normal mothers. Inversions also promote complex intrachromosomal re-arrangements and translocations as in p23.1, located on chromosome 8. A 4.5 Mb inversion can prompt the formation of a dicentric chromosome carrying a single copy of an inverted region. However, duplication and deletions of a substantial fraction of chromosome 8 are associated with developmental delay and mental retardation, along with other problems.

Apart from the generation of diseases, inversions play a pivotal role in the evolution of sex chromosomes. In humans, the Y chromosome is the only one blocked from recombining with the X chromosome along its length. A pattern of molecular variation reveals that a series of overlapping inversions might have progressively extended their size in the non-recombining region of the Y chromosome during its evolution.

3.3.3.2.1.6 Insertions Chromosomal insertion, also known as insertional translocation, occurs when a DNA segment of one chromosome gets inserted either into another location or another chromosome. Insertions are complex chromosomal rearrangements as they need a minimum of three breakage events, known as complex genomic rearrangements. They constitute more than one straightforward rearrangement with at least two DNA breakpoint junctions. Chromosomal insertions are of two types: interchromosomal insertion, where a DNA segment gets translocated and inserted into the interstitial region of another non-homologous chromosome, and intrachromosomal insertion, where a DNA segment gets inserted into a different region of the same chromosome. Through cytogenetic analysis, the incidence of microscopically visible insertions was 1 in 80,000 individuals, which was further reduced to one in 500 – 563, using array-comparative genomic hybridization (a-CGH) in conjunction with FISH. 2.1% of the interchromosomal insertions are the outcomes of imbalances generated from parents with balanced insertions. Balanced carriers will have occasional reproductive problems such as infertility, and pregnancy loss. An unbalanced insertion will have offspring with multiple congenital anomalies. Phenotypic consequences of the insertion vary depending on the size of the insert, orientation, and gene content. More than 3 DNA breaks generate complex insertions, and their joining events accomplish either a gain or loss of copy number at the inserted site. Recent advances in genome research have enabled us to understand the origin, mechanism, and detailed information on the breakpoints and the junction sequences of chromosome-specific recombination. Replication-based pathways such as fork stalling template switching and break-induced replication are the major pathways for chromosome-specific recombination (Takema et al. 2017). The products of these pathways manifest in complex structures, including duplication and triplication of the breakpoint proximity. Additionally, microarrays and next generation sequencing technologies have identified many complex chromosomal rearrangements that occur in one or few chromosomes, and this re-arrangement is known as chromothripsis.

3.3.3.2.1.7 Ring Chromosomes Ring chromosomes are unusual circular chromosomes in humans that result from breaks at the ends of both chromosome arms with the subsequent fusion of the broken ends, forming a continuous ring. Two types of ring chromosomes have been identified; Firstly, the non-supernumerary ring is the one that replaces one of the normal homologous chromosomes with a ring type chromosomal structure leading to a 46 (r) karyotype (Figure 3.3). This type of karyotype always leads to the loss of genetic material. Ends of broken chromosomes with the telomere of one chromosome arm fuse with the telomere of its opposite chromosome arm or even its sub-telomeric regions. Secondly, a supernumerary ring is a tiny extra chromosome formed from the main chromosome, composed of pericentromeric material of the parent chromosome. A supernumerary ring is a small chromosomal segment, separated at the middle of the main homologous chromosome, and its sticky ends are joined to form a ring, thus making the karyotype 47 + (r). To date, the exact mechanism of the supernumerary ring formation is not known. Ring chromosomes are formed in all human chromosomes, but their transmission from parent to children has been documented in rare cases (Mathew et al. 2012). 99% of the ring chromosomes arise sporadically and the phenotype associated with the ring chromosomes is highly variable, depending on the chromosome involved and the extension of the deletion. The transmission of the ring chromosome from parent to child (familial ring) is maternally originated.

One of the common complications of ring chromosome occurrence is their instability from sister chromatid exchanges during cell division. A ring chromosome is highly unstable due to its circular shape. The entry of a ring chromosome into mitosis and its participation in cell division depends on the breakage of its sister chromatids and joining in a crossed manner. In the absence of a sister chromatid exchange at the break, separation of the ring chromatids occurs during anaphase without any difficulty, producing equal sized rings like the original with a single centromere in each of them. However, the occurrence of a single cross over between sister chromatids within a replicated ring results in a single dicentric ring with two centromeres. During anaphase, these two centromeres move to opposite poles, producing an anaphase bridge and sometimes lagging the anaphase, leading

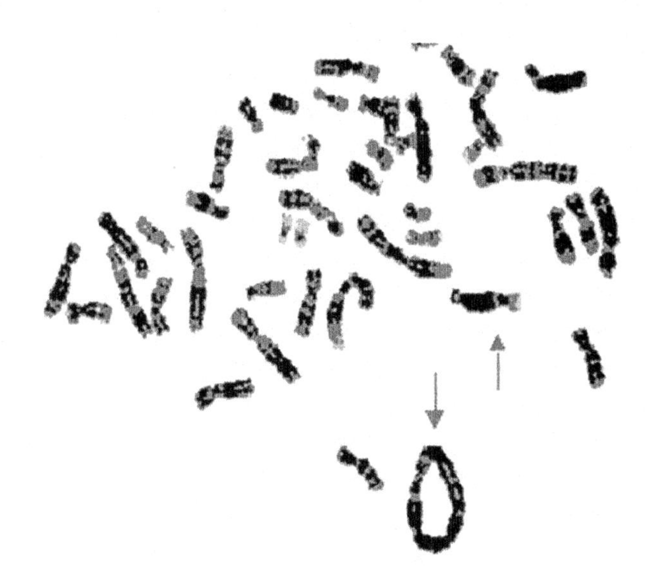

FIGURE 3.3 Karyotype showing Ring chromosome.

to the non-disjunction of the ring. Depending on breakage and reunion, unbalanced, unequal-sized sister rings will form with deleted or duplicated segments during telophase.

Cells of individuals carrying ring chromosomes with deleted, duplicated, or dicentric rings are demonstrated. The exchange of sister strands occurs in the ring. As a result, the replicated ring becomes intertwined, forming interlocked rings (Figure 3.4).

Breakage of these interlocked rings and their disruption during anaphase leads to more aberrations such as loss of the ring due to the reunion of the broken chromosomal ring ends of various sizes. The genetic imbalance in these cells with partial deletions, duplications, and ring loss will eventually make the cell non-viable and fail to survive in subsequent cell divisions. Cells that have managed to survive in a mosaic form would likely be unbalanced, contributing unfavorably to the phenotype.

Humans with ring chromosomes demonstrate ring mosaicism as an outcome of postzygotic ring instability and subsequent ring loss during the cell division, which continues the ring aberrations in their daughter cells. This process is known as dynamic mosaicism, as described by McDermott (1977). Dynamic mosaicism in subsequent cell divisions would lead to phenotype or lethal abnormalities. During meiosis, an unstable gamete is the outcome of the crossing over between the ring

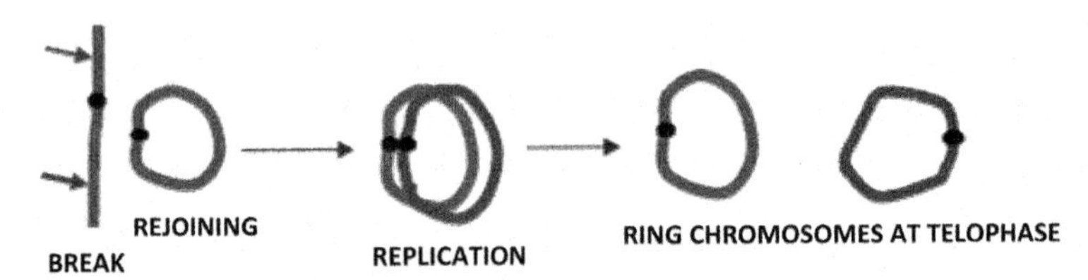

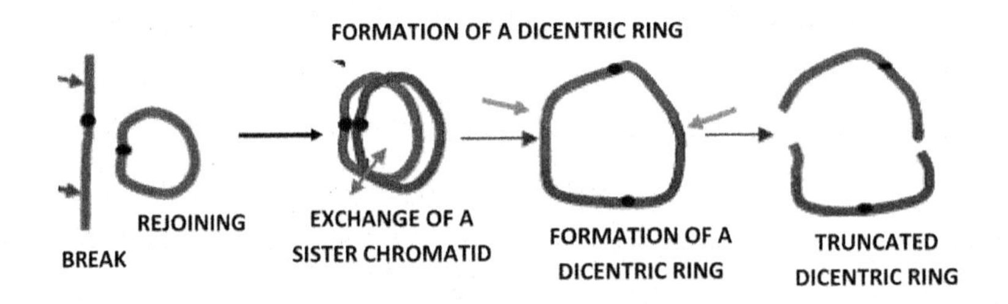

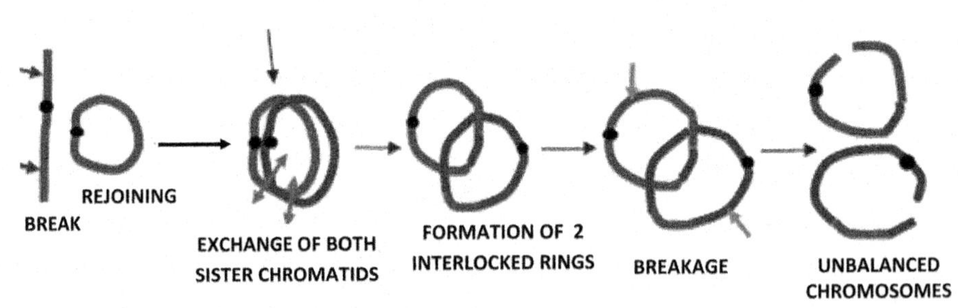

FIGURE 3.4 Ring chromosome during a cell division. Formation of dicentric and interlocked rings. ⟶ Denotes the points at which the initial breakage occurs for the ring formation. ⟶ Denotes points where the subsequent breakage occurs. ⟷ Denotes points where sister chromatid exchange occurs.

and its homolog with subsequent chromosome lagging from bridges at meiosis I and meiosis II, respectively. However, the recovery of the ring is possible if two-strand double crossovers occur. Failure of the unusual pairing between ring 21 and its homolog leads to univalents at meiosis metaphase I, leading to infertile males, due to the breakdown of spermatogenesis after this division. Impaired spermatogenesis in male heterozygotes might lead to total reproductive failure.

There is a high level of clinical manifestation in the familial ring chromosome compared with the sporadic ring. Inheritance of ring chromosomes for 8,13,14,17,18,20, 21 and 21 has been identified with abnormalities. An inherited ring chromosome 8 syndrome has been reported in a boy of age 6.5 years with short stature, microcephaly, mild mental retardation, and behavioral problems. The boy's mother had similar physical abnormalities with above normal intelligence. FISH of both boy and his mother have shown entire subtelomeric sequences with a short deletion in the euchromatic region. Their karyotypes have shown the loss of the ring in 8% of the cells. Transmission of ring 13 from short stature, mentally disabled mother to daughter involving the loss of a sub-telomeric q fragment has delayed the daughter's language development. Further, the transmission of complete ring 14 into two female children from a mother with far below normal intelligence has led to degenerative brain disease in the elder daughter and prolonged convulsions in the younger one. G-banding analysis has shown the ring as opened one as a satellite stalk associated with an active nucleolar organizing region (NOR). Ring 20 syndrome is chromosomal mosaicism that occurs due to the loss of the ring and the presence of dicentric and multiple rings, characterized by mental retardation and epilepsy (Moh-Ying, 2015). Epilepsy is the most distinctive feature of the ring 20 syndrome, usually diagnosed between 2 and 6 years of age and is the first manifestation of this syndrome. Overall, the ring chromosome patients fall into two categories, with severe clinical complications, and with no major clinical complications except infertility. Chromosomal imbalances occur in 68% of cases due to loss of the ring, 52% due to double ring formation, 36% due to ring opening, 16% due to ring doubling, and 16% due to other complex changes ring.

3.3.3.2.1.8 Isochromosomes Isochromosomes were first identified in the plant *Fritillaria kamtschatkensis* by Darlington in 1939. Isochromosomes are generated due to a misdivision of the centromere and structural rearrangement of chromosomes resulting in the attachment of two mirror image arms. The frequency of isochromosome formation is high in plants and insects compared with humans. Isochromosome formation in humans results in isodicentrics and monodicentrics without genetically identical arms. The formation of isochromosomes in the human X chromosomes is relatively more frequent, where the chromosomes are not dividing throughout their length but dividing transversely, leading to the formation of chromosomes with either two long arms or two short arms (Figure 3.5).

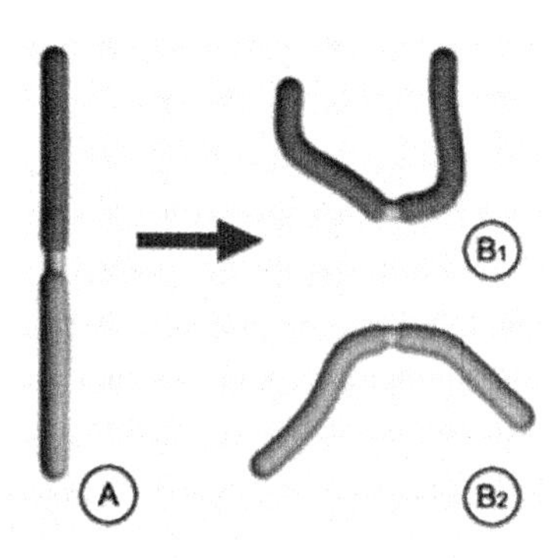

FIGURE 3.5 Formation of isochromosomes on the X chromosome. A. Normal X chromosome, B1. Isochromosome X (isoXp), B2. Isochromosome X (isoXq).

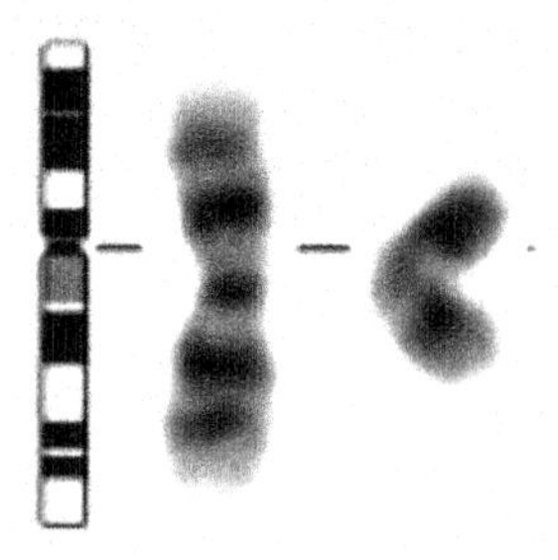

FIGURE 3.6 CTG banding of chromosome 9 isochromosome causing male infertility.

Each centromere of a dicentric chromosome (chromosomes with two centromeres) segregates independently towards opposite poles during the anaphase of the cell division. This segregation would result in anaphase bridges or chromosome breakage (sometimes both) in unstable chromosomes such as maize. Stable dicentric chromosomes such as humans will attain stability by inactivating one of the centromeres (structurally dicentric but functionally a monocentric chromosome) or by coordinating two functional centromeres with unknown stabilizing mechanisms. A centromeric misdivision is responsible for the formation of monocentric isochromosomes and the dicentric chromosomes. These chromosomes emerge from the breakage and reunion of the sister chromatids or homologous chromosomes. Their stability depends on the distance between the centromeres. A typical example of a stable human dicentric functional chromosome is the isochromosomes on the long arm of the X chromosomes, i Xq. These dicentric isochromosomes are found in 15% of Turner syndrome patients. Horizontal fusion at the centromere of the homologous chromosome 9 produces two metacentric isochromosomes i9p and i9q, leading to azoospermia in 15% - 20% of infertile men due to the failure of spermatogenesis (non-obstructive azoospermia) or obstruction of the seminal tract (obstructive azoospermia) (Figure 3.6).

3.3.3.2.1.9 Neocentric chromosomes Centromeres play a pivotal role in the inheritance of human chromosomes. These constrictions on the chromosomes separate the long and short arms and consist of heterochromatin DNA. The centromeric DNA consists of long stretches of repetitive DNA arranged in tandem arrays. Human centromeres consist of 2,000 to 2,400 Kb of a 171 bp α-satellite repeat. The presence of α- satellite DNA is a feature of all eukaryotic organisms except yeast. The presence of α-satellite DNA is a feature of all eukaryotic organisms except yeast, where the centromere is required to perform the functions of chromosomes, especially during cell division. The introduction of α-satellite repeats into cultured cells induced the de novo centromere formation, indicating its role in the centromere formation (David et al 2002). The generation of a centromere in another location of the chromosome is an extraordinary event, with potential insinuations to the cell and the organism bearing it. Neocentromeres are the new centromeres that appear on a chromosome other than the original existing centromere. The discovery of the human neocentromere without any α-satellite DNA was highly unexpected. The first human neocentromere was first detected in the karyotype of a boy suffering from learning difficulties. This neocentromere was the derivative of a *de novo* complex re-arrangement of chromosome 10, which has lost its original centromere and was named mardel 10. Despite the complete lack of α-satellite DNA, this neocentromere could be able to perform stable mitosis. To date, the location of more than 60 different neocentromeres on rearranged chromosomes that have lost their original centromeres has been reported. Figure 3.7 denotes the chromosomes with high neocentromeres.

Neocentromeres have also been detected in human cancer cells, providing valuable insight into their structure, function, and regulation. In addition, understanding the neocentromere properties has further opened the emerging fields for their utilization in artificial chromosomes and human gene therapy.

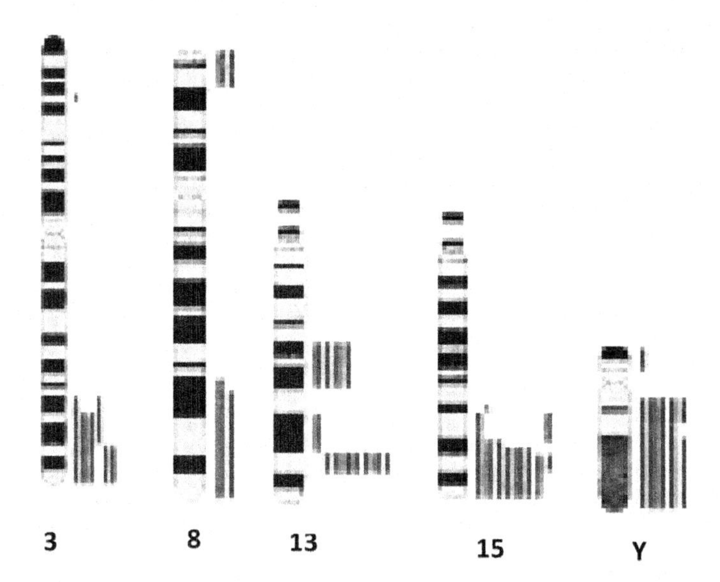

FIGURE 3.7 Chromosomes with high neocentromeres. Longer bars indicate the localization of neocentromere sites.

3.4 SINGLE GENE DISORDERS

Genetic disorders are the conditions which produce specific morphological and physiological changes as a response to a change in the genetic makeup of an organism. These changes are due to either excess or lack of genetic information in an individual, depending on multiple factors such as the genetic combination of the parents, individual, and affected chromosomal pairs. Studies on single gene disorders will provide invaluable insight into the underlying molecular mechanisms responsible for their increased understanding. Currently, these genetic disorders do not have any permanent cure. The treatment provided for these disorders is only palliative to improve their health condition. Even though the cytogenetic analysis is the fundamental beginning point for uncovering the mechanisms and etiology of genetic disorders, the amount of gene expression and its possible interaction with the environment is insufficient for disease diagnosis and its designation at the present stage of medicine. Not much knowledge on the nature of the genetic disorders and the mechanisms of the phenotypic outcomes and cellular pathways to disease is available. Single gene mutations are responsible for most chromosomal or genetic instability due to their manifestations in cellular and molecular pathways and alterations in their regulatory regions. The diseases associated with a single enzyme defect, or a pathway defect are attributed to a cascade of abnormal molecular and cellular events.

Single gene disorders are well understood and most studied sections of genetic disorders, with specific inheritance properties such as dominant and recessive and a very simple genetic etiology. Every individual carries two sets of 22 autosomes and either XX or XY allosomes, with each set inherited from each parent. It also carries two copies of every gene on these chromosomes, divided into autosomal, X-linked, and Y-linked inheritance. Autosomal and X-linked can be further divided into dominant and recessive, depending on whether one or two mutant alleles are required to cause disease. Even though a single gene primarily causes these diseases, several different mutations on the same gene can even lead to the same disease, with varying degrees of severity and phenotype. Sometimes, the same mutation can result in different phenotypes, mainly due to the environment in which the patient lives and the genetic variations, which influence the disease phenotype or outcome. Single gene disorders are described in the following sections.

3.4.1 Autosomal Dominant Disorders

One of the most common forms of Mendelian inheritance in human beings is the autosomal dominant, with a mutation in either one of the alleles sufficient for an individual to express the phenotype. Inheritance of this disorder is from parent to offspring, and those who do not inherit this disorder, cannot transmit it. This disease affects both males and females equally. In a family, the risk of its members carrying a mutation would be 50% for each offspring in a generation. A dominant trait may appear to skip an individual or a generation in some families, resulting from the individuals not exhibiting any symptoms for a generation (incomplete penetrance) or being unable to diagnose the disease due to very mild symptoms (variable expression). In some families, a child is born with a dominant genetic condition as the first child. The effect of the disorder might be due to a genetic change that occurred for the first time in either the egg or sperm involved in making a child. In this case, the parents are not affected, and it is unlikely that another child of the same parents will have the same conditions as the first child. However, the affected child will pass on this trait with an altered gene to his or her children. Some of the most common autosomal dominant disorders are Marfan syndrome, Huntington's disease, tuberous sclerosis complex, and neurofibromatosis

3.4.1.1 Marfan Syndrome

It is an autosomal dominant disorder with an estimated prevalence of one among 10,000 - 20,000 individuals. This rare hereditary connective tissue disorder affects multiple parts of the human body. It is also known as arachnodactyly as one of the significant symptoms is characterized by long, slender, spidery fingers and toes. This disease has no geographic, ethnic, or gender predilection. Liu Bei (161 - 223 A.D.), founder of the Shu Han dynasty, and Abraham Lincoln (1809 - 1865), former President of the United States of America, are believed to have had Marfan Syndrome (Pepe et al. 2016). This disease was identified for the first time in a 6-year-old boy by a French pediatrician Antonin Marfan. This disorder is present in the individuals belonging to 49% of families with this condition. In 20 - 30% of the patients, this disorder occurs without any prior family history, after considering gene mutation.

Symptoms of this syndrome will become more evident with age. One of the most common symptoms is myopia (an eye defect which cannot allow distant objectives to be seen clearly) and ectopia lentis (displacement of the eye lens from its normal location), retinal detachment, glaucoma, and early cataract formation. Other symptoms of Marfan syndrome include skeleton and connective tissue systems. Cardiovascular malformations are one of the most life-threatening phases of Marfan syndrome, with Aortic root dilation and mitral valve prolapse being significant in patients. Aortic root dilation is in 60% of patients suffering from Marfan's syndrome, with a differentiation of 74% males and 33% females. The aortic wall of the patient suffering from Marfan's syndrome is characterized by a widespread fragmentation of thin elastin fibers. Aortic dissection, a common feature in Marfan syndrome patients, is primarily due to an intimal tear in the proximal ascending aorta, with a dissection involving the sinotubular junction and aortic sinuses. Mitral valves prolapse in 91% of patients suffering from Marfan's syndrome with a differentiation of 74% males and 33% females. The principal changes in the mitral valve are annular dilation, fibromyxomatous changes to the leaflets and chordae, and elongation and rupture of chordae followed by calcium deposition.

3.4.2 Autosomal Recessive Disorders

The autosomal recessive disorder is inherited when both the copies of a gene on an autosomal chromosome carry a mutation, resulting in its loss of function. The condition at which both the copies of a gene have the same mutation is known as homozygous. Compound heterozygosity is the condition that arises due to the inheritance of each copy of a gene with a different deleterious mutation. Each parent of the affected patient is a typical heterozygous carrier with one regular and one abnormal copy of the gene. In most cases, the normal allele compensates for the defected copy, due to which the heterozygous carriers are asymptomatic. When two heterozygous carrier parents have four offspring, one among them (25%) should have the disease, two will be carriers, and one should be healthy. The frequency of autosomal recessive disorders increases with the increased frequency of consanguineous marriages. This condition might even occur due to founder mutations in isolated populations. Founder mutations might occur in an individual at a point in his pedigree and be subsequently propagated throughout the population. Cystic fibrosis and sickle cell disease are common diseases inherited in an autosomal recessive fashion.

3.4.2.1 Sickle Cell Disease

Sickle cell disease is a compound terminology for a group of autosomal inherited disorders such as Sickle cell anemia, hemoglobin sickle cell (HbSC), and HbSβ-thalassemia, characterized by mutations in the gene encoding for the hemoglobin β-subunit (HBB). Hemoglobin is a tetrameric protein composed of globin subunits in different combinations, with each globin subunit associated with a heme co-factor that can carry a single oxygen molecule. Gene Hb is expressed both in reticulocytes and erythrocytes. Several genes encode different globin protein types, and their various tetrameric combinations generate multiple Hb types, expressed at embryonic, fetal, and adult stages of life. Among them, Hb A is the most abundant Hb form expressed in adults comprised of two α-globin subunits encoded by HBA1 and HBA2 genes and two β-globin subunits. A single nucleotide change (substitution) in the hemoglobin β-subunit leads to sickle Hb allele (HbS) known as βS. The mutant protein generated from the βS is a sickle β-globin subunit under deoxygenation conditions (Hb is not bound to oxygen). Hb tetramers consisting of two mutant sickle β-globin subunits (βS) can polymerize and cause erythrocytes to assume a crescent shape or sickle shape, leading to sickle cell anemia. Sickle cell anemia is a lifelong disease characterized by chronic, hemolytic anemia, unpredictable duration of pain, and widespread organ damage. High levels of fetal Hb (HbF), combined with two α-globin and two γ-globin proteins encoded by HBG1 and HBG2 genes, lead to hemoglobin sickle cell (HbSC). Thalassemia occurs due to mutations in HBA1 and HBA2 genes, respectively.

The geographical distribution of the bS allele is under the influence of malaria endemicity and population movements. The overlap between the distribution of βS allele and the malaria endemicity in sub-Saharan Africa led to a hypothesis that individuals with Hb sickle might protect against malaria generated by *Plasmodium falciparum*, confirmed by the latest technological developments. This hypothesis explains that the individuals with HbAS provide a higher level of protection against *Plasmodium falciparum* malaria (90%) than the individuals with the normal Hb gene. This hypothesis also explains the occurrence of a high frequency of the βS allele in sub-Saharan Africa, the Mediterranean, the Middle East, and India regions. Population movements, especially the slave trade have provided a better platform for the wide distribution of βS allele in North America and Western Europe.

The incidence of birth rate with sickle cell anemia in the sub-Sahara African region was estimated to be more than 230,000 corresponding to 75% of the global births with sickle cell anemia. The West African region has reported the highest hemoglobin sickle cell disease, and the numbers are very likely to increase over the next 40 years. Even in countries such as the United Kingdom, where universal screening programs are under implementation, the estimated prevalence of the disease remains challenging, with more than 40,000 sickle cell disease cases confirmed among newborn babies. In the United States of America, 40,000 confirmed sickle cell disease cases were identified in 76 million newborn babies. In African Americans, 1 in every 360 newborn babies has sickle cell disease. Indian population consists of over 2,000 different ethnic groups, and most of them are practicing endogamy (the custom of marrying only within the limits of the local community) for centuries. Despite the βS allele identification in many ethnic groups, its prevalence has been identified only in some. Several hundred million of the Indian population are at risk of developing sickle cell disease. Among them, a majority belongs to historically disadvantaged groups (Kato et al. 2018). Screening of βS allele prevalence within these groups is in progress.

3.4.3 Sex-Linked Disorders

The traits associated with mutations in the genes located on humans X and Y chromosomes are known as sex-linked disorders. In humans, males have only one X chromosome and one Y chromosome, respectively, whereas females have two X chromosomes without any Y allele. Hence, the pattern of inheritance in this set of chromosomes is different from that of autosomes. For example, a mutant gene located on the X chromosome affects the father. Transmission of the mutant allele will be from the father to all his daughters but none of his sons. If a mutant allele on the Y chromosome affects the father, then its transmission is limited only to all his sons. If the mutant gene is on the X chromosome affecting the mother, transmission of the mutant allele will be both to her sons and daughters. Some of the sex-linked disorders are discussed here.

3.4.3.1 X-Inactivation

The human X chromosome consists of more than 1,000 genes, essential for the proper development and viability of the cell. Females carry two copies of the X chromosome transmitted from both the parents, resulting in a potentially toxic double dose of essential genes. This imbalance in the human X chromosome is corrected and controlled by a unique dosage compensation mechanism, known as X-chromosome inactivation (X-CI), which is distinct from other organisms. X-chromosomal inactivation also known as Lyonization is the inactivation process of one X chromosome copy during the early stages of fetus development (Ahn et al. 2008). X-chromosome inactivation is under the control of an inactivation center (XIC), which condenses the X-chromosome into high-density heterochromatins, visualized as compact structures in the nucleus known as the Barr Body. However, the ends of the X-chromosomes contain a small number of genes, which escapes the inactivation and are expressed. These regions are known as pseudoautosomal, with both the sexes having two copies of every gene from this pseudoautosomal region.

At the molecular level, X-chromosome inactivation occurs due to the transcriptional silencing of one X chromosome in the female mammalian cells, equalizing the gene products between XX females and XY males. X-chromosome inactivation occurs early during embryo development and both the X chromosomes in the female have an equal probability of being silenced (Barbara, 2008). However, the selection of the X-chromosome for the silencing makes it stable, and the same X chromosome inactivation occurs in all subsequent generations. The resulting female is a mosaic of cells with a silenced maternal or paternal X chromosome.

Two non-coding complementary RNAs X-Inactive Specific Transcript (XIST) and TSIX play a prominent role in X-chromosome inactivation. XIST has four specific properties.

1. It does not encode a protein but produces a 17 Kb functional RNA molecule
2. It is expressed only in the cells that contain two X chromosomes. Hence, not expressed in sperm cells. High XIST expression is in the cells with more X chromosomes.
3. XIST RNA is strictly nuclear specific, which can coat the chromosome from which it is produced
4. Paradoxically, XIST RNA is expressed from an inactive X-chromosome.

The expression of XIST RNA is sufficient for the inactivation of the X chromosome as it recruits various silencing protein complexes for labeling the inactive X chromosome. The mechanism of XIST expression on one X chromosome, with the other chromosome remaining silent, came into existence by discovering an XIST antisense partner TSIX. TSIX is a 40 Kb long non-coding RNA transcribed in the opposite direction across the entire XIST gene. Exactly like XIST, TSIX acts only on the chromosome that produces it. There is a reverse relationship between the expression of TSIX and SIXT. With the reduction of the TSIX transcription rate on one X chromosome, XIST expression will increase and leads to the inactivation of the same X chromosome. Overexpression of TSIX prevents the increase of XIST expression and blocks inactivation on the same X chromosome.

Inactivation of the X chromosome occurs in two possible ways: imprinted X-inactivation and random X-inactivation. During imprinted X inactivation, the silencing of the X chromosome transmitted from the father occur in the placenta. This type of silencing is present in the eutherians and early marsupial mammals such as Opposums and Kangaroos. For eutherians, both XIST and TSIX are essential. Marsupials do not have XIST homolog, and silencing occurs through the pre-silencing of the paternal X chromosome in the male germline cells. Here, both the X chromosomes are inactivated, and the inactive state of X is transmitted to subsequent generations during meiosis. This type of X-chromosome inactivation occurs only in female embryos as only females inherit the sX-chromosome paternally. Random inactivation of the X-chromosome occurs in the early female embryo, where both maternal and paternal X-chromosomes have an equal chance of getting inactivated. Each female cell has the task of distinguishing between two X chromosomes within the same nucleus, then designating one of them as an active and the other one as inactive. This silencing process is accomplished in each cell by XIST and TSIX along with specialized factors such as histone variants and chromatin modifiers. Along with the silencing, the cell also takes the responsibility of keeping another X chromosome active. There was a prediction that the two X chromosomes in a cell communicate to designate the fates mutually. Arbitration of this communication is by a protein and a transcription-dependent pairing between the X chromosomes during early development, making the random silencing more complex.

3.4.3.2 X-Linked Dominant Disorders

X-linked dominance is a genetic condition associated with mutations in the genes of the X chromosome. Here, a single copy of the mutated gene is sufficient to generate the disorder in both males and females. Under adverse conditions, the absence of a functional gene leads even to the death of the inherited person. If a father carries the abnormal gene on his X chromosome, all his daughters will inherit this disease, and none of his sons will have it. If a mother carries the abnormal X gene, 50% of their children (both sons and daughters) will inherit the disease. In X-linked dominant conditions, female phenotypes are generally milder than males.

3.4.3.2.1 Rett Syndrome

One of the well-studied examples of an X-linked dominant disorder is the classical Rett Syndrome. It is a neurodevelopment disorder that exclusively affects the female child. The disease is characterized by normal development until between 8 months and 2 years of age, followed by the gradual loss of purposeful hand movements and development of characteristic stereotyped hand movements, loss of speech acquired previously, neuromotor skill regression, and developmental stagnation. Other clinical features include screaming, inconsolable crying, and display of autistic tendencies. The growth of the head will begin to decelerate from 3 months of age. Brain size will be smaller than expected. Seizures such as generalized tonic-clonic and partial complex occur in 90% of the female children affected with Rett syndrome (Smeets and Pelc, 2011). Most males with Rett syndrome will not survive pregnancy, but some males are born with severe neonatal encephalopathy, which leads to death before they attain two years of age. Rett syndrome is frequently misdiagnosed as Autism, cerebral palsy, or non-specific developmental delay in a female child.

Rett syndrome occurs due to mutations in methyl CpG binding protein 2 (MECP2). This protein plays a vital role in the neuronal cell function by binding to the methylated DNA and interacting with other proteins for the formation of a complex, repressing the gene expression. Methylation of DNA usually occurs in the CpG islands, located near the promoter region of a gene. The binding of MECP2 will condense the DNA, making it inactive. Recent studies have shown that MECP2 forms a complex with histone deacetylase, responsible for removing the acetyl group on the histone, thereby blocking the gene's transcription. Diagnosis of Rett syndrome through confirmatory DNA testing is currently available in several labs. Treatment for this disorder mainly involves speech therapy and counseling.

3.4.3.3 X-Linked Recessive Disorders

They occur due to mutations on the X chromosome. As males are in a hemizygous situation (only one X chromosome), one altered copy of the gene is sufficient to generate the disorder, whereas, in females, the mutation should occur in both gene copies located on both X chromosomes to cause the disorder. It is improbable that females will have two altered copies of the gene. Males are highly susceptible to X-linked recessive disorders much more frequently when compared with females. Fathers affected with this abnormal X chromosomal trait will transmit this abnormal X chromosome to all their daughters and be considered obligate carriers. Female carriers will transmit this defective X-chromosome to half of their sons, who are born affected, and half of their daughters will become the carriers. The remaining half of their siblings will inherit normal X-chromosome copy. Another significant feature of the X-linked recessive inheritance is that this trait cannot be passed by an affected father to his sons as there is no male-to-male transmission possible in this condition Genetic disorders grouped into the X-linked recessiveness include hemophilia A, Duchenne/Becker muscular dystrophy, Lesch-Nyhan syndrome, customary conditions such as male pattern baldness, and less severe red-green color blindness.

3.4.3.3.1 Duchenne Muscular Dystrophy (DMD) and Becker Muscular Dystrophy (BMD)

DMD and BMD are the most prevalent neuromuscular disorders in male babies occurring at a rate of one per every 3,600 births worldwide. No clinical signs appear during birth. The first symptoms of the disease usually appear at 4 years. DMD is a progressive muscle degeneration and severe muscle weakness that progresses from proximal to distal and pseudo-hypertrophy of calves. During disease progression, loss of muscle and replacement of myocytes by fat and fibrotic tissue will occur. Before teenage, most of the patients will have difficulty in walking and are confined to a wheelchair. Skeletal deformities occur secondary to muscle weakness. The disease progresses pretty quickly and leads to death with cardiac and pulmonary complications at 20 years (Maria et al. 2015). Becker muscular dystrophy (BMD) is an allelic form of DMD, and a mild form of dystrophinopathy has one in every 18,518 male births. The disease occurs between 5 - 15 years of age. X-linked dilated cardiomyopathy is a different phenotype of dystrophinopathy, characterized by congestive heart failure between 10 and 20 years.

Muscular dystrophy occurs due to mutations in the dystrophin gene, the largest one identified in the human genome. Its mRNA is expressed in skeletal, cardiac muscles and in the brain. Three isoforms of the DMD gene are expressed in the brain, muscle, and Purkinje cerebellar neurons, respectively through three independent promoters. Other isoforms of the DMD gene are functional through alternate splicing. Dystrophin protein is localized in the sarcolemma intracellular surface of a healthy muscle along the length of myofibers. Its principal function is to stabilize the fibers during the muscle contraction by binding to F-actin at the N-terminal domain and to β-dystroglycan at its C-terminal domain, thus acting as a bridging and anchoring protein. DMD disease is associated with mutations in the form of deletions (65%), duplications (6 - 11%), small mutations, and other rearrangements (10%) that interrupt the ORF of the dystrophin gene. These interruptions lead to protein loss, muscle loss, pulmonary and cardiac failure, and ultimately death.

Additionally, the regenerative capacity of the myofibers is severely compromised in DMD patients due to chronic injury that induces satellite cell exhaustion and replacement of muscle with fibro adipose tissue. One of the most common molecular changes occurring to the dystrophin gene is the deletion of one or more exons, reported in 65% of the DMD cases, and exon duplication reported in

6 - 10% of the cases. The remaining 25% are due to minor mutations such as missense, nonsense, and splice variations, and small rearrangements such as insertions, deletions, and inversions. Rare cases with deep intron changes and complex rearrangements are also reported (up to 2%). Presently, multiplex ligation-dependent probe amplification (MLPA) is one of the sensitive techniques for diagnosing DMD. With this multiplex PCR-based technique, all 79 exons of the dystrophin gene can be screened for identifying the copy number variation (CNV). Similar results are also be obtained by oligonucleotide-array-based comparative genomic hybridization, which identifies copy number variation in the entire locus of the dystrophin gene. A variety of next generation sequencing technologies are also being used.

3.4.3.4 Y-Linked Disorders

The Y chromosome is the crucial determinant of male sex and transmits from father to son without undergoing any recombination. The traits associated with the Y chromosome are known as holandric or Y-linked traits. At least 14 such traits have been identified because of their transmission from father to son. One such holandric trait is Hypertrichosis Pinnae Auris responsible for abnormal long hair on the outer pinna. Several studies have also indicated the role of the Y chromosome as testes determining factor and three distinct loci azoospermia factors a, b, and c, involved in the spermatogenesis. Deletions of these factors are associated with varying degrees of spermatogenic failure.

The Sex determining region of the Y chromosome (SRY region), located on the distal region of the Y chromosome's short arm, is one of the essential genes that determine the potentiality of the gonads. SRY consists of a single exon with a well-conserved central motif known as a high mobility group box (HMG box) with DNA binding and DNA bending activities, suggesting that it serves as a transcriptional regulator for initiating testes development and differentiation of gonads into the testicular pathway. Mutations in the SRY region led to XY gonadal dysgenesis and sex reversal. Most of the mutations leading to these two complications are in the HMG box region of SRY. Gonad dysgenesis leads to the deficiency of antimullerian hormone (AMH) and testosterone resulting in under-masculinization, subsequently leading to infertility. Two-point mutations were identified in the upstream of the SRY region. The first one is A→G substitution, replacing glutamine with arginine at codon 57. This mutation leads to sex reversal in normal males. The second one is downstream of the HMG box region. It is an A→T substitution that replaces serine with cysteine at 143 codon, leading to gonadoblastoma. However, most of these mutations are *de novo*, affecting only one individual in a family. Similarly, mutations in the Azoospermia factor (AZF) region genes such as DAZ, RBMY, DBY, and USP9Y are known to cause spermatogenic dysfunction.

3.5 MITOCHONDRIAL DISORDERS

Mitochondria are the organelle present in all nucleated cells, whose primary function is the generation of ATP by oxidative phosphorylation. They are the only organelle with extra-chromosomal DNA in the human cells and are under the dual genetic control of nuclear and mitochondrial genomes. The mitochondrial genome in humans is 16.6 Kb, consisting of 37 protein encoding genes. 13 of them are involved in oxidative phosphorylation, 2 rRNAs, and 22 tRNAs for their translation within the organelle. Mitochondrial genetics is entirely different from Mendelian genetics in three main aspects: uniparental inheritance, multiple genome copies within a cell, replication, and transcription regulation mechanisms. Mutations are common in the mitochondrial genome, and a very high genetic diversity of the mitochondrial genome makes it more challenging to define the impact of mutations on human health. Mitochondrial DNA syndromes arising due to mtDNA mutations, their involvement in common disease phenotypes, and aging define the extent of damage posed by the changes in mtDNA (Table 3.2).

TABLE 3.2
Mitochondrial DNA Associated Diseases in Humans

S. No.	Disease	Features
1	Neurological	migraine, strokes, epilepsy, dementia, myopathy, peripheral neuropathy, ataxia, speech disturbances, sensorineural deafness
2.	Gastrointestinal	constipation, irritable bowel, dysphagia,
3.	Cardiac	heart failure, heart block, cardiomyopathy
4.	Respiratory	respiratory failure, nocturnal hypoventilation, recurrent aspiration, pneumonia
5.	Endocrinal: diabetes	thyroid disease, parathyroid disease, ovarian failure
6.	Ophthalmological	optic atrophy, cataract, ophthalmoplegia, ptosis
7	Gastrointestinal	vomiting, failure to thrive, dysphagia
9.	Renal	renal tubular defects
10	Liver	hepatic failure
11.	Endocrinal	diabetes, adrenal failure
12.	Ophthalmological	optic atrophy

3.5.1 Genetic Diseases due to the Mitochondrial Genome

The prevalence of mitochondrial genetic diseases in humans is estimated to be 1 among 5,000 across all ages. Affected children present pediatric complaints along with neurologic, muscular, cardiac, and gastrointestinal manifestations. Even though the diagnostic evaluation of a suspected mitochondrial disease is rapidly changing the scope of genetic information, the overall progress of diagnosis remains slow. To date, there is no cure for mitochondrial disease. A few available therapies clinically remain limited to symptomatic management, prophylactic antioxidant, and vitamin cofactor cocktails. Genetic diseases of human mitochondrial DNA were first described in 1988. The syndromes frequently reported in mitochondrial DNA inheritance are Kearns-Sayre Syndrome (KSS) and LHON syndrome. KSS syndrome occurs due to large scale deletions in the Mt DNA of muscle cells, leading to abnormalities in the muscles. Leber's hereditary optic neuropathy (LHON) syndrome is strictly maternal. It occurs due to point mutations in the ND gene family, leading to the degeneration of retinal ganglion cells and their axons, followed by vision loss in young adult males. Disorders due to MtDNA mutations are displayed in Table 3.3.

TABLE 3.3
Mitochondrial DNA Disorders in Humans

Gene	Mutation	Disorder
TRNL1	A→G at 3243 and T→C at 3271	mitochondrial myopathy, encephalopathy, lactic acidosis, and stroke (MELAS)
TRNK	A→G at 8344 and Myoclonic Epilepsy with Ragged T→C at 8356 Red	Fibers syndrome (MERRF)
ATP6	T→G at 8993	neuropathy, ataxia, and retinitis pigmentosa syndrome (NARP)
TRNL1	A→G at 3243	maternally inherited diabetes and deafness syndrome (MIDD)
ND1	G→A at 3460	optic neuropathy
ND4	G→A at 11, 778	optic neuropathy
TRNE	T→C at 14,709	myopathy, weakness, and diabetes
RNR1	A→G at 1555	deafness

Patients with a disease due to mutation(s) in mtDNA rarely have a phenotype. The mtDNA disease is into the differential diagnosis of many common clinical syndromes, making it difficult to diagnose by clinicians. One of the classic examples is diabetes, the best metabolic disorder affecting humans. Mutations in the mtDNA also lead to diabetes, but the proportion of the patients representing the mutations in mtDNA is less when compared with the patients associated with diabetes due to the changes in the nuclear DNA. It also implies that several mutations in the mtDNA cause diabetes, and it is uneconomical and impractical to screen all the patients with diabetes for the responsible mutations in the mtDNA. Another factor that makes the diagnosis of diseases due to the mutations in the mtDNA is the specific combinations of clinical symptoms. Mitochondrial diabetes is often accompanied by deafness, stroke, and migraine like symptoms. A clear association between a mitochondrial tRNA mutation and a metabolic syndrome characterized by hypertension, hypomagnesemia, and hypercholesterolemia have been identified. Many patients with diseases due to mutations in the mtDNA go undiagnosed for years, and probably many of them are never recognized. An increase in the awareness of these diseases among physicians, pediatricians, and investigations to diagnose the diseases at specialized centers can resolve these issues.

3.6 MULTIFACTORIAL DISORDERS

Modern genetics have further subdivided the genetic disorders into Mendelian and complex disorders. Mendelian disorders are determined by a mutation in a single gene, whereas complex disorders involve the combined contribution of multiple genes and different environmental factors interacting together, known as multifactorial inheritance. Francis Galton, a cousin of Charles Darwin and contemporary of Gregor Mendel, was the first person to identify and study multifactorial inheritance. Unlike Mendel, Galton has observed a blend of characters that describes a gradation in expression in which phenotypes do not fall into distinct categories. These characters are now known as continuous variations. When a trait exhibiting a continuous variation is plotted on a graph, its phenotypic distribution forms a bell-shaped curve. Accordingly, most of the individuals exhibiting a continuous variation form an intermediate phenotype. Traits exhibiting continuous variation are also known as quantitative traits, which involve a wide range of genetic and environmental factors responsible for producing a wide range of genotypes (Lobo, 2008).

Some traits with multifactorial inheritance exhibit no gradation. Mendel named these characters non-blending traits and placed them in a distinct category. Whenever there is an abrupt change from one phenotype to another, such as round and wrinkled peas in Mendel's inheritance, these traits exhibit discontinuous variation without any intermediate. Here the Pea shape is either round or wrinkled, falling only into one category. Another typical example of discontinuous variation is the human ABO blood antigen system. Many diseases are the result of discontinuous variation with phenotypes resembling continuous variation. This variation process might be due to a base of continuous variation on which the susceptibility process begins. The disorder is developed and expressed after a certain threshold of liability is reached. The more bestowing the threshold of liability, the more severe the disorder phenotype is. In contrast, an individual who does not reach the liability threshold will never develop this disorder. Irrespective of an individual having a disorder, it can be understood that it shows a discontinuous variation. Understanding of liability threshold would be better by considering the individuals with cleft lip and palate. It is a multifactorial disorder with discontinuous variation, in which an individual is born with un-fused lip and palate tissues. Individuals born with cleft lip and palate can have unaffected parents without any family history of the disorder but might have contributed some underactive genes required for a cleft lip and palate formation.

Additionally, nutritional deficiencies and cigarette smoking by mothers during pregnancy are also associated with this defect. When an individual is born with a cleft lip and palate, the contributing factors for this condition have surpassed the liability threshold. With further increase in

TABLE 3.4
Risk of Re-Occurrence of Cleft Lip and Palate in a Family

Relationship with person	Proportion of Common Genes	%
First Degree	½	
Child		4.20
Siblings		3.60
Second degree	¼	
Aunt or Uncle		0.70
Nephew or Niece		0.75
Third Degree	**1/8**	
First cousin		0.30

the threshold, the congenital disability increases its severity, forcing other family members to be affected. The risk of cleft lip re-occurrence in a family was mentioned in Table 3.4.

As illustrated in the table, the risk of cleft lip re-occurrence drops rapidly with a decrease in the degree of relatedness. This single gene pattern of re-occurrence cannot be demonstrated by Mendelian inheritance. As the risk of re-occurrence is similar in both siblings and children, this inheritance is unlikely to be a recessive one. A recessive inheritance implies a 25% risk for affected siblings and affected children, whereas an autosomal dominant inheritance involves a very high risk of 50%. Such high risk is not on the table. Instead, the risk of the cleft lip to the second-degree relatives correlates with the proportion of shared genes, indicating the involvement of genetic and non-genetic factors. First-degree relatives share half of their genes with one another (including the genes that contribute to the factors that alter the developmental processes responsible for the malformation), and their liability will be greater than that of distant relatives. As distant relatives will have a few genes in common, they will be less genetically viable. The proportion of common genes drops rapidly with a decline in the degree of relatedness. The risk of a more distant relationship also diminishes.

For decades, it has been speculative which type of inheritance theory exactly describes the inheritance of diseases in humans. Even though Mendelian inheritance describes a few disorders and traits, it is not fit for all of them. Similarly, Galtonian observations also do not fit all disorders. Currently, the description of a Mendelian or Galtonian pattern depends on the type of disease. A multifactorial disorder is a collective combination of certain characters, differentiated from Mendelian disorders. A multifactorial disorder can occur in children born to unaffected parents. There are chances of familial aggregation with multiple cases of the disorder in the same family without any Mendelian inheritance pattern. Environmental factors can increase the risk of a multifactorial disorder. These multifactorial disorders occur in one gender with higher frequency than in another gender. In addition, first degree relatives of the less frequent gender have a higher risk of bearing the disorder. The concordance rates of a multifactorial disorder in monozygotic and dizygotic twins (the rate at which both the twins bear a disorder) differs from Mendelian disorders. However, the multifactorial disorder is not a sex-limited trait. This disorder is reported to occur in one specific ethnic group such as Caucasians, Africans, Asians, and Hispanics with more frequency. For example, coronary artery disease, a multifactorial disorder, occurs more frequently in Afro-American males with high risk when compared to Caucasians and Asians.

A few complex disorders such as pyloric stenosis differs between the male and female sexes. This multifactorial disorder has a frequency of 5 per 1000 in males, whereas it occurs in only 1 in 1000 in females. The risk of re-occurrence for an affected male's son is 1/18 and for the son of an affected female is 1/5. The risk for an affected male's daughter is 1/42, whereas the risk for a daughter of an affected female is 1/14. Thus, it is evident that the risk of disease is high if the affected parent is female, and the risk of re-occurrence is less if the offspring is female. These findings suggest that the threshold of a multifactorial disorder is less in males. The effect of a multifactorial disorder will be high if the liability of a person falls beyond this threshold. If a female is affected, her liability, comprising genetic and non-genetic factors is high, generating a pattern of high re-occurrence risk among her children. They share one half of her genes, including those resulting in liability to the disorder. When both the parents are consanguineous, they have a great chance of sharing the deleterious genes. Parental consanguinity for a multifactorial disorder leads to a high risk of additional re-occurrence in their children. If consanguineous parents have a child with a neural tube defect, the re-occurrence risk is even greater for their subsequent children than for the children of the non-consanguineous parents. Suppose the parents are non-consanguineous, but a malformation has previously occurred in both the parental families, along with their child. In that case, the risk of re-occurrence will also be higher due to the liability condition of both parental families, which will produce a high additive re-occurrence risk to their subsequent offspring. In humans, many multifactorial disorders such as multiple sclerosis, diabetes type II, asthma, cancer, and congenital disabilities have been diagnosed due to the complex interactions of the genetic factors such as copy number variations, epistatic interactions, and environmental factors. In the cases associated with the discontinuous trait variation, the above said factors might not exceed the liability threshold, thus making it difficult to predict the chances of the disease occurrence. Genome Wide Association Studies (GWAS) is one of the most promising methods to identify major multifactorial disorders.

3.7 MOLECULAR BASIS OF CANCER

Cancer is a disease of uncontrolled growth and proliferation whereby cells have escaped the body's regular growth control mechanisms and have gained the ability to divide indefinitely. It is a multistep process that requires the accumulation of many genetic changes over time. These genetic alterations involve activating proto-oncogenes to oncogenes, deregulation of tumor suppressor genes and DNA repair genes, and 'immortalization'.

3.7.1 Cell Cycle Regulation and the Importance of Apoptosis

In normal cells, proliferation and progression through the cell cycle are through a strict regulation by a group of proteins interacting in a specific sequence of events. Checkpoints that ascertain those individual stages of the cell cycle are completed correctly and ensure that incompletely replicated DNA will not pass onto daughter cells. Core to this control system is cyclin-dependent kinases (CDKs). CDKs are 'master protein kinases' that drive progression through the different phases of the cell cycle by phosphorylating and activating other downstream kinases. CDK activity depends on activating subunits called cyclins that are synthesized and degraded in a cell cycle-dependent manner. CDK inhibitors further tightly regulate Cyclin-CDK complexes.

The re-entry of cells into a cell cycle is decided at the restriction point (R point). This decision is influenced by extracellular mitogenic signals transmitted via signaling pathways to key regulatory proteins, such as transcription factors (E2F). These regulatory proteins ultimately activate the S-phase CDKs, which trigger the start of DNA synthesis. In regular cells, transcription factor p53 is often referred to as the 'guardian of the genome'. p53 can impose cell cycle arrest and apoptosis (programmed cell death) through its ability to induce the expression of cell cycle inhibitors to prevent the proliferation of a cell until the repair of any damage. It can initiate apoptosis if the genomic damage is too high with no possibility of any repair.

3.7.2 The p53 Gene and Cancer

In the human body, cells are continuously exposed to various stresses responsible for genomic aberrations such as deletions, mutations, and translocations of the nuclear genome rendering genomic instability. Accumulation of genomic instability often leads to the development of cancers. P53 is a nuclear transcription factor responsible for maintaining genome integrity and protecting the cells from transforming into malignant forms by inducing apoptosis or cell cycle arrest. In >50% of all human tumors, the p53 pathway is aberrant (Toshinori and Akira, 2011). Inactivation of the p53 protein renders it unable to signal and activate the cell's apoptotic machinery resulting in increased survival of cancer cells. Under normal conditions, p53 is expressed at shallow levels due to proteasomal degradation mediated by RING-finger type E3 ubiquitin protein ligase MDM2, a functionally latent form. Upon the induction of DNA damage, p53 will accumulate in the nucleus through post-translational modifications such as phosphorylation and acetylation. Dissociation of MDM2 from p53 will further activate p53. Activated p53 transactivates a set of target genes for inducing apoptosis and or cell cycle arrest, depending on the extent and types of the DNA damage rendered. p53 mediated cell cycle arrest allows the cells to repair the damaged DNA and permits the re-entry of the repaired cells into the cell cycle. When the DNA of the cells is seriously damaged, p53 exerts its pro-apoptosis function to eliminate the cells, thereby inhibiting the transfer of damaged DNA to its daughter cells and maintaining the genome integrity.

P53 is a Knudson-type tumor suppressor. It has been demonstrated that more than 50% of human cancers carry loss of function mutations in the p53 gene. Mice with p53 deficiency mutations have developed spontaneous cancers. 95% of the mutations have been detected in the genomic region encoding the p53 DNA binding domain, and mutant p53 lacks the sequence-specific transactivation ability, which is highly associated with its pro-apoptotic function. Mutant p53 forms a hetero-oligomer with wild-type p53 through intact oligomerization domains and acts as a dominant negative inhibitor toward wild-type p53. Wild type p53 is short-lived for 20 minutes, whereas mutant p53 has a prolonged half-life of 2 to 12 hours, with an extreme oncogenic potentiality. Cancers occurring due to p53 mutations in humans have also displayed a chemo-resistant phenotype, indicating the critical role of p53 in the regulation of DNA damage response and forcing us to develop a novel strategy to eliminate the negative impact of mutated p53 on wild type p53 for efficient chemotherapy.

Mutational inactivation of p53 is one of the most common molecular mechanisms responsible for the dysfunction of p53. More than 50% of human cancers occur due to the loss of function mutations in p53. 90 - 95% of these mutations were identified in the exons 5 - 8, which encodes for the DNA-binding domain. Further analysis has revealed six amino acids most frequently mutated in human cancers Gly at 245, Arg at 175, 248, 249, 273, and 282, respectively. These mutations are found to disrupt proper confirmation of p53, thus making the p53 mutant defective in the sequence specific transcriptional activation. Mutant p53 also displays a dominant negative behavior with the wild type p53 through hetero-tetramer and has oncogenic properties. An increased amount of evidence demonstrates that cancer-derived p53 mutant forms will transactivate various target genes such as \Multiple Drug Resistance Gene 1(MDR1), c-myc, Proliferating Cell Nuclear Antigen (PCNA), interleukin-6(IL-6), Insulin Like Growth Factor 1 (IGFR), Fibroblast Growth Factor (FGF) and Epidermal Growth Factor (EGFR) (Figure 3.8). Cancer-derived p53 mutant transactivates Aspargin Synthetase and Telomerase Reverse Transcriptase, which confirms the role of p53 mutants in transactivation and oncogenes for the progression of aggressive cancers.

Along with the mutations in coding regions of the p53 genes, loss of function mutations was also found outside the DNA binding domain. Two-point mutations L344P and R337C in the COOH terminal oligomerization were found along with the R377C mutation, significantly reducing the transcriptional and pro-apoptotic activities. The R337H mutant of p53 has shallow stability when compared to that of p53 wild type. One of the p53 mutants, p53ΔC, whose oligomerization domain and nuclear localization signals are lacking in neuroblastoma-derived cell lines, changing the region of its expression from the nucleus to the cytoplasm with deficient pro-apoptotic ability when compared with wild type p53.

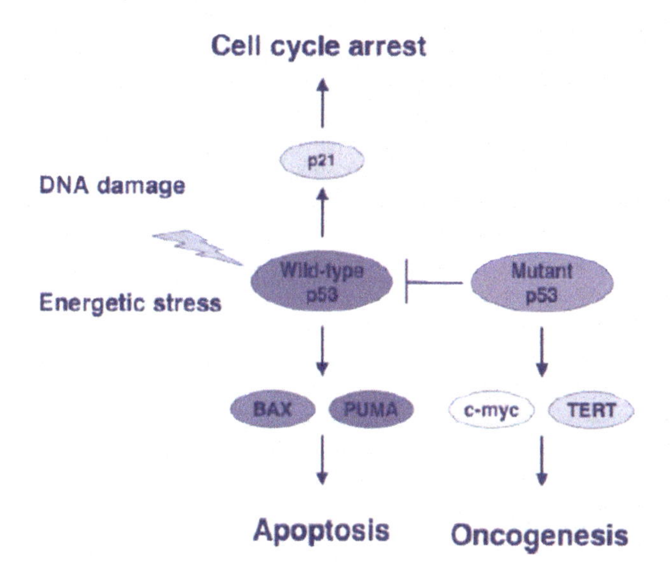

FIGURE 3.8 Dominant-negative effect of mutant p53 on wild-type p53. The pro-apoptotic function of p53 is significantly inhibited by specific p53 mutants, which induce malignant transformation through up-regulation of c-myc and TERT.

3.7.3 Telomeres, Cell Immortalization, and Tumorigenesis

Immortalization is the process of acquiring an infinite lifespan. Normal and healthy mammalian somatic cells proliferate a limited number of times before undergoing senescence. Senescent cells may remain metabolically active even though they have permanently ceased proliferation. Normal cells undergo genomic instability over a while and acquire the capability of replicative immortalization, leading to the transformation of normal cells into malignant ones. The attrition of telomere during successive cell divisions induces chromosomal instability and significantly contributes to the genome re-arrangements resulting in tumorigenesis. Telomeres are the DNA protein complexes present at the end of the chromosomes with a repetitive TTAGGG sequence, vital for the immortality of the cancer cells. The integrity of a telomere is maintained by telomerase, an enzyme responsible for maintaining telomeres at the ends of chromosomes. Telomerase can counter the progressive telomere shortening that would otherwise lead to cell death by extending telomeric DNA. Unlike normal cells that lack detectable levels of telomerase activity, approximately 90% of human tumors consist of cells that contain an active telomerase enzyme. The mechanisms responsible for maintaining the telomere length and telomerase expression involve transcriptional, post-transcriptional, and epigenetic regulation. Telomeres are involved in the protection of chromosome ends from fusion and DNA damage.

Dysfunctional telomeres arising from telomere shortening in normal somatic cells during progressive cell divisions elicit DNA damage responses, triggering cellular senescence. Cells that gain oncogenic changes will bypass senescence and will extend their life span. These cells will continue to divide until critically shortened telomeres initiate total replicative senescence, end-to-end chromosome fusions, and extensive apoptosis. These processes might lead to breakage-fusion-bridge cycles, in which two sister chromatids with lacking telomeres fuse, forming a bridge with a chromatin connection. During the anaphase of the mitotic cell division, the sister chromatids are drawn apart due to their movement towards the opposite poles, resulting in uneven derivative chromosomes, leading to genomic instability and extensive cell death. However, a few rare cells will escape these crises and maintain stability. They continue cell growth and division with short telomeres, eventually transforming into malignant cells. These malignant cells will achieve proliferative immortality by activating a set of silent genes, such as the TERT gene, which encodes telomerase for forming complexes with other proteins and functional RNA to make a ribonucleoprotein enzyme complex. Mechanisms involved in telomere length maintenance and expression of telomerase involve transcriptional, post-transcriptional, and epigenetic regulations. A rare DNA recombination mechanism known as alternative lengthening of telomeres reverses the telomere attrition to bypass the senescence. Further, an in-depth understanding of these mechanisms may help generate novel biomarkers and targets for early detection and diagnosis of disease, determination of its prognosis, and development of therapeutics.

3.7.4 Genes Frequently Mutated During Cancer

The genes implicated in carcinogenesis are divided into two broad categories oncogenes (cell accelerators) and tumor suppressor genes (cell brakes), and DNA repair genes (Carole et al. 2013).

3.7.4.1 Cellular Oncogenes

Genes that promote autonomous cell growth in cancer cells are called oncogenes, and their regular cellular counterparts are called proto-oncogenes. Proto-oncogenes are physiological regulators of cell proliferation and differentiation, while oncogenes promote cell growth without normal mitogenic signals. Their products, oncoproteins, resemble the regular products of proto-oncogenes except for oncoproteins, which are devoid of critical regulatory elements. Their production in the transformed cells becomes constitutive, not dependent on growth factors or other external signals. Proto-oncogenes can be converted to oncogenes by several mechanisms, including point mutation and gene amplification, resulting in overproduction of growth factors, flooding of the cell with replication signals, uncontrolled stimulation in the intermediary pathways, and cell growth by elevated levels of transcription factors. RAS oncogene is the most frequently mutated oncogene in human cancer. It encodes a GTP-binding protein Ras that functions as an on-off switch for several key signaling pathways controlling cellular proliferation. Ras is transiently activated in a normal cell and recruits Raf to activate the MAP-kinase pathway to transmit growth-promoting signals to the nucleus. The mutant Ras protein is permanently activated, leading to continuous stimulation of cells without any external trigger.

3.7.4.2 Tumor Suppressor Genes

Also known as anti-oncogenes, tumor suppressor genes are involved in the suppression of cell division or growth. As per Knudson's two-hit model, tumor suppressor genes are the ones whose both alleles are inactivated by genetic alterations such as chromosomal deletions, loss of function mutations, frame-shift mutations, insertions, missense mutations responsible for the inactivation, leading to the loss of functional activity of its protein, and development of cancer. Tumor suppressor genes encode proteins that are:

- Receptors for secreted hormones that function to inhibit cell proliferation
- Negative regulators of cell cycle entry or progression
- Negative regulators of growth signaling pathways (APC or PTEN)
- Checkpoint-control proteins that arrest the cell cycle if DNA is damaged or chromosomes are abnormal
- Proteins promote apoptosis DNA repair enzymes.

The transformation of a normal cell to a cancer cell accompanies the loss of function of one or more tumor suppressor genes, and both gene copies must be defective to promote tumor development. Recent studies have demonstrated that the epigenetic alterations of tumor suppressor genes, particularly transcriptional silencing of the hypermethylation of CpG islands in tumor suppressor genes, are an effective mechanism leading to tumorigenesis. Some tumor suppressors also encode for a haploinsufficient phenotype (a situation in which the total protein level produced due to the expression of a gene is half the average level and is not sufficient to perform the normal function of the cell thus leading to an abnormal phenotype).

3.7.4.3 DNA Damage Repair Genes

DNA damage is a change in the primary DNA structure. As a result, it is not replicated during the cell cycle, leading to damage. When the DNA carrying a damaged base undergoes replication, the incorrect base gets inserted into the opposite side of the damaged base in the complementary strand and thus becomes a mutation during the next round of DNA replication. Mutations can be avoided if the DNA repairing mechanisms recognize the DNA damage as abnormal structures and repair the damages before replication. DNA repair is a crucial protective process, which blocks the entry of cells into carcinogenesis. DNA damages occur in replicating, proliferating, in-differentiated and non-dividing cells such as neurons and myocytes. Cancers occur mainly in proliferating tissues. If the DNA damage is not prevented in the proliferating tissues by increased expression of a DNA repairing gene, the risk of cancer will increase, whereas the DNA damage in the non-proliferating tissues is not repaired due to the enhanced DNA repairing gene expression, the chances of premature aging will increase. Deficiency in the expression of *ERCC1*, *XPF1*, and *WRN* will increase the risk of cancer and pre-matured aging. Mutator genes are the ones that maintain genome stability and drive tumor progression by accumulating the mutations.

The accomplishment of the mutator phenotype is by inactivating the mismatch repair genes (MMR), maintaining an accurate repetition of microsatellites. During the S phase of the cell cycle, polymerases are prone to errors in the regions of high homology and high repeats after the replication of nuclear DNA. Mismatch repair genes code for the proteins that can repair such errors. Repetitive sequences such as mini-satellites and microsatellites are present throughout the genome, and mutations in the satellite repeats are due to mutations in the mismatch repair genes, making them non-functional, leading to genome instability. Well-studied examples of mismatch repair genes are *MSH2* and *MLH1*. People born with an inherited defect in these genes have a significant risk of developing hereditary non-polyposis colorectal cancer, also known as Lynch syndrome, which occurs due to the accumulation of microsatellite instability mutations in the coding regions of these two genes and mutations in the coding regions of oncogenes such as APC, KRAS, and the like. The APC gene has a microsatellite block in its coding region. Mismatch of this block leads to a frame-shift mutation, resulting in a non-functional APC protein. Polymerases are the protectors of stability and DNA fidelity in the human genome. These enzymes with proofreading activity, DNA-mismatch-repair protein complexes, and glycosylases are involved in the DNA excision repair pathway. When a cell is provided with an insufficient amount of these proteins (haploinsufficient phenotypes) or changed targets of these repeat proteins, it is subjected to mutations, leading to cancer.

3.7.5 Mutations in DNA and Cancer

A mutation is a change in the DNA sequence, in which the nucleotide bases are added, deleted, substituted, or even re-arranged. The mutated DNA sequence is copied during the replication of nuclear DNA. This mutation can prevent a gene from carrying out its regular function, activate an oncogene, inactivate tumor suppressor genes, or generate genome instability in replicating cells. Assembling such mutations in a proliferating tissue or a set of cells with a specific function will lead to cancer. Cancers arise from an assemblage of mutations leading to an expansion. Colon cancer occurs due to 15 driver mutations (they repeatedly occur in different colon cancers) and 75 passenger mutations (mutations that occur infrequently or with a low frequency). Colon cancers have an average of 28 deletions, 9 duplications of chromosome segments, 17 local amplification, and 10 translocations. Since mutations also have a structure like normal DNA, they cannot be recognized or removed by DNA repair processes occurring in the living cells. Removal of a mutation is possible only through cell death if it is deleterious to that cell.

Thousands of DNA damage events occur per cell every day in humans, mainly due to reactive molecules produced during different metabolisms or hydrolytic reactions occurring in various kinds of cells in the human body. Many DNA damages such as propano, etheno, and malondialdehyde derived DNA adducts, base propenals, estrogen-DNA adducts, alkylated bases, deamination of cytosine, adenine, and guanine to form uracil, hypoxanthine, and xanthine, respectively). Further, adducts formed with DNA by reactive carbonyl species are mutagenic, and the adducts are formed with DNA by reactive carbonyl species. Exogenous DNA damaging agents such as tobacco smoke severely delay the DNA damage repair mechanisms. Increased DNA damage will increase replication errors by translesion synthesis or increased error prone repair by non-homologous end joining, generating mutations. Increased mutations will activate oncogenes and inactivates tumor suppressor genes causing genomic instability. They also generate driver mutations in replicating cells, further increasing the risk of cancer. In most cases, cancer development occurs due to the accumulation of mutations in somatic cells over a lifetime rather than germline mutations. Under rare circumstances, mutations in the germline cells also contribute to the disease progression. Even though the rate of mutations is low in humans, they can be further enhanced by environmental carcinogens such as chemical mutagens, radiation, and tumor viruses, increasing the probability of cancer development. Chemical mutagens modify the DNA through various mechanisms, including alkylation or deamination of the bases or the intercalation between the bases. Oxidative damage of DNA is responsible for affecting DNA integrity. X-rays and UV radiation will induce double strand breaks in the DNA, lead to the formation of pyrimidine dimers by crosslinking the adjacent bases.

3.8 GENE EXPRESSION ANALYSIS

Gene expression analysis, also known as gene expression profiling, is a powerful technique applied in molecular biology, genetic engineering, and molecular diagnostics for checking the expression of thousands of genes simultaneously. In the context of cancer, gene expression profiling diagnoses and classifies tumors more accurately. The output of gene expression profiling will be able to predict the clinical outcome for a patient. As far as molecular biology is concerned, gene expression profiling is a technique for measuring the expression of thousands of genes at a single glance, which helps create a global picture of cellular function. Gene expression profiling will distinguish the actively dividing cells from the non-dividing cells or reveal the reaction of cells to a particular treatment. Many experiments based on gene expression analysis have measured the expressed gene in the entire genome simultaneously by studying the expression of every single gene located in a specific cell. Based on the data generated by gene expression profiling, several transcriptomics techniques have been successfully applied to generate the necessary data to analyze and measure the relative activity of many target genes previously identified. Additionally, sequence-based techniques, such as RNA-seq, provide information on the sequences of genes, in addition to their expression levels.

Expression profiling is a logical step implemented after sequencing the genome of an organism. The genome sequence of an organism will provide us with information regarding the number of genes that it contains, whereas its expression profiling will identify their function at any given point in time. Genes in an organism contain the instructions for making mRNA, but at any given moment, each cell makes mRNA from only a fraction of the genes it carries. Synthesis of mRNA makes a gene considered "on" or "off," which will also be determined by factors such as the time of day, whether the cell is actively dividing, its local environment, and chemical signals from other cells. An expression profile allows the deduction of a cell's type, state, and environment.

Expression profiling experiments often involve measuring the relative amount of mRNA expressed during two or more experimental conditions. Altered levels of a specific mRNA sequence under multiple conditions suggest a changed need for the protein coded by the mRNA, indicating cellular homeostasis or a response to a specific pathological condition. For example, higher mRNA levels for alcohol dehydrogenase suggest that the cells or tissues under study are responding positively to the increased levels of ethanol in their environment. Similarly, if breast cancer cells express higher levels of mRNA associated with a specific transmembrane receptor than normal cells do, it can be suggested that this receptor might play a pivotal role in breast cancer. A drug that interferes with this receptor may prevent or treat breast cancer. During the development of a drug, the gene expression profiling experiment's performance will help assess the drug's toxicity, probably by looking for a change in the expression levels of cytochrome P-450 genes, which may serve as biomarkers for the specific drug metabolism. Gene expression profiling can be considered an important diagnostic test. The human genome consists of 30,000 genes, which function in concert to produce 150,000 distinct proteins due to alternative splicing. Cells make essential changes to proteins through posttranslational modifications after making them in the ribosomes so that a given gene serves as the basis for many possible versions of a particular protein. Under any given condition, a single mass spectrometry experiment can identify around 2,000 proteins, representing 0.2% of the total. While knowledge of the number of proteins a cell makes is more relevant than knowing how much messenger RNA is synthesized from each gene, gene expression profiling provides a possible global picture in a single experiment. Gene expression profiling is used for studying the expression of thousands of genes simultaneously. Every cell contains the entire genome of that organism but expresses only a small subset of genes in the form of mRNA and proteins at any given time. Their relative expression levels are evaluated using gene expression profiling.

3.8.1 Gene Expression Profiling of Lung Cancer

Lung cancer occurs due to mutations in the genes located in the somatic chromosomes. The differences in gene regulation will also bring dramatic changes in gene expression leading to tumorous cells. Microarray technology can identify these changes, distinguish the highly aggressive tumor cells from the less aggressive cells, and provide better and more accurate treatment options. Gene expression profiling will facilitate the identification of a sub-type of lung adenocarcinoma with a relatively poor prognosis or a gene panel that can reliably predict the patient's survival for lung adenocarcinoma. The accuracy of the expression-based outcome prediction varies significantly among the studies and the accuracy of the outcome-based expressions prediction. The tumor-node-metastasis stage is the most vital outcome predictor for lung cancer, which functions as a surrogate for survival. However, identifying an accurate stage could be challenging as the micro-metastasis can be overlooked by conventional pathological examinations or by low sensitivity clinical imaging. As tumor cells undergo various changes in their gene expression from initiation to progression and metastasis, attempts have been made to perform a molecular tumor staging by analyzing the gene expression profiles of a primary tumor in the lung cells for achieving more accurate clinical outcome predictions and increased treatment options.

During molecular tumor staging, a correlation between un-supervised sample clusters and lung cancer stages was observed. More stage I tumors were clustered together. These tumors mainly were well-differentiated adenocarcinoma or BAC type. Advanced stage tumors of stage III tended to form another cluster. These features suggest that the genes dominating the expression pattern of a tumor are responsible for the histologic phenotypes and are less likely to measure a tumor's stage accurately. A pure histological examination of a primary tumor cannot reliably determine the presence or absence of a remote metastasis. However, noticeable differences in gene expression profiling have been made during metastasis, which can be better studied by comparing a metastatic tumor with its counterpart. Lymph node metastasis including the exact location and the number of lymph nodes involved is an important observation for separating an early-stage tumor from the later stage. Comparison of gene expression profiles of tumors with lymph node metastasis and those without will find a signature for predicting the lymph node status of a primary tumor.

For determining the association between gene expression profiling and patient survival, two common approaches were made: unsupervised clustering for identification of a tumor sub-class with a distinct clinical outcome, and supervised learning and testing to identify specific gene panels, which can be correlated with survival. The first approach does not consider the tissue annotations such as cell type and cluster formation. These annotations rely on the similarity of the gene expression patterns of the samples. In the second approach, a subset of genes will be primarily identified based on pre-defined classes such as long versus short survival, with metastasis versus without metastasis, and these genes are applied to a set of patients for outcome prediction.

3.9 HEMATOPOIETIC DISORDERS

Hematopoiesis is the formation, development, and maturation of blood cells in humans and other animals. In humans, blood and its components are formed in the highly specialized cells with a short life cycle of about 120 days for human erythrocytes, around 5 days for leukocytes, from several days to several months for lymphocytes, and 4 days for thrombocytes. Despite the continuous elimination of the blood cells from the bloodstream, the number of eliminated cells will be equal to the number of the newly synthesized cells throughout the life of an organism. Invertebrate animals carry hematopoiesis mainly in the blood and perivisceral fluids. In adult mammals and especially human beings, hematopoiesis takes place in the specialized hematopoietic organs. Integration of erythrocytes or red blood cells takes place in the bone marrow and the granular leukocytes, monocytes, macrophages, and thrombocytes. Production of lymphocytes occurs in the lymph nodes, spleen, thymus, and bone marrow. All blood cells, regardless of their differences, originate from a single parental hematopoietic stem cell. A strain of parental cells is maintained in the human body from its birth, ensuring the continuity of the hematopoiesis. After differentiation (maturity) these hematopoietic cells undergo complex changes and divide several times, leading to the development of many specialized cells from these parental cells.

Control of hematopoiesis occurs by regulating the qualitative and quantitative changes to blood cells according to the organism's needs and activities. Replenishment of the cells is needed in the cases of blood loss during injuries, surgeries, and the like. The accomplishment of this regulation is by several hormones, vitamins such as cyanocobalamin, folic acid, ascorbic acid, and special substances such as erythroproteins. Hematopoiesis is a complex process comprising various stages, which are quite sensitive. Most of the mechanisms regulating these stages and their regulations are unknown. Hematopoiesis starts in the yolk sac of human embryos, where the hematopoietic cells originate from mesenchymal cells. Foci of the hematopoietic tissue are formed initially in the embryonic mesenchyme and later in the liver. The formation of RBC and WBC also occurs in the liver at the embryo stage. Later hematopoiesis will be shifted to the bone marrow and the lymphocytes begin to develop in the thymus, spleen, and lymph nodes.

Hematopoietic disorders form the basis of pathogenesis in the blood. These disorders are either caused due to external factors such as physical, chemical, and infections or internal factors such as hormonal, metabolic, congenital, or hereditary factors. The factors responsible for disturbances in the blood are yet to be determined. Depending on the nature of the damage incurred by hematopoietic organs, disorders are classified into hyperplastic and hypoplastic. Hyperplastic disorders occur due to the excessive formation of hematopoietic elements in the tissue. Hypoplastic disorders occur due to the low formation of the hematopoietic elements in the tissue. Aplastic disorders result in the incapability of forming the elements or their maturation in the hematopoietic tissue. The category of the mutated, damaged, injured cells and their degrees of maturity such as granulo, erythro, thrombocyte, and lymphopoiesis at various maturity stages are other factors that determine the nature of the diseases. Leukemia and erythremia are the outcomes of hyperplastic hematopoiesis. During leukemia, bone marrow cells lose their capacity for differentiation or maturation, with a considerable reduction in proliferation rate. The life span of the immature elements will increase, leading to the accumulation of many cells with different degrees of maturity in the blood and hematopoietic organs. Variations in these elements will determine the type of leukemia, such as acute, chronic, myelocytic, or lymphocytic. Karyotypic studies of leukemia cells have revealed certain changes in the chromosomes, indicating the role of heredity in the hematopoiesis disturbances. During hypoplasia or aplasia, damage to hematopoietic stem cells leads to the depletion of hematopoietic cells in the bone marrow. This leads to a considerable decrease in the RBC count, hemoglobin levels, WBC count, and thrombocytes. A reduction in the blood cell count is responsible for hypoplastic and aplastic anemia, agranulocytoses, and metastases of tumors in the bone marrow.

Hematopoietic disorders assume a peculiar behavior, especially when the body suffers from a deficiency of vitamins, trace elements, or enzymes. The deficiency of vitamin B12 and folic acid in the human body will impair RBC formation, and the cells characteristic of embryonic hematopoiesis in the liver will start appearing in the bone marrow. Despite normal levels of erythrocyte forming cells in the bone marrow, RBC will have very little hemoglobin leading to iron deficiency anemia if the patient suffers from iron deficiency. If the structure of hemoglobin is impaired (leading to a disease hemoglobinopathies) or enzymes in the RBC are either absent or non-functional (enzymopathies), RBC will become abnormal and will be quickly destroyed in the bloodstream or spleen (also known as hemolytic anemia). In such cases, the bone marrow and the peripheral blood contain substantial amounts of immature cells of the erythrocytic series. Hematopoietic disorders involving lymphopoiesis will weaken immunity and generate significant protein changes in the blood.

3.9.1 Techniques for Studying the Hematopoietic Disorders

Fluorescence in situ hybridization (FISH) using cells arrested in the interphase is used to monitor the effect of interferon on chronic myelogenous leukemia patients. FISH is one of the most sensitive assays for identifying specific nucleic acid sequences and detecting numerical chromosome abnormalities, structural chromosomal rearrangements, and cryptic abnormalities (Discussed in Chapter 4). Bone marrow aspirate and peripheral blood smears are applied for FISH analysis. FISH uses fluorescent probes that hybridize only to complementary sequences and the resulting signals are examined under a fluorescent microscope. Southern blot hybridization is used for clonal analysis in diseases such as adult T-cell leukemia/lymphoma. PCR using genomic DNA is performed for limited diseases such as lymphoma with the bcl-2/IgH fusion gene. RT-PCR can detect the fusion gene transcripts with high sensitivity. This method is useful to detect minimal residual diseases after chemotherapy or bone marrow transplantation. For performing quantitative analysis, real-time PCR or competitive PCR will be of great support.

3.10 CONCLUSION

Biological systems function at the cellular, tissue, and organismal levels under the control of the genetic material embedded in the nucleus of every cell in an organism known as chromatin. During cell division in humans, the genetic material in the nucleus undergoes condensation to form 46 chromosomes, including 22 pairs of autosomes and one pair of allosomes (either XX or XY). Chromosomal disorders are the outcomes of either excess or deficiency of whole chromosomes or a portion of it in abnormalities. Documentation of many numerical and structural abnormalities during human karyotyping is associated with various disorders affecting human health and sometimes human life, causing mortality. In this chapter, aberrations that occur to human chromosomes and their adverse effects at the gene level, including single gene and multiple gene disorders, have been discussed. The human genome contains 30,000 genes, which build all proteins in the human body. Single gene disorders occur due to mutations in any of these genes responsible for the disorder. Studies of single gene disorders in humans have led to the identification of many unexpected findings such as trinucleotide repeats and their association with neurodegenerative disorders, including fragile x syndrome, myotonic dystrophy, and spinocerebellar ataxias. Studies on the transmission of monogenic disorders in identical twins have identified specific non-genetic mechanisms associated with a disorder. Online Mendelian inheritance in man (OMIM) is an online database established in 1997, which has reported over 387 human genes of known phenotypes over 2,310 human phenotypes of known molecular basis. This database has also reported 1,621 Mendelian phenotypes without any information about their molecular nature. After the elucidation of the human genome, there was a shift from monogenic disorders to polygenic disorders, involving multiple genes. Using SNP arrays, the SNP profile of an individual can be obtained and compared between affected and unaffected individuals for determining their segregation related to a disease. With an increased amount of information on human genetic diseases, many novel genetically inherent disorders are being identified at a rate of one disease per every 250 humans globally. Worryingly, this number is increasing due to rapid changes in the food and living habits of various human populations .

REFERENCES

Ahn, J., Lee, J. (2008) X chromosome: X inactivation. *Nature Education* 1(1):24.

Barbara, P. (2008). X-chromosome inactivation: the molecular basis of silencing. *J. Biol.* 7(8): 30.

Bezerra, P. K. M., Bezerra, P. M., Cavalcanti, A. L., Supranumerários, D. (2007). Revisão da literature e Relato de Caso. Rev. *Cien. Méd. Biol.* 6 (3): 349–356.

Carol, B., Anil, R. P., Valentine, N., Harris, B (2013). DNA damage, DNA repair and cancer. In: New Research Directions in DNA Repair. *Intech Publishers* 413–465.

Carole, B., Susan, H., Paul, Z., Albert, S., David, F. P. (1998). A chromosomal deletion map of human malformations. *Am. J. Hum. Genet.* 63:1153–1159.

Chamara, S. P., Nirmala, D. S., Kariyawasam, U. G. I. U., Dissanayake, V. H. W. (2019). The frequency and spectrum of chromosomal translocations in a cohort of Sri Lankans. *BioMed Research International* 2019: Article ID 9797104, 11 pages.

Clark, L. T., Emerole, O. (1995). Coronary heart disease in African Americans: Primary and secondary prevention. *Cleveland Clinic Journal of Medicine* 62: 285–292.

Cruz, R. M., Oliveira, S. F. (2007). Análise genética de problemas craniofaciais – revisão da literatura e diretrizes para investigações clínico-laboratoriais (parte 1). *Rev. Dent. Press Ortodon. Ortop. Facial* 12(5):133–40.

David, J. A., Andy Choo K. H. (2002). Neocentromeres: Role in Human Disease, Evolution, and Centromere Study. *Am. J. Hum. Genet.* 71(4): 695–714.

Goureau, A., Yerle, M., Schmitz, A., Riquet, J., Milan, D., Pinton, P., Frelat, G., Gellin, J. (1996). Human and porcine correspondence of chromosome segments using bidirectional chromosome painting. *Genomics* 36: 252–262.

Gregory, J. K., Frédéric, B. P., Clarice, D. R. (2018). Sickle cell disease. *Nature Reviews* 15(4): 18010.

He, W., Sun, X., Liu, L., Li, M., Jin, H., Wang, W-H. (2014). The prevalence of chromosomal deletions relating to developmental delay and/or intellectual disability in human euploid blastocysts. *PLoS ONE* 9(1): e85207.

Ivan, Y. I., Svetlana, G. V., Yuri, B. Y. (2019). Pathway-based classification of genetic diseases. *Molecular Cytogenetics* 12: Article No.52.

Kato, G. J., Piel, F. B., Reid, C. D., Gaston, M. H., Ohene-Frempong, K., Krishnamurti, L., Smith, W. R., Panepinto, J. A., Weatherall, D. J., Costa, F. F., Vichinsky, E. P. (2018). Sickle cell disease. *Nat. Rev. Dis. Primers*. 15(4):18010.

Kirkpatrick, M. (2010). How and Why Chromosome Inversions evolve. *PLoS Biol* 8(9): e1000501.

Korstanje, R., O'Brien, P. C., Yang, F., Rens, W., Bosma, A. A., van Lith, H. A., van Zutphen, L. F., Ferguson-Smith, M. A. (1999). Complete homology maps of the rabbit (*Oryctolagus cuniculus*) and human by reciprocal chromosome painting. *Cytogenet. Cell. Genet.* 86: 317–322.

Lobo, I. (2008) Multifactorial inheritance and genetic disease. *Nature Education* 1(1):5

MacDermot, K. D., Jack, E., Cooke, A., et al. (1990). Investigation of three patients with the "ring syndrome", including familial transmission of ring 5, and estimation of reproductive risks. *Hum. Genet.* 85:516–20.

Maria, S. F., Chiara, S. P., Alessandra, F. (2015). Duchenne muscular dystrophy: from diagnosis to therapy. *Molecules* 20: 18168–18184.

Marta, P., So`nia, C., Sergi, V., Mario, C (2015). Human inversions and their functional consequences. *Briefings in Functional Genomics* 14(5):369–379.

Martins, R. B., De Souza, R. S., Giovani, E. M. (2014). Cleidocranial dysplasia: report of six clinical cases. *Spec. Care Dentist.* 34(3):144–50.

Mathew, J. K. J., Prasad V. J. (2012). Chromothripsis: Chromosomes in Crisis. *Developmental Cell* 23: 908–917.

McDermott, A., Voyce, M. A., Romain, D. (1977). Ring chromosome 4. *J. Med. Genet.* 14:228–232.

Moh-Ying, Y. (2015). Autosomal ring chromosomes in human genetic disorders. *Transl. Pediatr.* 4(2):164–174.

Mossey, P. A., Little, J. (2002). Epidemiology of oral clefts: An international perspective. In: D. F. Wyszynski Eds. Cleft Lip and Palate: From Origin to Treatment, Oxford University Press, Pp127–158.

Myers, R. H. et al. (1990). Parental history is an independent risk factor for coronary artery disease: The Framingham study. *American Heart Journal* 120: 963–969.

Ohta, T., Gray, T. A., Rogan, P. K., Buiting, K., Gabriel, J. M., Saitoh, S., Muralidhar, B., Bilienska, B., Krajewska-Walasek, M., et al. (1999). Imprinting-mutation mechanisms in Prader-Willi syndrome. *Am. J. Hum. Genet.* 64(2): 397–413.

Pena, S. D. J., Sérgio, D. J. (1998). Molecular Cytogenetics II: PCR-based diagnosis of chromosomal deletions and microdeletion syndromes. *Genet. Mol. Biol.* 21(4): 3.

Pepe, G., Giusti, B., Sticchi, E., Abbate, R., Gensini, G., Nistri, S. (2016). Marfan syndrome: current perspectives. *The Application of Clinical Genetics* 9:55–65.

Sarnataro, S., Chiariello, A. M., Esposito, A., Prisco, A., Nicodemi, M. (2017). Structure of the human chromosome interaction network. *PLoS ONE* 12(11): e0188201.

Shen, G., Przemyslaw, S., Zeynep, C.A., Bo, Y., et al. (2016). Mechanisms for Complex Chromosomal Insertions. *PLoS Genet.* 12(11): e1006446.

Smeets. E. E. J., Pelc, K., Dan. B. (2011). Rett Syndrome. *Mol Syndromol.* 2:113–127.

Takema, K., Yuya, O., Hidehito, I., Yoshio, M., Seiji, M. et al. (2017). Genomic characterization of chromosomal insertions: insights into the mechanisms underlying chromothripsis. *Cytogenet. Genome. Res.* 153:1–9.

Toshinori, O., Akira, N. (2011). Role of p53 in cell death and human cancers. *Cancers* 3: 994–1013.

Wienberg, J., Stanyon, R., Nash, W. G., O'Brien, P. C., Yang, F., O'Brien, S. J., Ferguson-Smith, M. A. (1997). Conservation of human vs. feline genome organization revealed by reciprocal chromosome painting. *Cytogenet. Cell. Genet.* 77: 211–217.

William, J. M., Lutz, F., Stephen, J. O., Roscoe, S. (2011). The Origin of Human Chromosome 1 and Its Homologs in Placental Mammals. *Genome Research* 13:1880 1888.

Zakary, L. W., Casim, A. S., Marek, K., Seth, J. C. (2010). Hematopoiesis and its disorders: a systems biology approach. *Blood* 115(12): 2339–2347.

Zdravkovic, S., et al. (2002). Heritability of death from coronary heart disease: A 36-year follow-up of 20,966 Swedish twins. *Journal of Internal Medicine* 252: 247–254.

4 Techniques for the Diagnosis of Genetic Disorders

4.1 INTRODUCTION

The diploid human genome comprises of 23 chromosome pairs composed by 30,000 genes. The haploid human genome is estimated to be 3.2×10^9 bp. Each member of a chromosome is transmitted patrilineally, and another member is transmitted matrilineally. DNA base pairs consisting of genes are packed in the chromosomes. An allele is positioned on a locus, the specific location of a gene or DNA sequence on a chromosome, thus, the diploid genome contains two alleles of each gene. Chromosomes 1 to 22 are called autosomes, and the twenty-third pair X and Y are known as sex chromosomes. Steps in the transmission of genetic information include replication (DNA makes DNA), transcription (DNA makes RNA), RNA processing (capping, splicing, tailing, and RNA translocation to cytoplasm), translation (RNA makes protein), and protein processing, folding, transport, and incorporation. If any change in the DNA sequence due to either mutation or alteration is not repaired by the cell, subsequent DNA replications reproduce the mutation. Various mechanisms can cause these changes ranging from a single nucleotide alteration to the loss, duplication, or rearrangement of chromosomes. Genetic diseases that occur due to a single gene, chromosomal, and multifactorial disorders pose a threat to the health of human beings. With the changes in living and food habits, climate conditions impact human DNA, resulting in new combinations of genes, leading to the generation of novel genetic disorders needing diagnosis. The development of new technologies has provided an increased understanding of the changes in the human genome with more accuracy. Detection of mutations in human DNA occupies the central role in genetic diagnosis, including pre-implantation genetic diagnosis (PGD), prenatal diagnosis (PND), pre-symptomatic testing, conformational diagnosis, and forensic tests. The techniques for understanding the changes in the human DNA leading to genetic disorders and syndromes are broadly divided into two categories. Firstly, the indirect approach is based on results obtained by genetic linkage analysis using STR and VNTR DNA markers, flanking or within a gene. Secondly, the direct approach depends on the detection of genetic variations responsible for the disorder. The techniques for understanding these two categories are classified as either the molecular approach or the cytogenetic approach. The details of the techniques are discussed in this chapter.

4.2 MOLECULAR APPROACH

4.2.1 Nucleic Acid Isolation Methods

Recent advances in genetic disorders have necessitated the isolation of large quantities of good quality DNA from various sample sources under different conditions. One of the most traditional ways of isolating DNA is from the nucleated cells of peripheral blood. Due to the invasiveness of

DOI: 10.1201/9781003343790-4

this approach, other alternative sources of DNA isolation through non-invasive methods are being practiced.

4.2.1.1 DNA Extraction from Buccal Swabs

Isolation of DNA from buccal cells can be readily achieved using a buccal swab with a cotton swab or using a mouthwash. DNA isolation using a buccal swab has many advantages, including cost-effective processing, a lower sample volume requirement, long-term archiving, and suitability of self-collection. This isolation method is more comfortable for the patient, and the buccal swabs provide sufficient DNA for the PCRs, as they demand only a few nanograms of DNA.

The buccal swab samples are suspended in 500 µl lysis buffer [10 mM Tris (pH 8.0), 10 mM EDTA, and 2.0% SDS], and 50 µl 10% SDS, followed by the addition of 5 - 10 µl 20 mg/ml proteinase K. The samples are incubated for 1 - 3 hrs. at 56°C until the tissue is dissolved. The DNA is extracted from each sample with an equal volume of phenol: chloroform: isoamyl alcohol solution (25:24:1) and mixed gently by inverting the tubes for 3 min. The samples were centrifuged for 10 min with 10,000 g (4°C), and the upper aqueous layer was transferred to a fresh, sterilized microcentrifuge tube. 10 µl RNase A was added, and the solution was incubated at 37°C for 30 min. Equal volumes of chloroform: isoamyl alcohol solution were added and centrifuged, again with 10,000 g (4°C) for 10 min. The upper aqueous layer was transferred to a sterilized microcentrifuge tube, and double the volume of chilled isopropanol was added, along with the one-tenth volume of 3 M sodium acetate, and chilled at 20°C for 1 h for precipitation. After 1 h, the sample was centrifuged at 10,000 g (4°C) for 10 min. After decanting the supernatant, 250 µl of 70% ethanol was added and centrifuged at 10,000 rpm for 10 min, and the supernatant was decanted gently. The pellet was air-dried under laminar air flow, and the dried pellet was re-suspended in 50 µl nuclease-free water or 10 mM Tris-HCl, 1 mM EDTA, pH 7.6 (TE), buffer and frozen at 20°C or at -80°C for storage.

4.2.1.2 DNA Extraction from a Hair Sample

Human hair is one of the most common biological materials associated with legal investigations and is used for DNA-based forensic analysis. One of the most important applications of testing DNA isolated from the hair sample is the short tandem repeat analysis of nuclear DNA, possible with the root of the hair and or adhering tissue. However, telogen hair or shed hair often associated with a crime scene might not contain any nuclear material. Cellular mitochondria and their DNA (mtDNA) remain intact, while the nucleus degrades as the hair shaft hardens during keratinization. Hence, mtDNA analysis is feasible from the keratinized hair. Unfortunately, the protein-rich nature of the hair sample requires additional steps such as fragmentation using a microscopic glass grinder, followed by organic solvent extraction to break down the shaft and release the DNA.

DNA can be isolated from hair shafts using a modified microscopic glass-grinding and organic solvent extraction protocol. A digestion buffer (500 µl; 10 mM Tris-HCl, 10 mM EDTA, 50 mM NaCl, 20% SDS, pH 7.5) was added to a 1.5 - ml microcentrifuge tube, along with 40 µl of 1 M DTT (to a final concentration of 80 mM, 240 mM of sodium acetate, pH 5.2) and 151 of 10 mg/ml proteinase K (to a final concentration of 0.3 mg/ml; Himedia). A hair sample was added to this solution before vortexing and incubating for 2 hrs. at 56°C. After 2 hrs. of incubation, the sample tube was vortexed again, and an additional 401 of 1 M DTT and 151 of 10 mg/ml proteinase K were added, followed by gentle mixing and incubation at 60°C for 2 hrs. or until the hair was dissolved completely. The DNA was extracted from each sample with an equal volume of phenol: chloroform: isoamyl alcohol solution (25:24:1) and mixed gently by inverting the tube for a few minutes. The samples were centrifuged for 10 min with 10,000 g (4°C), followed by transferring the upper aqueous layer into a fresh, sterilized microcentrifuge tube. 10 µl of RNase A was added and kept for incubation at 37°C for 30 min. An equal volume of chloroform: isoamyl alcohol was added, and the tube was centrifuged at 10,000 g (4°C) for 10 min again. The upper aqueous layer was transferred into a fresh, sterilized microcentrifuge tube before double the volume of chilled isopropanol and one-tenth volume of 3M sodium acetate were added. The sample was chilled at -20°C for 1h for the DNA precipitation to occur. The sample was centrifuged at 10,000g (4°C) for 10min. The supernatant was discarded, 250µl 70% ethanol was added, and the pellet was tapped gently before further centrifugation at 10,000rpm for 10min. The supernatant was discarded, and the pellet was air-dried in a laminar airflow, re-suspended in 50µl nuclease-free water or TE buffer, and frozen at -20°C or -80°C for storage.

4.2.1.3 Extraction of DNA from Urine Samples

Forensic investigation of human urine stains is of great importance when identifying the exact crime location and the type of death. Human urine is a suitable sample for toxicological analysis in doping and drug screening tests. DNA found in urine is mainly derived from cells shed into the urine from the urinary tract. DNA isolated from urine may be used in many different applications in research including DNA methylation identification and cancer testing. DNA isolated from urine samples has unique advantages for diagnostics, such as that urine collection is wholly non-invasive. Technically, the isolation of DNA from urine is less labour intensive than isolating it from blood.

Frozen urine samples were thawed at room temperature and then placed immediately in ice before DNA isolation. The urine specimen was inverted or swirled in a specimen cup to create a homogenous suspension of cells. 1ml of the sample was transferred into an Eppendorf tube and centrifuged for 10min at 10,000g (4°C). The supernatant was removed, and a dry pellet containing cells was chilled at -20°C for 15min. Lysis buffer (5001; 10mM Tris, 1.2mM EDTA, 10% SDS, pH9.0) was added to the dry pellet, and the sample was vortexed to re-suspend the pellet. 20µl of Proteinase K was added, and the tube was incubated in a water bath at 56°C for 2h. Sodium acetate (601 of 3M) and 0.5ml cold isopropanol was added, mixed, and chilled at -20°C for 1h, followed by centrifugation at 10,000g (at 4°C) for 20min. The supernatant was discarded, 250µl of 70% ethanol was added, and the pellet was tapped gently, followed by centrifugation at 10,000rpm for 10min before the supernatant was discarded gently. The pellet was air-dried in a laminar airflow, and the dried pellet was re-suspended in 50µl nuclease-free water or TE buffer and frozen at -20°C or -80°C for storage.

4.2.1.4 DNA Extraction from a Blood Sample

Lymphocytes from whole blood were separated by lysing the red blood cells (RBCs) using a hypotonic buffer comprising ammonium bicarbonate and ammonium chloride with minimal lysing effect on lymphocytes. Three volumes of RBC lysis buffer were added to the blood sample, mixed by vortexing and inverting thoroughly for 5min, and centrifuged at 2000g for 10min. The supernatant was discarded, leaving behind 1ml to prevent loss of cells. 3vol RBC lysis buffer was added, and vortexing, inverting, and centrifuging steps were repeated two to three times until a clear supernatant and a clean white pellet were obtained. After the final wash, the supernatant was completely discarded. The pellet was re-suspended in 5001 PBS, followed by the addition of 400µl cell lysis buffer (10mM Tris-HCl, 10mM EDTA, 50mM NaCl, 10% SDS, pH7.5) and 10µl proteinase K. The sample was vortexed to dissolve the pellet completely and incubated for 2hrs. at 56°C in a water bath for lysis. An equal volume of phenol (equilibrated with Tris, pH8) was subsequently added to the tube and mixed well by inverting for 1min. The tube was centrifuged at 10,000g (at 4°C) for 10min, and the aqueous upper layer was transferred to a fresh tube containing equal volumes (1:1) of phenol and chloroform: isoamyl alcohol (24:1). The tube was mixed by inverting for 1min and centrifuged for 10min at 10,000g (at 4°C). The supernatant was then transferred to a fresh tube, and 10µl of 10mg/ml RNase A was added. The sample was incubated at 37°C for 30min before an equal volume of chloroform: isoamyl alcohol (24:1) was added and mixed by inverting the tube for 1min and centrifuging at 10,000g (at 4°C) for 10min. The supernatant was transferred to a fresh tube, and twice the volume of absolute alcohol was added and inverted a few times gently and chilled at -20°C, followed by centrifugation at 10,000g at 4°C for 20min. The supernatant was discarded, 250µl of

70% ethanol was added, and the pellet was tapped gently, centrifuged at 10,000 rpm for 10 min, and decanted the supernatant gently. The pellet was air-dried in a laminar airflow, and the dried pellet was re-suspended in 50 µl nuclease-free water or TE buffer and frozen at -20°C or -80°C for storage.

The concentration and purity of the isolated DNA are tested spectrophotometrically. The integrity of the isolated DNA can be tested by loading the DNA on 1% agarose gel and subjecting it to electroporation. The requirement of low sample volume and a non-invasive sample collection mechanism permits pediatric sampling that readily manifests in broader study recruitment in population-based case studies. Further, this simplified method can be applied as a diagnostic tool, and the DNA analysis is done in 8 hours, the desired requirement of the current diagnostic medical field.

4.2.2 Restriction Fragment Length Polymorphism

The discovery of the restriction enzymes in bacteria, which control the entry of foreign DNA by specific cleavage, has opened new doors in modern biology for specific cleaving of DNA and identifying the differences based on their cleavage sites. These enzymes are known as restriction endonucleases. They recognize the specific sites in foreign DNA and cleave them to prevent the invasion by viruses. However, bacteria will protect their DNA from attack by methylating the specific bacterial DNA sequence that the restriction enzyme would otherwise recognize. The specific sequences are 4 - 6 bp long. Since the discovery of Hind II from *Haemophilus influenza* in 1974, more than 400 enzymes have been isolated from different bacteria and are used in experiments related to molecular biology (Seshadri, 2011).

The usage of restriction endonucleases that cleave DNA at specific regions has identified polymorphic regions throughout the length of the targeting DNA sample. Different individuals have a particular number of bases in a specific area of the human genome. Depending on whether polymorphism falls within the restriction enzyme-specific cleavage site, short or long DNA fragments are obtained after their digestion by a specific enzyme. The generation of restriction fragment length polymorphisms (RFLPs) between two individuals is due to changes in the DNA, can be effectively used if the same is found close to a diseased gene or a mutated gene. RFLP can serve as a marker for the identification of a genetic disorder. Owing to the variations in the number of tandem repeats of a short DNA fragment, RFLPs serve as the basis for DNA fingerprinting. One individual may have 18 repeats of DNA segments on one chromosome and 12 repeats on the other, and their repeat pattern will differ. Tandemly repeated DNA sequences are very short, simple sequence repeats collectively known as (CA)n blocks. It is estimated that the human genome consists of 50,000 - 100,000 (CA)n blocks with n in the range of 15 - 30. A (CA)n block should be present in every 30 - 60 Kb segment of the human genome. These blocks play a pivotal role in the regulation of genes or the generation of recombinations.

4.2.2.1 Role of RFLPs as Markers for the Identification of Genetic Disorder

RFLPs are highly abundant in human DNA, which is at least one per every 500 bp. This number is high compared with the polymorphisms at the protein level and can be used as a genetic marker for the identification of a genetic disorder. To utilize RFLP as a genetic marker, the main criterion that should be satisfied is the genetic distance. The RFLP must be very close to the diseased gene. The closer the RFLP to the gene, the less chance there is of genetic recombination occurring. The RFLP and diseased gene will be linked, and the accuracy of the test as a specific marker would be very high. As these two get closer, the recombination fraction θ will be 0, making RFLP highly specific regarding its ability to screen for a specific disease. Accuracy can be further increased by using flanking markers, which are RFLPs located very close to both sides of a diseased gene. The rationale for using such flanking markers is that even if recombination occurs between one RFLP and diseased locus, it is very likely that the other RFLP would segregate with the disease. Genetic disorders can be detected directly by isolating an RFLP emerging from a change in a single base pair. For example, a mutation generated in the β-globin gene due to the substitution of adenine (A) for thymine (T) in codon 6, leading to valine for glutamine, thus generating a sickle cell hemoglobin. However, this approach of RFLP detection arising due to a point mutation is limited only to a few genetic diseases. In most cases, RFLPs arising due to the variations in the number of tandem repeats at a locus are used to study genetic disease. The sequences are short repetitive DNA, characterized as a hypervariable region. These hypervariable regions are cut using multiple restriction endonucleases and the differences in the lengths of resulting fragments can be used to establish a specific genetic disorder. The limitation of this approach is the inability of electrophoretic separation of fragments on the gel, which can resolve even minor differences in the sizes of restriction fragments.

4.2.2.2 RFLP Procedure

DNA located between two restriction sites is known as a restriction fragment. The size of a restriction fragment is determined by the DNA length between two restriction sites. Digestion of DNA with a restriction enzyme will lead to many fragments of varying lengths. As the nucleotide sequences vary from person to person, the number of restriction sites and the size of restriction fragment lengths also vary from person to person generating polymorphism. RFLP of human DNA, polymorphic for the restriction site *Pvu II*, is displayed in Figure 4.1. Individuals lacking the second Pvu II restriction site will have one restriction fragment of 5 KbP length, and the individuals with the second restriction site will have two fragments of 3 Kb and 2 Kb in length. Differences in the fragment lengths are known as RFLP.

RFLP is performed in two different ways: by Southern blot hybridization and through PCR analysis. Digestion of genomic DNA with a specific restriction endonuclease and separating resulting fragments according to their size by agarose gel electrophoresis is a significant part of this process. The resulting DNA fragments are denatured into single strands, transferred onto a nitrocellulose membrane that maintains the spatial orientation of the restriction fragments, and the band pattern on the membrane is identical to those of the gel. The fragments on the membrane are hybridized with a radiolabeled ^{32}P cDNA probe. Washing the membrane will remove any unhybridized probe and detect the hybridized DNA fragments by autoradiography (exposing the hybridized bands to an X-ray film). The blotting procedure is time consuming and will take at least 8 - 10 days. The amplification power of PCR is being exploited for the detection of RFLPs. In this technique, the fragments generated after digestion of DNA with the restriction endonuclease are amplified in a PCR thermal cycler using random primers, and the amplified fragments are separated by electrophoresis on 1% agarose gel. RFLP analysis through PCR is less time consuming and can be completed within a day. Alternatively, an amplified fragment can be transferred onto a membrane, detected by using a specific probe

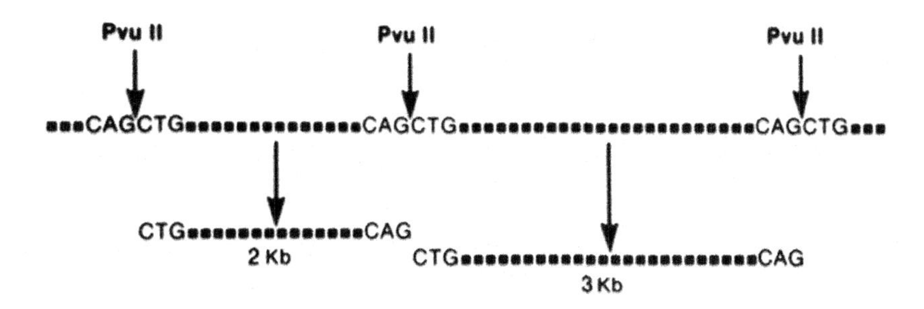

FIGURE 4.1 Generation of RFLP by Restriction endonuclease Pvu II, which recognizes the sequence CAGCTG and cuts between the G-C. The length of the restriction fragments generated by Pvu II is determined by the distance between the two sites, generating 2 kb and 3 kb fragments. In individuals lacking the middle Pvu II site, a 5 kb fragment only would result from digestion with this enzyme.

4.2.2.3 Applications of RFLP

RFLP technique is applied in various areas ranging from paternity testing to forensic sciences and detecting genetic disorders. RFLPs provide an unlimited source of genetic markers used for diagnosing genetic diseases in families. Enzyme exclusion of paternity with a cumulative probability of more than 99.9% is possible using a minimum of four probes. Similarly, the application of 3 - 5 probes can provide a completely individual fingerprint in forensic testing. (CA)n block markers are rapidly typed by PCR amplification followed by electrophoretic separation. This procedure permits the typing of hundreds of (CA)n block markers on a single gel using a small blood sample, thus making RFLP a powerful technique for detecting polymorphisms. Linkage analysis using an RFLP marker close to the defected gene is used for its detection. Genetic testing for a disorder involves screening several family members for detecting the disease, replaced by PCR amplification for studying genetic disorders such as hemophilia A and B, Duchenne and Becker muscular dystrophy, α-l-antitrypsin deficiency. PCR amplification of VNTR sequences in the 3' region of the apolipoprotein gene has led to the identification of 12 different alleles at a single locus in a single racial group. Similar studies are used for understanding the defects in other genes such as insulin. RFLP analysis of alcohol dehydrogenase loci (ADH 2 and ADH 3) helped in the identification of polymorphisms that led to liver cirrhosis in between 12 and 30% of heavy drinkers.

4.2.3 Hybridization-Based Methods

One of the central ideas for detecting changes in nucleic acids either due to mutation or other means was developed in 1970 as nucleic acid hybridization techniques based on the pairing of two complementary nucleotide strands to develop a hybrid DNA through hybridization. Here, the DNA or RNA of one strand known as a probe originates from a known organism. The other strand, known as the target, originates from an unknown organism that will be identified or detected. The hybrid DNAs are the results of DNA-DNA, DNA-RNA, and RNA-RNA interactions. During hybridization, single strands of DNA from two different species are allowed to join together to form hybrid double helices. The technique of DNA hybridization is based on two principles: that double strands of DNA are held together by hydrogen bonds between complementary base pairs, and that the more closely related the two species are, the higher the number of complementary base pairs in the hybrid DNA. In other words, the degree of hybridization is proportional to the degree of similarity between the molecules of DNA from the two species.

Hybridization of DNA is accomplished by heating the DNA strands of two different species to 86° C [186.8° F], which breaks the hydrogen bonds between all complementary base pairs. The single-stranded DNA from both species is mixed together and allowed to cool slowly. Similar strands of DNA from both species will begin to chemically join together, or re-anneal at complementary base pairs by reforming hydrogen bonds. The resulting hybrid DNA is reheated, and the temperature at which the DNA once again becomes single-stranded is noted. Because one cannot observe DNA separating, another technique must be used simultaneously with heating to show when separation has occurred. This technique employs the absorption of UV light by DNA. Single strands of DNA absorb UV light more effectively than double strands. Therefore, the separation of the DNA strands is measured by UV light absorption. As more single strands are liberated, more UV light is absorbed. The temperature at which hybrid DNA separation occurs is related to the number of hydrogen bonds formed between complementary base pairs. Therefore, if the two species are closely related, most base pairs will be complementary, and the separation temperature will be very close to 86° C [186.8° F]. If the two species are not closely related, they will not share many common DNA sequences, and fewer complementary base pairs will form. The separation temperature will be less than 86° C [186.8° F] because less energy is required to break fewer hydrogen bonds. With this information, a tree of evolutionary relationships based on the separation temperature of the hybrid helices is generated. DNA Hybridization can be carried out in two different ways depending on the platform holding the DNA.

4.2.3.1 Membrane-Based Methods

Membrane-based methods are blotting procedures performed either on nitrocellulose membrane (Southern and Northern) or PVDF membrane (Western blotting). Membrane-based nucleic acid hybridization is of two types:

1. DNA-DNA hybridization: a technique used to compare DNA from two different species, locate or identify nucleotide sequences, and establish the effective in vitro transfer of nuclear material to a new host. A single strand of DNA from one source is bound to a membrane to which a single strand of radioactively labeled DNA probe from a different source is added. Complementary base pairing between homologous sections of the two DNAs results in double-stranded hybrid sections that remain bound to the membrane, whereas single-strand sections are washed away. The amount of radioactivity remaining on the filter, compared with the amount washed away, gives a measure of the number of nucleotide sequences that the radioactive DNA and the original DNA share.
2. DNA hybridization: a method of determining the similarity of DNA from different sources. Single strands of DNA from two sources, such as different bacterial species, are put together, and the extent of double hybrid strand formation is estimated. The greater the tendency to form these hybrid molecules, the greater the extent of complementary base sequences, namely, gene similarity. This method is one way of determining the genetic relationships of species. Membrane-based hybridization is also carried out using mRNA (Northern blot hybridization) and protein (Western and Eastern blot hybridizations).

Saiki (1989) developed a fast, accurate, and convenient method for the detection of known mutations through reverse dot-blot and implemented the detection of mutations in b-thalassemia. This method utilizes the oligonucleotides bound to a membrane as hybridization targets for the amplified DNA. Advantages of the reverse dot blot are one membrane strip can be used to detect many different known mutations in a single individual (one strip-one patient assay), there is potential of automation, and it is easy to interpret the results using a classical avidin-biotin system.

The disadvantage is that this technique cannot detect unknown mutations. However, continuous development of the technique has given rise to allele-specific hybridization of amplified DNA.

4.2.3.2 Array-Based Methods

Microarray is a technique that involves the hybridization of DNA with targeted molecules both for the quantitative (gene expression) and the qualitative (diagnostic) analysis of large numbers of genes simultaneously. Each DNA spot known as a feature contains picomoles of a specific DNA sequence, known as a probe (or) reporter, which hybridizes with unknown target DNA or cDNA. Microarrays are differentiated based on characteristics such as the nature of the probe, the solid-surface support used, and the specific method used for probe addressing and target detection. DNA microarrays are successfully applied for testing multiple mutations. Here, single DNA strands, including sequences of different targets, are fixed on a solid support in an array format. Fluorescently labeled sample DNA or cDNA is hybridized into the chip. The amount of fluorescence emission is detected using a laser system, and the sequences and their quantities in the samples are determined. Protein microarray analysis quantifies the amount of protein present in biological samples. Similar to chromosome and DNA microarray analysis, the hybridization of labeled target proteins in a patient sample is measured against a reference sample. Also referred to as a biomarker, the presence, absence, increase, or decrease of a particular protein is used for indicating the occurrence of disease in a person, such as a patient's cerebrospinal fluid analysis for amyloid beta proteins to diagnose Alzheimer's disease.

Further, advancement in microarray technology has led to an improvement in the form of in-situ-synthesized arrays, which are extremely-high-density microarrays that use oligonucleotide probes for detection. An example of in-situ-synthesized arrays is Gene Chips, developed by Affymetrix, Santa Clara, CA, the most widely known. These in-situ synthesized arrays are made on the surface of a 1.2 cm^2 quartz wafer as these probes are typically short with a length of 20 - 25 bp. Multiple probes per target are included to improve sensitivity, specificity, and statistical accuracy. Bead arrays are other types of in-situ hybridized microarrays, developed by Illumina, San Diego, CA, that provide a platform for the high-density detection of target nucleic acids. Bead Arrays are manufactured on 3 μm silica beads that randomly self-assemble onto any of the two available substrates: the Sentrix Array Matrix (SAM) or the Sentrix Bead Chip. The most prominent application of bead arrays relies on passive transport for the hybridization of nucleic acids. Electronic microarrays utilize active hybridization via electric fields to control nucleic acid transport. Microelectronic cartridges such as Nano Chip 400, developed by Nanogen, San Diego, CA, use complementary metal oxide semiconductor (CMOS) technology for the electronic addressing of nucleic acids. Here, each Nano Chip cartridge is provided with 12 connectors that control 400 individual test sites. All these arrays mentioned above are two-dimensional (2-D) based. Suspension bead arrays are three-dimensional arrays, which use microscopic polystyrene spheres in either microspheres or beads as the solid support and rely on flow cytometry for bead and target detection.

4.2.4 Polymerase Chain Reaction-Based Techniques

The Discovery of PCR by Kary Mullis, which provides a quick optimization of DNA or cDNA using a thermostable Taq DNA polymerase, has greatly revolutionized molecular diagnostics. A robust technique consisting of many copies of the target sequence generated through its exponential amplification permitted the identification of a mutation within a single day. With the invention of PCR, molecular diagnostics has crossed the threshold to clinical laboratories for the provision of genetic services such as the screening of the population for known mutations, the prenatal diagnosis of inherited diseases, and the identification of unknown mutations. Each PCR cycle doubles the target DNA sequence, resulting in the exponential amplification of the DNA fragment. The DNA polymerase enzyme, the primers (both forward and reverse), and temperature, control PCR amplification. The PCR principle is based on the repetitive cycling of three simple reactions: Firstly, denaturation occurs at 95°C, which separates the template DNA double strand into two single strands. Secondly, annealing, where the temperature is reduced to 55°C, facilitates the attachment of two specific oligonucleotide primers to the DNA template complementarily. Thirdly, extension at which the temperature is raised to 72°C, facilitating the DNA polymerase to extend the primers at the 3´ terminus of each primer and synthesize the complementary strands along 5´ to 3´ termini of each template DNA using deoxynucleotides that are present in the reaction mixture. Each extension step leads to new double strand DNA copies, comprising two single template DNA strands and two synthesized complementary DNA strands. After extension, the reaction will be repeated at least 25 - 29 times with the above three steps, where each synthesized DNA copy serves as a template for further amplification. After completing the defined cycles, the DNA template is amplified 3 million times, and the whole process needs 2 - 5 hrs. depending on the number and types of nucleotides. With the advancement of technology, PCR has undergone many changes, and the modification of PCR used in molecular diagnostics to identify genetic disorders is discussed in this section.

4.2.4.1 Reverse Transcriptase PCR

Reverse transcriptase PCR is the technique of cDNA synthesis from mRNA, mediated by the enzyme reverse transcriptase through a process known as reverse transcription, followed by the amplification of a specific cDNA by PCR. This is one of the most valuable and sensitive techniques currently available for mRNA detection and quantization. The invention of RT- PCR has provided the foundation for many mutation detection schemes based on the amplified cDNA. RT-PCR generated cDNA fragments for their recognition and or examination. This was achieved by utilizing oligonucleotide probes through PCR, immobilizing onto the membrane, and hybridizing to spot genetic changes, particularly mutations. A significant number of mutation detection approaches has been built up and implemented in the following years. RT-PCR is performed as a single step or a two-step process. Single step RT-PCR combines reverse transcription steps involving cDNA synthesis and the subsequent PCR amplification of cDNA in a single reaction tube. This technique requires gene specific primers and has many high throughput applications. However, it is less sensitive and has lower efficiency since the reaction compromises reverse transcription and amplification conditions. Two-step PCR is performed in two steps, the first one for synthesizing cDNA from mRNA in a PCR tube, then amplifying a specific gene in the second step in a separate PCR tube. This technique is more sensitive, flexible and permits the analysis of multiple genes.

4.2.4.2 Real-Time PCR

Real-time PCR is a quantitative assay of an amplified cDNA sequence, illustrated for the first time by Higuchi in 1993. This technique utilizes fluorescent labeled probes for detection, confirmation, and quantification of the PCR products, generated in real time. Real-time PCR has the following novel features.

1. Temperature cycling occurs considerably faster than that of standard PCR assays.
2. Hybridization of specific DNA probes occurs continuously during the amplification.
3. A fluorescent dye is coupled to the probe and fluoresces only when hybridization is taking place.
4. The lack of post PCR processing of amplified products makes this technique more convenient as the production of amplified products can be observed automatically by real-time monitoring of fluorescence.
5. Since the tubes do not have to be opened at the time of reaction, the risk of carry-over contamination can be considerably reduced.

Depending on the transcription levels of the target gene, a small signal can be produced within 30 - 45 minutes. Some commercial automated real-time PCR systems such as Light Cycler, TaqMan, Smart Cycler, and the like, have become available in recent years. These systems perform real-time fluorescence monitoring using fluorescent dyes such as SYBR-Green I, which binds non-specifically to double-stranded DNA generated during PCR amplification. Machines such as TaqMan use florescent probes that bind specifically to amplification target sequences.

4.2.4.3 Allele-Specific Amplification (ASA) by Amplification Refractory Mutation System (ARMS) PCR

Determining the genotype of normal, heterozygous, and homozygous states to identify a point mutation, small deletion, or single nucleotide polymorphism (SNP) in a gene sequence is achieved by cutting the DNA sample with a specific restriction endonuclease. This technique was replaced by allele specific amplification (ASA) by the amplification of refractory mutation system (ARMS). ASA by ARMS uses two separate PCRs for the amplification of each allele with different primers. ASA PCR was introduced by Newton (1989). Here, the specific primers are designed to permit amplification only if the base at the 3′ end of the primer precisely complements the wild type, mutant, or variant sequences. Genotype amplification is achieved by using two complementary reactions in two PCR machines, one containing a specific primer for the amplification of normal DNA sequence at a given locus and the other containing a mutant-specific primer for the mutant-specific DNA amplification. After PCR amplification and electrophoresis, the patterns of specific PCR products will provide the differentiation of the products. The specific PCR product can be detected using either specific labeled probes or melting curve analysis through nucleic acid stains.

Shu (1992) performed ASA by using a single PCR with two annealing temperatures and four primers. Here, the tetra primers with a mismatched 3´ base were replaced by a pair of flanking primers (1, 2) and two internal primers (3, 4). The flanking primer's melting temperature (Tm) was at least 10°C higher than that of internal primers, achieved by designing long flanking primers and short internal primers. Primer 3 is complementary to the sense strand of allele 1 and primer 4 to the antisense strand of allele 2. The first block of cycles is performed with a higher annealing temperature and subsequent cycles at a lower temperature. During the first program, the DNA fragment will be produced only by utilizing the flanking primers as the temperature is too high for the internal primers to anneal. This DNA fragment acts as a concentrated template during the second program and facilitates the internal primers with low Tm for their specific annealing. Two heterozygote DNA fragments are generated from primers 1, 3 and 2, 4 respectively, but only the homozygote DNA fragment is amplified in allele1 (primers 1, 3) or allele 2(primers 2, 4). As the internal primers are not centered in the fragment generated by the flanking primers, two DNA fragments generated in the second program are of unequal length, which can be easily separated on an agarose gel electrophoresis.

ASA-PCR has a wide range of applications and is extensively used in studying genetic disorders. This PCR is used for checking the 35delG mutations in the GJB2 gene among deaf children. ASA-PCR is relatively a cheaper source for determining SNP than other available methods.

4.2.4.4 Multiplex PCR

Multiplex PCR is a widespread molecular biology technique for amplifying multiple targets in a single PCR experiment. Multiple primer pairs in a reaction mixture can amplify more than one target sequence in a multiplexing assay. As an extension to the practical use of PCR, this technique can produce considerable savings in time and effort within the laboratory without compromising the experiment's utility.

4.2.4.4.1 Types of Multiplex PCR

Multiplexing reactions are broadly divided into two categories:

1. *A Single Template PCR Reaction*: uses a single template that can be genomic DNA and several pairs of forward and reverse primers to amplify specific regions within a template.
2. *A Multiple Template PCR Reaction*: uses multiple templates and several primer sets in the same reaction tube. The presence of multiple primers may lead to cross-hybridization with each other and the possibility of mis priming with other templates.

4.2.4.4.2 Primer Design Parameters for Multiplex PCR

The design of specific primer sets is essential for a successful multiplex reaction. The important primer design considerations described below are key to specific amplification with high yield.

1. *Primer Length*: multiplex PCR assays involve the designing of a large number of primers. Hence it is required that the designed primer should be of appropriate length. Usually, primers of short length, in the range of 18 - 22 bases, are used.
2. *Melting Temperature*: primers with similar Tm, preferably between 55°C - 60°C, are used. Primers with a higher Tm (preferably 75°C - 80°C) are recommended for sequences with high GC content. A Tm variation of between 3° - 5° C is acceptable for primers used in a pool.
3. *Specificity*: it is crucial to consider the specificity of designed primers to the target sequences while preparing a multiplex assay since competition exists when multiple target sequences are in a single reaction vessel.

4. *Avoid Primer Dimer Formation*: the designed primers should be checked for the formation of primer-dimers, with all the primers present in the reaction mixture. Dimerization leads to unspecific amplification. All other parameters are like standard PCR.

4.2.4.4.3 Advantages and Applications of Multiplex PCR

Multiplex PCR has many applications in human genetics, linkage analysis, gene deletion analysis, template quantitation, RNA detection, mutation analysis, and forensic studies compared with the regular simple and uniplex PCRs.

One of the main problems in a simple PCR is false negatives due to reaction failure or false positives due to contamination. False negatives are often revealed in multiplex assays because each amplicon provides an internal control for the other amplified fragments. A further problem is that the expense of reagents and preparation time is less in multiplex PCR in comparison to other machines, whereas several tubes are used for uniplex PCRs. Additionally, a multiplex reaction is ideal for conserving costly polymerase and templates in short supply. The quality of the template may be determined more effectively in multiplex than in a simple PCR reaction. The exponential amplification and internal standards of multiplex PCR can assess the amount of a particular template in a sample. The number of reference templates, the number of reaction cycles, and the minimum inhibition of the theoretical doubling of product for each cycle must be accounted to quantitate templates accurately by multiplex PCR.

4.2.4.5 Digital PCR

Digital PCR is one of the third-generation PCR machines designed to meet absolute DNA or cDNA quantification demands, when they are being amplified. Here, the template is diluted to a specific concentration, and dispersed into several micro-reaction units, resulting in either a 0 or a 1 target sequence in each unit. After amplification, units consisting of target DNA or cDNA sequence copies show positive signals, denoted as 1, and the units without any target sequence are denoted as 0. The mean number and fraction of positive units will be quantified by applying a Poisson distribution, which minimizes the error rate generated by more than one copy of the target sequence in some units. The initial copy number and the final concentration of the target DNA or cDNA in each sequence unit are obtained (Debrand et al. 2015).

Template distribution in dPCR is mainly based on microwell chip, water-in-oil droplet, and microfluidic techniques. The water-in-oil droplet technique is most frequently used and is commercialized as droplet digital PCR (ddPCR). This technique distributes the template into 20,000 droplets before amplification, which provides a very high sensitivity for distinguishing the mutations in the template DNA with a detection limit of 0.001%. A flow chart concerning the methodology for ddPCR is provided in Figure 4.2.

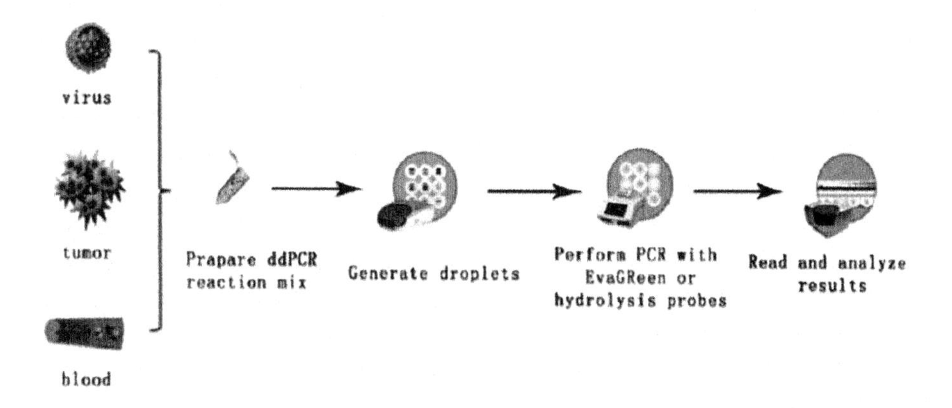

FIGURE 4.2 Flowchart determining the procedure of dd PCR.

4.2.4.5.1 Applications of Digital PCR

4.2.4.5.1.1 Detection of Mutations dPCR is being used extensively in medical applications to screen markers for different cancers, to detect pathogens, for environmental and food monitoring, for gene expression analysis, and to detect mutations. Gene mutation analysis is the most important among all of these fields. The detection of cancer-associated gene mutation comprising SNPs is more effective with dPCR when compared with the qPCR. dPCR is widely used to detect mutations in epidermal growth factor receptor (EFGR), non-small-cell lung cancer (NSCLC), KRAS mutation in colorectal cancer, and ESR1 mutation in breast cancer. ddPCR is a robust approach for detecting IDH1 mutation from a massive background of wild-type sequences in cerebrospinal fluid. ddPCR is also effective in diagnosing brain gliomas, where it is challenging to perform a biopsy. ddPCR is superior in detecting FGFR1-ITD mutations when compared to whole genome sequencing. It is a valuable tool for detecting dysembryoplastic neuroepithelial tumors (DNT) and low grade neuroepithelial tumors (LGNTs).

Gene mutation is considered to be one of the significant causes of drug resistance. Overexpression of the EGFR gene has been detected in NSCLC and EGFR tyrosine kinase inhibitors and cancer therapeutics. Mutation in EGFRT790M may induce drug resistance, and ddPCR is highly efficient in detecting the mutations in EFGR. ddPCR can also detect low mutation rates in the samples. dPCR can detect EGFR mutations at a level of 0.001%, which is far more sensitive than the sensitivity of ARMS-qPCR (1%), direct sequencing (20%), NGS (5%), and scorpion ARMS PCR (0.1%) respectively.

4.2.4.5.1.2 Prenatal Diagnosis of a Genetic Disorder Prenatal diagnosis offers excellent support to clinical genetics and gynecology, representing theoretical and clinical medicine integration. Invasive prenatal diagnosis, such as amniocentesis and chorionic villus sampling, involves certain risks that generate extensive psychological stress on parents, forcing them to follow non-invasive prenatal tests (NIPTs), based on the identification of cell-free fetal DNA (cffDNA) in maternal circulation. dPCR mainly contributes to the prenatal diagnosis of monogenic disorders. When a mother carries mutations, qualitative analysis of the concerned mutations is not sufficient to determine the mutational status of the fetus. This could be due to the difficulty in distinguishing the fetal alleles from the maternal DNA background in the presence of maternal and fetal DNA in maternal plasma. After pregnancy, the cffDNA is released into the maternal plasma DNA pool, causing a slight elevation of mutant or wild-type DNA ratio. The ddPCR will detect the subtle difference of mutant DNA in maternal plasma before and after pregnancy more precisely than other PCRs. Based on the ddPCR principle, a digital relative mutation dosage (RMD) approach has been developed to distinguish the balance between the mutant and wild-type causative genes of β thalassemia and hemophilia. Debrand and his associates in 2015 have used ddPCR to detect paternal CFTR mutations in cffDNA from the plasma of pregnant mothers. ddPCR has identified a ΔF508-MUT CFTR allele in cffDNA of all probed fetuses and excluded the unaffected fetuses with very high sensitivity and cost-effectiveness. Both regular and rare deletions in α-thalassemia were detected using ddPCR, which showed a detection limit of ~1 ng and rapid detection of α-thalassemia variants in a Malaysian population.

4.2.4.5.1.3 Detection of Mutations in Mitochondrial DNA Mutations in mitochondrial DNA (mtDNA) are associated with many pathological processes. However, detection of mutant mtDNA is still a challenge due to its heteroplasmy (coexistence of mutant and wild type mtDNA). Moreover, the diagnosis of the mitochondrial diseases remains challenging due to their high non-specific symptoms. Detection of mutations in the heteroplasmic mtDNAs with outstanding sensitivity and accuracy is the fundamental prerequisite for the early diagnosis and monitoring of disease progression. Taylor and his associates in 2014 have developed a digital deletion detection (3D) based on ddPCR, with an ability to quantify deletions in mtDNA critically. This method was also used to analyze the dynamic changes of age-related mtDNA mutations in the human brain. Through ddPCR, the heteroplasmy of human mtDNA has been validated, and the data confirmed. A combination of ddPCR with allele-specific probes and specific primers has detected G > A mutation in m.11778, responsible for Leber's disease. ddPCR meets all the requirements of data reproducibility and offers high sensitivity in detecting the heteroplasmy of mtDNA. ddPCR also contributes to the simultaneous detection and quantitative analysis of mutations in the mtDNA.

dPCR is superior in sensitivity and feasibility. Compared to qPCR, dPCR can accurately detect the copy number of target DNA, independent of Ct values with excellent sensitivity and high accuracy. dPCR can also reduce background fluorescence, which makes it less susceptible to inhibitors. These characteristics suggest the promise of the application of dPCR for the detection of rare mutations with precise quantification. Despite its high accuracy in detecting mutations, this technique has a few limitations, such as the requirement for highly allele-specific probes to reduce cross-reactivity and false positives. Large sample sizes must be used to cover the targeted mutations and maximize their application in clinical manifestations (Taylor et al.2014).

4.2.4.6 Nested PCR

Nested PCR is one of the modifications designed to enhance the sensitivity and specificity of the reaction. Nested PCR uses two sets of primers directed on the same target for running two successive PCR reactions. Here, the first set of primers is designed to anneal to the sequences located upstream to the second primer set, located internally or nested to the first primer set. The first set of primers, also known as outer primers, anneal to amplify a large fragment of the gene, as a template during the second round of PCR, further amplifies a small region of the amplicon by using the second set of primers known as inner primers or nested primers (Deepachandi et al. 2014). As the nested PCR primers will not find priming sites on any primer dimers or non-specific artifacts generated in the primary PCR, they will only anneal and amplify the specific product generated in the primary PCR, thus maintaining its specificity (Figure 4.3).

Traditionally, nested PCR is performed for many cycles using outer primers, opening the PCR reaction tube, and adding the nested PCR primers to run the nested PCR cycle. The major problem encountered with this approach is amplicon contamination, leading to the loss of specific assay. Single tube nested PCR reactions have been designed to address this issue, where both the outer and the nested PCR primer sets are added initially to the tube and an extended PCR is performed. Amplicon after PCR assay can be visualized by electrophoresing the reaction mixture on a 2% agarose gel along with an appropriate molecular weight marker.

4.2.4.6.1 Applications of Nested PCR

Nested PCR is also used for increasing the sensitivity of the PCR. The best standard way of ensuring the specificity is to sequence the PCR products and compare their sequences with the known target ones. Synthesis of non-specific products can also result in a false positive reaction product due to the non-specific binding of the primers (mispriming), which can be minimized by increasing the annealing temperature and maintaining stringent conditions. The amplification of the specific region can be detected by using Southern blot hybridization. Nested PCR is handy for suboptimal DNA samples, such as the samples isolated from tissues fixed in the formalin and those embedded in paraffin, and so forth. Nested PCRs are also very valuable in forensic sciences, and pathology for detecting microorganisms such as *M. tuberculosis*, present in low concentrations. However, this technique suffers from disadvantages such as having a high susceptibility to contamination, being time consuming, and being economically costly.

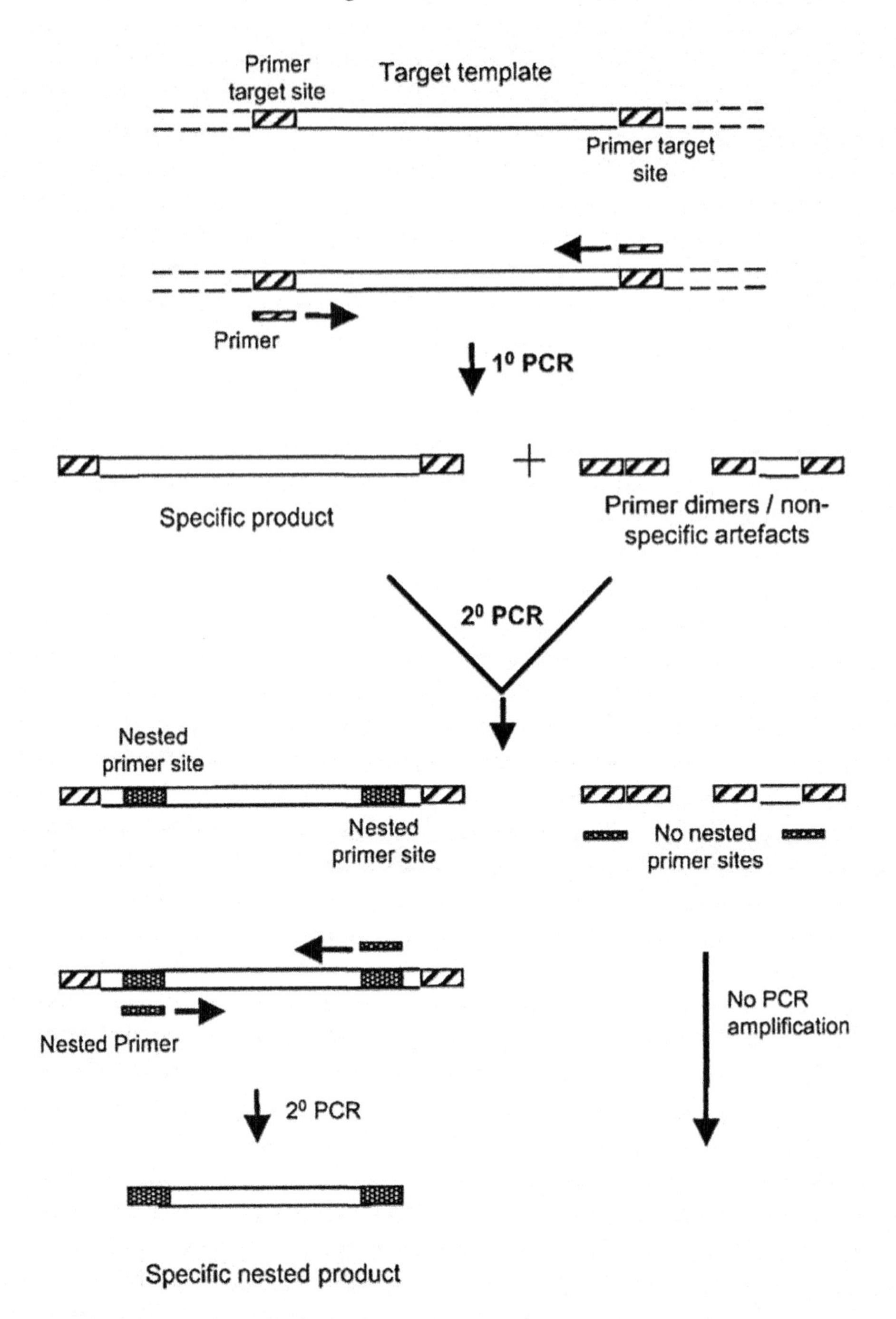

FIGURE 4.3 Flowchart determining the stages of a nested PCR.

4.2.4.7 Multiplex Ligation-Dependent Probe Amplification (MLPA)

Multiplex ligation-dependent probe amplification (MLPA) is a PCR technology that utilizes up to 45 probes. Each probe is specific for an exon of a specific gene of interest, which evaluates the relative copy number of each gene. Each probe of the MLPA is composed of two half probes, namely a 5′half probe consisting of a target-specific sequence and a 3′ half probe consisting of a universal primer sequence, which permits the simultaneous multiplex amplification of all probes at the same time. Additionally, one or two probes consist of a stuffer sequence, which provides differentiation during the electrophoretic separation of the amplified product, based on its length and size.

The MLPA reaction consists of five steps.

1. *DNA denaturation*: where DNA is denatured and is incubated with a mixture of MPPA probes.
2. *Ligation reaction*: where two half probes will recognize the contiguous target specific sequence only in the presence of a perfect match, without any gap. After hybridization, two half probes will be ligated for further amplification.
3. *PCR amplification*: where PCR amplification of the target sequences will be performed using a pair of fluorescently labeled PCR primers. Only ligated probes are amplified during the PCR reaction. The number of target sequences in the sample can be measured by counting the number of probe ligation products.
4. *Electrophoresis separation*: Amplified PCR products are separated in capillary gel electrophoresis under denaturation conditions.
5. *Data analysis*: The height or area of the PCR-derived fluorescent peaks is measured to quantify the PCR product after normalization and comparison with control DNA samples. The quality of the reaction is assessed by comparing the test samples with control samples, which provides information regarding the efficiency of the amplification and the exact amount of DNA used for the reaction. Here the MLPA assay does not amplify the target sequences but only the ligated probes using a single set of PCR primers, whereas a multiplex PCR requires specific primers for each target sequence.

Another critical aspect of MLPA assay utilization is interpreting its results for the molecular diagnosis of gene deletions or duplications. Homozygous or hemizygous deletions are evidenced by the absence of a specific peak for the target gene in the presence of a control probe, which shows regular amplification. In heterozygous deletions and duplications, the copy number variations produce a different height or different area of the relative peaks, whose interpretation is challenged. Consequently, different data analysis strategies have been developed for a correct interpretation of the data. The most widely used is Coffalyser software, an Excel-based program used for data normalization and corrections for signal slopping.

4.2.4.7.1 Applications of MLPA in the Detection of Genetic Disorders

Several Mendelian neuromuscular disorders include Duchenne muscular dystrophy, Becker muscular dystrophy, Spinal muscular atrophy, Charcot Marie Thoot disease, and hereditary neuropathy liability pressure palsies are inherited in humans due to deletions or duplications of specific genes. Most of these diseases are diagnosed by genetic testing, which identifies healthy carriers and evaluates the risk of recurrence. For this reason, the MLPA assay represents a powerful tool for the study of these different conditions.

Duchenne muscular dystrophy and Becker muscular dystrophy are X-linked genetic disorders that affect 1 in 3,500 and 1 in 18000 males. Both of the diseases occur due to mutations in the DMD gene located at Xp21.2 on the human X chromosome. 65% of Duchenne muscular dystrophy and 85% of Becker muscular dystrophy occur due to large deletions in the DMD gene. Duplications in the DMD gene are responsible for 5 - 10% of these two disorders and point mutations in the gene are responsible for 25 - 30% of the disorder. 98% of the deletions in affected males can be easily detected by using the multiplex PCR approach, which can analyze two hotspot regions located on exons 2 - 20 and 44 - 53, respectively. This approach will not detect heterozygous deletions in female carriers, crucial in determining the disease's recurrence risk within that family and a chance of preventing the birth of the affected children. About one-third of Duchenne muscular dystrophy is due to de novo mutations in the male children whose mothers are not healthy carriers and who have a low risk of disease recurrence. Moreover, multiplex PCR cannot detect duplications in the DMD gene of both affected males and female carriers. MLPA analysis can detect mutations in all exons of the DMD gene and several other control probes on sex chromosomes and autosomes to detect Duchenne muscular dystrophy and Becker muscular dystrophy both in affected male patients and female carriers. MLPA can analyze all exons of the DMD gene with high sensitivity, specificity, and quick identification of rearrangement breakpoints.

Spinal muscular atrophy (SMA) is a neuromuscular disorder characterized by symmetric proximal muscle weakness due to degeneration of the anterior horn cells of the spinal cord. This disorder, inherited as an autosomal recessive trait, has a prevalence of 1 per every 10,000 and a genetic frequency of 1 per 50. This disorder is classified as SMA I, II, and III according to the severity of symptoms. All three types are reported to occur due to homozygous mutations in survival motor neuron 1 (SMN1) gene located on the 5q13 locus of chromosome 5. 95% of this disorder occurs due to the functional absence of this gene, deletion, or its conversion as SMN2. SMA is diagnosed by using a PCR-RFLP, which detects a homozygous SMN1 loss. However, this method does not detect heterozygous SMN1 loss. Thus, it cannot be used to identify healthy carriers. MLPA assay for the molecular diagnosis of SMA is based on several probes for the SMA specific region, including specific probes for SMN1 and SMN2 genes. These probes will hybridize both the genes and other probes for sequences mapped within the SMA-specific region or other autosomal regions. Due to this highly specific set of probes, MLPA assay for SMA can detect the copy number of both SMN1 and SMN2 genes. Therefore, both homozygous and heterozygous SMN1 deletions and conversions to SMN2 can be detected, allowing the diagnosis of affected patients or healthy carriers.

4.2.4.7.2 Applications of MLPA in Prenatal Diagnosis

Prenatal diagnosis relies on the isolation, and culture of chorionic villi (CV) cells or amniotic fluid cells during pregnancy, followed by their chromosome investigation. It is one of the most standard assays developed to detect genetic alterations in the developing fetus. However, this analysis suffers from significant limitations, including the risk of abortion due to villocentesis or amniocentesis, and the time required for the culture and analysis of samples. Techniques employed to detect common aneuploidies on uncultured chorionic villi or amniocytes, such as FISH or QF-PCR, are expected to provide the first result within 24 - 48 hours, followed by the conventional karyotyping on the cultured cells. In recent years, these conventional methods were replaced by MLPA for the aneuploid screening of chromosomes 13, 18, 21, X, and Y, respectively. Sex determination using MLPA was also determined to be 100% accurate. MLPA is considered a rapid, flexible, sensitive, and robust test to detect prenatal aneuploidy.

4.2.4.7.3 MS-MLPA for the Detection of DNA Methylation

MLPA is resourceful in the identification of epigenetic alterations such as DNA methylation of specific genes. DNA methylation within the CpG islands of a promoter region is associated with its transcriptional silencing and is involved in several cellular processes, such as genomic imprinting, X chromosome inactivation, DNA repair, and so on. An aberrant DNA methylation of imprinted genes is associated with several inherited human diseases. Somatic "de novo" methylation of CpG islands in tumor suppressor genes has been implicated in tumorigenesis. Modified MLPA, known as methylation specific MLPA (MS-MLPA), has been developed for detecting epigenetic alterations. In this modified assay, the sequence targeted by specific probes contains a restriction site for the HhaI to recognize the unmethylated GCGC sequence. After hybridization, the annealed probe mixture is treated with HhaI endonuclease, which cleaves the probes hybridized to unmethylated DNA, leaving the undigested probes hybridized onto the methylated DNA.

Consequently, only the latter probes are amplified by PCR. Comparison of the peak size of a sample and control of methylated probes will provide information regarding methylation levels in the specific DNA regions targeted by the probes. Because of this ability, MS-MLPA is considered to be a golden technique for diagnosing several disorders due to abnormal DNA methylation. One of the most widely used applications of MS-MLPA is in the diagnosis of Prader Willi syndrome, Angelman syndrome, and genetic disorders that occur due to alterations in human genome imprinting. These diseases are linked to the 15q11-q13 locus of chromosome 15, involved in the paternal allele for Prader Willi syndrome and a maternal allele for Angelman syndrome. Prader Willi syndrome affects 1 in every 15,000 births and is characterized by hypotonia, hypogonadism, mental retardation, feeding difficulties in infancy, and excessive eating in the later stages of childhood. Deletion of the paternal allele mainly causes 70% of the Prader Willi syndrome in the chromosomal region 15q11-q13, 25% due to the maternal uniparental disomy (UPD) of chromosome 15, and 5% due to the abnormalities in the imprinting center sequence on the 15q11-q13 locus.

Even though MLPA is considered to be one of the efficient methods of diagnosing a genetic disorder, offering a rapid means of scanning up to 40 loci for gene dosage, it suffers from some limitations. MLPA works with 20 ng of the DNA sample. Nevertheless, 100 - 200 ng of the sample is the basic requirement for obtaining reliable and reproducible results. MLPA shows sensitivity to the type of sample used for DNA extraction, such as blood or buccal swab. MLPA is highly sensitive to contaminants such as small remnants of phenol and DNA degradation than conventional PCR. One of the crucial applications of MLPA is the detection of deletions involving a single exon. The deletion can change its exon sequence, which might hamper the stringency of hybridization with the specific probe. This sequence variation can be a point mutation or a polymorphism without affecting its gene function. These single exon deletions detected using MLPA should be further analyzed by an independent method.

4.2.5 Electrophoresis Based Methods

Technological developments for the identification and analysis of mutations are rapidly progressing. Over the past few years, the necessity of screening amplified DNA products to identify mutations or polymorphisms has gained increasing momentum in population genetics and molecular medicine to diagnose various genetic disorders. Mutation analysis refers to the identification of changes in DNA responsible for a disease or dysfunction. Detecting mutations either in DNA or in RNA requires physical mapping, cloning, or amplification of DNA, whereas screening methods to identify changes in DNA involves electrophoretic separation of DNA fragments followed by their sequencing. Various types of electrophoretic separation techniques used in human genetics to detect mutations are discussed in this section.

4.2.5.1 Confirmation Sensitive Gel Electrophoresis

Single-base changes are the mutations that occur most frequently in the human genome, leading to genetic disorders. However, these mutations occur in large, complex genes. Most of them are private in the sense that many unrelated individuals might consist of one out of several hundred different mutations in the same gene producing similar disease phenotypes. Detection of these mutations is a formidable technical challenge. One of the successful techniques is confirmation sensitive gel electrophoresis (CSGE), which offers a rapid and efficient mechanism for detecting such mutations. This method was developed by Ganguly and his associates in 1989 as a screening method to minimize DNA sequencing when screening large DNA fragments and complex genes for mutations. This method was initially developed to rapidly analyze multiple genes associated with collagen disorders, and has been highly sensitive in analyzing an extensive range of genetic disorders. CSGE was also proven to be highly useful for detecting sequence variations in cancer susceptible genes. This method is currently applied to detect single base changes such as insertions or deletions in a DNA sequence amplified by PCR. CSGE specifically detects the heterozygous changes in the sample DNA. CSGE is based on its ability to distinguish between homoduplex and heteroduplex DNA fragments when subjected to electrophoresis under denatured conditions. Homoduplex DNA consists of double stranded DNA fragments consisting of all bases paired correctly with their complementary bases on the opposite strand. Heteroduplex DNA consists of mismatched bases. PCR amplification of the DNA isolated from patients with a heterozygous mutation can form homoduplexes and heteroduplexes when double-stranded DNA dissociates and then re-anneals with the complementary strand coming from a different allele. The mismatched bases induce conformational changes in the heteroduplex DNA, as misaligned bases will not obey the Crick-Watson base pairing rules.

The methodology for CSGE is displayed in Figure 4.4, which involves generating heteroduplexes and or homoduplexes in PCR products through heating or slow annealing, followed by electrophoretic separation on a large polyacrylamide gel, crosslinked with BAP, which enhances the gel

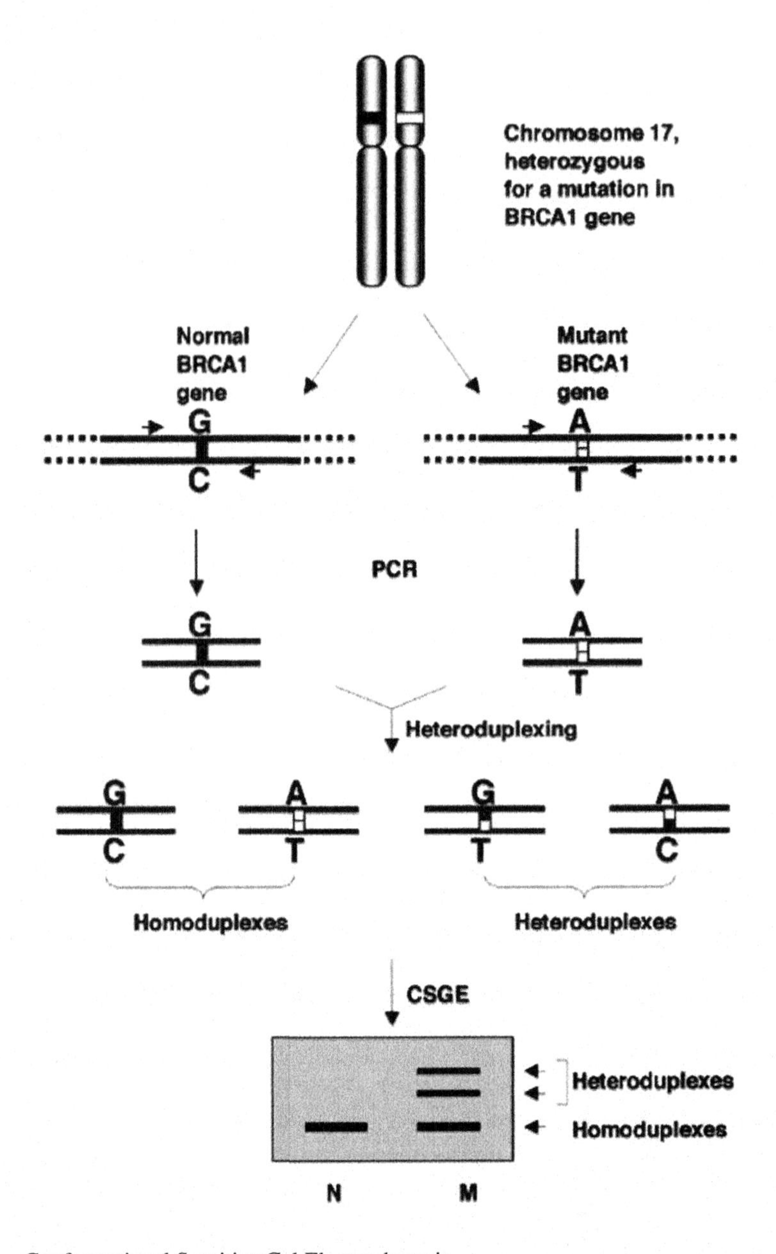

FIGURE 4.4 Conformational Sensitive Gel Electrophoresis.

strength and improves the conductivity. The gel also consists of ethylene glycol and formamide for mild denaturation. For the detection of homozygous changes, the PCR product will be mixed with a previously sequenced Y-linked sample at a 1:1 concentration. The PCR product is mixed with an X-linked sample at a 2:1 concentration of test and control to detect the hemizygous changes. Under these conditions, heteroduplexes can be separated from homoduplexes, which migrate more slowly through the gel matrix. Bands are visualized by staining the gel with ethidium bromide. Multiple bands are visible in the samples containing the heterozygous mutations, and a single band will be visible in samples without any sequence variation. CSGE provides information regarding the change in the sequence or the nature of the sequence variation, which can be further identified by sequencing the fragment. Conformational sensitive gel electrophoresis is a simple and efficient method for the separation of fragments up to a size of 450 bp. Detection does not require any radioactivity or other sources. Optimization of running conditions is also effortless with CSGE. Several samples of varying sizes can be loaded in a single lane, which is impossible with other techniques due to multiple bands in the samples. Using CSGE, one can detect 60 - 73 single base changes or mismatches in the PCR products ranging from 200 - 800 bp.

4.2.5.2 Denaturation Gradient Gel Electrophoresis

Denaturing gradient gel electrophoresis (DGGE) is a technique developed to separate DNA fragments after their PCR amplification. It facilitates the separation of DNA strands based on their base composition or the ratio of GC to AT base pairs that make up a particular segment. PCR products of a reaction are of similar size. Hence, the conventional separation on agarose gel results in a single DNA band, largely non-descriptive. DGGE can overcome this limitation by separating the PCR products based on their sequence differences, resulting in differential denaturing DNA characteristics. DNA is exposed to a denaturant gradient at elevated temperatures within a polyacrylamide gel. As the movement of the DNA sample progresses through the gel, from a low denaturant concentration to a higher one, it starts to melt at varying points. This is akin to the DNA "unzipping." The higher the GC content of the sample, the harder it is to melt. Thus, the DNA sample can progress further into the gel before stopping. Samples with lower GC content melt more rapidly in comparison. Therefore, they progress more slowly within the gel, thus becoming separated from the other fast-moving strands of DNA.

DGGE technique relies on the difference in the stability of G-C pairing with 3 hydrogen bonds instead of A-T pairing with 2 hydrogen bonds. A mixture containing DNA fragments of different sequences will be separated on an acrylamide gel with a linearly increased gradient of DNA denaturants. DNA fragments richer in GC will be more stable and remain double-stranded until they reach higher denaturation concentrations. Double-stranded DNA fragments migrate faster in the acrylamide gel, while denatured DNA fragments will either slow down or stop in the gel. In this manner, DNA fragments of different sequences separate on an acrylamide gel. DGGE gels are stained with DNA binding fluorescent dyes, such as SYBR Green, and are visualized under fluorescent light. Alternatively, they can also be stained with ethidium bromide and visualized under UV light. Samples on different gels are compared by using known standards. Ideally, a single band on the gel corresponds to one species. Therefore, the number of bands on the gel provides an idea of the diversity of the sample. DGGE is commonly performed for partial 16S rRNA but is also used for studying the differences in functional genes. The gene fragments can be eluted from the gel and can be further subjected to sequencing.

4.2.5.2.1 Applications of DGGE

There are advantages to using the DGGE technique.

1. It is very sensitive to variations in the DNA sequence.
2. It allows simultaneous analysis of multiple samples.

3. It is a useful method for monitoring shifts in community structure over time.
4. Community profiles can be analyzed with cluster analysis.
5. The use of universal primers allows the analyses of microbial communities without prior knowledge of the species.
6. The fragments separated by DGGE can be excised, cloned, and sequenced for identification.
7. It is possible to identify constituents that represent only 1% of the total community.
8. This technique can be applied to phylogenetic and functional genes (Muyzer and Smalla, 1998).

This technique also suffers from certain disadvantages.

1. DGGE analysis is time consuming.
2. DGGE analysis suffers from the same drawbacks as all PCR-based community analysis techniques, including biases from DNA extraction and amplification.
3. The variation in 16S rRNA gene copy number in different microbes makes this technique only semi-quantitative.
4. Microheterogeneity in rRNA encoding genes present in some species may result in multiple bands for a single species and, subsequently, overestimating community diversity.
5. Heteroduplexes can cause biases to the observed diversity.
6. No method for automated analyses is currently available.
7. It only works well with short fragments (<600 bp), thus limiting phylogenetic characterization.
8. Gels of complex communities may look smeared due to many bands.
9. Band position does not provide reproducible taxonomic information.
10. Results are challenging to reproduce between gels and laboratories.

4.2.5.3 Temporal Temperature Gradient Gel Electrophoresis (TTGE)

Temporal temperature gradient gel electrophoresis (TGGE) is an effective and sensitive technique to identify and characterize genetic polymorphisms of nuclear and mitochondrial genomes. Yoshino first introduced TGGE in 1991. TGGE and DGGE techniques have the same working principle, based on the melting behavior of DNA molecules. Compared to DGGE, TGGE can function even without using a chemical denaturing gradient. PCR-based TTGE is a non-radioactive-based technique. After detecting a mutation, the PCR product can be further processed for its subsequent sequencing analysis to identify the mutation. TGGE does not require specific GC clamped primers and can be efficiently applied to large PCR products of 1 kb. As the gradience in this technique is produced by the temperature, the pouring of a chemical gradient gel is eliminated.

TTGE uses an acrylamide gel that consists of a chemical denaturant such as urea, maintained at a uniform concentration. Here, the gradient will be generated by gradually and uniformly increasing the temperature by 0.5 - 3°C/hour during electrophoresis, resulting in a linear temperature gradience over the time course of the electrophoretic run. A denaturing environment is formed in the gel by maintaining a constant denaturant concentration in the gel with the temporal temperature gradient. As the denaturant is temperature, it is easy to modulate during electrophoresis to achieve a broader separation range, resulting in much higher detection sensitivity. TGGE can effectively detect single nucleotide changes, small size insertions, or deletions in the sample DNA. Even though the detection rate is not 100%, it is considered to be one of the most sensitive detection methods for identification of mutations and is one of the most accessible techniques. Further, the detection rate can reach 100% by optimizing gel running conditions such as temperature range, increment rate, percentage of the gel, and its running time.

The PCR-based TTGE mutation detection system is one of the most powerful methods available for detecting heterozygous nucleotide variants. This method differentiates the control and mutant DNA sequence melting behavior in linearly increasing temporal temperature gradience generated during the electroporation. Denaturation and re-annealing of the PCR products lead to the formation of homoduplexes along with the hybrid heteroduplex molecules. Four bands in a heterozygous mutation under optimal separation conditions are the result of two homoduplexes and two heteroduplexes, whereas a single band shifted either up or down is observed during the separation of a DNA fragment with homozygous mutation.

Using TTGE, 92 - 98% of all mutations in the 27 exons of the cystic fibrosis transmembrane regulator (CFTR) gene have been detected (Ozgul et al.2005). The specificity of this technique was estimated to be 99%, having a false negative rate of 1% and a false positive rate of 0%. DNA fragments with similar melting behavior can be separated under the same conditions using TTGE, which is time and cost effective. TTGE technique has also been applied to detect unknown mutations in nuclear genes such as CFTR, COL2A1, TP, FGFR2, and mitochondrial genes. TTGE has many other applications and cost-based advantages.

4.2.5.4 Single-Strand Conformational Polymorphism (SSCP)

SSCP is applied for the detection of mutations. Described for the first time by Orita (1989) and Hongyo (1993), this technique is used in many laboratories to detect mutations responsible for human genetic disorders. PCR will amplify the region of interest in the target DNA sample, and the resultant DNA is separated into single-stranded molecules on a non-denaturing polyacrylamide gel by electrophoresis. Single stranded DNA folds differently from other strands even if a single base change and mutation-induced changes in the tertiary structure of DNA have different mobility respectively for two single strands. Mutations responsible for changes in the DNA strands are detected as they appear as new bands on autoradiograms detected using radioactive probes. The tertiary structure of DNA will also change under varying physical conditions such as temperature and ionic environment. The operation of SSCP involves four steps.

1. *Amplification and labeling of DNA*: Here, 10 µl of reaction mixture contains 1 µg of DNA, 5 pM of each primer, 0.5 units of Taq polymerase, each dDNTP at 0.2 mM, 10 mM Tris-HCl (pH 8.3), 50 mM KC1, 1.5 mM $MgC1_2$, and 0.01% gelatin. DNA is labeled by adding 0.5 µCi of ^{32}p dCTP (3000 Ci/mmole) to the PCR mixture. PCR conditions are 30 cycles of 94°C for 30 sec, 55°C for 30 sec, and 72°C for 30 sec.
2. *Denaturation of the PCR product*: The amplified fragments are mixed with an equal volume of sample buffer comprising 95% formamide, 20 mM EDTA (pH 8.0), 0.05% xylene cyanol, and 0.05% bromophenol blue. The sample mixture is denatured at 85°C for 5 minutes and then cooled on ice.
3. *Gel conditions*: A 100 ml 5% non-denaturing gel is prepared, containing 17 ml of 30% polyacrylamide in lx TBE and cast in standard sequencing gel plates using 0.4 mm spacers and a shark's tooth comb. The casted gel is cooled in a cold room for at least 30 minutes before sample loading. 2 µl of denatured DNA are loaded and electrophoresed at 4°C at a constant voltage of 500 V for 12 - 15 hrs. in lx TBE electrophoresis buffer. After electrophoresis, the gel is transferred onto a sheet of Whatman 3MM paper and vacuum-dried before autoradiography.

4.2.5.4.1 SSCP Modifications

SSCP is modified into two subtypes.

1. *RNA-SSCP*: single-stranded RNA has an enormous repertoire of secondary structures as RNA base pairing is relatively more stable than both DNA-DNA and RNA-DNA base pairing and is expected to adopt a more conformational structure and hence be more sensitive to sequence changes. The efficiency of rSSCP is 70% compared with a 35% efficiency of SSCP as identified during the detection of factor IX gene of size 2.6 Kb.

2. *Restriction endonuclease fingerprinting-SSCP*: a modification of SSCP where a 1 kb segment is digested with 5 restriction enzymes designed to produce fragments of ~150 bp each. After digestion, these products are mixed, end-labeled with ^{32}P, denatured, and electrophoresed under non-denaturing conditions. Two 'components' are evident during this experiment, gain of restriction site and, loss of restriction site. Here, the efficiency of detection is 96%.

4.2.5.4.2 Applications

Detection of mutations for PCR-SSCP is more than 80% in a single run for fragments shorter than 300 bp. The sensitivity of PCR-SSCP decreases with an increase in the fragment length, with <300 bp being the optimum. Overlapping short primer sets are used to detect mutations in longer fragments, such as exons of size >300 bp and whole cDNAs. Alternatively, long PCR products are digested with appropriate restriction enzymes before SSCP. SSCP has been used most extensively to screen inherited mutations or somatic mutations in cancer cells. Due to its sensitivity in detecting single-base changes, SSCP has been used to search for polymorphisms in cloned or amplified DNA for its utilization as genetic markers (Gasser et al. 2006).

Despite its applications in the detection of mutations, this technique has certain limitations. Firstly, SSCP screening only tells whether a mutation exists or not. The mutation can be further confirmed by sequencing the subsequent DNA to determine the nature of the mutation responsible for a shift in the electrophoretic mobility of the sample. Secondly, not all point mutations in a given sequence will generate detectable changes in their electrophoretic mobility. However, the sensitivity of the SSCP can be further enhanced if the PCR conditions and gel running conditions are optimized. Several parameters such as type of mutation, size of DNA fragment, G C content of the fragment, percentage, and composition of the polyacrylamide gel, its size and potential, gel temperature during electrophoresis concentration of the DNA sample, running time of the electrophoresis, composition of the buffer, its ionic strength, and pH, and buffer additives, such as glycerol or sucrose also control the sensitivity of SSCP.

4.2.6 DNA Typing and Testing

Genetic identity testing involves identifying specific patterns in the genetic material namely, DNA, which is unique to every human being. The size of the haploid human genome is around 3 billion bp, consisting of ~35,000 genes. The coding region comprises only 10%, and non-coding and regulatory regions comprise most of the human genome. Genetic variation is limited in the coding region of the genome except for the HLA region. Any selection pressure does not control non-coding regions of the genome. Hence, mutations generated in the non-coding region will be transmitted to the offspring, leading to a tremendous increase in genetic variability. These regions are highly appropriate for forensic analysis in the genetic identification of humans. 30% of the non-coding region in the human genome consists of repetitive sequences, divided into two classes, tandem repetitive sequences and interspersed sequences. Tandem repetitive sequences exist as mini-satellite and microsatellite sequences, also known as short tandem repeats (STR). Most forensic testing is based on genetic loci with tandem repetitive sequences.

Mini-satellites are composed of sequence motifs with various sizes ranging from 15 - 50 bp, reiterated tandemly for a total length of 500 - 20 Kb. Microsatellite repeats range from 2 bp to 6 bp, reiterated for a total length of 50 - 500 bp. Mini-satellites are in the sub-telomeric regions of chromosomes, whereas microsatellites are distributed throughout the genome, with one repeat per every 6 - 10 Kb frequency of occurrence. The origin of the variability differs between micro and mini-satellite repeats. Unequal crossing over and gene conversion play a significant role in the variability of minisatellite repeats and replication slippage is mainly responsible for the origin of variability in microsatellite repeats.

The utilization of Y-chromosome-specific polymorphisms is used frequently in forensic analysis. One of the main applications of Y polymorphisms is deficiency paternity testing, especially when a male offspring is in question for criminal casework. Y polymorphisms are crucial for analyzing the male DNA fraction in the stains of male/female mixtures, which is the most common biological material available in crimes. Y-chromosome-specific markers are needed if the preferential sperm DNA isolation fails (which happens in 50% of the cases) and in rapes committed by azoospermic individuals (complete absence of sperm in the ejaculate, resulting in infertility but not sterility). The non-pseudoautosomal region of the Y-chromosome bears multiple types of polymorphisms such as biallelic markers, SNPs, mini and microsatellite repeats. Among the most used microsatellite repeats of the Y chromosome are trinucleotide repeat DYS392, tetranucleotide repeats DYS19, DYS385, DYS389-I, DYS389-II, DYS390, DYS391, DYS393, and recently added Y-microsatellite multiplexes, respectively.

Microsatellite repeats on the X-chromosome are used as markers for evaluating polymorphisms in mitochondrial DNA. Analysis of variations in the mitochondrial DNA control region is an efficient method for studying and comparing bones, old and degraded DNA, and in the analysis of telogenic hair effluvium (temporary hair fall). mtDNA analysis is considered to provide valid evidence in forensic genetics, which courts accept worldwide. However, mutation rate and heteroplasmy (presence of more than one type of organellar genomes within a cell or an individual) make the interpretation difficult.

STR analysis also plays a pivotal role in allogeneic bone marrow transplantation to treat hematological malignancies. During the reconstitution of the recipient's bone marrow with donor's stem cells by complete engraftment, a possibility of either rejection or relapse of the patient's original disease can occur, necessitating the recipient's peripheral blood and bone marrow. Microsatellite repeat analysis has amplified 9 microsatellite loci and the amelogenin gene located on X and Y chromosomes. To assess post-transplant engraftment status, microsatellite repeat profiling of the donor and recipient will be established before the transplantation. PCR analysis of the microsatellite repeat loci highlights the differences among pre-transplant donor and recipient leucocytes. Loci between the donor and the recipient are considered vital sources and monitored after transplantation to estimate the engraftment status. Microsatellite loci profiling post transplantation will be compared with that of the pre-transplantation profiling. The presence of only the donor's loci indicates complete engraftment, and mixed chimerism reveals both donor and recipient loci indicating partial transplantation. Microsatellite profiling can be the best support for the early diagnosis of chimerism in liver transplantation.

The utilization of identity testing has begun in forensic sciences with the analysis of blood groups, followed by identifying new markers. Based on the variations in serum proteins, RBC enzymes eventually lead to the utilization of the human leukocyte antigen system for fingerprinting. Sir Alec Jeffreys, a Professor at the University of Leicester, UK, was the pioneer of DNA fingerprinting. Professor Jeffreys identified the variations in the minisatellite repeats between two individuals, which could be used as highly informative genetic markers. His group developed a radioactive probe composed of short sequences that could latch on to the repetitive sequences and identify unique patterns for every individual, named DNA fingerprints. The steps involved in performing DNA fingerprinting are:

1. DNA is extracted from the sample (blood, semen, hair, and the like).
2. DNA is digested with restriction enzymes into tiny fragments that are different between two individuals.
3. These fragments are separated using electrophoresis on an agarose gel and are visualized by staining with ethidium bromide.
4. A Southern blot can also be performed by transferring the DNA onto the membrane, detecting the pattern using a radiolabeled probe, and exposing the membrane to an X-ray film. The

result of this fingerprinting is a pattern of 15 - 20 DNA bands different from each other, which appears like a bar code that is unique to every individual. However, the mother and father of a family will have their specific pattern. Their child is composed of both due to the inheritance of an allele, each from the father and mother. This composition can be applied in the forensic sciences for paternity identification.

4.2.6.1 Case study 1: Immigration Dispute and Paternity Testing A family was comprised of a mother, father, and their four children came to the United Kingdom and became its citizens. Eventually, one of their sons went back to Ghana. While returning to the UK, he was stopped because he had a forged passport. Their family lawyer contacted Professor Jeffreys to confirm that the boy was their son, not the mother's nephew, as she revealed sisters in Ghana. The absence of a father further complicated the situation. DNA samples were collected from the mother and son whose identity was disputed and the other three undisputed children. DNA patterns confirmed the relationship between the mother and the son in question, and the testing has confirmed that all four children had the same father. This immigration case opened new doors for the utilization of DNA fingerprinting in forensic cases and the determination of individual identity.

In 1986, the DNA fingerprinting technique was further refined by the addition of PCR technology to amplify microsatellite repeat sequences. Since then, microsatellite repeats have become the choice for forensic identification and analysis. Dinucleotide repeats are the most common microsatellite repeats present in the human genome, used for linkage analysis. For forensic analysis, tetra and pentanucleotide repeats are more suitable as they are less prone to slippage. Regarding their structure, microsatellite repeats range from most uncomplicated to highly complex. Simple microsatellite repeats have low mutation rate and are easy to standardize. Complex microsatellite repeats are hyper-variable. The selection of an ideal microsatellite repeat mainly depends on its size. In general, short sized repeats are preferred over the longer sizes as the short-sized fragments are amplified easily when isolated from degraded and decaying tissues.

Another vital factor is amplifying the multiple microsatellite repeats loci in a single multiplex reaction, coupled with their detection on polyacrylamide gels, making the microsatellite repeats profiling amenable to automation. Microsatellite repeats after amplification were initially analyzed on denaturing polyacrylamide gels, generating many variations (A-T rich and G-C rich), making the DNA typing prone to errors. The introduction of fluorescent-based technology and DNA sequencers has allowed the typing of large multiplex systems (more than 10) and the automation of DNA typing. The utilization of intelligent systems has further enhanced the quality of interpretation. Multiplex microsatellite repeat reaction units such as SGM Plus of Applied Biosystems, comprising of 10 systems to determine 10 loci, HUMFIBRA/FGA, HUMVWFA, HUMTH01, D18S51, D21S11, D6S477, D8S1179, D16S539, D19S433, and amelogenin are commercially available. Promega multiplex systems are in use for manual electrophoretic systems or monochromatic sequencing platforms. The latest highly discriminative 15-plex systems such as Powerpplex-16 from Promega and the identifier of Applied Biosystems are commonly used in forensic laboratories.

4.2.7 HLA Typing

Human leukocyte antigen (HLA) typing matches the recipients and donors for bone marrow and hematopoietic stem cell transplantation. HLA is the protein synthesized by most cells in the human body. The human immune system uses these proteins as markers for recognizing whether cells belong to its body or not. HLAs are inherited from parents to offspring, with one half from the mother and the other half from the father. Hence, brother and sister sharing the same parents have a 25% HLA match. Extended family members are very much unlikely to be close to HLA matches. In most cases (7 out of 10), patients who need a transplant will not have a fully matched donor in their family. An ideal match for the transplantation can be found by using HLA typing. Research has found that a donor must match a minimum of 6 HLA markers. Some HLA types are even found in certain racial and ethnic groups with high frequency (Sheldon and Poluton, 2006).

Superlocus 6p21, corresponding to the human major histocompatibility complex (MHC), also known as HLA, is located on the short arm of chromosome 6. HLA gene produces globular glycoproteins, each composed of two non-covalently linked chains. These proteins are ligand molecules, cell surface receptors, and other factors involved in the inflammatory response, recognition, processing, and presentation of foreign antigens to T cells as part of the adaptive immune response and the generation of innate immunity. In addition to the protein coding genes, MHC loci also contain pseudogenes, retrotransposon, and regulatory elements. HLA system comprises of ~ 220 genes, with 21 being of significant interest. These genes are located on 6p21.3 of chromosome 6 short arm, and their protein products mediate the response of humans to infectious disease and influence the outcome of cell and organ transplants. Human MHC loci are divided into three distinct regions class I, class II and class III. The class I region consists of genes encoding for HLA class I molecules HLAA, HLA-B, HLA-C, and non-classical HLA-E, HLA-F, HLA-G, and the class I-like molecules MIC-A and MIC-B. Class I loci are expressed on the surface of almost all nucleated cells, responsible for presenting intracellular derived peptides to CD8+ T cells. The Class II region consists of genes that encode for HLA class II molecules, HLA-DR (DRA, DRB1, DRB3, DRB4, or DRB5), HLA-DQ (DQA1, DQB1), and HLA-DP (DPA1, DPB1) molecules. These are expressed in professional antigen presenting cells (APCs), such as macrophages, dendritic cells, and B lymphocytes, to present extracellular derived peptides to CD4+ T cells. The location of Class III is between Class I and Class II. Class III contains non-HLA genes with immune function, such as complement components (C2, C4, factor B), cytokines, tumor necrosis factor (TNF), lymphotoxins, and heat shock proteins. HLA is one of the highly polymorphic genes in the human genome. HLAs corresponding to class I present peptides from inside the cell, produced from digested proteins broken down in the proteasomes. These peptides are small polymers of 9 amino acids in length. Foreign antigens attract killer T-cells (also called CD8+ or cytotoxic T-cells) that destroy cells. Class II presents antigens from outside the cell to T-lymphocytes. These antigens stimulate the multiplication of T-helper cells, which in turn stimulate antibody-producing B-cells to produce antibodies to that specific antigen. Suppressor T-cells suppress self-antigens. HLAs corresponding to Class III encode components of the complement system.

The diversity of HLAs in the human population is one of the critical aspects of defense against disease. As a result, the chance of two unrelated individuals with identical HLA molecules on all loci is meager. HLA genes can transplant organs between two HLA-similar individuals successfully. HLA complex is also responsible for the rejection of a transplanted organ or tissue. 6 HLA types are essential for stem cell transplants. For a bone marrow transplantation, the patient and donor must match at all 6 (100% match), whereas a cord blood transplant needs only 4 out of 6 (67%) between donor and patient. Cord blood donations are so important to help patients from minority or mixed racial backgrounds. Public banks always measure the HLA type of cord blood, and then the type is listed on a registry for search by the patients seeking a transplant. Family banks typically do not measure the HLA type at the banking time because it is an expensive lab test and can be done later from a testing segment of the stored cells.

HLA typing identifies the unique constellation of HLA antigens of an individual. Tests for HLA-class I (A, B, C) and class II (DR, DQ, DP) are performed by DNA-based molecular diagnostic techniques. HLA genes consist of 5 - 8 exons with varying lengths ranging from 4 to 17 kb. High-resolution four-digit HLA typing technologies such as sequence-based techniques (SBT) or probe-based hybridization techniques focus on deciphering the sequence of the antigen-binding groove due to the high cost of complete HLA genotyping and the limited time interval before HSCT. Before these techniques came to light, less informative serology methods were acquired for HLA typing. HLA typing by DNA technologies provides more robust, accurate testing that is reliable in resolving allele-level differences in HLA genes that serology cannot detect. Several approaches to HLA typing are used, offering a range of typing resolution levels from low (antigen-level) to high (allele-level). Tests used to identify HLA types rely on amplifying limited stretches of genomic DNA within the HLA genes. The genetic polymorphisms associated with the different HLA alleles are identified through hybridization with specific amplification primers (SSP) or probes (SSO). Next generation sequencing technologies constitute various strategies relying on a combination of template preparation, sequencing, and imaging followed by in silico genome alignment and assembly methods. Many companies are involved in building the fastest, least expensive, and more accurate sequencing systems, along with user friendly analyses, as the massive amount of data extracted from these machines, requires extensive bioinformatics knowledge. The most widely utilized sequencers for HLA typing are 454 GS FLX Titanium, 454 GS Junior of Illumina, and Roche 454. Next-generation sequencing technology offers many advantages. It allows setting the phase of linked polymorphisms within the amplicons produced during the first steps of the technique, which helps to determine the alleles and the identified groups of variants. All NGS systems share a higher level of coverage, leading to increased accuracy.

4.3 CYTOGENETIC TECHNIQUES

Cytogenetics studies changes in the chromosomal structure and properties, their behavior during mitosis and meiosis, their influence on the phenotype, and factors responsible for chromosomal changes. After Tijo and Leven reported the human chromosomes and their correct number in 1955, several studies were made to understand their abnormalities, such as monosomy of X and XXY, trisomy of the 21st chromosome leading to genetic disorders. These studies have paved the path for the continuous discovery of new techniques, improvements of existing techniques, or new combinations of well-established techniques, which lead to rapid progress in biosciences. Chromosomal abnormalities can be studied as classical cytogenetics and molecular cytogenetics.

4.3.1 Classical Cytogenetic Techniques

Traditionally, chromosomal aberrations were studied in human cells by banding techniques such as Q-banding, G-banding, R-banding, C-banding, T-banding, and banding of the nucleolar organizing region, high resolution banding, which allows precise identification of rearrangements in chromosomes. Casperson et al. in 1970 discovered the first chromosomal banding technique by staining the chromosome with a fluorochrome such as quinacrine mustard or quinacrine dihydrochloride and further examination using a fluorescent microscope, which produces bright and dark bands of chromosomes. This banding was further modified using antibiotics such as anthracyclins, which produce a similar banding pattern with more stability. G-bands are produced by staining the chromosomes with Giemsa stain, which produces a banding pattern like Q-banding with great resolution and high accuracy. However, G banding does not require fluorescent microscopy and allows permanent preparations. R-bands are the reverse of G-bands, produced by the thermal denaturation of chromosomes at a temperature of 87°C in Earle's balanced salt solution. As the staining ability of the chromosomes is lost due to heating, the usage of phase contrast microscopy provides a better contrast for the chromosomes with high resolution. Regions in the chromosomes with polymorphism can be visualized with C-banding, primarily localized in the acrocentric human chromosomes 1, 9, 16, and the distal portion of the human Y-chromosome. C-banding is used to study the chromosomes with multiple centromeres, molar pregnancies (abnormal pregnancy occurs due to a non-viable fertilized egg implanted in the uterus), true hermaphroditism, and distinguishing the donor and recipient cells during bone marrow transplantation. Telomeres of the chromosomes are stained by using T-banding.

Sometimes, combinations of banding techniques also help in obtaining the information necessary for chromosomal analysis. Scheres in 1974 developed CT banding to stain both centromeric heterochromatin and telomeric regions of chromosomes simultaneously by treating the slides with barium hydroxide. Chamia and Ruffie in 1976 further developed CT staining by incubating the slides in Hank's balanced salt solution and obtained complete C-T bands. NOR-banding is used to stain nucleolar organizing regions (NOR) of chromosomes located in the satellite stalks of acrocentric chromosomes and house-keeping genes of ribosomal RNA. NOR bands represent structural non-histone proteins linked explicitly to nucleolar organizing regions and bind to ammoniacal silver (replaced by silver nitrate, used widely). NOR banding is used to study the chromosome polymorphisms and identify satellite stalks, occasionally seen on acrocentric chromosomes.

The number of bands produced by these stains is ~500 per haploid genome with a resolution of 6 Mbp. This resolution is not sufficient to detect chromosomal rearrangements due to excessive condensation. The resolution was increased by adding high-resolution banding (from 500 to more than 1000 bands per haploid genome). High-resolution cytogenetics provides high accuracy in locating the chromosomal breakpoints, better than the earlier staining techniques. By applying high resolution staining, clinical syndromes such as Prader Willi and Angelman syndrome generated due to deletion at the proximal long arm of chromosome 15, Smith-Magenis, Miller-Dieker syndrome occurring due to deletions in the short arm of chromosome 17, and DiGeorge/ Velo Cardio Facial (VCF) syndrome with deletions in the long arm of chromosome 22 are linked to small chromosomal aberrations giving rise to the concept of the micro-deletion or contiguous gene syndrome.

4.3.1.1 Specialized Staining Techniques

Apart from the regular low resolution and high-resolution cytogenetic techniques, specialized staining techniques are also applied for determining the sex chromatin analysis, sister chromatin exchange, fragile sites, and chromosome breakage. The number of sex chromatin bodies will be one number less than the number of X chromosomes in the chromosome complement, achieved by fixing the buccal smears in ethanol on a clean slide and hydrolyzing with HCl, washing with distilled water, and staining with crystal violet. The presence of a chromatin mass known as a Barr body indicates the presence of a positive chromatin cell.

Sister chromatin exchange staining is performed by incorporating bromodeoxyuridine (BrdU) onto cells in the replication stage of the cell cycle consecutively for 2 cycles. As the DNA replication is semi-conservative, chromosomes will have one chromatid stained with BrdU on one strand of DNA and another chromatid will stain both strands, producing a staining pattern on a fluorescent microscope with one chromatid fluorescing better than the other. The exchange of sister chromatids appears as an interchange between the sister chromatids of bright and dull fluorescing segments.

Uncondensed regions of the DNA in a chromosome can be visualized as gaps during staining, and these gaps are responsible for chromosomal breakage. Gaps that are consistently noticed on a single chromosomal locus are fragile sites and can be visualized on chromosomes by either non-banding or Q-banding methods. Using these methods, fragile site Xq27.3 of the X chromosome that breaks more frequently than others has been identified and named fragile X syndrome leading to mental retardation, altered speech patterns, and other physical attributes in males (discussed in Chapter V).

4.3.2 Molecular Cytogenetic Techniques

Advancement in technology has witnessed the development of techniques that can identify re-arrangements in the chromosomes with a very high resolution, which was difficult to achieve with standard staining techniques. Techniques including fluorescent in situ hybridization (FISH), spectral karyotyping, and comparative genomic hybridization have provided an enormous advancement in the field of cytogenetics.

4.3.2.1 Fluorescent In Situ Hybridization (FISH)

Pinkel et al. in 1986 developed a technique for visualizing chromosomes using fluorescently labeled probes known as fluorescent in-situ hybridization, popularly called FISH. This approach allows chromosomal and nuclear locations of specific DNA sequences to be visualized through a fluorescent microscope. This technology enables the detection of specific nucleic acid sequences in morphologically similar chromosomes, cells, and tissues. Fluorescent probes are safe and straightforward to use. They can be stored for an indefinite amount of time and provide high resolution. FISH is performed on cells arrested either in metaphase or interphase. The process of FISH involves three steps.

1. *Preparation of sample*: The sample is fixed in a mixture of methanol and acetic acid for 10 minutes after application to a microscopic slide, dried, and then it undergoes denaturation and hybridization.
2. *Denaturation and hybridization*: A fluorescently labeled probe is applied to the test sample as an overlay, covered with a clean cover glass, and the cover glass's entire location is glued for proper sticking on to the slide. Denature the slides either by heating at 75°C for 5 minutes or by using formamide. Incubate the slide in a moist chamber for 12 - 14 hrs. at 37°C.
3. *Washing the unbound probe and detection using a fluorescent microscope*: Unglue the cover glass, wash the preparation in washing solutions under increasing stringency and dry the slide. Stain the slide with DAPI for its observation under a fluorescent microscope.

FISH probes are classified into two kinds, the location in the genome where they hybridize, and the type of chromosome aberration they detect. FISH is helpful in the diagnosis of patients with various congenital and malignant neoplastic disorders in conjunction with chromosome studies. Chromosomal abnormalities in small DNA segments can be detected by using FISH if the probe is either fortuitously situated or by virtue of its design in the targeted chromosome. FISH is used to establish the naturally occurring breakpoints relatively easier than banding. Karyotype analysis of nuclei in non-dividing cells can be performed by using FISH. This technique has been successfully applied to detect t2-5 and p23-q35 translocations in anaplastic large cell lymphoma for minimal residual disease in hematopoietic stem cell assays from peripheral blood stem cells of acute myeloid leukemia patients with trisomy 8 and for analyzing the chromosomal abnormalities of tumors in children.

4.3.2.2 Spectral Karyotyping (SKY)

This technique is also known as multicolor FISH (M-FISH), which allows the staining of all chromosomes with different colors, utilizing the concentrations and combinations of different fluorescent dyes. Staining each chromosome with a different color was used while analyzing the samples with complex aberrations such as solid tumors (Figure 4.5). Spectral karyotyping stains the entire chromosome with only a specific color, whereas M-FISH allows various fluorescent dyes to represent various probes simultaneously, offering a simultaneous presentation of all 24 human chromosomes with a single hybridization. The unequivocal color of each chromosome makes the analysis of complex chromosomal aberrations and the compositions of the chromosomes much more accessible, which can be applied for automated karyotype analysis (Figure 4.6). These two techniques are also helpful in the detection of chromosomal translocations.

4.3.2.3 Comparative Genomic Hybridization (CGH)

Comparative genomic hybridization is a technique that allows the detection of changes in the copy number without cell culture. It is one of the fastest screening techniques for pointing to the chromosome regions responsible for tumor progression. CGH was first reported by Kallioniemi et al. in 1992. In contrast to the staining techniques and FISH, this technique does not need the culture of

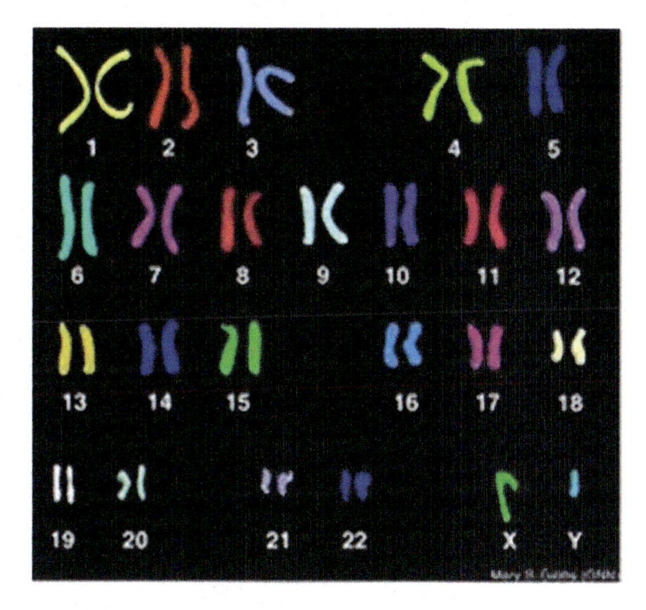

FIGURE 4.5 Spectral Karyotyping of Human Chromosomes.

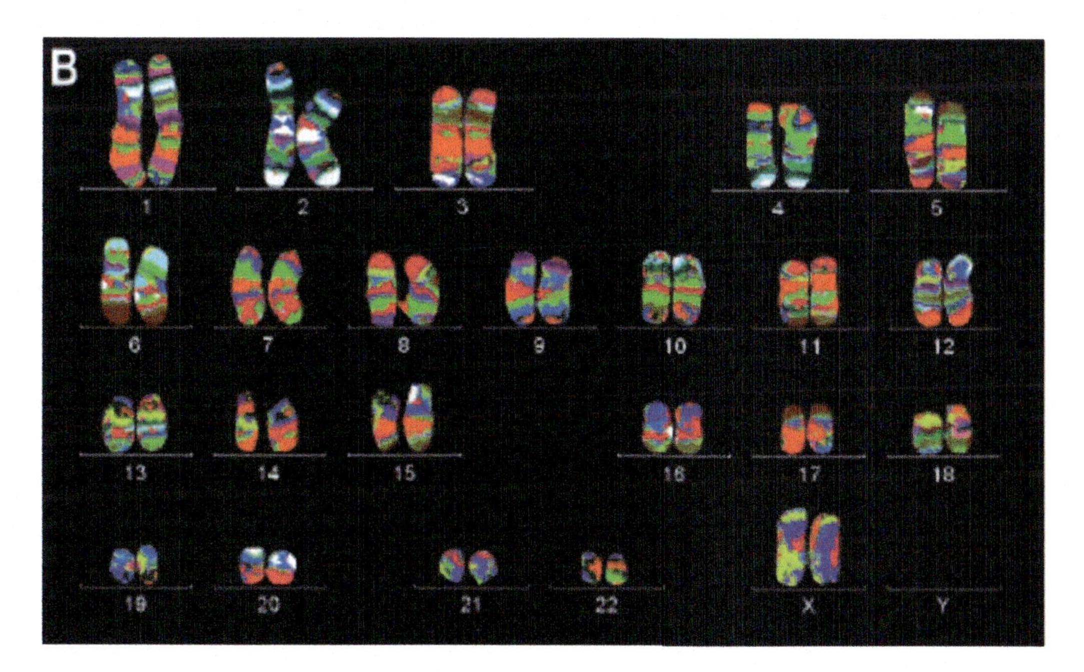

FIGURE 4.6 Multicolored FISH of Human Chromosomes.

cells at metaphase or interphase. Instead of hybridizing a labeled probe to human chromosomes on a glass slide, thousands of different probes are printed on a glass slide. The isolated DNA of the patient is labeled in a specific color, mixed with an equal amount of normal control DNA, and labeled in a different color. This DNA mixture is hybridized with the denatured DNA probe fixed on the glass slide. After subsequent washing, the fluorescence of each spot is analyzed by measuring the ratio of patient DNA over control DNA. This technique was further developed as array-based comparative genomic hybridization (Array-CGH). One CGH is equivalent to the conductance of thousands of FISH at one time. Array CGH provides a better picture of the copy number variations and breakpoints of segments, lost or added compared to traditional CGH. This technique also provides information regarding the changes and variations that occur within the human genome.

4.4 PRENATAL GENETIC DIAGNOSIS

Prenatal genetic diagnosis is a broad term referring to the techniques involved in generating information regarding the health of the developing fetus during pregnancy. Prenatal diagnosis is necessary in pregnancy cases, where the sonography tests fail to suggest chromosomal disorders, such as syndromes generated due to trisomy. Pedigree analysis is generally performed among individuals with a high risk of trisomic pregnancy to know their family history and couples suspected or known to be carriers of inherited genetic disorders.

4.4.1 Amniocentesis

Prenatal diagnosis of chromosomal aberrations requires cytogenetic analysis of the amniotic fetal cells through a process known as amniocentesis, a safe, reliable, accurate, and well-established technique for detecting chromosomal abnormalities and other genetic diseases. Amniocentesis is performed after 12 weeks of gestation. A small amount of amniotic fluid will be removed from the sac around the baby for detecting chromosomal problems and congenital disabilities during amniocentesis. The amniotic fluid consists of cells derived from the fetus's vital organs such as skin, kidney, bladder, gut, and other fetal tissues. Two methods are employed for culturing the amniotic cells, culturing and processing on a coverslip for retaining the individual colonies of the cells and culturing in flasks for mixing all colonies of cells. The cells are briefly centrifuged, and the amniotic fluid is decanted from the cell pellet into a sterile test tube. A suitable medium supplemented with fetal bovine serum, L-glutamine, and antibiotics are added to the tube, and the cultures are incubated at 37°C in a 5% CO_2 incubator for 10 days. After incubation, the cells are harvested and subjected to hypotonic and fixative treatments for chromosomal analysis.

4.4.2 Chorionic Villus Sampling

Chorionic cell sampling is considered to be one of the earliest methods for isolating fetal tissues for karyotype analysis. This method is generally performed after 8 - 11 weeks of gestation. Chorionic villus sampling is done broadly in two ways, transcervical sampling by inserting different types of catheters transcervically, monitored with ultrasound for the aspiration of chorionic villi, and transabdominal chorionic villus sampling by direct aspiration of placental villi transabdominally, monitored by ultrasound. Transabdominal sampling is routinely preferred as it can be performed even during the second trimester of pregnancy and is associated with a low risk of infection, bleeding, and concomitant pregnancy loss. Coelocentesis is the latest technique, which involves the puncture of the chorion and aspiration of the coelomic fluid from the extra-embryonic coelom and completely avoids the puncture of the amnion. Still, this procedure needs to be applied in clinical practice.

4.4.3 Fetal Blood Sampling

Fetal blood sampling is one of the techniques, which has gained a very low priority due to the difficulties and a very high rate of fetal loss of up to 0.4%. Fetal blood sampling is done by passing a fetoscope transabdominally after 20 weeks of gestation, under ultrasound guidance. Fetal blood sampling is replaced by cordocentesis, which allows the withdrawing of blood from the umbilical vein. Cordocentesis is applied to analyze prenatal karyotype disorders such as chromocutaneous albinism, Sjogren-Larsson syndrome, and prenatal genetic disorders to understand fetal physiology, development, and metabolism. Cordocentesis is also used for fetal liver biopsy to diagnose certain metabolic disorders and prenatal muscle biopsy to diagnose muscular dystrophy.

4.4.4 Prenatal Diagnosis of Genetic Disorders

Chromosomal abnormalities such as deletions, duplications, and translocations can be detected prenatally along with the chromosomal aneuploidy by using amniocytes, chorionic villi, or fetal blood. The result of the fetal karyotype is available in 72 hours for the chorionic villi or fetal blood samples and 2 - 3 weeks for amniotic fluid samples. FISH is being used to diagnose aneuploidy involving X, Y, 13, 18, and 21 chromosomes. Multicolored FISH is used to diagnose contiguous gene syndromes such as Prader Willi syndrome, Miller-Dicker syndrome, Angelman syndrome, and Di George syndrome, which involves the loss of functionally related genes contiguous along the chromosome. FIDH technique is also applied to detect Duchenne or Becker muscular dystrophy carrier status in females whose affected male relatives have a specific gene deletion. PCR technology aid in the detection of α-1-antitrypsin deficiency, Duchenne muscular dystrophy, phenylketonuria, and fragile X syndrome. RFLP detects thalassemia in prenatal cells.

4.5 DENATURING HIGH-PERFORMANCE LIQUID CHROMATOGRAPHY

Many sensitive high throughput techniques have been developed for the early diagnosis and analysis of genetic alterations. PCR-based techniques at both DNA and RNA levels are powerful, sensitive, and widely available. However, detection of the PCR products through agarose gel electrophoresis remains a rate-limiting step for high-throughput analyses using these techniques. In addition, most of the currently available techniques for mutation detection, such as single-strand conformation polymorphism (SSCP) analysis are cumbersome and use radioactivity for obtaining the highest degree of sensitivity, which is unsafe. Denaturing high-performance liquid chromatography (dHPLC) based analysis of DNA fragments is a novel technology that can completely replace the gel electrophoresis step for the analysis of PCR fragments. This technology can accommodate high-throughput analyses for quantitative PCR, genotyping, loss of heterozygosity (LOH) determinations, and detection of DNA mutations and polymorphisms. dHPLC is a recently developed specialized version of high-performance liquid chromatography (HPLC). The advantage of dHPLC over HPLC is its applications in DNA testing. Current dHPLC methods can detect DNA single nucleotide polymorphisms (SNPs) in less than ten minutes following PCR amplification.

In practice, dHPLC analysis begins with a PCR standard to amplify the fragment studied. If the amplified region exhibits hemizygous polymorphism, two kinds of fragments will be present in the PCR product, corresponding to the allele and the wild polymorphic allele. Denaturation-renaturation follows this step to create hetero-and homoduplexes from the two populations in the PCR product. Premixing a DNA wild population to a population of polymorphic DNA will be done to obtain heteroduplexes after the denaturation - renaturation step. Heteroduplexes are double strands of DNA containing a strand from the wild-type allele and another from the polymorphic allele. The formation of such DNA fragments causes the appearance of a "mismatch" or bad pairing where the polymorphism is located. These "mismatches" in the heteroduplex are based on polymorphism detection by dHPLC. Heteroduplexes are thermally less stable than their corresponding homoduplexes and will resolve in the chromatography column upon exposure to high temperatures. The consequence of this instability will be a mismatch of two DNA strands in the region of polymorphism when DNA is subjected to a high temperature. This mismatch will decrease the interaction with the column and reduce the retention time compared to the homoduplexes in chromatographic separation.

To achieve the separation of homo and heteroduplex strands, dHPLC uses a column of a non-grafted porous stationary phase composed of polystyrene – divinylbenzene alkyl. The stationary phase is electrically neutral and hydrophobic. The DNA is negatively charged at its phosphate groups and therefore can adsorb by itself at the column. Triethylammonium acetate is used for the adsorption, where the positively charged ammonium ion of these molecules interacts with the DNA and the alkyl chain interacts with the hydrophobic surface of the solid phase. When heteroduplexes are partially denatured by heating, negative charges undergo partial relocation, and the interaction force between DNA heteroduplexes and column decreases, compared to the strength of interaction of homoduplexes. As a result, the heteroduplexes will elute less rapidly by the mobile phase consisting of acetonitrile. Heteroduplexes are compared with the homoduplexes using a UV detector.

4.5.1 Applications

dHPLC is a novel technique that uses heteroduplex formation between wild type and mutated strands to identify mutations accurately. This method helps screen many samples for somatic as well as germline mutations. SNP analysis by dHPLC is a valuable method to screen SNPs present in the DNA samples before sequencing. By applying the dHPLC method, significant savings of time and money are realistically possible. The utilization of dHPLC for analyzing the DNA samples involved in a crime scene will save much time. Ease of instrument operation, high sample throughput, and shorter sample analysis time are the potential strengths of dHPLC. High throughput dHPLC is readily available and the cost per sample is low in comparison to the sequencing. dHPLC sample analysis time is less than ten minutes, which is much shorter than the PCR step in the analysis. dHPLC is a flexible platform for the development of new applications in DNA testing.

Despite so many advantages, this novel technique suffers from a few disadvantages, such as the absence of robust methods that can be routinely applied by crime labs and for training needs. The level of expertise and training needed to maintain and operate currently available dHPLC instrumentation is the same as that needed for HPLC. Standardized procedures are not available for instant use in forensic labs with existing equipment and varying levels of personnel qualifications. Intra- and inter laboratory studies using a standard method has not been conducted to validate the method.

4.6 CONCLUSION

Its more than 100 years since cytogenetics has been developed for diagnosing chromosomal aberrations. No system has been developed, which can classify the banded chromosomes as a skilled cytogeneticist can do. Current cytogenetics is paving the way to a molecular approach for understanding the micro changes. Millions of dollars are being invested in bringing automation and high throughput technology for performing the karyotype. Routine banding can be clubbed with SKY and M-FISH for increased detection of syndromes in children. Through the chromosome banding pattern analysis, thousands of chromosomal abnormalities associated with inheritance or de novo disorders can be diagnosed, generating many leads to the underlying molecular causes of these disorders. Present-day, high resolution genetic linkage analysis can be performed with much ease, and the gross chromosomal abnormality responsible for the disorder is heralded as a valuable resource for locating the disease gene. The challenge before us is to navigate from visible morphological alterations to the DNA sequence level at a low price, which a person with an average income can afford. Chromosomal abnormalities exist as nature's guide to many unexplained human disorders, which can be explored only by the rapid development of tools and techniques, which can accurately identify various genetic disorders, thereby providing an overall picture of the whole genome for analysis and paving the path for successful treatment of the chromosomal disorders.

REFERENCES

Andreas, G., Anastasios, G. K. (2017). Future perspectives in HLA typing technologies. In: umbilical cord blood banking for clinical application and regenerative medicine. Intech Publishers, Croatia, 2017, 45–68.

Angel, C., Paula, S-D. (2010). Forensic DNA-Typing Technologies. From: Methods in Molecular Biology, Edited by: A. Carracedo vol. 297, Humana Press Inc., Totowa, NJ 1–11.

Caspersson, T., Zech, L., Johansson, C. (1970). Differential binding of alkylating fluorochromes in human chromosomes. *Experimental Cell Research* 60: 315–319.

Chamla, Y., Ruffie, M. (1976). Production of C and T bands in human mitotic chromosomes after treatment. *Human Genetics* 34: 213–216.

Chen, X., Xing, Y., He, J., Tan, F., You, Y., Wen, Y. (2019). Develop and field evolution of single tube nested PCR, SYBRGreen PCR methods, for the diagnosis of leprosy in paraffin-embedded formalin fixed tissues in Vunnan province, a hyper endemic area of leprosy in China. *PLoS Negl. Trop. Dis.* 13(10): e0007731.

Darawi, M. N., Ai-Vyrn, C., Ramasamy, K., Philip P. J. H. et al. (2013). Allele-specific polymerase chain reaction for the detection of Alzheimer's disease-related single nucleotide polymorphisms. *BMC Med. Genet.* 14: 27.

Debrand, E., Lykoudi, A., Bradshaw, E., Allen, S. K. (2015) A non-invasive droplet digital PCR (ddPCR) assay to detect paternal CFTR mutations in the cell-free fetal DNA (cffDNA) of three pregnancies at risk of cystic fibrosis via compound heterozygosity. *Plos One* 10:45.

Deepachandi, B., Weerasinghe, S., Soysa, P. et al (2014). A highly sensitive modified nested PCR to enhance case detection in leishmaniasis. *BMC Infect. Dis.* 19: 623.

Dewald, G. W. (2002). Interphase FISH studies of chronicmyeloid leukemia. *Methods Mol. Biol.* 204:311–342.

Dodge, W., Cruz, J., Zamkoff, K., Hurd, D., Pettenati, M. J. (2004). Use of fluorescence in situ hybridization to detect minimal residual disease in hematopoietic stem cell assays from peripheral blood stem cells of 2 patients with trisomy 8 acute myeloid leukemia. *Stem Cells Dev.*13(1):23–26.

Dong, Y., Zhu, H. (2005). Single-strand conformational polymorphism analysis: basic principles and routine practice. *Methods Mol. Med.* 108:149–57.

Ercolini, D. (2004). PCR-DGGE fingerprinting: Novel strategies for detection of microbes in food. *J. Microbiol. Meth.* 56:297–314.

Ganguly, A., Rock, M. J., Prockop, D. J. (1994). Conformation-sensitive gel electrophoresis for rapid detection of single-base differences in double-stranded PCR products and DNA fragments: evidence for solvent-induced bends in DNA heteroduplexes. *Proc. Natl. Acad. Sci. USA.* 24 (91):5217.

Ganguly, A., Rooney, J. E., Hosomi, S., Zeiger, A. R., Prockop, D. J. (1989). Detection and location of single-base mutations in large DNA fragments by immunomicroscopy. *Genomics* 4:530–538.

Ganguly, T., Dhulipala, R., Godmilow, L., Ganguly, A. (1998). High throughput fluorescence-based conformation-sensitive gel electrophoresis (F-CSGE) identifies six unique BRCA2 mutations and an overall low incidence of BRCA2 mutations in high-risk BRCA1-negative breast cancer families. *Hum. Genet.* 102:546–549.

Gasser, R., Hu, M., Chilton, N. et al. (2006). Single-strand conformation polymorphism (SSCP) for the analysis of genetic variation. *Nat. Protoc.* 1: 3121–3128.

Gerard, M., Kornelia, S. (1998). Application of denaturing gradient gel electrophoresis (DGGE) and temperature gradient gel electrophoresis (TGGE) in microbial ecology. *Antonie van Leeuwenhoek* 73: 127–141.

Gisela, K., Arndt, H., James, M., Heinz, H. (2001). Denaturing high pressure liquid chromatography (DHPLC) for the Analysis of Somatic p53 Mutations. *Laboratory Investigation* 81 (12): 1735–1738.

Groppi, A., Begueret, J and Iron, A. (1990). Improved methods for genotype determination of human alcohol dehydrogenase (ADH) at ADH 2 and ADH 3 loci by using polymerase chain reaction—directed mutagenesis. *Clin. Chem.* 36:1765–1768.

Hagelberg, E., Sykes, B., Hedges, R. (1989). Ancient bone DNA amplified. *Nature* 342: 485.

Higuchi, R., Fockler, C., Dollinger, G., Watson, R. (1993). Kinetic PCR analysis: real-time monitoring of DNA amplification reactions. *Biotechnol. Nat. Publ. Co.* 11(9):1026–1030.

Hongyo, T., Buzard, G., Calvert, R., Weghorst, C. (1993). Cold SSCP: A simple, rapid, and non-radioactive method for optimized single-strand conformation polymorphism analyses. *Nucleic Acids Research* 21: 3637–3642.

Hosomichi, K., Jinam, T. A., Mitsunaga, S., Nakaoka, H., Inoue, I. (2013). Phase-defined complete sequencing of the HLA genes by next-generation sequencing. *B.M.C. Genomics* 14:355.

Jagtar, S., Niti, B., Shweta, S., Akshra, G. (2014). A critical review on PCR, its types, and applications. *Int. J. Adv. Res. Biol. Sci.* 1(7): 65–80.

Jeffreys, A. J., Tamaki, K., MacLeod, A., Monckton, D. G., Neil, D. L., Armour, J. A. L. (1994). Complex gene conversion events in germline mutation at human minisatellites. *Nat. Genet.* 6: 136–145.

Jeffreys, A. J., Wilson, V., Thein, S. L. (1985). Hypervariable minisatellite regions in human DNA. *Nature* 314: 67–73.

Johnson, W. M., Leek, J., Swin Bank, K., Angus, B., Roberts, P., Markham, A. F. et al. (1997). The use of fluorescent in situ hybridization for detection of the t(2;5)(p23;q35) translocation in anaplastic large-cell lymphoma. *Ann Oncol.* 8: S65–S69.

Kallioniemi, A., Kallioniemi, O. P., Sudar, D., Rutovitz, D., Gray, J. W., Waldman, F., Pinkel, D. (1992). Comparative genomic hybridization for molecular cytogenetic analysis of solid tumors. *Science* 258: 818–821.

Korkoo, J., Pihlajamaa, T., Prockop, D. J., Ala Kokko, L. (1996). Comparison of conformation sensitive gel electrophoresis with denaturing gel electrophoresis for detection of single base mutations in PCR products. Sixth International Conference on Matrix Biology, III-3.

Kulkarni, M. L., Sajeev, V. (1995). Prenatal diagnosis of genetic disorders. *Indian Pediatrics*. 32: 1229–1237.

Lee, T. Y., Lai, M. I., Ramachandran, V., Tan, J. A., Th, L. K., Othman, R., Hussein, N. H., George, E. (2016). Rapid detection of alpha Thalassaemia variants using droplet digital PCR. *Int. J. Lab. Hematol.* 38:435–443.

Liehr, T., Starke, H., Weise, A., Lehrer, H., Claussen, U. (2004). Multicolor FISH probe sets and their applications. *Histol. Histopathol.* 19: 229–237.

Mayor, N. P., Robinson, J., McWhinnie, A. J., Ranade, S., Eng, K., Midwinter, W. et al. (2015). HLA typing for the next generation. *PLoS One* 10(5): e0127153.

Muyzer, G., Smalla, K. (1998). Application of denaturing gradient gel electrophoresis (DGGE) and temperature gradient gel electrophoresis (TGGE) in microbial ecology. *Antonie van Leeuwenhoek* 73:127–141.

Newton, C. R., Graham, A., Hepstinstall, L. E., Powell, S. J., Summers, C., Kalsheker, N., Smith, J. C., Markham, A.F. (1989). Analysis of any point mutation in DNA. The amplification refractory mutation system (ARMS). *Nucleic Acids Res.* 17:2503–2516.

Orita, M., Iwahana, H., Kanazawa, H., Hayashi, K., Sekiya, T. (1989). Detection of polymorphisms of human DNA by gel electrophoresis as single-strand conformation polymorphisms. *Proceedings of the National Academy of Science* USA 86: 2766–2770.

Özgül, M. A., Güven, L., Lee-Jun, C. W. (2005). Mutation analysis by the use of temporal temperature gradient gel electrophoresis. *Turk. J. Med. Sci.* 35: 279–282.

Pinkel, D., Gray, J. W., Trask, B., Van D.E.G., Fuscoe, J., Van, D. (1986). Cytogenetic analysis by in situ hybridization with fluorescently labeled nucleic acid probes. *Cold Spring Harb. Symp. Quant Biol.* 51:151–157.

Pinkel, D., Straume, T., Gray, J. W. (1986). Cytogenetic analysis using quantitative, high-sensitivity, fluorescence hybridization. *Proc. Natl. Acad. Sci. USA*. 83:2934–2938.

Raimondi, S. C. (2000). Fluorescence in situ hybridization: Molecular probes for diagnosis of pediatric neoplastic diseases. *Cancer Invest.* 18:135–147

Reiss, J., Cooper, D. N. (1990). Application of the polymerase chain reaction to the diagnosis of human genetic disease. *Human Genet.* 85:1–8.

Saiki, R. K., Walsh, P. S., Levenson, C. H., Erlich, H. A. (1989). Genetic analysis of amplified DNA with immobilized sequence-specific oligonucleotide probes. *Proc. Natl. Acad. Sci. USA*. 86(16):6230–6234.

Scheres, J. M. J. C. (1974). Production of C and T bands in human chromosomes after heat treatment at high pH and staining with "stains-all". *Human Genetics* 23: 311–314.

Sheldon, S, Poulton, K. (2006). HLA typing and its influence on organ transplantation. *Methods Mol Biol.* 333:157–174.

Sheshadri, N. (2011). Applications of restriction fragment length polymorphism. *Annals of Clinical and Laboratory Science* 21 (4): 291–296.

Shiina, T., Hosomichi, K., Inoko, H., Kulski, J. K. (2009). The HLA genomic loci map: expression, interaction, diversity, and disease. *Journal of Human Genetics*. 54(1):15–39.

Shu, Ye., Steve, H., Fiona, G. (1992). Allele specific amplification by tetra-primer PCR. *Nucleic Acids Research* 20 (5): 1152–1152.

Souza, G., Almeida, A., Farias, A., Leal, N., Abath, F. (2007). Development and evaluation of a single tube nested PCR based approach (STNPCR) for the diagnosis of plague. In: Perry R.D., Fetherston J.D. (eds) The Genus Yersinia. Advances In: Experimental Medicine and Biology, vol 603. Springer, New York, NY.

Taylor, S. D., Ericson, N. G., Burton, J. N., Prolla, T. A., Silber, J. R., Shendure, J., Bielas, J. H. (2014). Targeted enrichment and high-resolution digital profiling of mitochondrial DNA deletions in human brain. *Aging Cell* 13:29–38.

Thirumulu, P. K., Zilfalil, B. A. (2009). Cytogenetics: past, present and future. *Malaysian Journal of Medical Sciences* 16 (2): 4–9.

Weiss, M. M., Hermsen, M. A. J. A., Meijer, G. A., Van Grieken, N. C. T., Baak, J. P. A., Kuipers, E. J., Van Diest, P. J. (1999). Comparative genomic hybridization. J. *Clin. Pathol. Mol Pathol.* 52:243–251.

5 Diagnosis of Genetic Disorders

5.1 INTRODUCTION

Human health is critically dependent on the identification of a disease and its diagnostics results. From the beginning of the cytogenetics for the identification of chromosomal disorders until the genetic tests performed for identification of cancer or culture of a microbe for identification of appropriate antibiotics for controlling infection, diagnostics provide critical insights at every stage of medical care for detection, treatment, prevention, and successful improvement of health conditions. There are five broad areas of diagnostics, are clinical chemistry, immunology, hematology, microbiology, and molecular diagnostics. Among these, molecular diagnostics has gained attention in recent years due to its deep insights enhancing diagnosis and treatment. Molecular diagnostics has successfully transformed diagnostics into a dynamic research area, leading to advancements in the treatment of many diseases, totally revolutionizing the health care system.

Genomic testing has begun a new era in the medical field. From the medicine, which is personalized, predictive, preventive, and participatory, genomics has triggered an evidence-based revolution, thus triggering a profound impact on clinical diagnostics. The American clinical laboratory association has defined genomic tests thus, *"A genetic or genomic test involves an analysis of human chromosomes, deoxyribonucleic acid (DNA), ribonucleic acid (RNA), genes, and gene products (e.g., enzymes, metabolites, and other types of proteins)."* As the cost of genome sequencing is dropping, it is becoming more affordable for a patient to use genomics for their health benefit. Clinical genomics is making use of sequencing technology for assessing the health status of a patient. Professionals in medicine and health care use diagnostics to decide on selecting an appropriate treatment for patients. Technical advancement and an expansion of molecular genetics and genomics have made them a thriving part of diagnostic services. The addition of two novel therapeutic approaches, namely pharmacogenomics and nutrigenomics has further strengthened diagnostics and provided a better treatment and management method.

Most of the molecular tests performed today are concerned with the detection of infectious diseases, including human immunodeficiency virus (HIV), hepatitis B virus (HBV), hepatitis C virus (HCV), human papillomavirus (HPV), cytomegalovirus (CMV), *Chlamydia trachomatis*, *Neisseria gonorrhoeae*, and *Mycobacterium tuberculosis*. A small proportion of molecular tests are used in forensic medicine for paternity testing, tissue typing, oncology, and food and beverage testing. Molecular diagnostic tests are performed with minimal blood or tissue obtained by small biopsies and fine-needle aspirates. Testing can even be performed on tissue collected in the past and stored in formalin. The molecular diagnostics approach for the quick diagnosis and treatment of genetic and pathogenic diseases has been discussed in this chapter.

DOI: 10.1201/9781003343790-5

5.2 THE HISTORY OF MOLECULAR DIAGNOSTICS

Advancement in molecular genetics has led to an increase in technology and its utilization in the laboratory. A wide variety of drugs, diagnostics, and therapeutic products developed for specific targeting of gene defects and protein mix expressions are in clinical trials and require the approval of regulating agencies. An increasing amount of learned people will demand information regarding the chances of getting or being susceptible to disease and the methodology employed to detect a disease followed by its treatment. Most pharmaceutical companies collaborate with diagnostic companies to develop more efficient and less toxic integrated personalized drugs in response to this increasing demand. This integration provides a new opportunity for clinical and pathology laboratories to invent new medicines, guide the selection, dosage, method of administration into the human body, and combinations with other drugs.

The term 'molecular disease' was introduced by Pauling et al. in 1949, based on their finding that change in a single amino acid at the β-globin chain is responsible for sickle cell anemia. In principle, their explorations have set the foundations of molecular diagnostics, even though significant transformations have occurred many years later. In 1980, a prenatal genetic test for detecting Thalassemia was performed based on restriction enzymes that cut DNA at specific short sequences and created bands corresponding to various sizes of DNA strands depending upon the presence of alleles (genetic variation). The term molecular diagnostics was utilized by companies such as Bethesda Research Laboratories in the 1980s. The 1990s have witnessed an era of novel gene discovery and application of novel techniques for their sequencing, which led to an emergence of molecular and genomic laboratory medicine, whose structure was determined by the Association for Molecular Pathology (AMP). Since 2002, the Hap Map Project combined information on single base genetic differences that recur in the human population, as single nucleotide polymorphisms (SNPs) and their relationships were mapped with diseases. In 2012, molecular diagnostic techniques using genetic hybridization tests to detect Thalassemia using specific SNPs were standardized (Sokolenko and Imyanitov, 2018). In 1998, the European Union's Directive 98/44/EC clarified that patents on DNA sequences can be allowed to be filed. However, in 2013, the US Supreme Court made a ruling that naturally occurring gene sequences cannot be patented.

5.2.1 Molecular Diagnostic Tests for the Detection of Changes in Genetic Composition

The focus of molecular diagnostics is to identify changes in nucleic acids. The rapid advancement of molecular diagnostics and basic molecular biology research has resulted in the development of diagnostic tests to identify genetic disorders. These tests are based on the analysis of DNA, RNA, and micro RNAs, and can placed into two categories, nucleic acid based diagnostic tests, and gene-based diagnostic tests. Proteomics and metabolomic array-based tests are also under development, and these might form a new sub-category.

5.2.1.1 Nucleic Acid-Based Diagnostic Tests

Molecular diagnostics is primarily represented by the nucleic acid test (NAT) to detect genomic DNA and RNA sequences and their characterization to understand aberrations and polymorphisms. Fundamentally, NAT is a hybridization-based assay, associated with detecting labeled molecules coupled to the probe sequence. Some NATs are also based on the direct hybridization detection of ribosomal RNA (rRNA). The high sensitivity of NAT is attributed to the fact that a cell contains thousands of rRNAs and the detection of less-abundant RNA or DNA molecules with increased sensitivity requires specific amplification to generate a detectable signal. Amplification of a target sequence mainly employs PCR technology. Roche Diagnostics Corporation, Indianapolis, developed the most notable PCR amplification tests, Strand-displacement amplification (SDA) tests developed by Franklin Lakes, NJ, transcription-mediated amplification (TMA) by Gen-Probe Inc., San Diego, and nucleic acid sequence-based amplification developed by Organon Teknika Corp. Durham, NC. All hybridization-based tests use either chromogenic, fluorescent, or optical reporter labels to produce the light detected by an optical device. The advancement of PCR technologies, multiplexing, sequencing, and the like, has provided diagnostic companies with novel capabilities. For example, Roche's PCR-based product lines have captured 47% of the molecular diagnostics market.

Providing better design and manufacture of probes and lines, for increased performance at a lower cost per test has prompted novel labeling methods. Homogenous labelling methods are widely used in laboratories for the detection of genetic disorders and cancer. Even though homogenous-based assays provide the best detection sensitivity, the availability of a limited number of fluorophores for the simultaneous detection of multiple analytes in a single sample is the major limitation of its usage in multiplex assays.

5.2.1.2 Gene-Based Diagnostic Tests

Novel research findings in molecular biology have further changed biological and biomedical research, which has become an indispensable tool in clinical diagnostics. Gene expression profiling, changes in protein levels, and information on the post translational modifications of proteins have become vital for diagnosing, classifying, and determining cancer stages. There has been a predominant increase in DNA microarrays in clinical research for the last few years. Even though microarrays are a method for testing thousands of gene expressions, Gene Chip of Affymetrix Inc. Santa Clara, CA, is an example of an advanced microarray platform designed for diagnostics. Flow-Thru Chip is a novel microarray platform developed by MetriGenix Inc., Gaithersburg, MD, which promises high assay performance, amenability to automation, and the flexibility to handle low-to-high throughput diagnostic applications. These microarrays target tens to hundreds of analytes rather than thousands. Platforms combine bioinformatics and microfluidic technologies with microarrays to miniaturize and simplify the tests. Novel technologies will extend the limits for the simultaneous analysis of thousands of genes for a better understanding of gene function and dysfunction. AmpliOnc I, a Vysis Geno Sensor microarray system from Abbott, was designed to detect abnormal increases in gene copies of 58 different genes.

The market potential for genetic disease and oncology testing is much smaller and is presently in the infant stage. The involvement of genetic markers in specific diseases and physiological pathways has begun to be biologically validated. In-house molecular tests offered as analyte-specific reagents (ASRS) will be vital in this testing. Only an estimated 25% of the 4000 known hereditary diseases can be analyzed through molecular testing techniques available today. Additional technical, patent, and ethical issues may further restrict genetic diagnostic test development and market penetration. Pharmacogenomic tests for a patient's susceptibility, response, and resistance to a specific drug hold exciting potential for future molecular diagnostics. Pharmaceutical medicines will be prescribed to targeted patient populations based on their individual genomic profiles. It is estimated that more than 370 biopharmaceutical drugs are in the late stages of clinical development. Management of several medical conditions, including diabetes, asthma, autoimmune disease, and cardiac disease, will significantly benefit from these gene-based diagnostics and therapeutic interventions

5.3 LAB ON A CHIP

A lab-on-a-chip (LOC) is a device, which integrates multiple analytical functions on a single chip of size ranging from a few square millimeters to centimeters. These devices can reciprocate the same functions as their full-scale counterparts. Processes customarily carried out in a clinical laboratory are miniaturized on a single chip to enhance efficiency, and mobility and reduce the sample and reagent volumes. The principle of LOC is based on microfluidics, a novel technology for manipulating and controlling fluids, particles of micron and submicron dimensions. The technology is associated with the development of devices that can undertake such operations. The utilization of microfluidic platforms supports the implementation of assay miniaturization. Such platforms, empowered by the addition of fluidic channels and chambers, will facilitate miniaturization, integration, automation, and parallelization, performing multiple tests simultaneously. Microfluidic-based LOC devices are specifically used in drug discovery, life sciences, ecology, and clinical diagnostics. Clinical diagnostics is one of the most significant LOC market segments and it has been subdivided into two categories, point of care testing, where a diagnostic test is performed near the patient without any clinical laboratory, and central laboratory-based testing which is performed in a clinical laboratory. Point of care testing ranges from simple immunochromatographic strips such as the kits used to determine pregnancy to complex systems requiring external machinery and expertise. Applications of point-of-care testing also include detection kits for nucleotides and peptides for the early detection of disease. LOC devices fulfil the requirements of a point of care testing diagnostic device by low consumption of sample and reagents, miniaturization of devices, and rapid turnout time for the analysis.

5.3.1 A Brief History of Lab on a Chip

A significant milestone in LOC was reached in 1979 by the development of a miniature gas chromatograph in a silicon wafer. At the beginning of 1990, several microfluidic devices such as microvalves and micropumps were developed by silicon micromachining, thus developing a platform for automating complex liquid handling protocols by microfluidic integration. Manz in 1990 developed the first microfluidic HPLC, which was marked as an emerging field of LOC devices, known as micro total analysis systems (μTAS). Simultaneously, much simpler microfluidic analysis systems based on wettable fleeces such as fabric or paper were discovered. The development of dipsticks for pH measurement based on a single fleece has paved the way for the development of complex strips for lateral flow movement, such as cardiac markers, pregnancy kits, and drug abuse. These strips are entirely automated and perform the biochemical analysis by microfluidic integration on a miniature paper. These strips are the first to obtain a vast market and remain one of the microfluidic systems sold in high numbers.

5.3.2 Materials Used for the Development of LOCs

The main issues with the manufacture of microfluidic devices are the formation of microfluidic channels, which are micro or even nanostructures. Here, the materials used in the manufacture of LOCs are of high prominence. Various materials such as silicon, glass, polymers, and paper are used to manufacture microfluidic channels. Since their inception, microfluidic channels were patterned directly into silicon due to their mechanical properties, good chemical resistance, and well characterized processing techniques with the capability of sensing circuitry. Glass is also preferred in the microfluidic channels due to high optical transparency and ease in the electro-osmotic flow. One of the most successful LOCs developed using glass is the capillary electrophoresis chip, manufactured using glass etching and fusion bonding techniques. Low cost, ease of manufacture, high reliability, and compatibility has made the manufacturers prefer polymers and paper. Polymers such as plastic are promising materials used for mass production using casting, hot embossing, injection molding, and soft lithography. Polydimethylsiloxane, a low cost, transparent, inexpensive, and elastomeric option with elastic properties at room temperature, is a workhorse material for the LOCs. Other polymer materials such as the derivatives of polyacrylate, polystyrene, polyethylene, and cycloolefin are also used to develop LOCs. Paper based LOCs are even cheaper and are simpler to manufacture. Commonly referred to as microfluidic paper-based analytical devices (μPADs), paper-based LOC devices can analyze a single liquid sample for multiple analytes. These devices are more functional than the traditional dipstick type paper devices as their functionality is achieved by creating channels and paths for the flow of reagents within the paper sheets, which allows the formation of distinct regions functionalized with chemical indicators.

5.3.3 Operation of a Microfluidic Unit

Operating a microfluidic system requires a micropump and microvalve, comprehensively controlling sample, buffer, reagent flow, and delivery. These two are the essential requisites for the operation of next generation LOC devices for integrating features such as sample separation, incubation, mixing, and separation. Two mechanisms were in use for the transportation of fluids, displacement pumps which work by exerting pressure and force on fluids through one or (sometimes) more than one moving boundary, and micropumps which work by reciprocating or rotary actuators. They also may contain piezoelectric, peristaltic, pneumatic, electrostatic, and electromagnetic moving units for displacing the fluids. Dilution of a sample, re-suspension, and reaction of reagents in LOC devices require rapid and efficient micro-mixing, achieved through active and passive mixing. External forces such as acoustic waves and magnetic beads are coupled with permanent magnets or actuated air bubbles during active mixing. During passive mixing, liquids are driven through microstructures, increasing the contact region between different streams to speed up either diffusive or chaotic mixing. Further, biochemical substances are separated by capillary electrophoresis, dielectrophoresis, isoelectric focusing, liquid (electro)chromatography, size-based filtration, magnetic fields, acoustic wave optical tweezers, and various combinations of flow, diffusion, and sedimentation-based phenomena (Chin et al. 2012).

5.3.4 Storage of Reagents on a LOC Device

Storing reagents on a LOC device for longer durations is one of the main properties, which increases their shelf life. Enzymes or antibodies can be stored in a wet or dry state. Wet storage is preferred if the dry storage will cause an unrecoverable loss of the activity. Most LOC devices prefer dry storage because the reagents, which are completely dried, have improved stability over those under wet conditions. Lateral flow assay strips consist of reagents such as antibodies maintained under dry conditions. Glucose sensors include dried glucose oxidase enzyme and electron transfer catalysts. Drying the reagents is generally achieved by either freeze-drying or lyophilization. The addition of sugars, such as trehalose, is one of the frequently implemented methods for improving the stability of reagents and retaining their activity.

5.3.5 Microfluidic Platforms Used as LOCs

A microfluidic platform provides a set of specific operations designed as a combination of manufacturing technology. A few microfluidic platforms are discussed here.

5.3.5.1 Lateral Flow Tests

Also known as test strips, where the capillary action drives the reagents stored in liquid form, the movement of the liquid is controlled by the wettability nature and feature size of the microstructured substrate. The readout of the test will be determined optically by the color change in the detection area, which the naked eye can detect. The most common example of this type is the pregnancy test strip.

5.3.5.2 Linear-Actuated Devices

These devices control fluid movement through a mechanical displacement such as a plunger. The liquid flow in this device will be linearly limited to one dimension without any branches or any alternate pathways. In these platforms, liquid calibrants and reaction buffers are pre-stored in pouches. The most common example of a linear actuated device is the i-STAT Analyzer manufactured by Abbott Point of Care Inc., USA. Using this portable hand analyzer, several blood parameters such as coagulation can be measured using different disposable cartridges. Once the blood sample is introduced into the cartridge, it is placed into the analyzer. Calibrant solution will be released first, which provides a baseline, and then the sample will be sent into the measuring chamber by displacing the calibrant solution. The analyzer will display the results of all determined parameters.

5.3.5.3 Centrifugal Microfluidic Devices

All processes in these devices are controlled by rotating a micro-structured substrate, which provides several forces such as centrifugal force, capillary force, Coriolis force, and Euler force by transforming Newton's laws of motion into a uniform rotating force for transport of the liquid. Assays are performed in a sequential manner of liquid operations arranged from radially inward positions to radially outward positions, respectively. Fluidic disks will transport the samples and reagents acted upon by the above mentioned forces. Fluids are pumped towards the rim of the disk with a wide range of flow rates controlled by the speed of the spin, channel dimensions, and surface energy. Temporary capillaries stop valves are opened to fluid passage by an increase in their rotational velocity. Microfluidic operations such as metering, switching, and aliquoting can be used for DNA and RNA extraction, plasma separation, and the like.

5.3.6 LOC Diagnostic Targets

Lab-On-Chip devices are being developed to identify the concentrations of biomolecules such as proteins, nucleic acids, metabolites, and sometimes whole cells. The biomolecules that are targeted by the diagnostic kits are mentioned below.

1. *Proteins*: LOC devices designed for targeting proteins will utilize immunoassay technology, including antigen-antibody binding. These assays are developed by targeting disease-specific protein markers such as glycated hemoglobin (HbA1c) for diabetes, C-reactive protein (CRP) for inflammation indicating cardiovascular disease, D-dimer for thrombosis, and troponin I or T for cardiac damage, prostate-specific antigen for prostate cancer, bacterial and viral infection-related markers such as human immunodeficiency virus (HIV), influenza, chlamydia, and hepatitis. One of the best protein-detection devices available in the market is the pregnancy test kit, a lateral flow strip, which measures the human pregnancy hormone chorionic gonadotropin. Currently, Bio-alternative Medical Devices Ltd., UK, is involved in developing a next generation pregnancy detection test using traditional chromatography-based lateral flow immunoassay tests by incorporating specific novel sensors, display, and power management capabilities for the detection.
2. *Metabolites and other small molecules*: Metabolites are the end products of chemical reactions and processes involved in the energy generation, and synthesis of nutrients and wastes. LOC assays are designed by targeting the metabolites, grouped as blood ions such as Na^+, K^+, Cl^-, non-protein hormones such as epinephrine, and cortisol, whose levels are the indicators of a specific disease. One of the best-known reagents for glucose determines the level of diabetes mellitus. Glucose biosensors account for ~85% of the global biosensor market. Most diabetic patients can regulate their sugar levels from home using handheld blood glucose meters such as the iSTAT system, manufactured by the Abbott Point of Care Inc, USA. These meters carry out a wide range of blood parameter analyses such as ions, carbohydrates including glucose and lactate, blood gases (pO_2 and pCO_2), peptides (brain natriuretic peptide), proteins (thrombin), and other blood indicators such as hematocrit. In less than two minutes, these analytes can be detected at clinically relevant levels using 65 μl whole blood samples.

3. *Nucleic acids:* Nucleic acid diagnostics, also known as molecular diagnostics, measures DNA, and various types of RNA for assessing the genomic and genetic details of a patient or to assay the DNA or RNA sequences, which are unique to pathogens. PCR and other sequencing methods for selectively amplifying the preselected nucleic acid sequences make such assays.
4. *Cells*: Identification and making a list of specific cells in the blood and other samples is one of the rapidly expanding fields in diagnostics. In addition to basic blood cell counting, POC cell assay-based devices conduct diagnostic and prognostic testing for infectious diseases, cancers, inflammatory responses, and hematological parameters.
5. *Pathogens*: Pathogens such as bacteria, viruses, and other parasites are vital targets for detecting infectious diseases. Rapid identification of the causative pathogen of infection can reduce treatment costs, reduce suffering, help systems against disease spreading, and save lives. As the identification of species and strain is a fundamental requisite, pathogens are often diagnosed by identifying their nucleic acid composition and sequence. In a few cases, immunoassays are used for the diagnosis of specific antibodies present in an infected host.

Detection for sensors on a microfluidic-based LOC device is of several types, designed based on optical, electrochemical, magnetic, and mass sensitive approaches. The development of detectors is based on two vital qualities, sensitivity, and selectivity that help in the minimization of false negatives and false positives.

5.3.7 Applications and Future perspectives of LOC

Presently 75 companies are manufacturing 154 LOC devices and 33 more devices are under development and trial. Most of the devices are involved in blood glucose and electrolyte analysis, HIV diagnostics, and the determination of cardiac markers. The leading companies in the LOC diagnostics market are Abbott, Alere, Arkray, Bayer, LifeScan, Menarini Diagnostics, Roche, and Siemens.

Miniaturization of LOC devices and their integration has enhanced their applications, especially for point of care. The degree of integrating a microfluidic technology in POC applications varies from having a disposable chip used with peripheral equipment such as pumps to having all functions needed for processing and analyzing a sample and reporting the results on the chip. The main selection criteria for developing LOC devices are portability, time to result, and costs per test. The size and weight of the LOC should be minimal, which affects the portability and energy consumption of the device. A hand-held device with low energy consumption is considered the most ideal for patient usage. The assay and results should be done between seconds and minutes, for which fully automated LOC devices would be ideal with the device analyzing the sample, calibrating the result, and recording and transmitting encrypted data wirelessly to an electronic record. The ability of a device to analyze multiple parameters from a single sample would be highly advantageous, as a small volume of samples would be needed for such devices (Gubala et al. 2012). Compared to conventional devices that typically use microliters of sample volume, microfluidic systems designed for using the nanoliter or picoliter of volume to fill the channels will reduce the reagent cost and ensure that the performance of the diagnostic tests is more efficient and accurate.

5.4 DIAGNOSIS OF GENETIC DISORDERS ASSOCIATED WITH THE CHROMOSOMAL ABNORMALITIES

5.4.1 Down Syndrome

Down syndrome is one of the well-known genetic disorders. This disorder occurs due to the trisomy of chromosome 21 (each cell has three copies of chromosome 21 instead of two). Down syndrome rarely occurs when a fragment or a portion of chromosome 21 gets translocated to

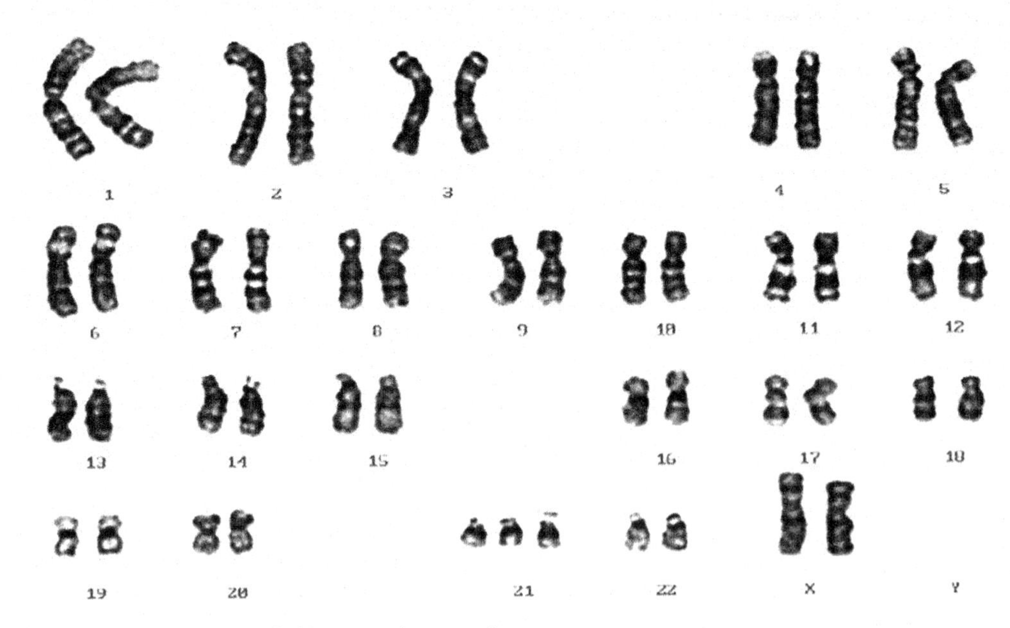

FIGURE 5.1 Karyotype of a Down syndrome patient.

another chromosome during spermatogenesis or oogenesis of the parents. Affected people have two regular chromosome copies and extra genetic material from chromosome 21, which is attached to another chromosome, resulting in three copies of chromosome 21 (Figure 5.1). The presence of an extra chromosome will disrupt normal development, leading to abnormalities in the body and brain development. In some cases, the physical development is slow and there might also be delayed mental development.

Down syndrome is reported to occur in 1 out of 800 newborn babies. In the United States, 5,300 babies are born with Down syndrome every year and more than 200,000 people have this condition. Even though women of any age can have a child with Down syndrome, their chances of having a child with Down syndrome will increase with their age.

5.4.1.1 Symptoms

Down syndrome leads to an intellectual disability with a characteristic facial appearance and a weak muscle tone due to hypotonia during infancy. All affected individuals experience cognitive delays, but their intellectual disabilities vary from mild to moderate. Children born with Down syndrome may experience various congenital disabilities such as heart problems, digestive abnormalities, and intestine blockage. With the child's growth, increased risk of medical conditions such as gastroesophageal reflux (backflow of acid and other stomach contents into the esophagus), celiac disease (intolerance of wheat protein gluten) will develop. 15% of people with Down syndrome have an underactive thyroid gland (hypothyroidism). Individuals with Down syndrome may also have an increased risk of developing hearing and vision problems. It was reported that a small percentage of children with Down syndrome develop leukemia.

Children with Down syndrome develop behavioral problems. Speaking and learning skills will be developed much slower in children with Down syndrome compared with healthy ones, and the speech of affected individuals will be challenging to understand. Behavioral issues such as attention problems, obsessive or compulsive behavior, stubbornness, or tantrums are likely to be developed in individuals with Down syndrome. Developmental conditions such as autism, which adversely affect their communication and social interaction, may also develop. People with Down syndrome experience a gradual decline in their thinking ability upon reaching 50 years of age. Down syndrome is also associated with the development of Alzheimer's disease, leading to gradual loss of memory, judgment, and functional disabilities. Even though Alzheimer's disease occurs in the old, people with Down syndrome develop this in their early fifties or sometimes late forties.

5.4.1.2 Etiology

In most cases, Down syndrome is not genetically inherited. The trisomy of chromosome 21 occurs as a random genetic event during germinal cell formation in the parents. An abnormality known as non-disjunction occurs in egg cells, and occasionally in sperm cells resulting in cells with an abnormal number of chromosomes. If these cells are involved in a child's genetic makeup, they will have the extra chromosome 21 in each cell of their body. People with translocation Down syndrome can inherit this condition from an unaffected parent.

In some cases, parents carry a chromosome 21 genetic material re-arrangement with other chromosomes, and this process is known as a balanced translocation. During this process, no genetic material is either gained or lost. Hence, the balanced translocation changes do not cause any health issues. However, this translocation can become unbalanced during inheritance to the next generation, leading to an extra chromosome 21, causing Down syndrome. A small percentage of people have an extra copy of chromosome 21 in some of the cells in their bodies. This condition is known as mosaic Down Syndrome. Trisomy 21 and mosaic Down syndrome are not genetically inherited. Mosaic Down syndrome occurs as a random event during cell division in the early stages of fetal development. As a result, some cells of the same human body have two copies of chromosome 21, while other cells have three copies of the same.

5.4.1.3 Diagnosis

Down syndrome can be detected during pregnancy through prenatal tests and after birth through karyotype analysis. Prenatal tests are performed based on the screening tests for determining Down syndrome. Two types of screening tests are being conducted, blood tests for measuring protein and hormone levels in pregnant woman, indicating their genetic condition, and the ultrasound imaging technique, which uses sound waves for generating the image of the fetus. An ultrasound scan can identify congenital heart conditions and other structural changes in the fetus, such as extra skin at the base of the neck, which might occur due to Down syndrome. Clinicians combine the results of both blood tests and ultra-scan, matching the mother's age for estimating the chances of a fetus developing Down syndrome. Further, the MaternaT21 or Panorama test for checking the DNA of the fetus in the mother's blood will indicate the exact probability of the fetus developing Down syndrome.

Diagnostic tests can determine the occurrence of Down syndrome in the developing fetus with 100% accuracy. Two diagnostic tests are in practice. The first of these is chorionic villus sampling, which is a diagnostic method performed during 10 - 14 weeks of gestation by conducting a biopsy of the placenta for a specific genetic test to detect any condition that involves chromosomal abnormalities such as Down syndrome. The second is amniocentesis which is a prenatal diagnosis method conducted after 15 weeks of gestation by inserting a needle into the amniotic sac that surrounds the fetus. Amniocentesis is one of the most frequently tests used to detect Down syndrome and other chromosomal abnormalities. Down syndrome can also be diagnosed after the birth of a baby either by fluorescent in-situ hybridization (FISH) which confirms the extrachromosomal material of chromosome 21, or by karyotyping which provides more intense information on the type of Down syndrome, which is essential for determining the chance of Down syndrome in a future pregnancy.

5.4.1.4 Treatment

As of now, there is no complete cure for Down syndrome. However, various treatments and therapies are in practice to address the medical and behavior issues due to Down syndrome. 50% of children with Down syndrome are born with congenital heart defects. Pediatric cardiologists treat these children depending on the intensity of the disease. A variety of therapies are available for addressing physical and behavioral issues, such as occupational therapy, physical therapy, speech therapy, and behavioral therapy. The average life span of people with Down syndrome has increased from 55 - 70 years compared with early 1900. With the help of specialized educational programs and support, many children with Down syndrome are living, growing, and successful in their lives.

5.4.2 Turner Syndrome

Also known as gonadal dysgenesis or gonadal agenesis, this is one of the chromosomal disorders that occur exclusively in females. This genetic disorder is named after Henry Turner in 1938, one of the first to report this disorder in medical literature on seven women with short stature, sexual immaturity, cubitus valgus, webbed neck, and low posterior hairline. This disorder occurs due to either partial or complete loss of one of the sex chromosomes (monosomy). In most cases, the missing second X chromosome in females leads to various symptoms due to the malfunction of different organs (Figure 5.2).

Generally, women with Turner syndrome are infertile. This disorder is not inherited in the families and appears to occur sporadically without any apparent reason. In the 1950s, advances in cytogenetics have allowed the identification of one normal X chromosome and one missing or altered X chromosome in the females with Turner syndrome. In 1965, Dr. Malcom Ferguson-Smith proposed the short stature of Turner Syndrome. Other clinical findings were due to deletions in the genes located on the X chromosome, which were validated in subsequent years.

In normal females, one copy of the X chromosome is inactivated for achieving a balanced gene expression between males and females. Thus, the absence of one X chromosome would not have any effect on female health. However, if the inactivation of the X chromosome is incomplete, 15% of the genes on the silenced X chromosome escape inactivation and are expressed on both chromosomes.

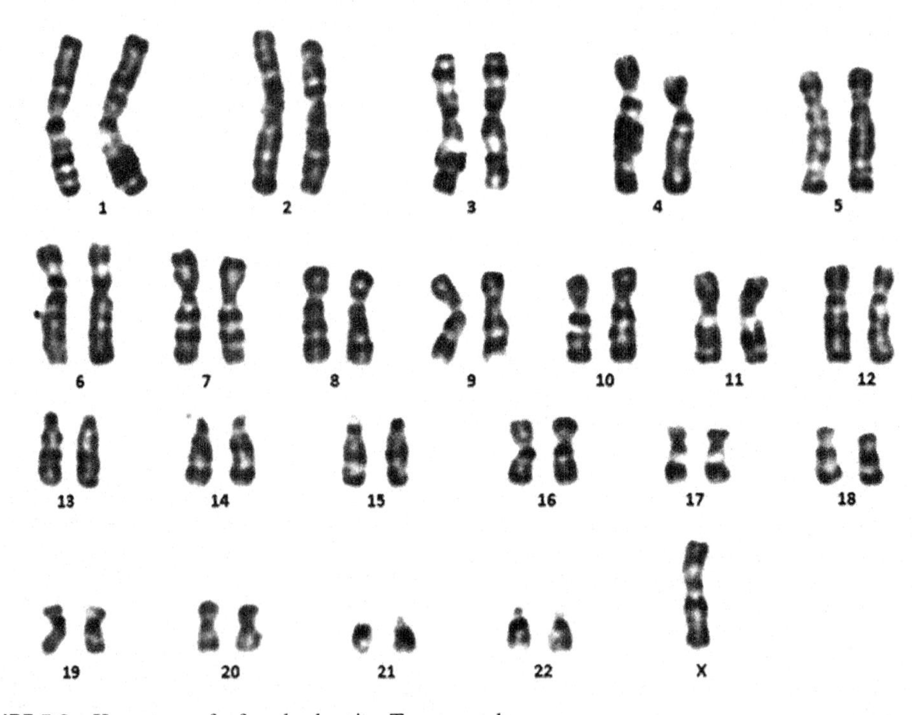

FIGURE 5.2 Karyotype of a female showing Turner syndrome.

Most of the genes that escape X-chromosome inactivation are pseudo autosomal genes located on the tip of both X and Y chromosome short arms tip inside the telomere of the pseudo autosomal region (PAR1) (Rao et al. 1997). Most of the pseudo autosomal genes located on the X chromosome have homologous genes on the Y chromosome. Hence, abnormalities in Turner syndrome might be due to the haploinsufficiency of the genes expressed by both the X chromosomes (Ellison et al. 1997).

5.4.2.1 Symptoms

Symptoms of Turner syndrome vary with the severity of the disorder. All females with this disorder exhibit growth failure and attain a shorter height than average, less than 5 feet. They develop distinctive physical features such as a short neck with a webbed appearance, low hairline at the back of the head, low set ears, narrow fingernails, and toenails turned upward. A broad chest with widely spaced nipples referred to as shield chest might occur. The hands and legs of some individuals will be swollen and puffy due to lymphedema. Additionally, few individuals with Turner syndrome may also develop a receding jaw (retrognathia), crossed eyes (strabismus), lazy eyes with drooping eyelids (amblyopia), and narrow high arched roof of the mouth. Some individuals may also develop tiny colored spots (pigmented nevi) on the skin.

One of the main features of Turner syndrome is the gonadal dysgenesis that occurs due to the failure of ovaries to develop, leading to the loss of ovarian function early during childhood. Ovaries are essential to produce estrogen and progesterone at puberty, necessary to develop secondary sexual characteristics, needing hormonal therapy. Intelligence levels in individuals suffering from Turner syndrome will be less. Individuals might also develop learning disabilities such as visual-spatial relationships, light-left disorientation, and directional sense. A few females may also face difficulty in social situations.

Some individuals with Turner syndrome may also develop skeletal malformations such as short bones of the hands, especially the fourth metacarpals, arms turned out at the elbows, and flat feet (pes planus). Under rare circumstances, individuals may also develop abnormal sideways curvature of the spine (scoliosis). Individuals with Turner syndrome may also develop congenital heart diseases due to lymphedema. Renal abnormalities such as horseshoe kidneys, and the absence of a kidney (agenesis) may also occur, increasing the risk of urinary tract infections and hypertension. Liver abnormalities such as fatty liver also can occur. Some affected individuals might show hypothyroidism. Problems with the immune system failing to recognize the thyroid gland leads to a condition known as Hashimoto syndrome, also known as autoimmune thyroiditis.

5.4.2.2 Etiology

Turner syndrome occurs due to either complete (monosomy) or partial loss of the second allosome. The reason for the monosomy is still unknown and is believed to be the result of a random event. In some cases, it might be due to an error in the cell division of the parent's germinal cells (spermatozoa or egg cells), resulting in a genetic error. Rare chromosomal abnormalities such as ring chromosomes or isochromosomes lead to Turner syndrome. Most of the symptoms of Turner syndrome are developed when cells receive only one copy of the X chromosome. In rare cases, some human body cells will have one copy of the X chromosome, while other cells will have one copy of the X chromosome and some Y chromosomal material. The amount of Y chromosomal material is not sufficient for the development of male features but will be associated with gonadoblastoma development, a rare form of cancer.

5.4.2.2.1 SHOX Gene and Turner Syndrome

The short stature homeobox containing gene, also known as the SHOX gene, plays a pivotal role in Turner syndrome (Gerhard, 2011). This gene was discovered by the deletion mapping on the short arm of the X-chromosome by Rao and his associates in 1997. Similar studies were also performed by Ellison in 1997 to discover the SHOX gene but called the PHOG gene, which was published much later. The SHOX gene is located on the tip X and Y chromosome short arms inside the telomere of the pseudo autosomal region (PAR1). PAR1 is around 2.6 Mb, which contains the genes that can escape X inactivation. Due to this, SHOX is expressed on both the X and Y chromosomes, and there is no difference between the SHOX X and SHOX Y. The SHOX gene is a hot spot for highly frequent recombination events between the X and Y chromosomes that occur during the meiosis pairing of spermatogenesis. The deletions in the SHOX gene on the short arm of the X chromosome or small terminal deletions in the short arm of the Y chromosome are associated with a short stature in human beings.

mRNA of SHOX gene comprising 7 exons named 1 - 5, 6a, and 6b encompassing up to 40 Kb of genomic DNA. This gene consists of two motifs. The homeobox motif encodes for a homeodomain, which enables specific DNA binding and the transactivation of the protein. The protein domain is involved in the nuclear translocation and dimerization of SHOX. The second motif encodes for a C-terminal located OAR domain involved in the transactivational activity of the SHOX gene. Alternate splicing of the SHOX gene results in the generation of two products, SHOXa and SHOXb, a shortened version of SHOXa, lacking the information for the OAR domain, hence is not an active transcriptional activator. The SHOX gene is a member of the paired-like type of homeobox containing genes, which play vital roles during embryogenesis and development by regulating the pattern formation in time and space (Elizabeth et al. 2017).

The occurrence of mutations in the SHOX gene and short stature ranges from 2 to 15%. Turner syndrome is associated with losing one SHOX gene due to either numerical or structural aberration of the X chromosome. Mutations in the SHOX gene cause the short stature of the family dominantly. A decrease in the SHOX gene protein is responsible for a decrease in linear growth. SHOX haploinsufficiency is a major cause of growth failure leading to Turner syndrome. A loss of the SHOX alleles results in a complete lack of SHOX, leading to an extreme phenotype osteodysplasia, termed Langer syndrome. A gain of one or two additional copies of the SHOX gene is associated with tall stature.

5.4.2.2.2 UTX Gene and Turner Syndrome

Ubiquitously transcribed tetratricopeptide repeat on the X chromosome (UTX) gene is vital in Turner syndrome. This disorder is also oriented with alterations in the immune response. Epigenetic regulation of T cells immune alterations, and their relationship with the haploinsufficiency of the UTX gene is also responsible for Turner syndrome. A gene expression microarray analysis has been performed on the peripheral blood mononuclear cells to identify pseudo autosomal X-linked genes associated with the immune alteration in the Turner syndrome. 1169 genes have shown a differential expression between Turner syndrome and control peripheral blood mononuclear cells. Among them, 35 are located on the X chromosome. UTX gene, located on the Xp11.3 locus has shown the most significant decrease in its expression. UTX is the only candidate gene that escapes X chromosome inactivation, and a histone H3 lysine 27 dimethylase regulates gene expression epigenetically. The UTX gene is required for optimal $CD4^{+}$ T cell differentiation to Tfh cells during chronic viral infection. Females with Turner syndrome have demonstrated a reduced UTX gene expression and reduced numbers of CD4+ T cells compared with the normal ones. Decreased UTX gene expression in individuals with Turner syndrome might increase their predisposition to viral infections due to deficiency in the Tfh cells with subsequent reduction in antibody levels. Recently, the roles of the TIMP1 and TIMP3 genes in developing bicuspid aortic valve and aortic abnormalities in female individuals with Turner syndrome were identified (Francisco and Roberto, 2018).

5.4.2.3 Diagnosis

Turner syndrome is diagnosed using both postnatal and prenatal tests. This disorder should be suspected in girls with short stature and growth deficiency without any appropriate reason. Further, karyotype analysis of the suspected blood sample will confirm the occurrence of the disorder. Turner syndrome is also diagnosed by using imaging techniques such as magnetic resonance imaging (MRI) to assess abnormalities in vital organs such as the liver, heart, kidney, and so forth. MRI produces cross sectional images of specific organs and body tissues by using a magnetic field and radio waves. Individuals with Turner syndrome undergo complete cardiac checkups through echocardiogram for assessment of heart structure and function. Evaluation of thyroid function, bone growth, and age, and hypertension screening will also determine the outcome of Turner syndrome in an individual. Infants showing Turner syndrome also undergo nose, throat, and auditory examinations.

Prenatal diagnosis of Turner syndrome is being achieved. Increased prenatal diagnosis is reported for Turner syndrome as there is no specific pattern for understanding the pedigree of this disorder. A non-invasive prenatal test for Turner syndrome on a maternal blood sample can be confirmed by amniocentesis after 16 - 18 weeks of pregnancy. The fetal ultrasound can determine specific physical changes in the fetus due to Turner syndrome, such as accumulation of lymph fluid around the neck region. If a prenatal karyotype shows Turner syndrome with a routine fetal ultrasound, the symptoms develop after birth, which will be difficult to predict.

5.4.2.4 Treatment

Treatment for Turner syndrome depends on the symptoms and varies from individual to individual. For the treatment of an infant suffering from Turner syndrome, a team of specialists comprising pediatricians, surgeons, cardiologists, endocrinologists, speech therapists, otolaryngologists, ophthalmologists, psychologists, and other healthcare professionals gather for systematic, comprehensive planning and implementation. Correction of hearing loss with a hearing aid and special services such as psychological support, and speech therapy will improve their social interaction. To date, there is no treatment for Turner syndrome. However, a few therapies are in practice for the improvement of physical development.

With appropriate medical care and therapy, females with Turner syndrome should be able to lead a complete and productive life. Growth hormone therapy and estrogen therapy are the primary therapies for the ones affected by Turner's syndrome. Individuals suffering from Turner syndrome will benefit from growth hormone therapy for normalizing their height. Food and drug administration has approved the usage of recombinant growth factor hormone for patients suffering from Turner syndrome, under the consultation of a pediatric endocrinologist. Most females with Turner syndrome might require sex hormone replacement therapy to undergo normal development associated with puberty. Estrogen and progesterone replacement therapy will promote puberty and the development of secondary sexual characters. The average age for females to undergo sex hormone replacement therapy is 12 years.

Most of the individuals with Turner syndrome are unable to conceive children. For them, in vitro fertilization with a donor egg and implanted pregnancy is sometimes possible. Nevertheless, these pregnancies carry risks and require close consultation and monitoring. Females with Turner syndrome and Y-chromosome mosaicism are at an increased risk of developing a tumor in the gonads, which can be attended by removing the non-functioning gonadal tissue through surgery. Thyroid hormone replacement therapy may be helpful in the treatment of individuals whose thyroid gland is affected due to Turner syndrome. Reports suggest that Turner syndrome affects one among every 2000 - 2500 live female births. The is an estimation of more than 75,000 women suffering from Turner syndrome in the United States of America. No ethnic group or racial factors are involved in the frequency of this X-chromosome-specific syndrome. Sometimes, genetic counseling will help control the spread of Turner syndrome into a family.

5.4.3 Klinefelter Syndrome

Klinefelter syndrome is one of the most common sex chromosomal genetic disorders that occur in male human beings. This disorder is a clinical manifestation that occurs due to an extra X

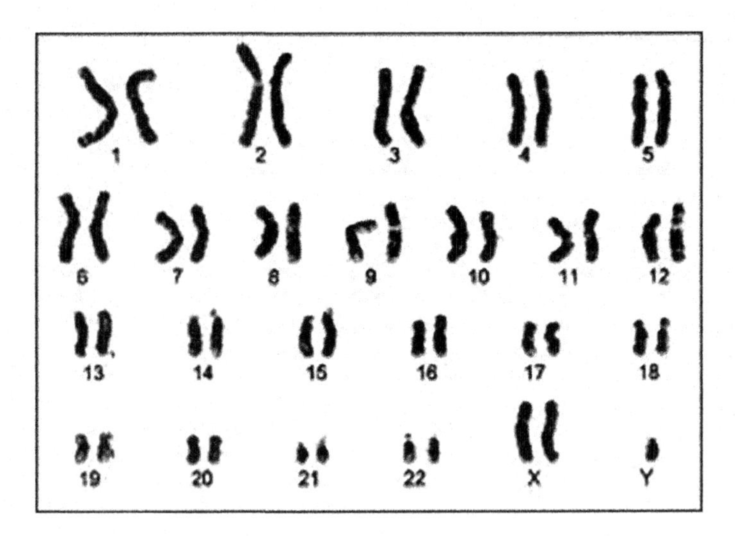

FIGURE 5.3 Karyotype showing Klinefelter syndrome (47, XXY).

chromosome in males. Klinefelter syndrome was first discovered by Harry Klinefelter in 1942 while demonstrating the elevated urine follicle stimulating hormone excretion among 9 patients with small testes, and gynecomastia, without the capability of spermatogenesis. Plunkett and Barr, in 1956 have studied the buccal mucosa cells of males with Klinefelter syndrome and described their specific condition and appearance known as Barr body. Jacobs and Strong, in 1959 correlated the Klinefelter syndrome with the chromosomal abnormality of possessing an extra X chromosome (47, XXY), responsible for this situation.

The majority of patients with Klinefelter syndrome (80%) have a karyotype of 47, XXY, referred to as a classic type (Figure 5.3). The other 20% of the patients have 46, XY and 47, XXY mosaic forms, with high level aneuploidies and severe X chromosomal structural abnormalities.

5.4.3.1 Symptoms

One of the significant targets of Klinefelter syndrome is the development of hypogonadism. However, the symptoms vary from individual to individual. Congenital malformations such as clinodactyly, cleft palate, and inguinal hernia were noticed more frequently in male infants suffering from Klinefelter syndrome when compared to healthy male children. These congenital malformations will increase in patients with a high degree of chromosomal abnormalities 70-100% of congenital malformations occur in individuals with chromosomal abnormalities 48, XXYY, 48, XXXY, and 49, XXXXY. The most common malformations are clinodactyly, cleft palate, inguinal hernia, cardiac abnormalities, and radioulnar synostosis. Genitourinary malformations such as hypospadias and undescended testes are the malformations that occur quite rarely.

The physical changes and developmental characteristics of children suffering from Klinefelter syndrome are not different from normal healthy children. However, the size of the penis and the volume of the testes are smaller in children suffering from Klinefelter syndrome, and these children are taller than normal children. Children with Klinefelter syndrome also develop problems while learning, such as delay in speech and movement, learning difficulties, delay in reading, and abnormalities in cognition such as aggressiveness and non-compliance. The degree of these problems increases with an increase in chromosomal aberrations. IQ levels of infants suffering from the classic Klinefelter syndrome are expected to decrease with increasing malformations. Most of these features are not detected in individuals with classic Klinefelter syndrome till they reach the age of adolescence. To date, only 10% of individuals with Klinefelter syndrome are diagnosed in their pre-pubertal period. The features will start emerging once they reach their adolescence. One of the most important physical changes in Klinefelter syndrome patients is gynecomastia, an increase in breast gland tissue in boys and men due to aromatization or deficiency of androgen. It is a permanent physical change in 50% of patients suffering from Klinefelter syndrome and will not change back with treatment. Androgen deficiency in Klinefelter syndrome patients will cause character disorders, personality disorders, psychological disorders, and behavioral problems.

Patients suffering from Klinefelter syndrome mosaicism will not display any symptoms and only 25% of the patients can get diagnosed during adulthood. Individuals with mosaic Klinefelter syndrome are more variable, and also display weaker phenotype properties than classical cases of the condition. The most common mosaic forms are 47, XXY/46, XY, 47, XXX/46, and XY. Mosaicism varies among individuals with this chromosomal arrangement, and the symptoms also vary. 14 - 62% of individuals with mosaic forms of 47, XXY/46, XY have normal XY karyotype in their testicular tissue, and their fertility is preserved. The bone mineral density of the patients suffering from Klinefelter syndrome will be average from childhood to puberty and will lower with age. Hypogonadism is responsible for bone mineral deficiency. Patients suffering from Klinefelter syndrome have a 20 - 50-fold risk of getting breast cancer compared with healthy men. 6% of all breast cancer cases in men are due to Klinefelter syndrome. The genetic risk of breast cancer will also increase with the existence of two X chromosomes. The probability of disease occurrences such as autoimmune disorders, type 2 diabetes mellitus, and leg ulcers is high in patients suffering from Klinefelter syndrome.

5.4.3.2 Etiology

An extra X chromosome occurs during gametogenesis if either sperm or egg cells carry an extra chromosome along with a normal sex chromosome. Klinefelter syndrome rarely occurs due to errors in mitotic divisions (3.2%) after zygote formation. 53.2% of the Klinefelter syndrome cases occur due to errors in the paternal division, and 43.7% in the maternal division. 34.4% of Klinefelter syndrome occurs due to errors in the first meiotic division, and 9.3% occurs due to second meiotic division errors. Most of the patients with Klinefelter syndrome have 47, XXY chromosomal arrangement. 10 - 20% of the patients with Klinefelter syndrome display abnormality and the rate of abnormality increases with an increase in the X chromosome number. Along with the numeric aberrations, structural chromosomal aberrations such as isochromosome Xq are also responsible for Klinefelter syndrome. These structural chromosomal abnormalities constitute from 0.3% to 0.9% of all patients suffering from Klinefelter syndrome.

A few theories explain the impact of the second X chromosome on the phenotype of Klinefelter syndrome. In healthy females, random inactivation of one X chromosome expression compensates for the Y chromosome's minimal gene content in males. These inactivated X chromosomes are visualized via microscope as Barr bodies (also known as sex chromatins) (Figure 5.4). X-chromosome

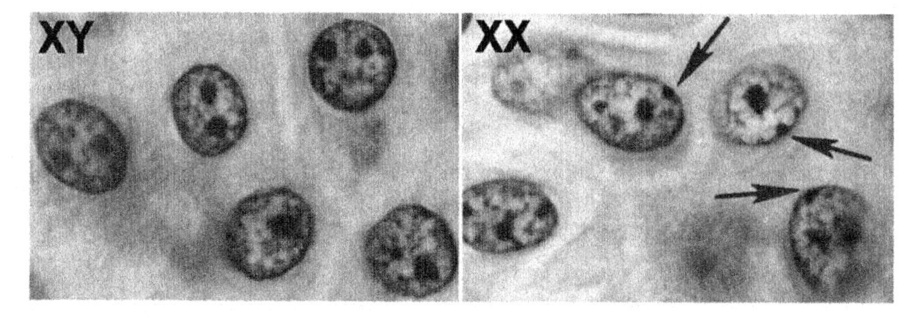

FIGURE 5.4 Nuclei of cells obtained from male (XY) and female (XX). The dark stained area identified by the arrow marks represents Barr bodies.

inactivation occurs due to the activity of the X inactive specific transcript (XIST) gene, located on the X chromosome and which expresses a long non-coding RNA. Expression of the XIST gene coats the X chromosome that was selected for inactivation.

Similarly, the X chromosome in the somatic cells of Klinefelter syndrome will be inactivated and result in the synthesis of the Barr body, used to diagnose this disorder (Figure 5.4). This inactivation of the redundant genetic material is one of the reasons that Klinefelter syndrome is not characterized by severe abnormalities, unlike other aneuploidy disorders. 65% of the genes on the X-chromosome of healthy females are inactivated, and the other 35% of the genes will escape this transcriptional inactivation either globally (20%) or randomly (15%). Hence, it can be pointed out that tissues in female cells are mosaics of cells with different X chromosome inactivation patterns. This mechanism does not exist in normal males with XY chromosomes but in males with XXY Klinefelter syndrome, responsible for genetic peculiarities such as gynoid proportions, autoimmune disorders, and breast cancer.

The androgen receptor gene undergoes inactivation in individuals with Klinefelter syndrome. This gene contains a stretch of CAG repeats coding polyglutamine. The length of the CAG repeats is correlated with the chromosomal arm length and arm span. Both X and Y chromosomes contain terminal regions exhibiting identical haplotypes. They recombine during the meiosis and are inherited in an autosomal mechanism. Hence, these regions are also known as pseudo autosomal regions (PAR). Most of the genes in the PAR will escape X chromosome inactivation, resulting in the generation of two active copies in males and three active copies in Klinefelter syndrome patients. One of such genes is a short stature homeobox on chromosome X (SHOX), located on PAR1, responsible for the tall stature and long legs observed in the Klinefelter syndrome phenotype. Many of the deregulated genes in Klinefelter syndrome patients are found to lie outside the X-chromosome. This fact implies that a supernumerary chromosome might affect the expression levels of multiple genes throughout the genome. Accordingly, differential methylation of multiple loci belonging to both male and female controls has been demonstrated in Klinefelter syndrome, where most of the genes have been related to energy balance and immune regulation.

5.4.3.3 Diagnosis

Characteristic features of patients suffering from Klinefelter syndrome will vary according to age. A reduction in testicular size is a significant feature of this disorder, with a testis volume of 3 - 4 ml. Attaining puberty is delayed in a majority of Klinefelter syndrome patients. The development of gynecomastia is another manifestation of Klinefelter syndrome. In patients with Klinefelter syndrome, serum gonadotropin and follicle stimulating hormone will begin with puberty. Levels of serum testosterone will start decreasing. However, normal serum testosterone levels are present in patients with mosaic form and much lower levels with more chromosomal defects. In classic Klinefelter syndrome, serum testosterone levels gradually decrease over time. Patients with Klinefelter syndrome also show an increase in their estradiol levels due to increased testosterone aromatization.

Another common clinical manifestation of Klinefelter syndrome patients is a severe reduction in the sperm number. Even though most patients will have normal ejaculation, sperm is rarely found in the ejaculate. Patients with mosaic form might contain sperm in their ejaculate. Histology of the patient's testes demonstrates the fibrosis of seminiferous tubules. Chromosomal analysis of lymphocytes isolated from peripheral blood is the most accurate diagnosis for Klinefelter syndrome. This disorder can also be diagnosed quickly and reliably by Barr body analysis of a buccal mucosa biopsy. The sensitivity and reliability of the Barr body analysis of buccal mucosa are 82% and 95%, respectively. Klinefelter syndrome is also detected prenatally by karyotype evaluation of the pleural fluid in amniocentesis as the fetus with Klinefelter syndrome does not show any noteworthy features in ultrasound scanning during pregnancy. Factors such as maternal age have no relationship with the increasing probability of Klinefelter syndrome. Nothing specific can be observed either during the delivery or in the infants with classic Klinefelter syndrome

5.4.3.4 Treatment

Testosterone replacement for patients suffering from hypogonadism is one of the regular treatments. Testosterone therapy will help increase muscle mass, strength, toughness, hair growth, bone mineral density, and libido during puberty. Testosterone enanthate and testosterone cypionate are the most used testosterone replacement therapy agents, injected intramuscularly, within 3-week intervals. Depending on the seminiferous tubule fibrosis in patients, testosterone levels decrease with the course of age. Hence, patients must follow this up periodically. Patients with gynecomastia should undergo cosmetic surgery for the resection of breast tissue.

If there is no spermatogenesis in patients with the Klinefelter syndrome, then the possibility of fertility is exceptionally low. In patients with sperm in their ejaculate, pregnancy can be possible with human chorionic gonadotropin therapy. However, if the patients are azoospermic, fertilization is not possible with gonadotropin treatment. In these cases, fertilization can be achieved by isolating the detected sperm through the testicular sperm extraction method (TESE) and ejaculation of the sperm into the ovum through intra-cytoplasmic sperm injection (ICSI). In patients with Klinefelter syndrome, a 40 - 50% chance of sperm extraction with the TESE method and up to 25% possibility of pregnancy with ICSI was achieved. Children of a father with Klinefelter syndrome will mostly have a normal karyotype. However, they have high rates of chromosomal hyperploidy (a chromosomal abnormality characterized by the addition of chromosomes resulting in a chromosome number that is not the exact multiple of its haploid number) and need genetic follow-up.

5.4.4 Huntington's Disease

Huntington's disease is one of the treatable monogenic neurodegenerative disorders, first identified by Waters in 1842. In 1872, George Huntington described and analyzed this disease as hereditary chorea, later named Huntington's disease. Huntington described the hereditary nature of the disease, which was associated with psychiatric and cognitive symptoms. Huntington had identified the manifestation of the disease in adult life between 30 and 40 years of age. He also mentioned the progressive nature of the disease. He stated, "*Once it begins, it clings to the bitter end.*" This chorea is a neurodegenerative disorder inherited from generation to generation, characterized by choreatic movements, and behavioral and psychiatric disturbances leading to dementia. In 1980, complete information on the non-motor signs and symptoms of the disease was obtained. A linkage with the loci on human chromosome 4 was established in 1983. By the end of 1993, the gene corresponding to the generation of the disorder was identified in the human genome. Identifying the gene has provided information for the premanifest diagnosis of this disorder and other similar disorders due to mutations in the genes responsible for an increase in the number of repeats. Huntington's disease also serves as a model for many studies on neurodegenerative disorders in medicine. If the symptoms of Huntington's disease start before 20 years of age, it is known as Juvenile Huntington's disease (JHD). In 75% of the cases, juvenile Huntington's disease is genetically transmitted through the father.

5.4.4.1 Symptoms

Major symptoms of Huntington's disease include neuromotor, cognitive, and psychiatric disturbances. Changes in neuromotor function are involuntary, leading to unwanted movements. These movements occur in the distal extremities such as fingers, and toes and will spread to facial muscles. Most of the facial muscle twitches are invisible. Slowly, unwanted movements spread to all other muscles from distal to proximal and axial. Choreatic movements are present all the time. Walking will be unstable, and the person's walk looks as if he or she is drunk. Talking and swallowing will become more problematic, leading to choking. Dystonia, characterized by slow movements with increased muscle tone, leading to an abnormal posture such as torticollis, is the first major motor sign of Huntington's disease. Hyperkinesia and hypokinesia generate difficulties in walking, and standing, leading to ataxic gait and frequent falling. Daily activities such as getting out of bed, toileting, showering, and so on, will become difficult for the patient.

Psychiatric and behavioral symptoms occur very frequently during the early stage of the disease, mainly with the onset of motor symptoms. Generally, 33 - 75% of Huntington's disease patients show psychiatric signs. These signs will profoundly impact the patient's daily life, functioning, and family life. The most frequent symptom is depression, and its diagnosis is difficult due to weight loss, apathy, and inactivity. The patient develops low self-esteem, feelings of guilt, and anxiety and may commit suicide in some extreme cases. Obsessions and compulsions may disturb the patient greatly, leading to irritability and aggression. Loss of interest and increasing passive behavior can be noticed. Psychosis may also appear during the later stages of Huntington's disease, leading to schizophrenia with paranoid and acoustic hallucinations. Hypersexuality during the initial stages of the disease can also pose problems in a relationship, which will slowly transform into hyposexuality during later stages (Raymund, 2010).

The first signs of juvenile Huntington's disease begin with behavioral disturbances and learning difficulties at school. Motor behavior is often hypokinetic and bradykinetic and chorea can be noticed only in the first decade of the disease and disappears in the second decade. The life of a Huntington patient transmitted from his father can be divided into three stages, at risk, preclinical, and clinical. During the at-risk stage, the stage determines whether a person carries the increased trinucleotide repeats on the chromosome or not. If the number of repeats supports the chance of the disease, he or she will undergo preclinical and clinical stages, which indicate a decrease in the patient's independence and increased care for him. Many scales were developed to follow a Huntington patient for research purposes. The best-known scales are the united Huntington disease rating scale, consisting of a history, motor, behavior, cognitive function, medication units, and the European network for Huntington's disease comprising of a whole assessment scale, used by more than 6,000 patients in Europe.

5.4.4.2 Etiology

Huntington's disease is a monogenic disorder, which is autosomal dominantly inherited due to increased CAG repeats in exon 1 of the Huntingtin gene located in the 4p16.3 locus on the short arm of chromosome 4, producing Huntingtin protein. The control gene in a healthy person consists of CAG repeats in a range of 6 - 26, where each trinucleotide repeat codes for a glutamine amino acid. If the number of repeats raises to 27 - 35, then it is considered potentially unstable with intermediate alleles with a chance of getting subjected to changes during reproduction. If the number of CAG repeats is 36 - 39, it is abnormal with incomplete penetrance of the disease or very late onset. CAG repeats above 40 is a definite clinical manifestation of Huntington's disease (McColgana and Tabrizia, 2018). An inverse correlation between the repeat length and age of onset has been observed. The first motor imbalance determines the correlation. Increasing repeat length will denote early onset of the disease. In the case of juvenile Huntington's disease, which begins at the age of 20 years, the length of the repeats often crosses 55. The length of the repeats will determine 70% of the variance in the onset's age and provide no hint or symptom about the illness before that. The only correlation that has been identified is the weight loss associated with the repeat length. The longer the repeat length, the fast will be the weight loss of the patient.

In healthy humans, the protein Huntingtin plays a vital role in synaptic function. It also plays an anti-apoptotic function, much needed in the postembryonic period, to protect against the toxic mutant huntingtin. The mutant form huntingtin will lead to a loss of function as well as a gain of function. The role of mutation in the generation of the disease is still not clear. Even though intranuclear and intra-cytoplasmic inclusions are found in the neurons of the patients, their roles are still not clear. One prediction that can be made while studying Huntington's disease is that when this disease is passed paternally from a father to a son, the length of the CAG repeats will be expanded, which might be due to a more significant repeat variability and larger repeat size shown by the sperm cells.

Genetic factors independent of the CAG length on the mutated Huntingtin gene are also reported to be modifying Huntington's disease. Genome wide association studies (GWAS) on Huntington's disease have identified genes involved in DNA repair, with the capability of altering the age of motor onset. Two such genes, FAN1 (Fanconi anemia FANC1/FANCD2 associated endonuclease) and MTMR10 (myotubularin related protein 10) have been identified on chromosome 15 to be the most significant. Considerable associations were also noticed with RRM2B (a subunit of DNA damage p53-inducible ribonucleotide reductase M2 B) and URB5 (a HECT domain E3 ubiquitin-protein ligase) genes located on chromosome 8.

Huntington's disease is a neuropsychiatric disorder with a prevalence of 5 - 10 per every 100,000 in the Caucasian population, and 1 per 10,000 in the United States population. 5 at-risk individuals for every affected person are existing. About one tenth prevalence of this disease in the Japanese population. Taiwan and Hong Kong have a much low prevalence of 1-7 per every million populations. In South Africa, black populations show less disease prevalence when compared with the white and mixed populations. The difference in the prevalence among different ethnic groups is correlated with differences in the Huntingtin gene. Populations with a high frequency of Huntington's disease have a long CAG repeat average as identified in European ancestry with 18.4 - 18.7 and Asian ancestry with 16.9 - 17.4. Recently, several Huntington's disease phenocopies with lower prevalence have been identified and studied. These phenocopies are identified and defined by diagnosing the disease with chorea, psychiatric and cognitive symptoms, inherited in an autosomal dominant pattern in family history.

5.4.4.3 Diagnosis

In most cases, the diagnosis of Huntington's disease is based on a person's clinical symptoms and psychiatric, and behavioral signs. Before 1993, this disease diagnosis was mainly on the family history with clinical and behavioral verification, supported by medical records and biopsy reports. The current way of determining the disorder is counting the CAG repeats in the Huntingtin gene on chromosome four, which is considered the gold standard for diagnosing the disease. Studies are underway to determine the biomarkers from clinical tests such as the blood test, DNA test, and MRI scan. Changes in function and brain imaging through MRI scan before clinical manifestation are also under trial because brain volume and brain connections are subjected to changes several years before the clinical symptoms begin. Prenatal diagnosis of Huntington's disease through a DNA test is being practiced for the earliest detection. This diagnosis is done by chorionic villus sampling during 10–12 weeks and amniocentesis during 15–17 weeks of pregnancy for the DNA test. In the last few years, pre-implantation diagnostics are also practiced. This pre-implantation starts with in vitro fertilization, and when the embryo is in eight cell stage, one cell is separated for DNA testing. Without any long CAG repeats, the embryo is placed in the mother's womb and develops pregnancy. Before initiating the pre-implantation process, one must make sure that the genetic status of the parent is thoroughly known.

5.4.4.4 Treatment

As not much information is available regarding the pathogenesis of Huntington's disease, a permanent cure is not available. Despite many therapies for the treatment of symptoms, they only help improve the patient's condition temporarily. Many of the drugs prescribed for symptoms are personalized and based on expert opinion and daily practice. Psychiatric signs such as depression and aggressive behavior are treated with medication and non-medical interventions such as physiotherapy, occupational therapy, speech therapy, and the like, which will improve the quality of the patient's life. Unfortunately, no permanent cure is available for Huntington's disease to date.

5.4.4.4.1 DNA and RNA Therapies for Huntington's Disease

Gene therapy is one of the most promising approaches for the permanent cure of Huntington's disease, achieved by decreasing the expression of the mutant Huntingtin gene in humans by targeting either its DNA or RNA (Edward and Sarah, 2017). Two different approaches can target the DNA of the mutated Huntingtin gene, the zinc finger proteins approach and the clustered inter-spaced short palindromic repeats (CRISPR/Cas9) system. Zinc finger proteins are the structural motifs that bind to DNA. Synthetic zinc finger transcription factors that can target and bind to CAG have been used to reduce the levels of mutated Huntingtin protein in animal models. Further work is in progress to control the immune reactions. Bacteria use CRISPR/Cas9 to cleave foreign DNA. This approach is used as a tool for genome editing with a multitude of applications in human disorders. CRISPR has been used in the fibroblasts of a patient, with Huntington's disease, for excising the promoter regions, transcription start site and a patch of CAG repeats in the mutated Huntingtin gene, resulting in the permanent and mutant allele specific inactivation of the mutated Huntingtin gene. This approach has been successfully tested in a Huntington's disease rodent model, affirming the feasibility. However, much work needs to be done for bringing this technology to a functional state.

RNA of the mutated Huntingtin gene can be targeted using three approaches, antisense oligonucleotides, RNA interference, and small molecule splicing inhibitors. Antisense oligonucleotides are under their first human trial (phase 1b). They are delivered for catalyzing the degradation of mutated Huntingtin gene mRNA by RNAse H, thereby reducing the production of the mutated Huntingtin protein. 80% reduction of mutated Huntingtin gene mRNA levels has been achieved today through this approach. RNA molecules are bound to mRNA in the cytoplasm in the RNAi approach, prompting its removal by argonaut 2, the RNAse enzyme present in the RNA-inducing silencing complex (RISC). Approaches using this technology are in the preclinical stage. Delivery of RNA into cells is more invasive than antisense oligonucleotides, which require an intracranial injection into the striatum. Small molecule splicing inhibitors have shown a promising result in animal models with small muscular atrophy, and a similar approach is underway to identify the small molecule splicing modulators of mutated Huntingtin gene RNA.

5.4.5 Fragile X Syndrome

Fragile X syndrome is a clinical and cytogenetic entity responsible for several intellectual disabilities after Down's syndrome. Fragile X syndrome is also known as Martin-Bell syndrome after identifying the pattern in decreasing intellectual ability and linking it with X-chromosomal inheritance in 1943. Lubs (1969) was the first to report a specific fragile site on the X-chromosome, segregated into three generations in a family and responsible for the intellectual disability. He also observed a marker on the X chromosome of a mother and her three sons in a family, who expressed a constriction at the end of the long arm of the X-chromosome's and named it a fragile site, hence giving the name fragile X syndrome. In 1991, the locus responsible for the fragile X syndrome on the X chromosome was identified. Currently, fragile X syndrome is responsible for 2.5% of all intellectual disabilities and is the most prevalent cause of intellectual disabilities in males, with a range of 1 per every 5000 - 7000 men globally. Fragile X syndrome is not a chromosomal disorder inherited through Mendelian genetics but occurs due to an increase in trinucleotide repeats in the 5′ UTR region of a gene located on the X chromosome.

5.4.5.1 Symptoms

Men affected by fragile X syndrome are characterized by a long face, intellectual disabilities, large protruding ears, and macroorchidism (large testicular size) in post pubertal males. Females, heterozygous for the fragile X syndrome have a 30% chance of normal phenotype and 25% chance of IQ levels <70, presenting learning deficits and emotional difficulties. This phenotype in females represents an X-inactivation pattern.

Fragile X syndrome shows a multi-systemic condition with the capability to affect any human body organ. Even though the child looks normal at the time of birth, facial characteristics such as a long narrow face, prominent ears, prominent lower jaw, a high arched palate, puffiness around eyes, long palpebral fissures, closely spaced eyes, epicanthal folds, flat nasal bridge, broad nose, broad philtrum, and facial hypotonia will become more distinctive during early childhood. Not all features are noticed in children with this disorder, but 30% of young children will show most of these characteristics. One of the most prominent defects noted in patients with fragile X syndrome is a global developmental delay in intellectual abilities. A delay in psychomotor abilities occurs at walking age and first wordage. Patients have various learning disabilities, classified as normal, mild, and severe intellectual disabilities. Females with fragile X syndrome (scarce) present various phenotypic characters, mostly involving cognitive impairment.

A significant co-morbidity of fragile X syndrome is epilepsy. Children with this disorder show both partial and complex seizures and the prevalence of seizures are 10 - 20% in boys. The onset of seizures is 2 - 10 years. Patients with fragile X syndrome have 74% of electroencephalogram (EEG) abnormalities. They also develop atrophy of cerebellar vermis, thinning of the corpus callosum, hippocampal anomalies, enlarged fourth ventricle, and lacunar infarction of the basal ganglia. A study by Hall and his team members in 2016 has identified an increased fractional anisotropy in patients with fragile X syndrome in the left, right inferior fasciculus, left cingulum hippocampus compared with that of healthy humans. Most the patients with fragile X syndrome show connective tissue disorder. The skin will be soft with hypermobility of the small joints (metacarpal and phalangeal). Skeletal abnormalities may include a high-arched palate, scoliosis, pectus excavatum, and flat feet.

5.4.5.2 Etiology of Fragile X Syndrome

Inheritance of fragile X syndrome is according to the changes in fragile X mental retardation-1 gene (FMR1), located on the Xq27.3 locus of the X chromosome. 99% of the disorder occurs due to the loss of function caused due to an increased number of (more than 200) CGG trinucleotide repeats in the 5′ UTR region. This type of allelic constitution is known as a full mutation, which generates the expression of the cytogenetic fragile site (FRAXA). Expression of FRAXA leads to the hypermethylation of the promoter followed by the inhibition of FMR1 transcription. All this process results in a drastic reduction of the fragile mental retardation protein (FMRP). The fragile X syndrome phenotype is the consequence of the absence of FMR Protein (Fu et al. 1991). The other 1% of the disorder is due to alterations in the FMR1 gene such as deletions, duplications, and single nucleotide variations.

In a healthy human, the number of CGG repeats in the FMR1 gene ranges from 5 - 44. An increase in the repeat number ranging from 45 - 54 is considered a grey zone, characterized by peripheral neuropathy, ataxia, anxiety, depression, clinical symptoms such as bradykinesia, rigidity, memory complaints, and a positive response to dopaminergic medications. Repeat range from 55 - 200 is known as pre-mutation, associated with premature ovarian failure in females (significantly very less frequent) and fragile X-associated ataxia syndrome in males. Repeats above 200 are considered a full mutation. Transcription silencing, the absence of FMRProtein due to hypermethylation of the CpG islands adjacent to the expanded trinucleotide repeats are the outcomes of full mutation. Heterochromatin confirmation of the FMR1 gene promoter region may also lead to the disorder. The transition of a pre-mutation into full mutation can occur due to the expansion of a maternal X chromosome (or a paternal X chromosome which is very rare) carrying a pre-mutation to her children. The individuals with pre-mutation allele frequency are one in eight hundred and fifty males and one in three hundred females.

Fragile X syndrome occurs due to a rare, highly fragile, unsteady site on the Xq27.3 locus, named FRAXA. This site is visualized as a non-staining gap in the metaphase chromosomes under selective culture conditions, such as folic acid or thymidine deprivation. The FMRProtein is involved in various functions such as regulating post-transcriptional RNA metabolism, synaptic plasticity, dendrite, and axon development of nerve cells, in underlying learning and memory. FMRProtein acts as a shuttle for RNA transport and RNA binding protein during translation. FMRProtein is also involved in a feedback loop by controlling its protein levels. The absence of FMRProtein leads to a variable expansion of the trinucleotide CGG repeats in the 5′UTR region of the FMR1 gene. A low trinucleotide CGG repeat number may also have some important clinical implications. CGG number below 26 is also considered a risk factor for cognition disability and mental health problems. Hence, it can be understood that a delicate homeostatic equilibrium exists between the expression levels of FMRP and CGG repeat numbers, which ultimately alters brain function.

Chances of a pre-mutation allele becoming a full mutation allele can be correlated with the number of CGG repeats. When a mother transmits the permutation, alleles with repeats of more than 100 have a chance of becoming full mutation alleles in the next generation. If the father transmits the permutation, a slight increase in the repeats might occur but not result in full mutation. The paternal alleles with <70 repeats have a possibility of instability. Maternal alleles with a large repeat size of more than one hundred can cause instability. For example, grey zone alleles with repeat number 54 are unstable and can evolve into pre-mutation alleles.

Another major factor that influences the stability of the FMR1 gene is the presence of AGG tri-nucleotide repeats interspersed within the FMR1 repeated region. One or two AGG repeats are interspersed with CGG repeats, in the 5′ end of the FMR1 gene. These CGG interspersions are present in 94% of the population alleles. Maternal alleles without any AGG interspersions have a significant risk of becoming full mutations for the disorder, whereas the presence of even a single AGG repeat will significantly reduce the risk until repeat number 70. If the repeat number exceeds 70, the allele shows instability even interspersed by the AGG repeats. If the length of the CGG repeats goes beyond 90, AGG interruptions will not have any ability to control the instability. A rare decrease in the CGG repeat number retracts the allele from full mutation to partial mutations and partial mutations to the normal length of the repeats. The retraction from full mutations to partial mutations appears to be sporadic. This retraction occurs post-zygotically due to the excision of a variable number of trinucleotides leading to a pre-mutation or grey zone or a normal size of the repeats. Excision of repeats may also lead to mosaicism, a significant source of phenotypic variability observed mainly in males. The prevalence of mosaicism in fragile X syndrome varies between 12 and 45% among male patients. The variations might be due to varying degrees of methylation from tissue to tissue due to deletions during embryonic cell divisions, leading to two distinct subpopulations of cells carrying both deletions and full mutation alleles.

In a few cases, this disorder occurs through mutations in the coding regions or deletions in the FMR1 gene. Mutations such as point, missense, nonsense, and frameshift are responsible for the fragile X syndrome in 4% of patients with a normal range of CGG repeats. X chromosomal inactivation is another crucial factor responsible for this disorder. Inactivation occurs when one of the X chromosomes gets silenced in mammalian females. Under normal conditions, X-chromosome inactivation occurs randomly. However, preferential X-chromosome inactivation has also been observed. X chromosome carrying FMR1 mutation is preferentially inactivated. Hence, FMRProtein produced by the normal allele results in a less severe phenotype.

5.4.5.3 Disease Diagnosis

Fragile X syndrome is postnatally and prenatally diagnosed using cytogenetic and molecular analysis. Initially, Fragile X syndrome was detected through cytogenetic analysis for the presence of the FRAXA site in peripheral blood lymphocytes. However, this procedure was time consuming and required specific technical skills. This technique was also incapable of distinguishing between FRAXA and other neighboring fragile sites on the X chromosome. These limitations were overcome by utilizing FISH, using DNA probes for hybridization. Cytogenetic analysis was replaced by Southern blot analysis of DNA isolated from the peripheral blood cells by digesting with specific restriction endonucleases. Using Southern blot hybridization, all FMR1 normal, pre-mutation, and full mutation alleles can be detected. One can also determine the methylation status of the FMR1 gene promoter. However, Southern blot hybridization is time consuming and expensive. Standard PCR and southern blot hybridization were the gold standard for detecting the FMR1 gene for a long time, despite its low-resolution estimation of the CGG repeat number (up to 300 repeats in males and 160 repeats in females). To increase the resolution, triplet primed PCR (TP-PCR) was designed (Claudia et al. 2017). This technique comprises a forward primer located upstream of the CGG region. Another primer overlaps the CGG repeat and the adjacent unique sequence region. After PCR amplification, the CGG repeat can be determined by separating the amplified product on capillary gel electrophoresis. Using TP-PCR, FMR1 alleles are amplified by using primers flanking the repeated region, and CGG repeats are simultaneously amplified using a third primer complementary to the FMR1 triplet repeat region. TP-PCR is the best technique for the first level diagnosis of fragile X syndrome due to its ability to detect expanded alleles even in mosaic regions. During the second level analysis, the silencing of an FMR1 gene promoter is evaluated by methylation testing. Specific kits such as the methylation-sensitive long-range PCR kit measure the methylation level in each FMR1 allele, using DNA digested with methylation-sensitive restriction endonucleases. This approach is adapted to identify a rare full mutation in unmethylated males, and in asymptomatic carriers. Fragile X syndrome is prenatally diagnosed by using LR-PCR amplification of DNA isolated from either chorionic villi or amniocytic cells under the American College of Medical Genetics and American Congress of Obstetricians and Gynecologists guidelines.

5.4.5.4 Treatment

There is no permanent cure for fragile X syndrome. However, a wide variety of treatments are available for minimizing the effect of this disorder. Pharmacological agents such as anti-depressants and anticonvulsants are used under the direction of medical experts. A ray of hope for the permanent cure of this disorder is the development of specific treatment. Two possible approaches are being considered, reactivation of the affected gene, and compensating for the lack of the FMR1 promoter. The strategy for restoring the activity of the FMR1 gene is based on the coding sequence and targeting the reversible epigenetic changes, mainly DNA methylation. The first drug tested on the cells derived from fragile X syndrome patients was 5-aza-deoxycytidine, a methyltransferase inhibitor, which restored the transcription and translation of the FMR1 gene. Further, treatment with histone deacetylase inhibitors such as butyrate and 4-phenylbutyrate has enhanced DNA demethylation and generated the changes in the epigenetic code of histones H3 and H4, resulting in the reactivation of the inactive and methylated full mutation allele. Even though the main reason for this demethylation is unknown, pharmacological conversion of a methylated full mutated allele into a normal control allele is one of the most targeted approaches for curing this disorder.

5.4.6 Thalassemia

Thalassemia (s), also known as Mediterranean anemia, is a genetically inherent disorder that affects hemoglobin synthesis in humans. Physicians, Cooley and Lee first discovered this disorder in 1925 in a series of profoundly anemic infants who had developed splenomegaly and bone change during the first year of their birth. The pathological conditions of the disease were first described by George and William in 1932 in the patients, belonging to the Mediterranean region and named the disease thalassemia. The genetics of this disorder have been studied since 1940.

A healthy hemoglobin molecule is a heterotetramer during normal conditions, consisting of two α-globin chains and two non-α-globin chains known as β-chains, each of them carrying a heme molecule with a central iron. This structure enables the hemoglobin molecule to have maximum oxygen carrying capacity. α-globin chains are coupled with β-globin chains to form a major fraction of adult human hemoglobin, whereas α-globin chains coupled with δ-chains form a minor fraction of adult hemoglobin. α-globin chains are coupled with γ-globin chains to form the hemoglobin in fetal cells. The physiology is controlled by the balanced production of the α, β, γ and δ globin chains, ensuring a reciprocal pairing of these chains into normal tetramers. This equilibrium is disturbed by the reduced production of globin chains due to mutations in the genes responsible for their synthesis, leading to thalassemia(s). Reduced α-globin chains lead to an excess synthesis of g-globin chains in the developing fetus and the newborn. This will further enhance the levels of b-globin chains in children and adults. Excess accumulation of these chains will accelerate the destruction and damage of RBC. Reduced production of these globin chains within the RBC will lead to an imbalance in the tetramer, which makes it incapable of carrying oxygen to its fullest potential, resulting in anemia. This anemia is known as hereditary anemia as it occurs due to mutations in the DNA that can cause severe mortality and morbidity. Further, excess production of β-globin chains leads to the formation of either β-4 or Hb-H soluble tetramers, and this form of hemoglobin is precarious and tends to precipitate within the cell, forming insoluble inclusions (also known as Heinz bodies), damaging the RBC membrane (Hamidreza, 2018).

Thalassemia(s) is a heterogeneous group of inherited disorders affecting hemoglobin synthesis. All forms of thalassemia are characterized either by the absence or reduction of one or more globin chains. Thalassemia is classified into α, β, δβ, γδβ, and ϵγδβ types. Classification is based on a specific globin chain, whose synthesis is hampered in the phenotype (Michael and Stephan, 2019). However, thalassemia occurs due to an imbalance in the synthesis and production of α and β globin chains. The most prevalent clinical forms of thalassemia are α-thalassemia and β-thalassemia. α-thalassemia occurs due to a decrease in the production of α-globin chains, resulting in an excess number of unpaired β-globin chains. α-thalassemia is responsible for a diminished hemoglobinization of individual RBCs, resulting in damage to the erythrocyte precursors and ineffective erythropoiesis in the bone marrow, hypochromia, and microcytosis in circulating RBC. β-thalassemia occurs due to a reduction in β+ or absence of β0 globin chains, leading to accumulation and precipitation of α-globin chains in the erythroid precursors and circulating RBC. Precipitation of α-chains results in insufficient and ineffective erythropoiesis enhanced erythropoietin, and proliferation of the bone marrow (25-30 times more normal than expected). This enhanced bone marrow proliferation can lead to further bone abnormalities. Prolonged, severe anemia and increased erythropoiesis result in hepatosplenomegaly and extramedullary erythropoiesis, leading to ineffective erythropoiesis and RBC pre-mature death. The degree of globin chain reduction depends on the intensity of mutations occurring in the genes responsible for their synthesis.

5.4.6.1 Symptoms

The production of hemoglobin starts in the pro-erythroblast cells, which further increases during the maturation of erythroid cells through basophilic, polychromatophilic, and orthochromatic phases of RBC maturation. Excess α-globin chains will precipitate the cell membranes of erythroblasts, leading to oxidative membrane damage and pre-mature death through apoptosis, leading to ineffective erythropoiesis. Some of these immature RBCs will enter the blood circulation. Due to their membrane defect, they are fragile and prone to hemolysis. These immature RBCs also exhibit altered deformability and are trapped by the spleen and destroyed by the macrophages. This continuous removal of RBC, WBC, and platelets by the spleen leads to massive spleen enlargement and it eventually develops functional hypersplenism. Overall, ineffective erythropoiesis, removing abnormal cells by the spleen, and hemolysis will contribute to severe anemia. The physiological response to anemia is twofold. Firstly, there is enhanced erythropoietin secretion by the kidneys. Erythropoietin is a cytokine that targets RBC precursors in response to the oxygen requirement of tissues. Erythropoietin secretion will enhance defective RBC production. Defective RBC makes the erythropoiesis worse, leading to various complications such as the expansion of hematopoietic tissue in bone marrow, destruction of bone architecture, bone diseases, and extramedullary hematopoietic mass development in the liver, spleen, and reticuloendothelial system. Secondly, there is the production of hepcidin, an iron absorption regulator by the liver cells. Hepcidin regulates the expression of ferroportin protein, facilitating enterocytic iron absorption in the gut region. Severe anemia suppresses hepcidin production, resulting in enhanced iron absorption and leading to a severe iron load.

5.4.6.2 Etiology

Thalassemia is a genetic disorder inherited in an autosomal recessive manner through the impaired synthesis of either α or β globin chains. According to their order of activation and expression, genes involved in the synthesis and regulation of globins are arranged in two different clusters, in a sequential mode of 5′ to 3′ direction. α-globin genes are located as a cluster at 30 Kb in the telomeric region of chromosome 16 short arm. The cluster includes an embryonal gene, three pseudogenes, two α-genes, named α1 and α2, and a pseudogene θ (Figure 5.5).

Gene clusters corresponding to the synthesis of β, γ, and δ globins are located in the 60 Kb region on the short arm of chromosome 11. The cluster consists of genes corresponding to ε, Gγ, Aγ, pseudogene Ψβ, δ, and β arranged in 5′ to 3′ order (Figure 5.6). All globin genes share a standard structure comprising of three exons separated by two introns. Regulatory regions required for gene expression are highly conserved and located at the proximal promoter regions, intron, exon boundaries, and 5′ and 3′ UTR regions.

Globin chains in the developing fetus are encoded by Gg and Ag genes. Sequences of both these genes are almost similar except for the replacement of glycine residue by alanine in the Ag gene at codon 136 in the proximal promoter region. Along with the promoter regions, both α and β gene clusters are equipped with an upstream regulatory region, which plays a pivotal role in the erythroid specific gene expression and the developmental regulation of each gene in the clusters. In the α-gene cluster, this regulatory region is known as HS-40, located upstream of the embryonal α-globin gene. This region in the β-cluster is the locus control region (LCR), a significant element of size 20 Kb, located at the 25 Kb upstream of the ∈-globin gene. This region consists of fine DNAse I hypersensitive erythroid specific sites, HS-1,2,3,4 and 5, which define the sub-regions of open-chromatin bound by multiprotein complexes. The expression of globin genes is regulated by refined and complex transcriptional, post-transcriptional and post-translational mechanisms to ensure the equal production of globin chains for the correct and proper hemoglobin assembly (Daniel, 2017).

α-thalassemia has both absence (known as α^0 thalassemia) and reduction (known as α^+ thalassemia), resulting in an imbalance of globin chains in the hemoglobin. Consequently, the excess β and γ globin chains aggregate to form homotetramers Hb Bart's and HbII, respectively. These homotetramers are responsible for severe hemolytic anemia. α-thalassemia forms are further subdivided into deletional and non-deletional according to the defect. α-thalassemia forms arise due

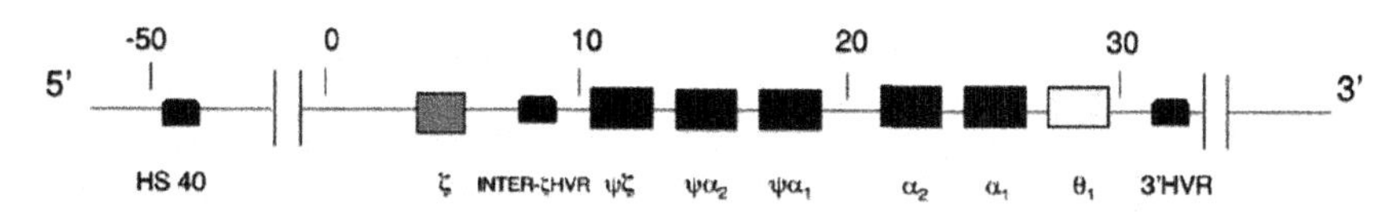

FIGURE 5.5 Structure of an α-globin gene cluster located on chromosome 16.

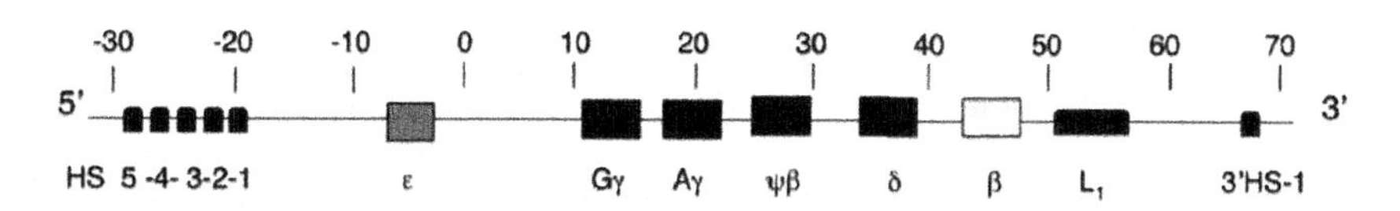

FIGURE 5.6 Structure of β-globin gene cluster located on Chromosome 11.

to deletions in one or both α-globin genes located on chromosome 16. The non-deletional form of α-thalassemia arises due to point mutations that affect the functional expression of one α globin gene. However, non-deletional forms are less common. The main reason for the increased susceptibility of the α-globin genes deletional defects is highly homologous regions scattered within the cluster, predisposing to events such as unequal recombination. A healthy normal individual has four α-globin genes as each chromosome 16 carries two linked α-globin genes. Hence the normal genotype can be written as αα/αα. Deletions in the genes result in the loss of one (-α) or both (--α) of the duplicated α-globin genes from the same chromosome. The number of remaining functional genes determines the clinical significance of α-Thalassemia. The deletional loss of either α-1 (α3.7 deletion) or α-2 (α-4.2 deletion) is the most common molecular reason responsible for α-thalassemia. Unequal crossing over events responsible for the origin of these genes also gives rise to triplicated or quadruplicated gene rearrangements. In α^0 thalassemia, large deletions responsible for removing the entire α-globin cluster region have been identified.

Non-deletional α-thalassemia occurs due to point mutations in the genomic regions critical for regular expression of the α-genes. Point mutations affecting the α-2 gene will drastically impair the expression of the α-globin gene as the α-2 gene expression is three times more than that of the α-1 gene during normal conditions. Point mutations in the α-thalassemia genes have an adverse effect on RNA processing (α^{hph} mutation), RNA translation (α^{nco} defect), and protein instability (Hb Suan Dok or Hb Evanston). Point mutations in the termination codon will give rise to long α-globin chains (Hb Constant Spring).

β-thalassemia is subdivided into β^{0-} (complete), β^{+}(mild), and β^{++} (severe) groups depending on the deficiency in the β-globin chain synthesis, leading to an imbalance. β-thalassemia molecular characterization is highly heterogeneous, with more than 200 mutations identified and described to date. Most of these mutations are point mutations with single base changes, insertions, and deletions. Very few mutations involve large deletions encompassing the β-globin gene cluster (Tangvarasittichai, 2011). These mutations occur in the exon, intron, promoter, 5´ and 3´ UTR regions, affecting the expression of β-globin gene at different stages as mentioned below.

1. *Transcription Efficiency*: Mutations occurring in the promoter region will alter the recognition sequences of proteins such as TATA, CCAAT, and CACCC boxes. Generally, these mutations are β^{+} and β^{++} types, resulting in the mild and severe forms of β-thalassemia.
2. *Pre-mRNA Maturation*: Occurs if the mutations fall in the splicing or polyadenylation (polyA) regions. Mutations in the RNA splicing region are the most commonly occurring and represent a large portion of thalassemia mutations. These mutations will affect the splicing process, and the effect depends on the position of the mutation. Mutations affecting the dinucleotide at the intron-exon junction (GT motif at the 5´ region and AG motif at the 3´ region) will abolish the splicing process altogether, resulting in β^0 thalassemia. Mutations occurring in the splicing consensus sequences will lead to variable degrees of defective splicing, resulting in mild β-thalassemia. Mutations in the intron and exon sequences may also activate a cryptic splicing site, leading to abnormal mRNA processing.
3. *RNA Stability*: Mutations occurring in the 5 ´ cap site, 5´ UTR region, 3´UTR region, and the Poly A site are associated with the phenotypes of mild β-thalassemia. In particular, mutations occurring in the 5´ UTR region are too mild to be considered silent β-thalassemia alleles, showing normal hematological phenotypes among heterozygotes.
4. *mRNA Translation*: Premature nonsense codons generated due to mutations will result in the premature termination of globin chain synthesis leading to the production of the short non-viable β-globin chains and nonsense-mediated decay of abnormal mRNA. All these cases result in severe thalassemia.
5. *Protein Instability*: Mutations give rise to either truncated or elongated globin chains, forming insoluble tetramers.

Thalassemia genes are distributed from the Mediterranean basin and Sub-Saharan Africa through the Middle East to the Far East including South China and the Pacific Islands. These genes are a little rare in the endogenous populations of northern regions. Hb constant spring point mutations that affect the stop codon and give rise to a long α-globin chain are widespread in Southeast Asia.

5.4.6.3 Diagnosis

α-thalassemia mainly occurs due to deletions in the α-globin genes, which are common in affected populations. Hence, diagnostic techniques are concentrated on the identification of these common deletions. Earlier, DNA blot analysis was applied to identify deletions in the genes, which were completely swapped using PCR technology. GAP-PCR is one of the most used PCR technologies for the identification of deletions. PCR primers flanking the common breakpoints can specifically amplify the deletion regions by generating PCR products of different specific sizes. Alternatively, multiplex ligation dependent probe amplification (MLPA) is also considered. This technique can quantify the gene copy numbers and detect deletions based on the reduction in gene dosage. MLPA detects unusual deletions in the genes that cannot be detected by using standard PCR technology. This technique successfully characterizes the novel deletions in the α-globin gene locus responsible for the α-thalassemia. Array-comparative genomic hybridization (Array-CGH) is being used to identify deletions in the gene locus quickly. By creating a custom oligonucleotide array that densely tiles to the HBA locus, heterozygotes for the common Filipino and Southeast Asian cis deletions in the α-globin gene were detected. Heterozygotes with single gene trans – α 3.7 and –α 4.2 deletions on the α-globin gene were also detected using Array-CGH.

In most cases, β-thalassemia occurs due to point mutations in the β-globin gene. Sanger's sequencing is considered the most applied method for the detection of all mutations. DNA sequencing of the β-globin gene is less complex than that of α-globin. PCR amplification of the β-globin gene is much easier as there is only a single β-globin gene on chromosome 11 with less GC content when compared with of α-globin gene. Multiple mutations in the β-globin gene can be detected simultaneously using melting curve analysis. Sanger's sequencing has identified 24 mutations responsible for β-thalassemia in the Southern Chinese Population. Alternatively, resolving DNA fragments can also be detected using MALDI-TOF mass spectrometry. Deletions in the β-globin gene responsible for δβ-thalassemia have been detected by using GAP-PCR (Daniel, 2017).

5.4.6.4 Treatment

Blood transfusion is regularly practiced from childhood as a treatment for severe anemia due to the thalassemia disorder. However, regular transfusions from childhood will have severe iron overload, necessitating extra medication. The first drug available for treating iron load during thalassemia was deferoxamine (also known as desferrioxamine B), which binds to iron, and decreases the toxic reactions, and iron uptake by tissues. In the case of problems with deferoxamine therapy, deferiprone, the oral iron chelator, can be used. Alternatively, deferasirox is used as it has good bioavailability with a long half-life. It can remove iron from the blood by coordinating two dederasirox molecules with a single iron ion.

Young people with thalassemia may undergo bone marrow transplantation. Patients without matched compatible donors can be offered bone marrow transplantation from haploidentical mother to child. Stem cell transplantation is another option for children born with severe thalassemia. Stem cell transplantation involves high doses of drugs and radiation to destroy the diseased bone marrow before receiving infusions from the donor. These clinical procedures are risky and sometimes cause even death.

5.4.7 Sickle Cell Disease

It is a blood disorder, also known as sickle cell anemia or depanocytosis. Acquisition of a rigid, abnormal sickle shape by the RBC during deoxygenation due to abnormal hemoglobin, leading to hemolysis and anemia, is the characteristic feature of this disorder. Sickle cell disease was first reported and explained by a Chicago based cardiologist and professor in medicine, K. James B. Herrick, in 1904. This discovery occurred after identifying peculiarly elongated sickle-shaped cells in the blood of a 20-year-old Canadian, Walter Clement Noel, admitted to the Chicago Presbyterian Hospital, suffering from anemia. Vernon Mason named sickle cell Anemia in 1922. Linus Pauling, 1949 was the first to demonstrate that sickle cell disease occurs due to an abnormality in hemoglobin. Linus Pauling was the first to demonstrate that sickle cell disease occurs due to an abnormality in hemoglobin. It was the first incident linking a genetic disorder to the mutation of a protein, which is a milestone in the history of molecular biology.

Hemoglobin A (HbA), hemoglobin F (HbF), and hemoglobin A2 (HbA2) are three different types of hemoglobin found in human beings. Each hemoglobin molecule consists of 4 polypeptide chains, whose structure varies according to the type of hemoglobin. Hemoglobin A comprises of 2 α-globin chains and 2 β-globin chains, consisting of 95 - 97% of normal hemoglobin. 2.5 - 3.5% of hemoglobin A2 is normal with 2 α-globin and 2 γ-globin chains. Hemoglobin F consists of 2α-globin chains and 2 Δ-globin chains with <1% normal hemoglobin. The genes coding for the α, b, g, and D globin chains are located on chromosomes 11 and 16, respectively.

5.4.7.1 Symptoms

Sickle cell disease is a multi-organ, multi-system disorder responsible for chronic and acute complications when the fetal hemoglobin levels drop towards the adult level within 5 - 6 months of age. Sickle cell anemia (HbS) is the most predominant type of sickle cell disease, followed by HbSC, the result of the co-inheritance of HbS and HbC. Co-inheritance of HbS with β-thalassemia results in sickle β-thalassemia (genotype represented as HbS/βo or HbS/β+). When RBC with HbS or the combination of HbS with other abnormal b alleles is exposed to the deoxygenated environment, they will undergo polymerization, and become rigid, leading to hemolysis. Rigid RBC will also increase their density affecting the blood flow and the integrity of the endothelial vessel wall. Dense and rigid RBC will also lead to vaso-occlusion, tissue ischemia, and infarction. Hemolysis leads to a cascade of events, including nitric oxide consumption, and endothelial dysfunction, leading to complications such as leg ulceration, stroke, pulmonary hypertension, and priapism. Increased hemolysis will reduce the life span of sickle RBC to 10 - 20 days from 120 days. During deoxygenation, healthy hemoglobin re-arranges into a further confirmation, enabling the binding of CO_2 and reverting to normal upon release. In contrast, sickle celled hemoglobin (HbS) gets polymerized into rigid insoluble strands and tactoids (gel-like substances containing Hb crystals). Intravascular hemolysis results in free hemoglobin with increased Na^+, Ca_2^+ concentrations, and decreased K^+. An increase in the Ca_2^+ blood concentrations leads to dysfunction of the calcium pump.

5.4.7.2 Etiology

Sickle cell is an autosomal recessive disorder, that occurs due to a transversion mutation in the β-globin gene, located on the short arm of chromosome 11. Individuals receiving the mutated genes both from their father and mother will develop this disease. After receiving one mutated gene and one normal gene, a person does not show any symptoms but carries this diseased allele into the next generation. If two carrier parents have 4 children, one will develop the disorder, and one will be the carrier. As this gene is incompletely recessive, carriers may produce very few sickle-shaped RBCs, not showing any symptoms but can resist malaria. In the carrier, RBC with defective hemoglobin ruptures prematurely, not permitting the plasmodium parasite to reproduce.

Further, the polymerization of hemoglobin will not permit the parasite to digest. Due to this feature, heterozygotes have better fitness than both homozygotes, and this feature is known as heterozygotes advantage. Due to this advantage, Sickle cell disorder is still prevalent in malaria-stricken regions such as India, Africa, the Mediterranean, and the Middle East. Heterozygotes will have minor polymerization problems as one normal allele produces 50% of the hemoglobin. In the blood of HbS homozygous people, long chain polymers will distort the shape of RBC from smooth count like to ragged and full of spikes, making it fragile enough to break within capillaries. Carriers show symptoms only if deprived of oxygen or severely dehydrated (chances are 0.8 times a year per carrier patient).

Sickle cell disease occurs due to the benign mutation of a single nucleotide from the GAG codon to the GTG codon of the β-globin gene, resulting in the substitution of glutamic acid by valine at position 6. Hemoglobin with a substitution mutation is known as HbS, whereas the normal one is HbA. This mutation does not cause any apparent loss to the hemoglobin secondary, tertiary, or quaternary structure but undergoes polymerization under low oxygen concentrations, exposing a hydrophobic patch on the protein between E and F helices. The hydrophobic residues of the valine amino acid at position 6 of the β globin chain in the hemoglobin will become associated with the hydrophobic patch, causing HbS to aggregate and form a precipitate of fibers.

Sickle cell disorder is one of the most common genetically inherited disorders in human beings that specifically affect people from Africa, India, the Caribbean, the Middle East, and the Mediterranean regions. More than 80% of the annual births in Sub-Saharan Africa are with sickle cell disorder, with the highest contribution from Nigeria and the Republic of Congo. The gene frequency of sickle cell disorder is also high in West African Countries where 1 among every 3 to 4 are carriers compared with 1 among 400 African Americans. This frequency is highly variable among European populations. The prevalence of sickle cell disorder is increasing alarmingly in European countries due to the migration from high prevalence countries. More than 14,000 people are living with sickle cell disorder in the United Kingdom and France. With the advancement in medical technology, the age distribution of sickle cell disorder is also changing from infants and childhood to adults and old age. More than 94% of the people born with sickle cell disorder in the United States, United Kingdom, and France can survive into adulthood, whereas 50 - 90% of the patients in the Sub-Saharan Africa region have succumbed to death in the first five years after their birth. Lack of early diagnosis, low education, and preventive therapies are the main reasons for the high number of deaths in this region.

5.4.7.3 Diagnosis

A complete blood count is the efficient way of diagnosing sickle cell anemia, with hemoglobin levels falling in the range of 6 - 8 g/dL due to the occurrence of reticulocytosis and bone marrow hyperplasia. In other forms of sickle cell disorder, Hb levels are high. A patient's blood film may reveal many sickle shaped cells and features of hyposplenism. The addition of sodium metabisulfite can induce the presence of sickle shaped cells in the blood film. The presence of abnormal hemoglobin types in the blood is diagnosed by hemoglobin electrophoresis. Hemoglobin S and Hemoglobin SC are the two most common forms found in the blood of sickle cell disorder patients, detected by using this gel electrophoresis separation technique. The presence of sickle hemoglobin is demonstrated by using the sickle solubility test, which gives a turbid appearance when dissolved in a reducing solution such as sodium dithionite. HPLC can also confirm the diagnosis. Chest X-ray detects occult pneumonia. Genetic tests are rarely performed for sickle cell disorder. Fetal blood sampling and sometimes amniocentesis is done for detecting the disorder in the developing fetus. Amniocentesis is safe for detection as fetal blood sampling might lead to miscarriage.

5.4.7.4 Treatment

Manifestations that occur due to sickle cell disease can be minimized if diagnosed early and followed up regularly. Nutrient supplements such as omega 3-fatty acids purified from fish oil and folic acid are prescribed for patients with sickle cell disease. Along with the nutritional supplementation, oral application of prasugrel drug inhibits platelet aggregation. Oral intake of apixaban will prevent the activation of prothrombin and thrombin. Niprisan, a traditional herbal product, is being used in Nigeria, showing promising results, and is shown to have interactions and significant inhibition of cytochrome CYP3A4 activity in sickle cell disorder patients. However, niprisan is still under clinical trials.

Individuals suffering from sickle cell disorder with a baseline of anemia mainly due to chronic hemolysis are treated with blood transfusion. Transfusion of HbS negative blood will reduce the proportion of sickled hemoglobin in circulation and considerably reduce vessel occlusion and hemolysis. Bone marrow transplantation is one of the most successful methods for the complete cure of sickle cell disease with a success rate of 91% and has reduced the mortality rate to 5%. However, it carries significant risks, such as newly synthesized leucocytes might attack host tissue cells of the skin, liver, gastrointestinal tract, and eyes through a process known as graft-versus-host disease generating symptoms such as nausea, and jaundice. Hence, bone marrow transplantation is recommended if the symptoms and complications are severe.

Gene therapy for sickle cell disease is in the infancy stage. This approach is based on stem cells. Instead of isolating the embryonic stem cells, host stem cells are derived by manipulating and re-programming the cells from the patient's blood cells through genetic engineering, where the patient's inborn genetic error can be rectified. As the cells are derived from the patient, there is no need for a donor. Here the patient's blood cells will be transformed into pluripotent stem cells, which replace the defective fragment of the gene. These stem cells will be coaxed to become hematopoietic cells, explicitly regenerating the entire RBCs. This process of gene therapy is in the clinical stage with different lentiviral vectors.

5.4.8 Alzheimer's Disease

Alzheimer's is a complex heterogeneous neurodegenerative disorder that occurs due to genetic and epigenetic factors. Alzheimer's disorder is an aggressive form of dementia, manifesting in memory, language, and behavioral deficits. The dementia condition leading to the disorder was described for the first time by a German physiatrist and neuropathologist Dr. Alois Alzheimer. As predicted by the World Health Organization, the global prevalence of Alzheimer's will increase tremendously in the next few decades, reaching up to 114 million patients by 2050. To date, there are no practical option is available for the treatment of Alzheimer's disease, which is of two primary forms, sporadic Alzheimer's and, familial Alzheimer's. The sporadic form is the most common one without any family history. The development of sporadic Alzheimer's disease is strongly dependent on genetic, environmental, vascular, and psychical factors and their interactions. Familial Alzheimer's disease is a rare form with strong family history. This Alzheimer's form occurs in 1% of patients due to deletions or alterations in the specific genes, inherited from one generation to another. Alzheimer's is a complex disease that cannot be treated by using a single drug or intervention. Approaches focusing on the maintenance of mental function, and behavior symptom management might slow or delay the symptoms to some extent. The development of therapy, targeting its genetic, molecular, and cellular mechanisms, might provide the primary basis for understanding the actual underlying cause of the disease and its prevention.

5.4.8.1 Symptoms

The progression of Alzheimer's disease comprises four main stages, predementia, mild, moderate, and severe. Each stage has significant symptoms and challenges. The predementia stage is characterized by the deterioration of episodic memory, verbal and visuospatial functions without declining sensory or motor performance. An individual at this stage is highly independent and is not diagnosed with suffering from Alzheimer's. A mild stage of Alzheimer's will increase memory loss and drastically affect the recent declarative memory, compared with short-term and implicit memories. Recent declarative memory further continues to deteriorate in the moderate stage. Due to their inability to generate new memories, Alzheimer's patients would live in the past. They will need help in dressing, grooming, and for every daily practice. Complete loss of insight occurs at this stage of the disease, and the patients will become delusional. Cognitive decline, aggression, depression, and incontinence in patients will be reflected in this stage. The patient might even lose his early memories during the severe stage, with a gradual decline in basic activities of daily living. Communication will deteriorate to single words or phrases, significantly impairing the language. Behavioral disturbances will be more frequent, disrupting the caregivers. One of the most common causes of death for Alzheimer's disease patients is pneumonia, followed by myocardial infarction and septicemia.

5.4.8.2 Etiology

Alzheimer's disease is categorized into two sub-types, early onset, and late onset.

5.4.8.2.1 Early Onset Alzheimer's disease

Early onset Alzheimer's disorder is identified in fewer patients (5 - 10% of cases). Early-onset Alzheimer's is inherited via a Mendelian pattern of inheritance, known as familial early onset Alzheimer's disease. However, early onset Alzheimer's disease is also reported without any family history, known as sporadic early onset Alzheimer's disease. Three genes amyloid precursor protein (APP), presenilin 1 (PSEN1), and presenilin 2 (PSEN2) are identified as the main factors responsible for the early onset of Alzheimer's disease. Mutations in these genes might alter the production of amyloid beta (Abeta 40 and Abeta 42), leading to apoptosis of neurons and dementia. The APP gene is located on the q21 locus of chromosome 21, consisting of 19 exons for encoding the APP protein. Among them, exons 16 and 17 encode abeta peptides. Following transcription and alternate splicing, a minimum of five APP protein isoforms with abeta peptide sequence have been identified. At least three mutations were identified on exons 16 and 17, suggesting that APP is a rare risk factor for early onset Alzheimer's disease (Bertram et al. 2001).

PSEN1 and PSEN2 are two genes with a homology of 67% and a highly similar structure has been identified based on linkage analysis. These genes contain 12 exons. Among them, 10 coding exons (numbered 3 - 12) make a protein of 450 amino acids. Both PSEN1 and PSEN2 are transmembrane proteins with a minimum of seven transmembrane domains. PSEN1 gene is located on the q24.2 locus of chromosome 14. Two transmembrane aspartate residues at 257 and 385 of PSEN1 are involved in the synthesis of gamma secretase. More than 70% of Alzheimer's disease risk factor mutations were detected on the PSEN1gene locus. More than 180 mutations have been described in the PSEN1 gene that is associated with familial Alzheimer's disease. Individuals with mutations in the PSEN1 gene might develop Alzheimer's disease symptoms during their 40s or early 50s. However, few individuals have reported the development of symptoms even in their late 30s. Several missense mutations in the PSEN1 gene might increase or decrease the production of abeta 40 and 42, responsible for dementia. Gene PSEN2, located on the q42.13 locus of chromosome 1, is considered a significant risk factor for early onset Alzheimer's disease among the European population. Alzheimer's disease generated from PSEN2 gene mutations is high, variable occurring between 40 - 75 years.

In addition to these three genes, a polygenic inheritance is also reported to exist, responsible for early onset Alzheimer's disease. Recent whole genome and whole exome studies have identified the involvement of more than 20 loci of genes in various metabolic pathways, such as sortilin related receptor 1 (SORL1), anti-inflammatory microglial triggering receptor 2 (TREM2) expressed on myeloid cells, ATP binding cassette subfamily A member 7 (ABCA7), genes involved in endocytosis or endolysosomal transport, immunologic reactivity, and lipid metabolism (PLD3, PSD2, TCIRG1, RIN3, and RUFY1) with each of them contributing to the risk of developing early onset Alzheimer's disease. Combinations of these genes lead to an increased risk of early onset Alzheimer's disease with 72.9% - 75.5% accuracy.

5.4.8.2.2 Late onset Alzheimer's disease

Several genetic and environmental factors are responsible for the progression of late onset Alzheimer's disease. The apolipoprotein gene (APOE) located on the q13.2 locus of chromosome 19 is the primary risk factor for late ongoing Alzheimer's disease (Coon et al. 2007). APOE protein is one of the major cholesterol carriers in the human brain. This protein is also involved in neuron maintenance and repair. This protein binds to the lipid delivery and transport receptors, glucose metabolism, neuron signaling, and mitochondria function. APOE protein binds to the abeta peptide and plays a crucial role in its clearance.

Two single nucleotide polymorphisms (SNPs) have been described in the human APOE gene (Kim et al. 1999). These SNPs are located at the codons 112 and 158, respectively, responsible for at least 3 main APOE gene variations named E2, E3, and E4 alleles or isoforms. E3 is a common allelic form consisting of cysteine at codon 112 and arginine at codon 158. E2 allele carries cysteine at codon 158, and the E4 allele carries arginine at codon 112. A total of six different genotypes are distinguished with the homozygous E4/E4, E3/E3, E2/E2, and heterozygous E2/E3, E2/E4, E3/E4 combinations. E3 is the most common allele detected with high frequency (77%) followed by E2 (8%) and E4 (15%) alleles. A high frequency of E4 alleles has been found in Alzheimer's patients, and an increased risk of the disease has been identified in both the homo and heterozygous alleles. Pathogenicity of the E4 allele is associated with a structural change in the APOE protein, consisting of a 22 KDa N-terminal domain and a 10 KDa C-terminal domain, connected by a hinge region (Liu et al. 2013). Arginine at codon 112 interacts with glutamate at codon 255 in the C-terminal domain, resulting in the structural changes of APOE protein, leading to neuronal death and neurodegeneration. A mutation of arginine at codon 61 to threonine and glutamate at 255 to alanine might also reduce the domain interactions. Prevalence of the E2 allele is less among individuals with dementia. The E2 allele acts as a protective agent against Alzheimer's disease. Genome-wide association studies (GWAS) have identified novel genes such as clusterin, complement receptor 1 (CR1), located on chromosome 1, Phosphatidylinositol binding clathrin assembly protein (PICALM) located on chromosome 11, sortilin-related receptor (SORL1) located on chromosome 11, and bridging integrator 1 (BIN1) located on chromosome 2 that are associated with late onset Alzheimer's disease. Betram and his associates have created a database by summarizing all potential genes related to Alzheimer's disease, available at www.alzgene.org. This database has a collection of potential candidate genes involved in late onset Alzheimer's disease, which can be used as powerful tools for understanding the genetics related to this disorder (Bertram et al. 2007).

5.4.8.3 Diagnosis

Diagnosis of Alzheimer's follows a logical sequence comprising two stages. Firstly, to distinguish dementia syndromes from other similar conditions such as depression, delirium, and mild cognitive impairment which mimic them. Secondly, after recognizing dementia syndrome, the subtype diagnosis should be performed to determine the kind of possible treatment. For the detection of Alzheimer's, neuroimaging is promising and is one of the highly developing areas. It can be achieved through multiple brain imaging procedures such as PET, MRI, and CT scans used to identify abnormalities in the brain. Each scanning procedure involves a unique technique that can detect specific abnormalities in the brain and its associated parts. Even though brain imaging is not the standard procedure for Alzheimer's testing, current clinical studies have shown promising results while diagnosing the disease. Despite extensive research, there is no effective treatment currently for Alzheimer's disease. For effective treatment, this disease should be diagnosed well before the symptoms will start becoming evident. There is a great need for effective diagnostic methods so that the treatment will be in place to slow down the process or symptoms. One of the main pathological characteristic features of Alzheimer's is the development of insoluble amyloid plaques in the brain and neurons. The formation of these plaques in the brain is measured by using position emission tomography (also known as PET cameras), which aids in the visualization of radioactive trace molecules that bind to the plaques by creating a 3-dimensional color image of the human brain. Amyloid plaques are also formed in cerebrospinal fluid, whose number reduces with an increase in the intensity of the disease. A computed tomography scan (CT scan) generates a series of cross-sectional images of the human brain or other parts of the human body integrated and incorporated into a single detailed image. CT scans the density of the tissues in various parts of the brain. Magnetic resonance imaging (MRI) creates two or three-dimensional images of the body to diagnose illness. MRI can identify structural changes in the brain of Alzheimer's disease patients due to cellular death. Hippocampal atrophy is a frequently seen pathogenic condition in Alzheimer's disease even before the appearance of clinical symptoms. Using MRI, the volume of hippocampal atrophy can be detected, which can be used as an indicator for the development of Alzheimer's disease.

Apart from the image scans, a diagnosis of Alzheimer's disease is also made, by means of neurological examination, and neuropsychological tests. According to the Diagnostic and Statistical Manual of Mental Disorders (Edition IV), criteria for diagnosing dementia requires the loss of at least two or more of the features involving memory, language, calculation, orientation, and judgment. Mini-mental state examination for Alzheimer's can help in evaluating changes in the cognitive abilities of the patient. The exact diagnosis of Alzheimer's requires the assessment of major histopathological features such as neurofibrillary tangles and amyloid plaques, which allows the experts to predict 80 - 90% of the time. Even though the formation of plaques and tangles is common among individuals with age, the plaques' density and the tangles' distribution are more severe in patients with Alzheimer's. Accumulation of neurofibrillary tangles in neurons is another significant feature for the diagnosis of Alzheimer's. The temporal and spatial appearance of these tangles containing hyperphosphorylated tau reflects the disease severity rather than plaques. Tangles are formed by the hyperphosphorylation of a microtubule-associated protein tau, responsible for their aggregation in an insoluble form.

The presence of apolipoprotein E4 –allele on chromosome 19 has been correlated with the development of Alzheimer's disease later in many individuals' lives. Diagnosis of cognitive function is performed for major depression, anxiety disorder, and dementia due to illness. PCR-based methods monitor the mutations in the risk factor genes by isolating the DNA from blood and bone marrow cells (Mahley et al. 2006). Specific primers are designed to amplify risk factors genes APP, PSEN1, PSEN2, and APOE. Several methods have been developed to detect mutations such as restriction fragment length polymorphism (RFLP), single-strand conformation polymorphism (SSCP), and denaturing gradient gel electrophoresis (DGGE), temperature gradient gel electrophoresis (TGGE), and heteroduplex analysis. Among these methods, denaturing gradient gel electrophoresis is a rapid, commonly used method to detect mutations (Waring and Rosenberg, 2008). RFLP is based on recognizing specific cleavage sites, used for genetic mapping and linkage analysis. Their specific sequencing achieves the detection of polymorphisms in PCR products.

Prion diseases such as familial Creutzfeldt-Jakob disease (CJD) and fatal familial insomnia are also aggregated in families, sharing the clinical features with early onset familial Alzheimer's disease. Familial prion disease occurs due to mutations within the prion protein gene (PRNP). A mutation in codon 200 of the prion protein gene is associated with familial Creutzfeldt-Jakob disease, and a mutation in codon 178 is associated with fatal familial insomnia. Cognitive impairment, myoclonus, periodic sharp-wave complexes in electroencephalogram (EEG), and hyperintensity in the cortex or striatum are the characteristic features of Familial Creutzfeldt-Jakob disease identified by the diffusion-weighted images of MRI.

5.4.8.4 Treatment

Two types of medications are use in practice for the treatment of Alzheimer's disease, acetylcholinesterase inhibitors, and N-methyl D-aspartate antagonists. Patients with Alzheimer's will have low levels of acetylcholine in their brain, responsible for decreased communication between the neurons. Acetylcholinesterase inhibitors will increase the acetylcholine availability in synaptic neurotransmission for the treatment of memory-associated disturbances. Three acetylcholinesterase inhibitors such as donepezil, rivastigmine, and galantamine are utilized in the first-line treatment of mild and moderate Alzheimer's disease. Modulation of N-methyl D-aspartate antagonists receptors results in reduced glutamate-induced excitotoxicity. Memantine is an effective non-competitive NMDA receptor antagonist in the treatment of moderate to severe Alzheimer's disease. Depression is ubiquitous during the early and late courses of Alzheimer's, minimized by using anti-depressants, especially serotonin reuptake inhibitors such as citalopram, fluoxetine, paroxetine, sertraline, and trazodone. Alternatively, tricyclic agents and combined serotonergic and noradrenergic inhibitors are also used. Antipsychotic drugs such as olanzapine, quetiapine, and risperidone are prescribed to treat psychosis and agitation. Delaying neurodegeneration by targeting the plaques and tangles is one of the potential mechanisms for the future treatment of Alzheimer's disease.

5.4.9 Schizophrenia

Schizophrenia is a word used to describe a mental condition with a spectrum of varied conditions and symptoms such as perception alterations, thought alterations, self-decrease in violations, slowing of psychomotor skills, and showing antisocial behavior. It is a chronic psychiatric disorder, occurring due to a heterogeneous genetic and neurobiological combination. It heavily influences the early brain development of the fetus, reflected in a variety of psychotic symptoms, including hallucinations, delusions, disorganization, motivational and cognitive dysfunctions. Schizophrenia is not a common mental sickness, and the result can be highly disabling, with 7 - 8 of every 1000 individuals suffering from this disorder. This disease is one of the seven costliest diseases due to frequent hospitalizations, the need for psychosocial services, and poor productivity (Mc Grath et al. 2004).

5.4.9.1 Symptoms

The psychological symptoms of this disorder are divided into three main categories, positive, negative, and cognitive. Even though positive and cognitive symptoms are ubiquitous during schizophrenia, no correlation or relationship has been observed between these two categories. Positive symptoms of Schizophrenia are the ones, which are easily identified in the patients and are not found in healthy individuals. These symptoms include diverse delusions such as thought, persecutory, delusions of control, somatic delusions and granulose, and hallucinations such as auditory, visual olfactory, and gustatory, with auditory hallucinations most common. Negative symptoms of Schizophrenia are the reduction of features that are generally present in healthy individuals. These symptoms cannot be easily identified in patients and are associated with a high morbidity rate. These symptoms include the blunted effect of immobile facial expression and monotonous voice, anhedonia (lack of pleasure), avolition or apathy (diminished ability to initiate or follow through ongoing plans), and alogia (a reduction in the quantity of speech). Negative symptoms are more persistent and less fluctuating when compared with positive symptoms. As these symptoms are less apparent, patients are often perceived by their relatives, and others close by to be lazy and willfully unengaging. Cognitive symptoms are the most deteriorating, disabling, and tenacious features present in the long term in schizophrenia patients. Cognitive symptoms include attention, concentration, lack of psychomotor speed, impaired memory, and impaired executive functions such as abstract thinking, and problem solving. Among all cognitive disabilities, visual and verbal declarative memory, working memory, and processing speed are known to have pronounced effects on Schizophrenia patients. These cognitive symptoms will severely impair the patient's communication skills by drastically reducing speaking ability and attention. Impaired cognitive symptoms are noticed in acute and chronic cases of Schizophrenia (Cardno and Gottesman, 2000).

5.4.9.2 Etiology

Studies that have been carried out over the past year have identified multiple factors responsible for the development of Schizophrenia. Schizophrenia is also genetically inherited and runs in families sometimes. This disease occurs in more than 1% of the population, and this ratio will increase to 10% in individuals with first-degree relatives such as parents, brothers, and sisters suffering from this disease. The combination of genetic and environmental factors such as exposure to viruses, malnutrition, problems during birth, and other unknown psychological factors might also lead to Schizophrenia (Kenny et al. 2014).

The genetic component of Schizophrenia is very high, with an estimated heritability of ~80%. However, the genetic transmission of this disease is complex and will not follow Mendelian single gene inheritance patterns. More than 1000 susceptible genes and at least 30 risk factor genes are predicted to interact with epigenetic and environmental factors responsible for Schizophrenia (Fromer et al. 2014). Despite the replicative linkage, several putative genes located on chromosomes 1q, 2q, 5q, 6p, 6q, 7q, 8p,10p,11q,12q,13q,14q,15q and 22q are also involved in the disease. Several candidate genes such as AKT1, BDNF, CHRNA7, COMT, D2, DAAO, DAOA, DISC1, DTNBP1, ERBB4, FEZ1, G72, GAD1, GRM3, GSK2B, HTR1A, HTR6, LEP, MUTED, MRDS1, NCS1, NRG1, PAFAH1B1, and RGS4 are responsible for Schizophrenia (Zafar et al. 2013). However, the specific target gene responsible for the disease is yet to be identified.

The molecular mechanisms of schizophrenia across diverse populations have been elucidated based on twin studies and genetic discoveries. The loci containing the common variants for DRD2 encode the dopamine D2 receptor, glutamate receptor components GRM3, GRIN2A, and GRIA1, encoding metabotropic glutamate receptor 3 (mGluR3), GluN2A, and GluA1, SRR encoding serine racemase, an enzyme responsible for the biosynthesis of NMDA receptor allosteric site ligand have been identified (Clinton and Meador-Woodruff, 2004). A large portion of chromosome 6 consisting of the major histocompatibility complex region has also been identified as a target site for the disease. Each locus is responsible for a small amount of individual risk with a difference of >20% between diseased and normal individuals. The mechanisms by which the genetic signals are translated into molecular mechanisms are not known. These genes' associations are clues for understanding molecular mechanisms, which are being explored by using molecular and bioinformatics strategies (Lips et al. 2012, René et al. 2015).

Most of the genes associated with schizophrenia have shown a preferential expression during the development stage of the fetus, suggesting that the genetics of schizophrenia correlate with the genetics of brain development, which supports the hypothesis that schizophrenia has its origins in early life (Fatemi and Folsom, 2009). This hypothesis is consistent with the epidemiological data, which establishes a link between the intensity of obstetric complications and the increased risk of developing the disorder. Individuals with schizophrenia who evidence this hypothesis during adulthood have compromised early neurodevelopmental milestones, and cognitive development is compromised in patients long before they manifest this condition during their adult stage. It is unclear why such developmental antecedents manifest into cognitive and social difficulties during the first 20 years of life and emerge as a profound psychotic illness during adulthood. The brain structure of schizophrenia patients is different when compared to healthy individuals. Fluid filled ventricles at the center of the brain are larger in schizophrenia patients. Patients with schizophrenia have increased dopamine levels, and the reasons for the increased dopamine levels leading to schizophrenia are still unknown. Three possible mechanisms have been hypothesized for schizophrenia.

1. *Neurodevelopment hypothesis*: postulates that the effects of environment or genetics on brain development during the embryonic stage will result in defective neural activity and altered neuronal functioning later in their life. Neurodevelopmental disturbances lead to hippocampal formation in the superior temporal lobe, one of the alterations observed in a schizophrenic patient (Fatemi and Folsom, 2009).

2. *Dopamine hypothesis*: is one of the most widely contemplated neurochemical hypotheses for schizophrenia. This hypothesis is based on the excess neurotransmission of dopamine in the mesolimbic and striatal brain regions that might lead to positive symptoms, ultimately turning to schizophrenia (Jentsch and Roth, 1999). The primary evidence for this hypothesis came from individuals who used amphetamine (drugs that stimulate the central nervous system), which produce increased amounts of dopamine, showing schizophrenia symptoms (Gründer and Cumming, 2016).
3. *Glutamate hypothesis*: states that the dis-functioning of the human brain due to enhanced dopamine levels is associated with the dis-functioning of glutamate in the thalamocortical loop, leading to changes in dopamine concentrations, ultimately developing psychotic symptoms. Two groups of glutamate receptors have been identified in the human brain whose differential functions might lead to the appearance of schizophrenic symptoms. Among them, N-methyl, D-aspartate is a primary receptor responsible for schizophrenia in many patients by changing the dopamine level from normal to high (Hu et al. 2015).

Schizophrenia is a cosmopolitan disorder that occurs throughout the world. The prevalence of this disease is 1% globally, with an incidence of 1.5% per 1000 individuals. Males are more susceptible to this disease than females during early life, but women are more susceptible during later stages. The age of the onset of the disease is 18 - 25 years for males and 25 - 35 years for females. The average lifetime prevalence of schizophrenia is >1% globally and is 0.87% in Finland. However, there is a variation in prevalence rates geographically of up to fivefold. Schizophrenia patients will have a shorter life than healthy individuals. The standardized mortality rate is 2.6, with suicide being the main reason during the early course of illness and cardiovascular disorders during the later stages.

5.4.9.3 Diagnosis

No clinical or laboratory tests are available for the diagnosis of schizophrenia. No biological marker has been identified for defining the onset of the disease or following its progression. Diagnosis of schizophrenia is primarily based on the clinical history of the patient and his/her mental status examination. Currently, two clinical diagnosis systems are commonly employed, the international classification of diseases (ICD- version11), and the Diagnostic and Statistical Manual of Mental Disorders-IV Text Revision (DSM-IV TR). Even though there are no clinical diagnostic tests, the diagnosis is on the comparative clinical criteria. Alternatively, schizophrenia is diagnosed through functional neuroimaging. Recent functional MRI studies have identified reproducible abnormalities in the brains of schizophrenia patients. However, this technique with current technology might not be sensitive enough to identify the differences in more than 40 - 50% of patients. Moreover, this disease is non-specific, as seen in 30% of the patient's first-degree relatives and 10% of other normal controls. The search continues to find appropriate tools for the accurate diagnosis of schizophrenia.

5.4.9.4 Treatment

As the primary symptoms of schizophrenia are still unknown, treatment for this disorder is dependent on the symptoms, and antipsychotic drugs are the primary sources. Chlorpromazine is a first-generation antipsychotic drug used for reducing the intensity of schizophrenia. Other first-generation drugs such as loxapine, fluphenazine, perphenazine, and haloperidol were also made available for the same purpose. Nevertheless, all these drugs have a significant side effect on extra-pyramidal systems, which cannot be minimized. Hence, these drugs are no longer used. Clozapine is a second-generation antipsychotic drug that has proved to be better than the first-generation drug in respect of its side effects. This drug blocks dopamine receptors to some extent (Shahid et al. 2018). After the success of clozapine, several second-generation antipsychotic drugs were released and are serving schizophrenia patients with better efficiency.

Therapies have been developed as an alternative treatment option for schizophrenia. Schizophrenia therapy is provided to patients incapable of receiving their medication and may increase the risk of relapse. The intensity of the therapy will change according to the severity of the disease, as evidenced by the clinical reports. Augmentation therapy is provided to patients in combination with second-generation antipsychotics and electroconvulsive therapy with mood-stabilizing agents, such as lithium. The efficiency of clozapine will increase if prescribed in combination with other therapies. Even though patients are being treated by various pharmacological and non-pharmacological methods, proper treatment for this disease is still not available. The development of research in this field will address the gaps existing for the potential treatment of this disorder.

5.4.10 Cystic Fibrosis

Cystic fibrosis (also known as mucoviscodosis) is the most severe life-shortening multisystem monogenetic disorder caused due to the dysfunction of chloride channels in exocrine glands. The first clinical description of the syndrome was made in the year 1939. Much progress has been made in the diagnosis of the disease and in the improvement of the life expectancy of patients suffering from this disease. 70,000 individuals globally and more than 1,000 new individuals per year succumb due to abnormal viscous secretions in the lung's airways, and ducts of the pancreas, causing obstructions, leading to inflammation, tissue damage, and destruction of both organ systems.

5.4.10.1 Symptoms

Clinical manifestation of cystic fibrosis is subject to changes in the individuals and might appear in the neonatal period or even in the later stages of life. The most common clinical symptoms of this disorder include chronic cough, diarrhea, and malnutrition. This disease also affects multiple systems and organs. Mutation(s) in the CFTR gene causes the absence or dysfunction of cystic fibrosis transmembrane conductance regulator (CFTR) protein, which functions as a canal of chloride in the apical membranes of epithelial cells.

CFTR protein is a member of the ABC (adenosine triphosphate (ATP)-binding cassette) transporter, whose function is to hydrolyze ATP for transporting ions, amino acids, sugars, drugs, and proteins against a concentration gradient. The primary function of CFTR protein is the transportation of chloride and bicarbonate ions. This protein consists of two membrane spanning domains and two nucleotide binding domains. The transportation process begins with binding an appropriate substrate to its specific site on the membrane spanning domains, which promotes the ligation of ATP onto the nucleotide binding domains and their subsequent dimerization. The binding of ATP will provide energy to release the bound substrate across the cellular membrane. Hydrolysis of ATP will de-stabilize the dimer of the nucleotide binding domain, allowing the release of inorganic phosphate, ADP, and the protein will regain its basal conformational state. This process is known as the ATP switch model. Even though CFTR protein belongs to the ABC transporter's group, it possesses distinctive characteristics such as, it is the only protein capable of transporting the ions across the cell membrane, that this protein is equipped with a unique regulatory domain and two terminal (both N and C) extensions, hypothesized to regulate the gating mechanisms and modulate interactions between the ion channel and other cellular proteins. Also, CFTR protein is in the apical surface of the airway, and intestinal and exocrine epithelial cells. Functional CFTR protein maintains optimal volume, electrolyte composition, and pH of the airway surface liquid in human lungs, a thin fluid layer that protects the epithelium from inspired air. Airway surface liquid levels are essential as the lipid contains antimicrobial properties and phagocytic cells for the innate immune system. Insufficient or dysfunctional CFTR protein will impair chloride secretion and epithelial sodium channel regulation, leading to a reduction in anion secretion, increased sodium and water re-absorption from airway surface liquid, through the apical cellular membrane. This situation augments the viscosity of the airway surface liquid and drastically reduces mucociliary clearance efficiency, the first line of defense in the respiratory system against pathogens. Mutated CFTR protein will also affect mucous production, secretory granules, and intracellular organelle, impacting multiple organs and their clinical and physiological responses, leading to death in 90% of the patients. Death in most patients occurs due to the involvement of the respiratory tract.

Another most common and crucial symptom of cystic fibrosis affecting the digestive tract is exocrine pancreatic insufficiency, characterized by chronic diarrhea with undigested food. A reduction in the secretion of sodium bicarbonate is responsible for reducing the efficacy of pancreatic enzymes and the precipitation of bile salts, resulting in an increased pH in the duodenum, leading to malabsorption. In addition, obstruction of pancreatic canaliculi by mucous plugs prevents the release of enzymes into the duodenum, causing poor digestion of fat, proteins, and carbohydrates, which leads to malnutrition, supported by inadequate food digestion and increased energy needs due to anorexia and recurrent respiratory disease. Cystic fibrosis patients also show gastrointestinal symptoms such as nausea, vomiting frequently, gastroesophageal reflux disease, esophageal adenocarcinoma, distal intestinal syndrome, and cholelithiasis rarely. Abdominal pain is the common symptom in cystic fibrosis patients with distal bowel obstruction syndrome and fibrosing colonopathy as the main characteristic features of the complications. Gastroparesis is another major complication of the lung, predominantly observed in cystic fibrosis patients, especially in children and young individuals with severe deterioration of the pulmonary tract.

5.4.10.2 Etiology

Cystic fibrosis is an autosomal recessive disorder, occurring if both parents carry the mutated or diseased alleles responsible for the disorder in their genomes. Carriers will have one normal and one mutant CFTR gene, due to which their health is not affected. Carriers have a 50% chance of transferring the mutated gene to the next generation. If both the parents are carriers, there is a 25% chance of having cystic fibrosis in every pregnancy for the child to have cystic fibrosis, a 25% chance that the child will have two normal CFTR genes, and a 50% chance of having a child, as a carrier.

Cystic fibrosis occurs mainly due to the mutation(s) in the CFTR gene, located on the q31.2 locus of the chromosome 7 short arm. The length of this gene is 230 Kb with 27 exons and 26 introns. Upon translation, it yields CFTR protein of length 1480 amino acids. As per the Cystic Fibrosis Mutation Database, more than 2,000 mutations have been identified and described. Altogether, 2,000 mutations have been identified on the CFTR gene, with 40% being substitution mutations of single amino acids, 36% are nonsense, missense, frameshift, and mis-splicing mutations involved in RNA processing, 3% in large re-arrangements, 1% affecting the promoter region. 14% of the mutations are neutral, and the effect of the remaining 6% is unknown. Based on molecular and functional defect, mutations responsible for cystic fibrosis are divided into six classes. Class I mutations such as Arg553X, Gly542X, and Trp1282X mainly consist of nonsense mutations, arising due to the generation of premature stop codons. The resultant truncated and unstable RNA segment completely precludes the production of functional CFTR protein. Canonical splice mutations and chromosomal deletions lead to the total absence of CFTR protein and belongs to class I mutations. Class II mutations are the most frequently encountered CFTR mutations observed among different ethnic groups. 90% of European and American cystic fibrosis patients harbor class II mutation on at least one CFTR allele. Phe508del (popularly known as ΔF508) is the major mutation found in most of the population. In this mutation, the deletion of three nucleotides leads to missing phenylalanine at position 508 of the amino acid sequence, resulting in misfolding and abrogation of trafficking to the cellular membrane. The cellular machinery recognizes this faulted protein as abnormal and degrades it well before reaching the plasma membrane. Other class II CFTR mutations identified are Gly85Glu, Arg560Thr, Ile507del, and Asn1303Lys, which are known to reduce transportation efficiency. Mutations of this class result in an insufficient amount of CFTR reaching the surface of apical cells.

Class III mutations, also known as gating mutations, will deregulate the opening function of CFTR protein by reducing the opening time of the ion channel gate and blocking the passage of chloride ions. One of the most common class III mutations is Gly551Asp. Class IV mutations comprise structurally normal CFTR protein with defective conductance capacity, leading to inefficient electrolyte transport, even when the channel is open. Arg117His and Arg347Pro mutations belong to this class. Class V mutations are mainly found in introns, adversely affecting splicing and ultimately reducing CFTR synthesis. Class VI mutations reduce CFTR protein stability at the plasma membrane. This class comprises missense mutations such as 432delTC, resulting in unstable protein.

An estimate of one individual per every 3,000 has this disease. Ireland has the highest disease incidence, with 1 per 1400 live births. The prevalence of cystic fibrosis varies with ethnic group. 1 per every 1800 Caucasians born in Europe, one per every 5000 born in the United States, 1 per every 40,000 Afro Americans, and 1 per 4,000 - 10,000 Latin Americans suffer from cystic fibrosis. This disease has a low incidence rate in Asian and African populations. Some small European populations, such as Albania, have a high disease incidence of 1 per every 555 individuals, especially Albanian immigrants to northern Italy. However, several countries have reported that >50% of their cystic fibrosis patients fall under 18 years of age.

5.4.10.3 Diagnosis

95% of patients showing the symptoms of cystic fibrosis are diagnosed with the sweat test by measuring the chloride concentration after applying pilocarpine, a muscarinic receptor antagonist, which stimulates sweat production. The optimum chloride concentration in the sweat of an average individual is 30 $mmol^{-1}$ at any age with an upper limit of 40 $mmol^{-1}$. Any chloride concentration above 60 $mmoll^{-1}$ is abnormal at all ages. However, there is some uncertainty regarding screening by using a sweat test. A few rare cases with chloride concentrations are <30 $mmol^{-1}$, which might be due to a heterogeneity in CFTR expression or could be due to the influence of other genes such as ENaC. The United States Cystic Fibrosis Foundation has coined the term 'CFTR-related metabolic syndrome' for characterizing the sweat chloride values of individuals <60 $mmol^{-1}$ and <30 $mmol^{-1}$ and two mutations on the CFTR gene with one of them, for the disease to be occurring. Alternatively, the European Cystic Fibrosis group has proposed cystic fibrosis screen positive with inconclusive diagnosis (CFSPID) under similar diagnostic conditions. Genetic testing also plays a vital role in the diagnosis of patients with rare cystic fibrosis mutations. Measurement of potential differences across the nose or respiratory epithelium using a peripheral electrode is the adjunct test for diagnosing cystic fibrosis. In vitro measurement of the CFTR activity on excised rectal biopsy tissue can also be carried out using open or closed-circuit currents.

5.4.10.4 Treatment

Patients with chronic lung infection are provided with nebulized tobramycin, an aminoglycoside antibiotic for improved lung function, reduced exacerbations, and increased weight. Mucolytic, hydrating, and bronchodilator agents clear mucous in the lungs. Antibiotics control chronic respiratory infections. Prescription of pancreatic enzymes and vitamin supplements for patients with pancreatic insufficiency will optimize their nutritional status. Three categories of therapies are available for Cystic fibrosis therapies, modulator drugs, corrector molecules, and gene therapy

5.4.10.4.1 Modulator Drugs

Drugs that control cystic fibrosis are differentiated into correctors or potentiators. Correctors aim to repair the molecular defect such as a folding defect in Phe508del, while potentiators enhance the channel activity without correcting the genetic or structural abnormality. The first CFTR potentiator, ivacaftor (formerly known as VX-770), was commercialized in 2009 by Vertex Pharmaceuticals Inc., Boston, MA, USA. Ivacaftor will be administered to patients of 12 years age, carrying at least one Gly551Asp allele, presenting a forced expiratory volume of 1 second between 40% AND 90% of the predicted value. There was significant, rapid, and sustained improvement in pulmonary function among patients supplemented with ivacaftor along with reduced pulmonary exacerbations. In February 2014, the US Food and Drug Administration approved a supplemental new drug application for ivacaftor, extending its use for patients with non-Gly551Asp class III mutations. Ivacaftor also benefits the patients with Arg117His class IV mutations.

5.4.10.4.2 Corrector Molecules

Corrector drugs overcome the pre-mature stop codons of Class I mutations by reducing the translational fidelity. Premature termination codon suppressors such as aminoglycosides are capable of binding to the ribosome and alter the recognition of premature termination codons, which triggers the recruitment of a tRNA rather than a termination complex. This tactic will restore the premature stop codons into sense codons and permits the ribosomes to read through the disease-causing premature stop codons without interrupting protein synthesis, allowing the production of full length CFTR proteins to the normal chloride secretion level. Ataluren (formerly known as PTC124, where PTC stands for premature termination codon), an orally supplemented premature termination codon suppressor, has been evaluated for phase I and phase II clinical trials, where the results were promising. Unfortunately, the phase III trial failed to reach its endpoint.

5.4.10.4.3 Gene Therapy

Enormous interest in gene addition therapy might be due to its potential in all classes of cystic fibrosis mutations. Gene addition strategies with viral and non-viral vectors have been pursued but remain unsuccessful. An initial strategy comprising of an adenoviral vector has been designed. However, this vector has been shown to induce potent cellular and humoral immune responses that preclude subsequent re-administration. Subsequently, an Adeno-associated viral vector strategy has surpassed the adenoviral vectors. The advantage of this vector system is that it does not contain viral genes and remains episomal inside the nucleus. The initial phase I trial has proved that adeno-associated vectors are safe gene transfer reagents.

Non-viral gene transfer agents such as cationic lipid or polymer carrier molecules, which bind to plasmid DNA and promote cellular integration, were also generated. Even though this strategy is less efficient, it can escape the immune response and enable repeated administration. The UK gene therapy consortium's monthly aerosolized lipid vector-based gene therapy is in phase III of the clinical trial. This novel approach of transforming premature stop codons into a sense codon using a recombinant enzyme is successful. A genetically faulty mRNA has been corrected using a genetically encoded enzyme editase and has successfully restored its function. However, the homogenous gene transfer of cystic fibrosis agents into the patient's lungs remains challenging. For achieving success, certain obstacles such as penetration of the target vector into the thick mucus layer for reaching the target region, need to be resolved.

5.4.11 Thrombophilia

Thrombophilia is a hypercoagulable state characterized by an increased tendency to develop thrombosis and an obstructive clot in arteries (arterial thrombosis) or veins (venous thrombosis). Thrombosis induces changes in vessel walls, blood flow, and blood composition. Even though these changes contribute to thrombosis, arterial thrombosis is mainly regulated by the changes in the vessel wall, particularly atherosclerosis, whereas prothrombotic blood abnormalities regulate venous thrombosis (Maurizio et al. 2008). Thrombophilia is of two types, venous thrombo embolism (VTE), also known as deep venous thrombosis, pulmonary embolism, and inherited thrombophilia. VTE occurs due to a combination of congenital pro coagulant defects and environmental factors, including age, male sex, obesity, exposure to "risk periods" of immobilization, trauma, cancer, pregnancy, the use of exogenous hormones, and chemotherapy. Hereditary thrombophilia occurs in individuals with a family history of thrombosis

5.4.11.1 Symptoms

Thrombophilia leads to blood clots forming mainly in the veins of the legs known as deep vein thrombosis, in the lungs known as pulmonary embolism, and rarely in the arteries, causing heart attacks, strokes, and blocked blood flow in the arm or leg. Inherited thrombophilia occurs due to mutations in the genes leading to a deficiency in the blood clotting proteins such as protein C, protein S, and anti-thrombin. Individuals with thrombophilia will never develop a blood clot. Nevertheless, their blood gets clotted due to the combination of risk factors such as surgery, hospitalization, extensive mobility, estrogen therapy for either birth control or hormone replacement after menopause, pregnancy, obesity, smoking, having a family history, and old age (Renu and Monica, 2009). Thrombophilia plays a significant role in hyper blood clotting, but it might not be the sole reason for the clot.

5.4.11.2 Etiology

Inheritance of thrombophilia applies to conditions such as mutations affecting the protein's amount or function in the coagulation system. These mutations comprise loss of function mutations affecting antithrombin, protein C, protein S, and gain of function mutations comprising factor V Leiden and prothrombin gene 20210 A/G (Jody, 2011).

5.4.11.2.1 Antithrombin Deficiency mutations

Antithrombin is a major physiological inhibitor of coagulation, exhibits anti-inflammatory properties, and accentuates a crosstalk between both. It is a serine proteinase inhibitor (also known as serpin) superfamily, with a highly conserved structure composed of three β-sheets and eight α-helices. Antithrombin deficiency in humans will lead to a substantial risk of developing venous thrombosis, classified into type1 and type 2. Type 1 involves quantitative defects such as low activity levels and antithrombin levels. Type 2 occurs due to qualitative defects in the gene, such as mutations in the thrombin binding site, producing a dysfunctional molecule of normal antigen levels with reduced antithrombin activity. Type II is further classified into Type II reactive site (RS), Type II heparin binding site (HBS), and Type II pleiotropic effect (PE). There are not many clinical differences between type 1 and type II.

Antithrombin deficiency mutations rarely occur in an autosomal dominant manner, with a prevalence of 1 in 2,000 - 5,000 individuals. Heterozygotes will have an 8.1 times higher probability of developing thrombosis. Mutations on the SERPIN C1 gene responsible for antithrombin deficiency are located on the q23-25 locus of chromosome 1. 60% of the patients with antithrombin deficiency exhibit recurrent thrombotic episodes, and 40% exhibit pulmonary embolism (Ming and Stephan, 2015).

5.4.11.2.2 Protein C Deficiency

Protein C deficiency is one of the most common in the human population, especially among Caucasians. This deficiency is prevalent in heterozygous populations ranging from 1 out of 500 in asymptomatic patients to 1 out of 5000 symptomatic patients. Heterozygous protein C deficiency is inherited both in an autosomal dominant and severe autosomal manner. Two kinds of protein C deficiencies have been described, type 1 deficiency comprising of reduced enzymatic and immunological activity, and type II deficiency involving dysfunctional protein C. Heterozygous protein deficiency produces a 7.3-fold increased risk of thrombosis over homozygous deficiency. Rare cases of purpura fulminans (thrombotic disorder that manifests as blood spots, bruising, and discoloration of skin due to coagulation in small blood vessels) have been reported in newborn babies due to deficient blood vessel protein C activity (less than 1%) resulting from homozygous mutations.

5.4.11.2.3 Protein S Deficiency

Protein S deficiency leads to the generation of a robust thrombophilic abnormality in humans. Protein S deficiency arises due to congenital genetic defects as well as from acquired plasma perturbations. Patients with protein S deficiency tend to have an increased risk of developing venous thrombo embolism. Heterozygotes have an 8 - 9 times higher risk of developing thrombosis. Homozygous individuals have developed recurrent venous thrombo embolism at a young age. Under normal conditions, 60% of the protein S binds to C4b binding protein, and only 40% of the protein S is functionally active. Increased C4 binding protein levels will further reduce the protein S quantity (Sandra, 2014). Acquired protein S deficiency is observed during pregnancy, overutilizing oral contraceptive pills, acute thromboembolic disease, anticoagulant therapy, and liver diseases.

5.4.11.2.4 Factor V Leiden Mutation

This mutation is one of the most common and frequently inherited prothrombic conditions accounting for 40 - 50% of the cases. The occurrence of this disorder due to a mutation in factor V is known as Dutch investigators were the first to report this mutation from the city of Leiden. It is a genetic disorder characterized by an inadequate anticoagulant response to activated protein C (APC), a natural anticoagulant protein, which cleaves and inactivates procoagulant factors such as Va and VIIIa, leading to the downregulation of thrombin. Factor V Leiden mutation is guanine to adenine substitution at 1691 nucleotide of the factor V gene, which eventually substitutes glutamine instead of arginine at Arg 506 APC cleavage site. APC inactivates factor Va by cleavage at three different amino acid positions R306, R506, and R679. Because of this single base substitution, Factor Va will become resistant to APC, inactivated at least 10-fold more slowly than for normal and healthy individuals, increasing thrombin. Factor V cleaved at amino acid 506 will also function as a co-factor for the APC mediated inactivation of factor VIIIa.

The prevalence of this mutation is higher in the Caucasian population, ranging from 1% to 7%. It is sporadic among Asian, African, and Australian populations. Heterozygous patients of factor V Leiden mutation have a 7-fold increased risk of venous thrombo embolism, and homozygous patients with a prevalence of 1 in 500 will have a 50 - 80-fold increased risk of developing venous thrombo embolism. A very high incidence of venous thrombo embolism with factor V Leiden mutation (35%) has been found in the women who take more oral contraceptives. The presence of factor V Leiden mutation might lead to an increased risk of fetal loss. The frequency of homozygosity for the Factor V Leiden in the European population is around 1 in 5000. Haplotype analysis of the gene strongly suggests that the Factor V Leiden mutation was a single event that must have occurred 20,000 - 30,000 years ago after the evolutionary separation of Europeans from Asians and Africans. The high prevalence of Leiden among Europeans suggests a balanced polymorphism with some survival advantage associated with a heterozygous state. Speculation of a mild prothrombotic state conferred by the mutation could have reduced the mortality from bleeding associated with childbirth or trauma in pre-modern times. Several cases have identified that heterozygotes of Factor V Leiden had significantly less blood during menses, childbirth, and cardiac surgery. Hemophiliacs who are heterozygous for Factor V Leiden have less severe bleeding and reduced clotting factor.

5.4.11.2.5 G20210a Mutation

Poort and his associates in 1996 performed DNA sequencing of the prothrombin gene in patients with VTE symptoms. His gene is located on chromosome 11 with single missense guanine to adenine substitution at position 20210 in the 3' UTR region, known as G20210A mutation. As the mutation site is not in the ORF of the gene, it will not affect the actual structure and function of the prothrombin. However, this mutation leads to elevated prothrombin levels, up to 133% more than regular, sufficient to develop venous thrombo embolism. G20210A mutation leads to an increased prothrombin mRNA and protein expression (Mehrez, 2011). Prevalence of G20210A mutation is high among Caucasians, with 3 - 17% of the patients with VTE and 1 - 8% healthy normal ones.

The prevalence of the prothrombin G20210A mutation is higher in southern European countries when compared with northern European countries. The prothrombin G20210A mutation is rare and is even absent in the Asian, African, native populations of America (Amerindians) and Australians. The only exception to this observation is the high prevalence of the G20210A mutation in Hispanics and Mexican Mestizos (Ceelie et al. 2004). The latter are the descendants of mixed marriages between Europeans and Amerindians.

5.4.11.2.6 MTHFR Mutation

The MTHFR gene is located on the short p arm of chromosome 1 at locus 36.22 from 11,785,73 to 11,806,103 bp. The expression of the MTHFR gene produces an enzyme known as methylenetetrahydrofolate reductase, which plays a prominent role in processing amino acids. The specific function of this enzyme is to convert methylenetetrahydrofolate (Vitamin B9) to 5-methyltetrahydrofolate, which is found in the blood and is necessary for the conversion of amino acid homocysteine to another amino acid methionine. At least 40 mutations have been identified on the MTHFR gene. Among them, the C677T mutation with cytosine to thymine substitution at 677 nucleotide position leads to an elevation in the levels of homocysteine known as homocystinuria, increasing the risk of thrombosis.

5.4.11.2.7 Deficiency of Factor XII

Factor XII, also known as the Hageman factor, is an enzyme, which initiates the coagulation cascade. Hageman factor was identified in 1955 while examining the blood sample of a 37-year-old rail brakeman named John Hageman found to have prolonged clotting time without any hemorrhagic symptoms. Dr. Oscar Ratnoff examined Hageman and found that he lacked an unknown clotting factor and studied the genetics of this deficiency. Factor XII plays a vital role in a cascade of reactions involved in coagulation. It activates Factor XI and subsequently factor IX. Factor XII is also crucial for the conversion of plasminogen into plasmin and the initiation of fibrinolysis. Factor XII deficiency leads to an asymptomatic prolongation of activated partial thromboplastin time (Mohammad et al. 2015). As coagulation and fibrinolysis are constantly in a dynamic equilibrium, deficiency of Factor XII can potentially increase the risk of thrombosis. Factor XII deficiency is rare and is inherited as an autosomal recessive or acquired disorder.

5.4.11.2.8 Dysfibrinogenemias

Dysfibrinogenemias is a group of disorders characterized by qualitative abnormalities of fibrinogen. Fibrinogen consists of two half molecules, with each half consisting of one aα, one bβ, and one γ chain. Each half of the fibrinogen is associated at the N terminus to form a central E domain. Each half also consists of a terminal D domain containing the C terminus of one bβ and one γ chain and a terminal α-C domain, containing the C terminus of one aα chain. Fibrinogen and its proteolytic cleavage product fibrin are critical plasma proteins, which perform multiple functions in blood clotting, including fibrin clot formation, factor XIIIa-mediated fibrin cross linking, non-substrate thrombin binding, platelet aggregation, and fibrinolysis. Many structural abnormalities occur in fibrinogen, which interferes in its hemostatic functions collectively known as dysfibrinogenemias (Mark et al. 2002). Based on the mechanism of occurring, dysfibrinogenemia disorders are of two types. Firstly, inherited dysfibrinogenemia, which occurs due to mutations in the coding region of the fibrinogen aα, bβ, and γ gene. To date, more than 250 fibrinogen abnormalities have been reported, named after the city, where the patient lives, or the city of the hospital, which evaluates the patient. Roman numerals are added after the city name if more than one dysfibrinogen is reported from the same city (Figure 5.7).

Secondly, there is acquired dysfibrinogenemia, caused due to liver disease or biliary tract disease associated with cirrhosis, chronic active liver disease, acute liver failure, and various causes of obstructive jaundice. The prevalence of dysfibrinogenemia is more severe in patients with liver

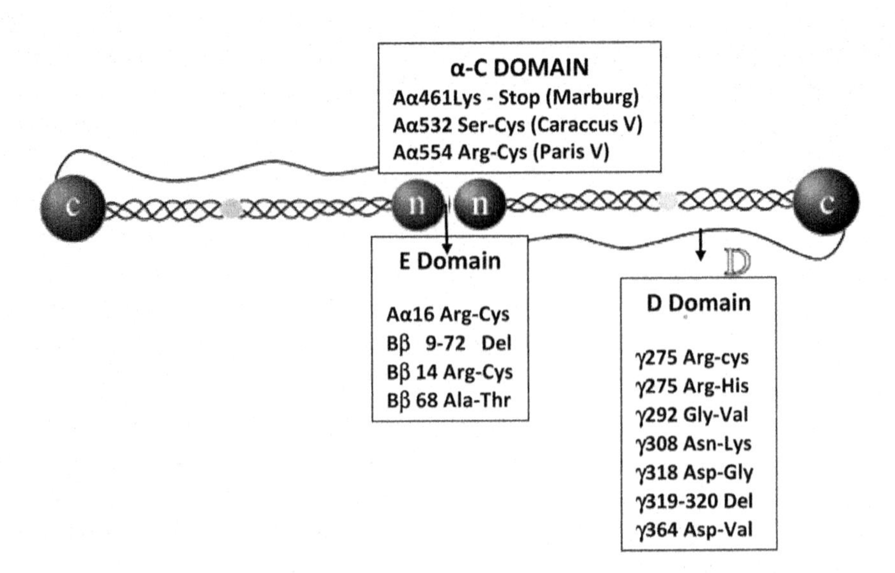

FIGURE 5.7 Structure of fibrinogen and location of mutations that are associated with thrombophilia.

disease (76 - 86%) than in obstructive jaundice (8%). Acquired dysfibrinogenemia is considered a paraneoplastic marker for malignancies such as hepatoma and renal cell carcinoma. The dysfibrinogen is found during the diagnosis, disappears during the tumor remission, and reappears after tumor relapse.

5.4.11.3 Diagnosis

Physical examination is the first step performed for the diagnosis of thrombophilia. Even though a family history of thrombosis is an essential factor for testing, adverse family history does not exclude hereditary thrombophilia due to the low penetrance of thrombophilic defects and new mutations that might occur in the individual. Preliminary laboratory tests include the prothrombin time activated partial thromboplastin time, thrombin time, and fibrinogen. A single base mutation in the factor V gene is determined by using inherited APCR. Over 90% of the cases concerned with this mutation are identified by using APCR. Despite its simplicity and inexpensiveness, APCR is not sufficiently sensitive and fully specific for the factor V mutation, which is replaced by new second generation APTT based assays, where plasma of the patient is pre-diluted with factor V deficient plasma, thus improving the sensitivity and specificity of the factor V Leiden mutation. The G20210A mutation is the second most inherited risk factor for thrombophilia, diagnosed by genotyping for the G20210A mutation. Hyperhomocysteinemia or homocystinuria, characterized by high homocysteine concentrations in plasma, is associated with arterial and venous thrombosis. The MTHFR gene C677T polymorphism is a common variant of homocysteinemia that appears in the heterozygous form in 30 - 40% and in the homozygous form in 10 - 13% of Caucasians. This polymorphism leads to a thermolabile MTHFR enzyme variant found in high concentrations in the patients' plasma. This polymorphism is also identified by genotyping. However, a large-scale population-based case-control mega study conducted by Bezemer and his colleagues in 2007 has shown clear proof that no association exists between MTHFR C677T polymorphism and VTE. Hence, it is not considered to be a tool for the diagnosis of thrombophilia.

5.4.11.4 Treatment

Family testing is one of the most applied procedures useful for women from thrombophilia families who intend to marry. Females and the relatives of thrombophilia families with antithrombin, protein C, protein S deficiency, and the carriers of Factor V Leiden mutations and G20210A mutation carriers have a high incidence of VTE at 4% per pregnancy. Women homozygous for factor V Leiden mutations have a high risk of up to 16% per pregnancy without prophylaxis. Anticoagulant therapy with low heparin injections is prescribed for these patients. Genetic testing is also helpful for the women from thrombophilia families with factor V Leiden, G20210A mutations or antithrombin, protein C, and protein S deficiency. The risk of VTE is increased in women using oral contraceptives. The advantages of genetic testing are higher for women undergoing postmenopausal hormone therapy due to the high risk of developing VTE during middle age.

The rapid expansion of pharmacogenetics and pharmacogenomics has allowed rapid progress in personalized therapy for individual patients by understanding the genetic variations related to drug efficacy and toxicity. Pharmacogenetic information has the potential to improve the efficacy and safety of antithrombotic drugs. One of the best examples of pharmacogenomic testing is warfarin, a widely prescribed oral anticoagulant used for decades to treat thrombosis. Sizeable inter-individual variation with 20-fold differences in sensitivity among patients is one of the significant problems with warfarin, having severe consequences of either over or under anticoagulation. In individuals with extreme sensitivity, the exact dosage may lead to adverse effects such as hemorrhage. Genetic variation has momentously impacted the dose variation along with age, and dietary vitamin K intake. Pharmacogenetic studies have identified polymorphisms in two genes corresponding to cytochrome 450 enzyme complex gene CYP2C9 and VKORC1 as principal genetic determinants of warfarin dose. CYP2C9 is involved in the metabolic clearance of warfarin, responsible for its therapeutic effects. Based on non-synonymous SNPs that result in Arg144Cys (*2) and Ile358Leu (*3), two alleles of CYP2C9, CYP2C9*2, and CYP2C9*3 were identified. Both these variants are associated with the reduced metabolic clearance of warfarin, lowering the dose requirements. Carriers of these variants display a high sensitivity to warfarin and increased risk for hemorrhagic complications compared with the individuals homozygous for allele*1. SNPs on the CYP2C9 gene account for more than 12% of the total variance, and SNPs on gene VKORC1 account for 30% of the variance among required warfarin dosage. VKORC1 gene encodes for vitamin K epoxide reductase complex, the target enzyme inhibited by warfarin. Vitamin K epoxide reductase complex is essential for recycling vitamin K and subsequent activation of several clotting factors. One significant SNP for VKORC1 is 1639 G>A, a common polymorphism of the promoter sequence, which defines two haplotypes A and B. Haplotype A is associated with high warfarin sensitivity. Hence, a lower dosage of warfarin drug is required. In contrast, haplotype B has low sensitivity to warfarin, requiring high dosages of the drug.

Clinical manifestations and morbidity are associated with thrombophilia during pregnancy, including pregnancy loss and other adverse outcomes. The effect of preventive anticoagulant therapy during pregnancy in women inherited with thrombophilia is still controversial. One large study encompassed 1011 pregnancies in 416 woman carriers of factor V Leiden mutation or prothrombin gene variant G20210A (Kevin et al. 2014). The outcome was evaluated according to the low molecular weight of heparin or aspirin and the pregnancy period when the treatment was started. Results have shown that low molecular heparin had a protective effect on miscarriages and venous thromboembolism, whereas the administration of aspirin has shown no effect on preventing complications and venous thromboembolism. Finally, it is imperative to provide genetic counseling to the patients and their asymptomatic family members. Adequate genetic counseling is essential for educating family members to increase awareness of risk factors and interventions for preventing thrombophilia.

5.4.12 Tay Sachs Disease

Tay Sachs disease, named after Warren Tay in 1881, was the first to identify and describe an asymptomatic red spot on the eye retina. Bernard Sachs, in 1887, described and studied most of the symptoms and etiology of the disease. This disorder, also known as gangliosidosis, occurs due to the deficiency of an enzyme β-Hexosaminidase, which degrades the GM2 ganglioside (ganglioside monosialic acid 2), leading to lysosomal swelling, neuronal dysfunction, and progressive neurologic degeneration. Three major isoenzymes for hexosaminidase, HexA, HexB, and HexS were identified. Isoenzyme HexA consists of two α and β subunits. HexB is a homodimer, composed of two β subunits and HexS consists of two α subunits. Both α and β subunits are synthesized in the endoplasmic reticulum (ER), where glycosylation, intramolecular disulfate bond formation, and dimerization occur. After dimerization, b-hexosaminidase is transported to the Golgi complex for post-translational modification, such as adding mannose-6-phosphate to the oligosaccharide side chains. Residues of the phosphorylated mannose can be considered as a marker for its recognition by the specific receptors located on the inner surface of the Golgi complex membranes. Due to the presence of phosphorylated mannose, a lysosome recognizes the enzyme and will absorb it. Inside the lysosome, α and β subunits are proteolytically processed into a mature form. HexA and HexB isoenzymes hydrolyze a wide range of substrates such as glycoproteins, glycosaminoglycans, and glycolipids upon maturity. Hex A also interacts with the GM2 ganglioside complex and hydrolyzes GM2 ganglioside. In the absence of GM2A, these isoenzymes also hydrolyze synthetic substrates such as 4-methylumbelliferone-GlcNAc fluorescent substrate (4MUG), which can be used for the diagnosis of GM2-gangliosidosis and detection of HexA and HexB gene mutations.

GM2-gangliosidosis occurs due to mutations in three genes:

1. Tay Sachs disease due to mutations in the HexA gene located on chromosome 15, disrupting the HexA activity
2. Sandoff disease due to mutations in HEXB gene located on chromosome 5, disrupting the activities of both HexA and HexB
3. GM2 activator protein deficiency occurs due to the mutations in the GM2A gene located on chromosome 5.

In patients with Tay Sachs disease, GM2 ganglioside accumulates in the lysosomes leading to their enlargement and forms inclusions within the cells known as membranous cytoplasmic bodies. As the maximum concentration of GM2 is present in the neurons, deficiency of HexA affects the nervous system, causing a motor developmental delay in the patients.

5.4.12.1 Symptoms

Tay Sachs disease is a genetic disorder characterized by the deficiency of HexA, leading to progressive damage and proliferation of microglia. HexA deficiency also leads to the accumulation of complex lipids in macrophages of the brain tissue, in the cerebellum, basal ganglia, brain stem, spinal cord, spinal ganglia, and in the neurons of the autonomous nervous system. Ganglion cells in the retina also get swollen and contain GM2 gangliosides along the macula edges, resulting in the appearance of cherry red spots.

Two forms of Tay Sachs exist, the infantile form and the late onset form (Figure 5.8). The infantile form occurs more frequently than the late onset form, characterized by an onset around 6 months of age with a deficient HexA activity of <0.5%, which can be correlated with developmental delay in mental and motor function. Neurodegenerative symptoms such as hypotension, inability to sit or hold their head unsupported, eye movement abnormalities, dysphagia, spasms, and hypomyelination are noticed in infants. In the severe forms, symptoms will start a few months after birth, and infantile Tay Sachs disease patients may not survive more than 3 - 5 years of age.

The late onset form of Tay Sachs disease comprises of two sub forms, juvenile onset form, and adult-onset form. Juvenile onset is characterized by ataxia, dysarthria, dysphagia development, hypotension, and spasm progression with affected individuals entering this stage between 5 and15 years

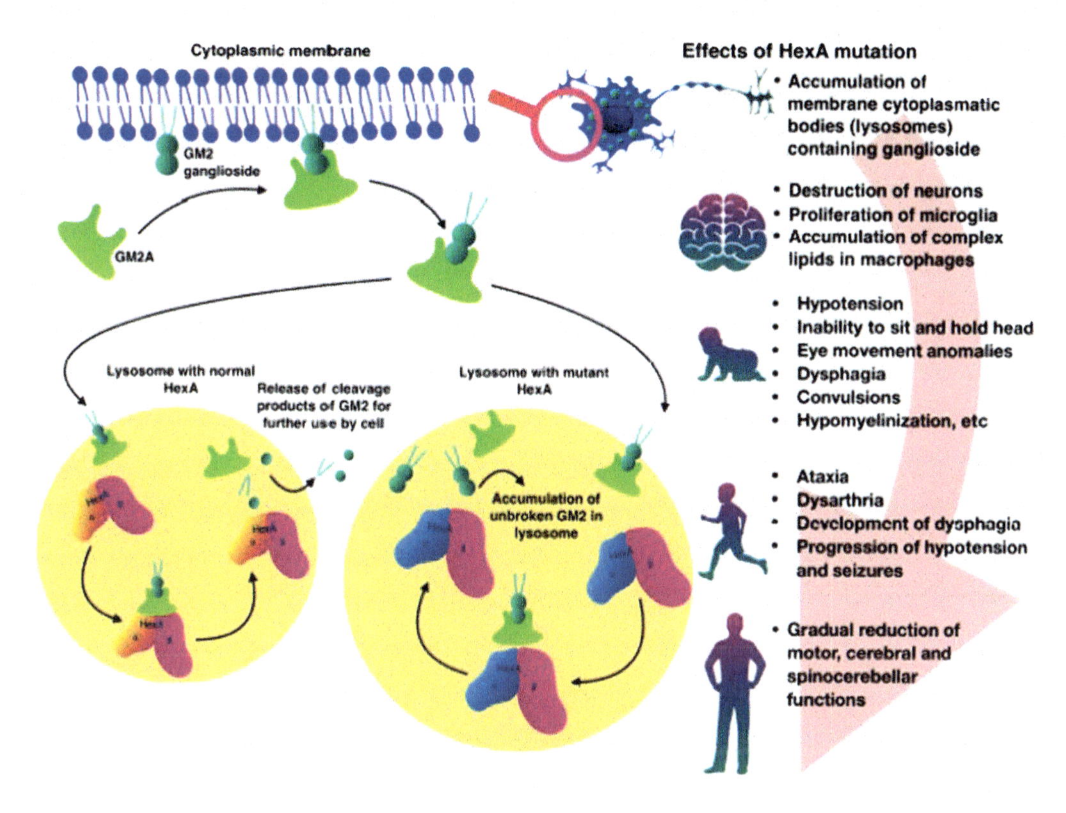

FIGURE 5.8 Biochemistry and symptoms of Tay Sachs disease.

of age. Most of the patients do not live past the age of 15 years. In contrast to the infantile and juvenile forms, the adult-onset form is less aggressive, characterized by fewer mutations and comparatively high quantities of HexA (5 - 20% of the normal activity). Here, symptoms appear during adolescence, early or late adulthood. Progress of neurodegeneration is slowed, manifested with the signs of cerebellar involvement such as dysarthria, dysmetria, ataxia, and anterior horn cell involvement such as proximal muscle weakness, fasciculations, and atrophy. In some patients, neuropsychiatric problems such as bipolar disorder and depression are also observed.

5.4.12.2 Etiology

Infantile Tay Sachs is a genetic disorder inherited in an autosomal recessive fashion. It occurs due to mutations in the HexA gene, located on the long q arm of chromosome 15 at the locus 23 - 24, coding for the α-subunit of the HexA enzyme. More than 125 different mutations were identified in the HexA gene, including single base insertions and deletions, splice phase mutations, missense mutations, altering or severely inhibiting the protein function. Late onset Tay Sachs is a rare variant phenotype whose symptoms appear after 30 - 40 years of age. The severity in other forms of Tay Sachs is low due to residual enzyme activity. In most cases, infantile Tay Sachs disorder is inherited in a homozygous or compound heterozygous for severe chronic deleterious null alleles. In contrast, late onset Tay Sachs combines one severe allele and another associated residual enzyme activity.

The incidence of Tay Sachs disease is concentrated on specific populations such as the Ashkenazi Jewish population comprising Jews of Central and Eastern European descent. Incidence of the disease is at the rate of 1 in every 3,500 newborns, 100 times higher. Every 1 among 29 individuals is heterozygous or asymptomatic and is also the carrier for Tay Sachs disease. Among the Ashkenazi Jewish population, three mutations are responsible for 98% of incidence of the disease.

TABLE 5.1
Mutations in the HEXA Gene of Different Population Groups, Responsible for Infantile, Juvenile, and Adult Tay Sachs Disease

	Mutation	Change	Population Group
Infantile			
	509 G→A	170 Arg → Gly	Japanese
	532 C → T	178 Arg → Cys	Czechoslovakian
	533 G → A	178Arg → Hist	Diverse
	1260 G → C	420 TRP → CYS	Irish and German
Juvenile			
	570 G → A	Abnormal splicing	Tunesia
	749 G → A	250 Gly → Asp	Lebanese
	1496 G → A	499 Arg → His	Scotland and Ireland
	1511 G → A	504 Arg → His	Assyrian
Adult			
	805 G→ A	269 Gly → Ser	Diverse

A 4 bp insertion at exon 11 of the HEXA gene (insertion of TATC at 1278 bp), found in 80% of the Ashkenazi Jewish patients with Tay Sachs disease. Further, transversion of guanine to cytosine in the first base of intron 12 is found in the 16% of Ashkenazi Jewish patients. Additionally, transition of guanine to adenine in the last nucleotide of exon 7, upon translation, would substitute glycine for serine at 269 amino acid position in the α-subunit of HexA enzyme (also known as Gly269Ser mutation). This mutation is found to be present in 2% of the Ashkenazi Jewish patients. The first two mutations are responsible for the infantile onset whereas the third is responsible for the late onset Tay Sachs disease. A few mutations responsible for Tay Sachs of different population groups are displayed in Table 5.1

5.4.12.3 Diagnosis

Tay Sachs disease is diagnosed by an initial enzyme assay involving the measurement of HexA enzyme activity in the patient's blood serum, fibroblasts, and leucocytes. HexA activity is reduced in the individuals with late onset Tay Sachs and is almost absent in the infantile Tay Sachs form. Further, molecular genetic testing can diagnose Tay Sachs disease by detecting mutations in the HEXA gene. In some population groups such as Ashkenazi Jews, the prenatal diagnosis of Tay Sachs disease has been made successfully since the 1970s by monitoring the fate of subsequent pregnancies using the diseased child of a couple in a family. With the development of technology, carrier screening programs using enzyme testing were initiated widely in high risk populations for detecting and informing carrier couples about the family history of the disease. By using enzyme testing, more than one million individuals have been tested for the carrier status. More than 36,000 carriers have been identified, and more than 1000 couples have been informed of their risk of having a child with Tay Sachs. These programs led to a 90% decrease in the Tay Sachs disease among the Ashkenazi Jews in Canada, Israel, and the USA. Other options available to carrier couples are adoption, donation of sperm or egg, pre-implantation diagnosis, and reproductive abstention (taking 25% risk). With technology development, prenatal diagnosis is being made either by amniocentesis or chorionic villus sampling. During amniocentesis, a sample of amniotic fluid surrounding the developing fetus is removed and studied for the presence of HexA enzyme, whereas for chorionic villus, tissue samples from a portion of the placenta are used for the test. Prenatal tests also identify the mutations in the HEXA gene of the DNA isolated from amniotic cells. Patients with infantile Tay

Sachs disease have a cherry red macula in their retina, diagnosed using an ophthalmoscope. Blood tests are available for determining the carrier status of individuals for Tay Sachs disease. Couples with Jewish ancestry are strongly encouraged to undergo carrier screening before proceeding with pregnancy.

5.4.12.4 Treatment

There is no permanent cure for this disease, or no specific treatment is available for this disease. Current treatment is based on the symptoms that are apparent to the individual, which might require the coordinated efforts of a team of specialists to work with pediatricians to plan for the treatment systematically. However, with advancements in technology, a series of therapies are available, providing us with a cure to this disorder. Five therapies have passed experimental stages for the cure of Tay Sachs.

5.4.12.4.1 Substrate Reduction Therapy

This therapy utilizes small molecules as agents for reducing the rate of glycolipid biosynthesis. Bembi and his colleagues in 2006 have successfully demonstrated the efficacy of N-butyldeoxynojirimycin (NB-DNJ), a small iminosugar and competitive inhibitor of glucosylceramide synthase, in the prevention of GM2 ganglioside accumulation in murine models. NB-DNJ catalyzes the first step of glycosphingolipid synthesis and can penetrate the blood brain barrier. They also assessed its clinical efficacy in two infantile Tay Sachs disease patients and identified a significant increase in NB-DNJ concentrations in their cerebrospinal fluid. Similar results in a clinical trial on five patients (NCT00672022) and the usage of NB-DNJ have prevented n-macrocephaly.

5.4.12.4.2 Enzyme Replacement Therapy

Enzyme replacement therapy is a promising option for the treatment of lysosomal storage diseases. Enzyme replacement is clinically approved for Gaucher, Fabry, Pompe, Type I, Type II, and Type VI mucopolysaccharidosis. However, enzyme replacement therapy has been less effective in preventing neurodegenerative disorders, as the intravenous administration of the enzyme does not permit the enzyme to cross the blood-brain barrier. The major challenge for creating the HexA based enzyme replacement therapy is the need to synthesize both subunits. Tropak and his associates in 2016 created a hybrid micro subunit by combining the critical characteristics of both α and β subunits of the HexA enzyme. The hybrid micro subunit consists of an active α-subunit site and a stable β-subunit interface and unique regions of each subunit necessary to interact with GM2A. To purify the micro-homodimer of the resulting HexM, HEK239 cells with CRISPR deleted HexA and HexB genes, and the stably expressing micro subunit was used. Hydrolysis of GM2-ganglioside derivative by combining HexM and GM2A was demonstrated in cellular and in vitro conditions.

5.4.12.4.3 Bone Marrow Transplantation

A combination of bone marrow transplantation and substrate reduction therapy is implemented to treat Tay Sachs diseased patients. The application of bone marrow transplantation and Zavesca R (miglustat) usage has increased HexA activity in leucocytes and plasma 23 months after transplantation but failed to prevent neurological dysfunction development. In vivo studies on bone marrow transplanted Tay Sachs diseased mouse models have shown an increased survival rate from 4.5 to 8 months with improved neurological manifestations. Without suitable bone marrow donors for transplantation, the patients can transplant hematopoietic stem cells from umbilical cord blood obtained from partially HLA-matched donors. Transplantation of umbilical cord blood cells is a promising approach for treating neurodegenerative diseases concerned with ischemic or traumatic spinal cord injury.

5.4.12.4.4 Gene Therapy

Choices of the vector and delivery method with minimal side effects are the significant challenges for the Tay Sachs disease gene therapy. Gene and cell engineering to correct mutations in the HEXA gene were begun during 1990 using adenoviruses. Using Herpes Simplex Viral Vector (HSV-1) vector system, enzyme secretion has been promoted in the serum, and enzyme activity restoration has been achieved in the peripheral tissues. However, the main limiting factors for the successful utilization of adenoviruses are their inability to overcome the blood brain barrier, their immunogenicity, and their tropism, leading to a high accumulation of vector and overexpression of the transgene. Injection of HSV-1, consisting of HEXA gene into the inner capsule of the left cerebral hemisphere of Tay Sachs diseased mice could restore HexA activity, leading to the disappearance of the GM2 ganglioside in the hemispheres, cerebellum, and spinal cord within a month after injection. However, given the large size of the human brain, this approach might not achieve uniform distribution of the vector in the central nervous system, necessitating many injections.

Considerable progress has been achieved in controlling the GM2 gangliosides through gene therapy using Adeno-associated virus-based vector systems. Intracranial administration of recombinant Adeno associated vector serotypes encoding HEXA and HEXB cDNAs into Tay Sachs diseased mice has widespread HexA in the nervous system without much cytotoxicity, increasing their survival rate. The incapability of Adeno-associated virus based-vectors to carry genes of size above 4.5 Kb was minimized by designing the self-complementary AAV9.47 encoding a hybrid μ subunit ($_{SC}$AAV9.47-HEXM), whose intracranial injection has decreased GM2 ganglioside accumulation in the brain of Tay Sachs diseased model mice. Its intravenous administration has led to a long-term decrease of GM2 ganglioside accumulation in the central nervous system of newborn Tay Sachs diseased mice models.

The therapeutic efficacy of AAVrh8 based gene therapy has been studied on 2 - 4 months old Tay Sachs diseased Jacob sheep models by intracranial injection of AAVrh8-HEXA. All injected sheep have shown a delay in the appearance of disease symptoms. An excellent distribution of HexA in their brains has been observed. However, the distribution of HexA in their spinal cord was low. There was also a decrease in the activation and proliferation of microglia in their brains. Similar data was obtained upon the intracranial injection of AAvrh8 on Tay Sachs diseased cats, which demonstrated a widespread HexA in their central nervous system. Further studies are in progress ensuring the safety of AAvrh8 on other animals.

5.4.12.4.5 Genetically Modified Multipotent Cells

Transplantation of ex vivo modified multipotent neural cells in the central nervous system is another strategy for controlling Tay Sachs disease. A multipotent neural cell line with HEXA gene overexpression has been produced by retroviral transduction, which stably secreted biologically active HexA enzyme and cross-corrected the metabolic defect in Tay Sachs diseased patient-derived fibroblast culture under in vitro conditions. Intracranial injection of a multipotent neural cell line with overexpressed HEXA to mice has significantly increased the HexA transcription and active enzyme production. Even though these therapeutic methods could partially restore HexA activity in the human central nervous system and reduce GM2 ganglioside accumulation in cells, their efficacy was low in preventing neurodegeneration. The therapeutic effect can be achieved to its maximum only by starting the therapy from the time of its early manifestation.

5.4.13 Muscular Dystrophy

Muscular Dystrophy is a progressive motor paralysis that occurs primarily in males due to the decrease or absence of dystrophin protein, leading to a breakdown of muscle fibers and their consequent replacement by a fatty tissue that gradually weakens the muscle. Dr. Charles Bell first reported this muscular paralysis in 1830. Dr. Edward Meryon described the clinical and pathological findings of this disorder and its inheritance in 1864. Over the next 100 years, different forms of muscular dystrophy were identified. The genes that encode for structural proteins associated with sarcolemma, nuclear membrane proteins, and myofiber metabolism proteins have been isolated, sequenced, and mutations leading to specific forms of dystrophy have been studied. Over 10 different forms of muscular dystrophy are known in humans, including Duchenne muscular dystrophy, Becker muscular dystrophy, myotonic dystrophy, oculopharyngeal muscular dystrophy, Emery-Dreifuss muscular dystrophy, facioscapulohumeral muscular dystrophy, muscle-eye-brain disease, female dystrophinopathy carriers, and manifesting female dystrophinopathy carriers (Strehle and Straub, 2015). Among them, Duchenne muscular dystrophy is more prevalent. Congenital muscular dystrophies (CMDs) are clinically and genetically heterogeneous neuromuscular disorders that onset during birth or children's infancy. Fukuyama's congenital muscular dystrophy and Collagen Vi deficient CMD represent the common types of CMD due to mutations.

5.4.13.1 Symptoms

Deficiency in the synthesis of dystrophin protein forms the primary functional abnormality for muscular dystrophy. Dystrophin is the main protein, forming a dystrophin-associated protein complex for connecting the actin cytoskeleton with an extracellular matrix. Dystrophin is in the cytoplasm of sarcolemma, whose function is to maintain the structural integrity of skeletal muscles and heart. The absence of dystrophin leads to the loss of DAPC from sarcolemma. As a result, muscles will not withstand the stress generated due to regular muscle contractions. This might lead to an extra influx of calcium and activation of proteases and hydrolases in the cell, inducing myocyte apoptosis, inflammation, and fibrosis. DAPC loss will kickstart a cascade of deleterious processes that cause muscle fiber regeneration to fail, replace fat and connective tissues. An increase in calcium levels will activate calpains, myocardial cell degeneration, apoptosis, and fibrosis. Excess fibrosis leads to a decline in cardiac function. In the brain, the absence of dystrophin will disrupt synapse integrity and adversely affect the interneuron transmission. Cognitive impairment may also occur due to a reduction in glucose metabolism in the cerebellum.

Duchenne muscular dystrophy causes disability, respiratory disorders, and cardiac dysfunction, eventually leading to mortality. Patients look normal during their birth, but the proximal muscle gets weak at the age of four years as the first symptom of this disorder. Patients will develop difficulty rising from the floor at the age of 7 - 9 years and will have a severe inability to walk from 13 years. Walking inability is followed by a decline in the development of neuro motors, leading to reduced physical endurance and frequent falls. Cardiac abnormalities will begin at the age of 10, the severity of which depends on the individual. Fibrosis of respiratory muscles leads to respiration insufficiency (Carsten et al.2014). Death occurs either due to cardiac or respiratory failure. Increased plasma creatinine kinase levels above 1000 IU/L can reach up to 30,000 IU/L in patients.

5.4.13.2 Etiology

After the identification of the inheritance of muscular dystrophy by Dr. Edward Meryon in 1864, various genetic forms of muscular dystrophy have been recognized, and the mutations leading to these distinct forms have been characterized. Duchenne muscular dystrophy occurs due to a mutation in the gene responsible for the reduction or modification in the synthesis of the protein dystrophin. This gene is located on the locus 21, P arm of the X chromosome. Duchenne muscular dystrophy mainly occurs in males. As females have two X chromosomes and males have only one, in more than half of the cases, the diseased condition is inherited from the mother serving as a carrier. It can also occur due to novel mutations arising in the child's genes in rare cases. Female carriers under normal circumstances are not affected as the dystrophin gene on the second X chromosome can be the source for the synthesis and production of protein. A small number of female carriers have a degree of muscle weakness known as manifesting carriers. The chance of a carrier's son having a disorder is 50%, and each daughter of a disordered father has a chance of 50%of being a carrier.

The dystrophin gene was identified in 1987. This gene is the longest one discovered in the human genome accounting for 0.1% of the total human genome. Mutations in the dystrophin gene such as deletions, duplications, re-arrangements, and even point mutations will adversely affect the synthesis of dystrophin protein, responsible for the disorder. Most of the mutations are deletions that occur between exons 44 and 55, causing interference in the ORF of the dystrophin gene, resulting from a truncation in the formation of dystrophin protein, thus developing Duchenne muscular dystrophy.

Becker muscular dystrophy is an inherited muscle wasting disease, caused by mutations in the dystrophin gene located on the 121st locus of the X chromosome, resulting in the deficiency or abnormal production of dystrophin protein leading to a condition known as dystrophinopathy. Becker muscular dystrophy is also inherited by X-linked inheritance, predominantly affecting males. Under rare conditions, female carriers are also affected. Limb girdle muscular dystrophy is a rare kind of muscular dystrophy, characterized by atrophy and weakness in the voluntary muscles of hip and shoulder areas, which will slowly spread to other parts of the body. Based on the mutations, 15 different subtypes of Becker muscular dystrophy have been identified so far, inherited as autosomal dominant or autosomal recessive traits. Autosomal dominant is known as LGDM1, comprising of five subtypes (LGDM A, B, C, D, and E), whereas the autosomal recessive type is known as LGDM2 consisting of 10 different subtypes (LGDM A, B, C, D, E, F, G, H, I, J). Spinal muscular dystrophy is a progressive neuromuscular disorder characterized by the degeneration of nerve cell groups in the lower brain stem region and motor neurons in the spinal cord. A deletion in the SMN gene located on 13 loci of the chromosome 5 q arm and the inheritance of a mutated gene in an autosomal recessive pattern causes spinal muscular dystrophy. Congenital muscular dystrophy and congenital myopathies occur due to mutations in the RYR1 and SEPN1 genes. The incidence and prevalence of congenital muscular dystrophy range from 0.68 - 2.5 per 100,000 individuals. The relative frequency of individual types also varies among different population groups, with Fukuyama's congenital muscular dystrophy being the most diagnosed in the Japanese population due to mutations in the fukutin and COL6-RD gene. Fukutin gene mutations are sporadic among other population groups.

5.4.13.3 Diagnosis

The diagnostic process for Duchenne muscular dystrophy begins during early childhood after noticing definite signs and symptoms such as weakness, clumsiness, difficulty in stair climbing, and toe walking. The disease is diagnosed by examining the increased concentrations of alanine aminotransferase, aspartate aminotransferase, lactate dehydrogenase, and creatine kinase in the serum. Values of creatine kinase in the plasma of healthy humans are 1000 IU/L which might even exceed 30,000 IU/L in patients suffering from this disease. In 70% of individuals, Duchenne muscular dystrophy occurs due to a single and multi-exon deletion or duplication in the dystrophin gene (David et al. 2018). The first confirmatory test to diagnose Duchenne muscular dystrophy is a deletion or duplication testing of the dystrophin gene by multiplex-ligation dependent probe amplification or comparative genomic hybridization array, indicating that the mutation is preserving or disrupting the open reading frame of the gene. If the result of this test is negative, then the gene should be isolated and sequenced from the patient to detect other mutations such as missense, nonsense, small deletions, duplications, and insertions involved in the generation of this disorder. If the gene test does not confirm the diagnosis of Duchenne muscular dystrophy, then a muscle biopsy test should be done using immunostaining, Western blot, and immune fluorescence to detect the presence of dystrophin protein.

Female family members of the individual diagnosed with Duchenne muscular dystrophy may undergo carrier testing for establishing their risk of being a carrier. If identified, they can consider pre-implantation genetic diagnosis or prenatal genetic testing through chorionic villus or amniotic fluid sampling. If the relative is a child, then the ethical guidelines prescribed by the American Medical Association should be followed for their genetic testing. In addition, newborns are screened with the measurement of creatine kinase concentrations from their dried blood spots.

5.4.13.4 Treatment

Presently, there is no permanent cure for muscular dystrophy. However, due to promising research on this disease, a few therapies are emerging.

5.4.13.4.1 Steroid Therapy

Steroid therapy is provided by using corticosteroids for various applications such as reducing muscle necrosis, inflammation, modulation of cell response to inflammation, and increased muscle regeneration and growth due to anabolic effects. A reduction in the rate of muscle breakdown and intake of corticosteroids can slow down the decline in function and muscle strength, improve lung function, delay the onset of cardiomyopathy, and prolong independent ambulation. Corticosteroid therapy is also reported to reduce the mortality rate and lower the incidence of cardiomyopathy (Beytia et al.2012). One of the most prominent adverse effects of using corticosteroids is a reduction in the patient's height and an increase in weight. Other adverse effects are cataract formation, vertebral fractures, cushingoid facies, acne, hirsutism, arterial hypertension, behavioral disorders, and delayed puberty. Corticosteroid therapy is not recommended for patients under 2 years, as their motor skills are still developing. Corticosteroids prednisone, prednisolone, and deflazacort are administered to Duchenne muscular dystrophy patients to improve muscle strength and immunosuppressive properties.

5.4.13.4.2 Gene Therapy

Restoration of the expression lost due to the mutations in the gene resulting in the disorder is one of the challenges faced during gene therapy for muscular dystrophy, followed by the size of the gene product and the origin of the gene expression. Myostatin is a gene therapy target for regulating muscle size. Inhibition of myostatin can be a possible route for gene therapy, achieved by administering follistatin, a myostatin receptor blocker and inhibitor of TGF-β during muscle growth (Moh et al. 2017). Utrophin, another potential target for gene therapy is an autosomal homolog of dystrophin, sharing 74% homology at the amino acid level and possessing a very similar structure, binding to the proteins in the protein complex attached to dystrophins. Nabumetone is a potential activator of utrophin with anti-inflammatory properties.

Exon skipping is another approach to gene therapy, used since the discovery of gene expression and manipulation using antisense oligonucleotides. These antisense oligonucleotides are designed to produce mRNA, whose translation gives several folds of truncated and functional dystrophin. Antisense oligonucleotides are short DNA sequences that selectively bind to the mRNA or pre-mRNA, inflicting a small double–helix region on the target mRNA. By binding to the target region and the formation of the double helix, the mutated exon is skipped. This process excises the exons that carry premature stop codons and terminator codons in the exons located within the frameshift mutations (intra-exonic mutations). Drisapersen is an antisense oligonucleotide that is responsible for the skipping of exon 51. A significant improvement in walking distance in the patients has been noticed when the patients receive 6 mg of drisapersen per week. Eteplirsen is involved in the removal of exon 51 during the RNA splicing process. Patients receiving 30 - 50 mg of etiplirsen per week have shown an increase in the dystrophin protein content in their muscles and improvement in their walking abilities. Recombinant Adeno-associated vectors carrying the miniature functional dystrophin gene (also known as minidystrophin) reduce muscle failure and repair it (Malerba et al. 2012). However, the transfer of minidystrophin has low efficiency and triggers an immune response.

5.4.13.4.3 Stem Cell Therapy

Stem cell therapy treats Duchenne muscular dystrophy by in vitro transduction of isolated stem cells by retroviral or lentiviral vectors. These genetically modified cells are developed ex vivo and are systemically injected into the target muscle (Sienkiewicz et al. 2015). Transplantation of autologous genes through stem cell modification can prevent the side effects related to immune response and cell rejection. Several clinical trials have shown an improvement in the patient's muscle strength after receiving stem cell therapy.

5.4.14 Ataxia

Ataxia is an impaired coordination condition due to poor voluntary muscle movement. The term, Ataxia is quite often used to describe a highly non-coordinating, unsteady, and parabolic walk. This disease manifests in various neurological conditions such as stroke, brain tumor, multiple sclerosis, traumatic brain injury, infection, and congenital cerebellar defects, adversely affecting the coordination of fingers, hands, arms, eye movements, and even speech. Two broad categories of Ataxia, genetic, and acquired or degenerative have been identified. Genetic ataxia occurs in an individual with a family history. It is inherited by autosomal recessive, X-linked, maternal inheritance through mutations in mitochondria. Sporadic ataxia occurs in an individual without any family history. Metabolic disorders such as Niemann-Pick type C, and Tay-Sachs are late-onset ataxia without any family history. Paraneoplastic spinocerebellar degeneration, gluten, immune-mediated, and degenerative are acquired forms of ataxia. Acquired ataxia occurs due to Vitamin B12, Vitamin E deficiency, alcohol toxicity (alcohol-related atrophy), phenytoin, infections (HIV), and progressive leukoencephalopathy (Marie et al. 2017). Chronic progressive ataxia occurs due to neurodegenerative neoplasms, and chronic infections to the cerebellum.

5.4.14.1 Symptoms

Ataxia occurs either due to cerebellar dysfunction, or impaired vestibular dysfunction, even due to proprioceptive afferent input to the cerebellum. Its onset can be insidious with acute, sub-acute, episodic, or chronic evolution. Symptoms and signs of the disease are more related to lesions in the cerebellar region of the human brain. Lateralized cerebellar lesions are responsible for ipsilateral symptoms, signs and diffused cerebellar lesions are responsible for general symmetric symptoms. Lesions in the cerebellar hemisphere are responsible for limb ataxia. Lesions in the vermis cause truncal and gait ataxia, causing minimal damage to limb movements. Lesions in the vestibulocerebellar region led to disequilibrium, vertigo, and gait ataxia. Patients with ataxia show clumsiness, unsteadiness, non-coordination, and slurred speech. They also experience oscillopsia (a sensation that the surrounding environment is in constant motion jumping and jiggling while it is stationary) under rare conditions. Patients' reflexes get considerably reduced in Friedreich's, vitamin E deficiency, oculomotor apraxia, and spinocerebellar ataxia. Abnormal visually enhanced vestibule-ocular reflexes are also the characteristic features of cerebellar ataxia.

Ataxia may appear during an individual's infancy due to congenital, developmental, or sometimes genetic causes. Early onset ataxia is a genetic disorder that develops before 20 years of age. In most cases, it is autosomally recessive or maternally inherited through mitochondria. Late onset ataxia occurs during the late 30s to early 40s is autosomal dominant. Friedreich's and Cerebellar ataxia, neuropathy, and vestibular areflexia syndrome (CANVAS syndrome) are inherited during the early 40s in an autosomal recessive fashion (David et al. 2015). Inheritance of spinocerebellar ataxias is autosomal dominant and occurs during the 20s. Rapid progression is the characteristic feature of paraneoplastic spinocerebellar degeneration and sporadic Creutzfeldt-Jakob disease, which spreads in a few weeks. In contrast, multiple systems atrophy type C and congenital neurodegenerative ataxias might progress over many years.

5.4.14.2 Etiology

Ataxia, in most cases, is genetic, inherited from one generation to the other, and is passed on to the families. Genetic ataxia is inherited in autosomal dominant, autosomal recessive, and X-linked patterns. The progress of genetic ataxias is slow and is associated with atrophy of the cerebellum.

5.4.14.2.1 *Autosomal Dominant Ataxia*

The incidence of autosomal dominant disorders is 5 amongst every 100,000 individuals. Spinocerebellar ataxias, including dentatorubral-pallidoluysian atrophy (DRPLA), Gerstmann-Straussler-Scheinker disease (GSS), and episodic ataxias, are autosomal dominant disorders. So far, around 40 types of spinocerebellar ataxias have been identified and partially characterized (Table 5.2).

TABLE 5.2
Genetics of Spinocerebellar Ataxias and Their Epidemiology (Courtesy NCBI, Pub Med)

Type	Locus	Mutation	Epidemiology
SCA1	6q23	CAG repeats exceeding >44	worldwide
SCA2	12q24	CAG repeats exceeding >33	worldwide
SCA3	14q24.3	CAG repeats exceeding >41	worldwide
SCA4	16q22.1	C-T substitution in 5' UTR	Utah, Germany, Japan
SCA5	11p11 - q11	deletions and missense Leu253-Pro	US, Germany, France
SCA6	19p13	CAG repeats exceeding >19	worldwide
SCA7	3p21.1	CAG repeats exceeding >28	worldwide
SCA8	13q21	CTG repeats exceeding >80	worldwide
SCA9	Unknown	Unknown	Unknown
SCA10	22q13	ATTCT repeats 800 - 4500	Mexican
SCA11	15q15.2	Stop, frame shift, insertion, deletion	British
SCA12	5q31-33	CAG expansion in 5'UTR	Germany, USA
SCA13	19q13.3	2 missense mutations R420H and F448L	French, Philippines
SCA14	19q13.4	missense mutations	Japanese, Dutch, British
SCA15	3p26.1	Deletions and missense mutations	Australia, France
SCA16	3p26.1	Deletions and missense mutations	Australia, France, Japan
SCA17	6q27	CAG/CAA repeats exceeding >45	Japan, Germany, Italy
SCA18	7q22	DNA polymorphisms	Ireland, USA
SCA19	1p13.2	Missense	Dutch
SCA20	11q12	260 Kb duplication at the locus	Australia
SCA21	1p36.33	Missense and stop	France
SCA22	1p13.2	Missense	Dutch
SCA23	20p13	Missense	Dutch
SCA24	Unknown	Unknown	Unknown
SCA25	2p21	DNA polymorphisms	France
SCA26	19p13.3	Missense	Norway
SCA27	13q34	Missense and Frame shift	Dutch, Germany
SCA28	18p11.21	Missense and Frame shift, Deletions	Unknown
SCA29	3p26.1	Deletions and missense mutations	Australia, France
SCA30	4q34.3	Unknown	Australia
SCA31	16q21	TGGAA repeat insertion into the locus	Japan
SCA32	7q32	Unknown	China
SCA33	Unknown	Unknown	Unknown
SCA34	6q14.1	L168F, incomplete penetrance	France, Canada
SCA35	20p13	Missense	China
SCA36	20p13	GGCCTG repeat exceeding 2500	Japan, Spain
SCA37	1p32	Unknown	Spain
SCA38	6p12.1	Leu72Val and Gly230Val	Italy and France
SCA39	Unknown	Unknown	Unknown
SCA40	14q32.11	Missense	Hong Kong and China

All these spinocerebellar ataxias are inherited in an autosomal dominant fashion. Each son and daughter of a patient has a 50% chance of inheriting this disorder. Spinocerebellar ataxia affects males and females equally. Each spinocerebellar ataxia occurs due to a mutation or a change in the gene responsible for normal brain function. CAG repeat expansion is the genetic change responsible for 1, 2, 3, 6, 7, 12, and 17 spinocerebellar ataxia types. In these types, CAG, a trinucleotide, is repeated too many times (up to 45), disrupting the gene's normal function. SCA8 occurs due to a CTG trinucleotide repeat expansion of more than 80 times. SCA10 and SCA31 are disrupted by pentanucleotide insertions ATTCT and TGGAA, respectively. Most of the spinocerebellar ataxias are characterized by a phenomenon known as anticipation, which refers to an early stage of the disorder with increasing symptoms from one generation to another. The molecular basis of anticipation is associated with the repeat number and the change in its size while passing from one generation to another. Anticipation cannot be predicted accurately in spinocerebellar ataxia as there is an association between the repeat size, age of the onset, and severity of the symptoms. Under general conditions, the larger the repeat size, the younger a person will develop its symptoms. Repeat numbers cannot predict the exact age of onset or predict the severity of the symptoms. Penetrance is another major factor, which determines the proportion of individuals who develop symptoms. All spinocerebellar ataxias show a very high penetrance rate. Individuals subjected to this disorder develop symptoms at some point during their life span.

5.4.14.2.2 *Autosomal Recessive Ataxia*

Autosomal recessive ataxias are neurological disorders characterized by degeneration or abnormal development of the cerebellum and spinal cord. Autosomal recessive ataxias onset before 20 years. This group encompasses many rare disorders such as Friedreich ataxia, Joubert syndrome, Cayman cerebellar ataxia, and ataxia with isolated vitamin E deficiency. All these disorders occur due to mutations in specific genes. A few of them are even heterozygous due to mutations in more than one locus.

Friedreich ataxia is one of the most common forms of progressive autosomal recessive ataxia, which onsets during childhood and leads to ambulation after 10-15 years. It considerably reduces the life span as it often affects the heart. Friedreich ataxia occurs due to an abnormal GAA trinucleotide repeat expansion in the frataxin gene. This gene is in the mitochondria of highly metabolically active tissues. Point mutations are also responsible for the expansion of the GAA repeats under rare circumstances. The age of disease onset is inversely correlated with the number of GAA repeats on the small allele. Joubert syndrome is a rare disorder formed by the abnormal configuration of superior cerebellar peduncles connecting the midbrain and thalamus. The most common clinical manifestations of Joubert syndrome include early onset of cerebellar ataxia at an infant stage, nystagmus, vertical gaze paresis, mental retardation, retinopathy, and episodic hyperpnea. Five loci of chromosomes 9q34.3, 11p11.2-q12, 6q23, 2q13, and 12q21 have been mapped with 1, 2, 3, 4, and 5 types of Joubert syndrome. Hypotonia, delayed psychomotor development, non-progressive cerebellar dysfunction (truncal and limb ataxia, nystagmus), and intention tremor are the characteristic features of Cayman cerebellar ataxia. This autosomal recessive disorder has been discovered in the isolated population of Grand Cayman Island, and it occurs due to the mutations in the ATCAY gene that encodes a CRAL-TRIO domain located on chromosome 19.

5.4.14.2.3 *X-Linked Ataxia*

X-linked ataxia is an emerging group of ataxia disorders characterized by a defect in the cerebellum due to mutations and imbalances in the genes located on the X chromosome. Classification of this group of disorders is little known as most of the pathogenesis is yet to be understood. X-linked cerebellar ataxia, associated with cerebellar hypoplasia due to mutations in OPHN1 or CASK genes, fragile X, and oral digital type 1 syndromes have been best characterized.

5.4.14.2.4 Mitochondrial Ataxia

Mitochondrial diseases occur at any age stage of human life and result from mutations affecting the mitochondrial respiratory chain function. Genetically characterized mitochondrial disorders occur in a combination of clinical features, as in the case of Kearns-Sayre syndrome, chronic progressive external ophthalmoplegia, mitochondrial encephalopathy, lactic acidosis, and stroke-like episodes, myoclonic epilepsy with ragged red fibers, neurogenic weakness with ataxia and retinitis pigmentosa, and Leigh syndrome. Neurodegeneration is the vital cause of neurological disability in patients suffering from mitochondria disorders, primarily arising either due to point mutations or large deletions in the mitochondrial DNA. Sometimes it may also occur due to mutations in the nuclear genes involved in maintaining mitochondrial DNA. The central nervous system has a vast energy requirement and depends on mitochondria for ATP generation. In patients with mitochondria disorder, neurons are highly vulnerable to degeneration, as reported in cerebellar ataxia (Nichola et al. 2012). A considerable amount of work is done to characterize neurodegeneration using neuroimaging techniques for the diagnosis of mitochondrial disorders, as mitochondrial DNA mutations cannot be identified by using regular genetic tests.

5.4.14.3 Diagnosis

Individuals suffering from ataxia are evaluated with primary care (also known as the first lane of diagnosis) such as thyroid function, serum B12, folate, homocysteine, and secondary care by a neurologist. Along with cranial and spinal imaging, patients may also need genetic testing. Rapid progression of ataxia can be diagnosed by serological testing for paraneoplastic antibodies. Frequently, neurologists conclude that alcohol toxicity causes progressive cerebellar ataxia in patients, which needs further investigations. Blood alcohol levels above 300 mg/dl will affect coordination and balance. Therapeutic drug levels should be examined for the diagnosis of drug-induced ataxia, as the overdose of phenobarbital and butobarbital drugs might also lead to ataxia. For predicting hereditary ataxia, the following laboratory tests are performed:

1. High or low vitamin E levels,
2. High abnormal lipoprotein for the detection of abetalipoproteinemia,
3. High fetoprotein,
4. Low immunoglobulin for the detection of ataxia telangiectasia,
5. High serum cholesterol,
6. Low albumin for detecting ataxia with oculomotor apraxia type 1,
7. High fetoprotein in ataxia with oculomotor apraxia type 2.

Genetic tests are available for the identification of trinucleotide CAG repeats involved in SCA1, SCA2, SCA3, SCA6, SCA7, SCA12, and SCA17 ataxia. CTG repeats responsible for SCA8, GAA expansion in the development of Friedreich ataxia, pentanucleotide repeat ATTCT expansion in SCA36, and conventional mutations in SCA5, SCA13, SCA14 are tested by specific mutation analysis. Detection of mutations in mitochondrial DNA associated with the ataxias such as myoclonic epilepsy with ragged red fibers, mitochondrial encephalomyopathy, lactic acidosis, stroke-like episodes, neuropathy, ataxia, retinitis pigmentosa syndrome are commercially being carried out. However, genetic tests for the diagnosis of inherited ataxia have shown a considerable improvement mainly due to the recent advances in molecular diagnostics. With the advent of next generation sequencing technology, targeted searches for known gene mutations in individuals other than family members may also become obsolete. Next generation sequencing permits the search for a wide array of Ataxia genes. Whole-exome sequencing is also being considered the diagnosis. Next generation parallel sequencing technology shows great promise for diagnosing patients with ataxia by setting up panels for the known ataxia genes. Exome sequencing and whole genome sequencing (adapted by the 100,000 genomes project) are extensively used for the clinical diagnosis of ataxia. Still, next generation sequencing technology has not reached full reliability on large scale deletions, genome rearrangements, and nucleotide expansions for accurately diagnosing ataxia.

Prenatal genetic testing can also be performed to find the chances of spinocerebellar ataxia in the developing fetus. Diagnosis techniques such as amniocentesis and chorionic villi sampling are applied for the prenatal diagnosis of ataxia. Genetic testing can be performed on an embryo to detect the inheritance of spinocerebellar ataxia through pre-implantation genetic diagnosis. This method performs a genetic test on the embryos generated through in vitro fertilization at 8 cell stage for the inherited spinocerebellar ataxia gene. If the embryo is found to consist of the inherited gene, it will not proceed with implantation in the uterus. Only non-inherited embryos will proceed with their implantation into the uterus for pregnancy.

5.4.14.4 Treatment

Except for gluten ataxia and immune mediated ataxias, other ataxias are rarely curable. Treatment is provided for a few types of ataxias only for control. Patients with antigliadin ataxia are recommended to follow a gluten-free diet and the antibody titers should be repeated every 6 months to confirm their elimination until stabilization of symptoms. After confirmation by a genetic test, low vitamin E ataxia is provided with replacement doses of up to 1500 mg per day. CoQ10 ataxia is a recessive disorder, resulting in a low concentration of CoQ10 in skeletal muscle, leading to seizures and mild mental retardation. Supplementation of CoQ10 can potentially improve ataxia. Ataxias that are treatable among children include CoQ10 deficiency, hypobetalipoproteinemia, Hartnup disease, biotinidase deficiency, and pyruvate dehydrogenase deficiency. However, these are being treated by pediatricians. Few strategies based on the symptoms are derived for the treatment of ataxia. Cerebellar dysfunction, including the subcortical frontal impairments, will affect personality, behavior, judgment and generate mental health complications such as anxiety, and depression, exacerbating people's sense of isolation and fear of the future. Sleep disorders and fatigue quite often accompany these symptoms. This symptomatic assessment and management will be entirely handled by a multidisciplinary team involving therapy specialists, significantly enhancing patient care. Speech and language therapy is essential for infants throughout their treatment, from monitoring the swallowing function during the initial stages to planning and implementing percutaneous gastrostomy feeding. It is noteworthy that all these interventions to manage ataxia patients are non-evidence based and will never be included in high quality randomized controlled trials.

5.5 DIAGNOSIS OF RARE GENETIC DISORDERS

The term rare disorders refers to a group of disorders whose prevalence is so low and is considered to be non-viable for the therapeutics market in the absence of full support and incentives, too rare for full investigation, and is appropriately managed by the professionals. The term emphasizes rare disorders which are infrequent, with 6,000 rare disorders affecting an estimated 300 million people globally. A few rare genetic disorders are discussed in this section.

5.5.1 Emanuel Syndrome

Also known as derivative 22 syndrome, derivative 11;22 syndrome, partial trisomy11;22, and supernumerary der (22)t(11;22) syndrome, Emanuel syndrome is a rare genetic disorder that occurs due to the inheritance of additional genetic material on chromosomes 11 and 22. This disorder is named after Dr. Beverly Emanuel, a cytogeneticist in Philadelphia, US in 2004. Dr. Emanuel is a founding member of chromosome 22 disorder. Earlier, this disorder was known as derivative 22 syndrome.

5.5.1.1 Symptoms

Emanuel syndrome occurs due to the inheritance of genetic material from a parent having a balanced translocation between chromosomes 11 and 22. Patients with this rare disorder have a distinctive phenotype consisting of facial dysmorphism, microcephaly (tiny head), abnormal facial features, small jaw and ear anomalies, arched palate, and cleft palate. This syndrome also causes severe mental retardation, delay in development during childhood, kidney malformations, cardiac defects, and genital abnormalities in males (Choudhary et al. 2013). Most of the children with Emanuel syndrome have developmental delay and intellectual disability. Most children cannot ambulate independently and learn to walk with support. Speech and expression are severely impaired, and introductory speech acquisition was also noticed in 20% of children.

5.5.1.2 Etiology

Emanuel syndrome is an inherited chromosomal abnormality whose etiology is associated with supernumerary marker chromosomes (sSMCs). sSMCs are structurally abnormal chromosomes that are present as an addition to the average number of chromosomes as an extra chromosome. sSMCs can originate from any human with 23 pairs of chromosomes and may also replace any one of the 46 chromosomes present in the human cells. The frequency of sSMCs in newborns is up to 0.044%, in prenatal cases is 0.075%, in mentally retarded cases is 0.288% and in subfertile cases is 0.125%. The adverse effect of sSMCs depends on their size, genetic content, and degree of mosaicism. Buckton and his colleagues have identified that 86% of the sSMCs were the derivatives of acrocentric chromosomes. A majority of the sSMCs (up to 65%) originated from chromosome 15, 7% were from chromosomes 13,14, 21, and 9% were from 22, respectively. 70% of the sSMCs are new *de novo* mutations, and 30% get segregated within a family.

Emanuel syndrome results from a 3:1 meiotic segregation of a balanced translocation between chromosomes 11 and 22. It is the most common recurrent reciprocal translocation in humans. This abnormality occurs due to an imbalance occurred by the derivative of chromosome 22known as [der (22)] as an sSMC with a karyotype

47, XY, +der (22)t(11;22) (q23; q11) in males (rare) and
47, XX, +der (22)t(11;22) (q23; q11) in females.

The breakpoints of t(11;22) translocation are in the palindromic AT-rich repeats of chromosomes 11 and 22. These breakpoints make hairpin or cruciform structures for mediating double-strand breaks during meiosis. These breaks permit recombination between 11q23 and 22q11, leading to a recurrent translocation. Emanuel syndrome is the derivative of a partial trisomy 11;22, adding one extra chromosome and making the number 47 (Figure 5.9). The extra chromosome is the combination of chromosomes 11 and 22 with the top and middle regions of chromosome 22 and the bottom of chromosome 11, indicating the presence of many extra copies of the genes present on chromosomes 22 and 11 (Ikbal et al. 2015).

Emanuel syndrome is a rare disorder with only 392 reported cases to date ((http://ssmc-tl.com/chromosome-22.html). Male and female balanced carriers have up to 0.7% and 0.3% the risk of having children with supernumerary der(22). Carriers of Emanuel syndrome are examined for multiple miscarriages, infertility, or after the birth of a child with Emanuel syndrome.

5.5.1.3 Diagnosis

The diagnosis of Emanuel syndrome is made with a blood test to analyze the chromosomes. Babies with birth defects or delays in their development have chromosome analysis performed. This disorder can also be diagnosed through clinical tests, including FISH (Figure 5.10), whole chromosome paint (WCP), array genomic hybridization, or multiplex ligation-dependent probe amplification (MLPA)

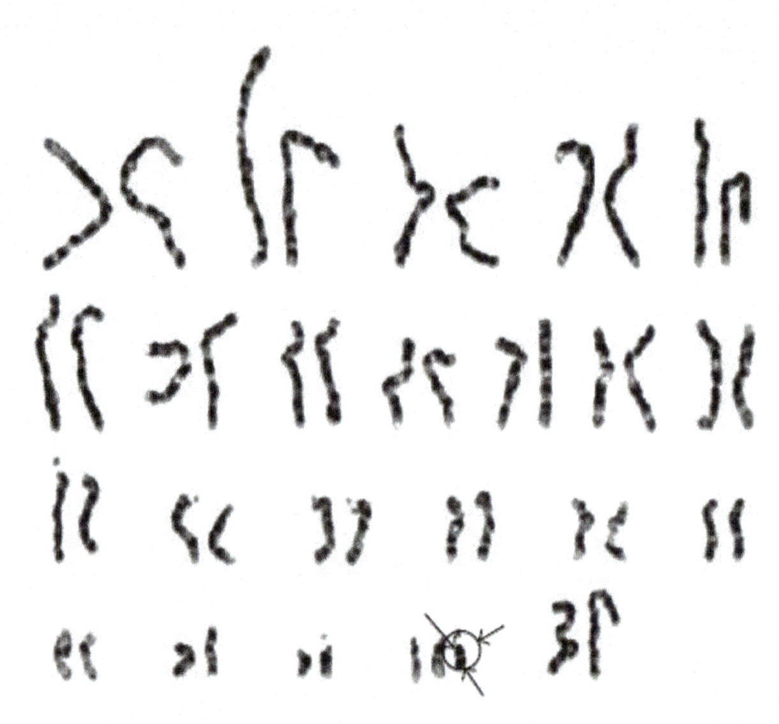

FIGURE 5.9 Karyotype of Emanuel Syndrome patient showing extra supernumerary chromosome.

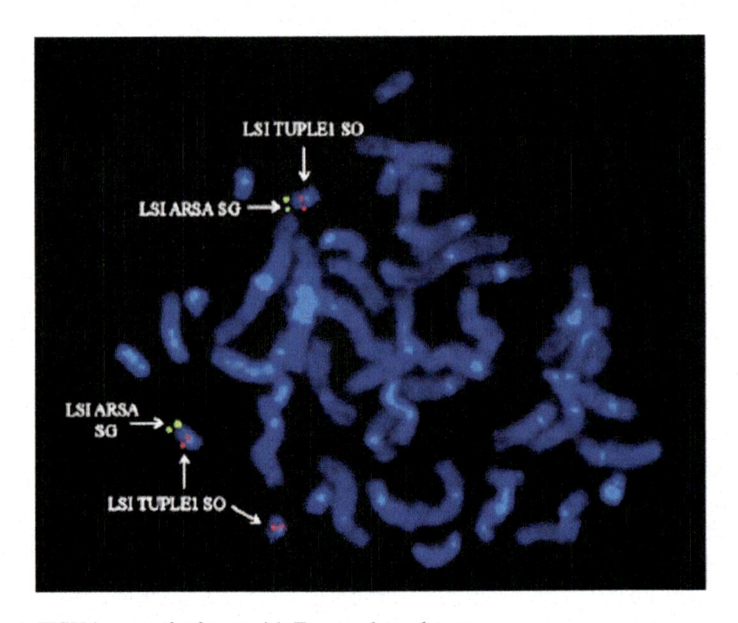

FIGURE 5.10 A FISH image of a fetus with Emanuel syndrome.

assay. The past few years have witnessed the rapid development of next-generation sequencing technology. The discovery of cell-free fetal DNA (CfDNA) and non-invasive prenatal screening (NIPS) has brought profound changes in detecting genetic conditions during early pregnancy. NIPS was found to have a detection rate of 99.7% for trisomy of chromosome 21 with a false positive rate of 0.04%. For trisomy of chromosome 18, the NIPS detection rate was 97.9%, with a false positive rate of 0.04%. For trisomy of chromosome 13, it was 99% with a false positive rate of 0.04%. NIPS coverage aids the detection of deletion syndromes, which was applied for fetuses with Emanuel syndrome (Guven et al. 2006). The results are successfully confirmed by chromosomal karyotyping, SNP array, FISH, and MRI scan. Cat-eye syndrome is one of the most differential diagnoses of Emanuel syndrome. However, iris coloboma, a cardinal feature of CES, is not reported in Emanuel syndrome.

5.5.1.4 Treatment

Treatment for Emanuel syndrome needs to be provided by a multidisciplinary team comprising a pedodontist, a pediatrician, geneticist, gastrologist, speech therapist, surgeon, and ophthalmologist. The mortality rate is high in the first few months of birth. Patients with cleft palate might face difficulties while feeding, requiring a surgical closure (Luo et al. 2017). Genetic counseling is conducted for the parents and family members of patients suffering from Emanuel syndrome with two issues. Firstly, if either of the parents is a carrier, future pregnancies are at increased risk. Prenatal cytogenetic testing should be offered for all future pregnancies. Secondly, carrier testing of unaffected siblings will be offered if they cannot understand the complications of being a carrier after attaining adulthood.

5.5.2 Cat Eye Syndrome

Cat Eye Syndrome, also known as Schmid-Fraccaro syndrome, chromosome 22 partial tetrasomy (22pter-22q11), chromosome 22 Inv dup (22pter-22q11), chromosome 22 partial trisomy (22pter-22q11) is a rare genetic disorder in humans with an incidence of 1 per 150,000 live births (Mears et al. 1994). This genetic disorder is named cat eye syndrome due to vertical colobomas (fissures in the lower part of the eye) in the eyes of the patients suffering from this disorder. So far, this disorder has been reported in 100 patients. This syndrome occurs due to the tetrasomy in chromosome 22 producing a supernumerary dicentric marker chromosome.

5.5.2.1 Symptoms

Cat eye syndrome is caused by a partial tetrasomy of chromosome 22. Persons suffering from cat eye syndrome have a specific defect known as coloboma (fissure in the lower part of the eye, which fails to close during the early stages of development), leading to an elongated pupil and a hole in the iris (colored part of the eye) that resembles a cat eye. The symptoms of cat eye syndrome vary in severity, including cardiac defects, anal atresia (absence of anal canal), pre-auricular skin tags (small skin growth), or pits (depressions) in front of the ears. Reproductive tract defects and skeletal defects are the other symptoms.

5.5.2.2 Etiology

Cat eye syndrome is a rare developmental disorder that occurs due to a supernumerary dicentric marker chromosome in the cells resulting from a partial tetrasomy of chromosome 22. Patients will have a partial tetrasomy of a region involving the entire short (p) arm and a part of the long (q) arm of chromosome 22 and far band 11 including low copy repeat regions. These low copy repeats mediate meiotic unequal nonallelic homologous recombination events, resulting in chromosomal rearrangements (Belien et al. 2008) Chromosomal rearrangements lead to the formation of supernumerary dicentric marker chromosome, responsible for cat eye syndrome. A large subset of supernumerary dicentric marker chromosomes is derived from chromosome 22, conferring tri or tetrasomy to the proximal part of the q22 known as cat eye syndrome critical region (CESCR). This critical region is of 2 Mb in size. Different supernumerary marker chromosomes, including bisatellite small, large cat eye syndrome chromosomes, and a small ring like supernumerary marker chromosome containing the CECSR have been identified.

Based on the location of the breakpoints, cat eye syndrome chromosomes are classified into type 1 and type II. Type I are the small cat eye syndrome chromosomes symmetrical with both the breakpoints located within the proximal interval. Type 2 are the oversized cat eye syndrome chromosomes, further divided into asymmetrical type IIa with one breakpoint situated on each of the two intervals or symmetrical and type IIb where both breakpoints are in the distal interval. Based on the low copy repeats in the q11 region where the re-arrangement occurs, supernumerary marker chromosomes derived from chromosome 22 might have a different size (Edelmann et al. 1999). It has been observed that the cat eye syndrome phenotype might have a variable phenotype. No correlation has been found between cat eye phenotypes and supernumerary regions.

5.5.2.3 Diagnosis

Prenatal diagnosis of cat eye syndrome is achieved with an ultrasound test, which reveals certain defects in the fetus that are characteristic of cat eye syndrome. Following an ultrasound test, amniocentesis can diagnose the presence of extrachromosomal material from chromosome 22q11. A genetic test involving the karyotyping, FISH, can confirm the diagnosis (Leihr et al. 1992; Bartsch et al. 2005). After the detection of cat eye syndrome, other abnormalities regarding the heart and kidney defects can be detected using imaging tests such as X-rays, electrocardiography, echocardiography, hearing tests, and cognitive function tests.

5.5.2.4 Treatment

Treatment for cat eye syndrome essentially depends on the symptoms, involving a team of pediatricians, surgeons, heart specialists, gastrointestinal specialists, eye specialists, and orthopedists. As of now, there is no permanent cure for cat eye syndrome. Treatment is done according to specific symptoms, including medication, surgery for anal atresia, skeletal abnormalities, physical and occupational therapy, hormone therapy, and focused training for individuals with intellectual disabilities (Berends et al. 2000). The life expectancy of people with cat eye syndrome varies widely from individual to individual. Treatment based on the symptoms may prolong life. In most cases, cat eye syndrome will not reduce life expectancy.

5.5.3 Uniparental Disomy

Uniparental disomy is when both the chromosomes or homologous regions or chromosomal segments are inherited from a single parent, either, mother or father. Inheriting two copies of a single parental homolog associated with a reduction to homozygosity in the offspring is known as isodisomy, and inheritance of both homologs from a single parent is known as heterodisomy. Homozygosity of autosomal recessive mutations might be a possible feature of isodisomy. This was evidenced in 50 patients with recessive disorders due to isodisomy (Amazawa et al. 2010).

Humans inherit two gene copies, one from the mother and one from the father. Both these copies are active and considered to be turned on. In some individuals, only one gene copy is turned on. Which among the two depends on the origin of the parent. Some genes are active only if paternally inherited, and some are active if inherited from the mother. This phenomenon is known as genomic imprinting. Among the genes that undergo genomic imprinting, their parent of origin is methylated during the formation of egg and sperm cells. These methyl groups can identify the inheritance of the gene copies from mother and father. The addition and deletion of these methyl groups regulate their activity. Only a small number of human genes are reported to undergo genomic imprinting. These imprinted genes tend to cluster in the same regions of chromosomes. Two significant clusters of imprinted genes have been identified in the human genome. One is located on the short p arm of chromosome 11 at position 11p15, and the second cluster is on the long q arm of chromosome 15 in the region 15q11 to 15q13. Still, it is not clear why only specific genes will undergo imprinting.

5.5.3.1 Symptoms

Uniparental disomy doesn't always turn out into a disease. Its clinical outcome is directly related to the genetics of the disease, the amount of the affected chromosome region, and mosaicism. As most genes are not imprinted, uniparental disomy doesn't impact human health. It makes no difference whether both the copies of a gene are inherited from one parent instead of inheritance from each parent. It makes a difference whether the gene is inherited from the mother or father. A person with uniparental disomy may lack active copies of essential genes for undergoing genomic imprinting, resulting in loss of function. This might lead to delayed growth and development, intellectual disability, and other health problems. Uniparental disomy can lead to genetic disorders such as Prader-Willi syndrome, characterized by uncontrolled eating and obesity, Angelman syndrome, responsible for intellectual disability, and delayed speech. Both the syndromes occur due to the errors in genomic imprinting of the genes located on the long q arm of chromosome 15. Beckwith-Wiedemann syndrome, characterized by accelerated growth, is associated with the abnormalities of impaired genes located on the short p arm of chromosome 11.

5.5.3.2 Etiology

The origin of uniparental disomy can be studied by haplotype studies involving the whole chromosome. Complete isodisomy implies three conditions, the total homozygosity of all markers arising from postzygotic chromosome duplication, monosomy rescue when a monosomic gamete is fertilized by a nullisomic gamete, and a postfertilization error in the normal zygote such as mitotic disjunction. Mosaicism is unlikely to exist during monosomy rescue due to the lethality of the cell line. In postfertilization, mosaicism of the trisomic cell line is caused by the mitotic disjunction. The resulting isodisomic cell line generated by the subsequent trisomy rescue can be detected. Somatic recombination is another process where a normal zygote leads to isodisomy. Reciprocal exchange of chromosomal segments results in segmental isodisomy between the recombination point and telomere, contributing to the loss of heterozygosity. This is associated with some cancers like imprinted disorders. Paternal segmental uniparental disomy for 11p15.5 due to mitotic recombination has been found in the mosaic forms of Beckwith-Wiedmann syndrome, a congenital disorder characterized by overgrowth, abdominal wall defects, and tumor predisposition.

Abnormalities in the germline meiotic segregation are also responsible for uniparental disomy, which occurs through fertilization between a disomic gamete and a nullisomic gamete through a process known as gamete complementation. In these cases, uniparental disomy occurs from the intercrosses between the parents, harboring balanced chromosomal translocations. Heterodisomy exists along with isodisomy and occurs through mitotic rescue after the formation of a trisomic zygote or gamete complementation. These mechanisms occur during crossing over and recombination at meiosis I. The pericentromeric regions of the chromosomes display heterozygous haplotypes if the trisomic or a disomic gamete arises from an error in meiosis I. In contrast, they display homozygous haplotypes if the trisomic or a disomic gamete arises from an error in meiosis II. Investigations of the markers in the pericentromeric regions will help in elucidating the onset of uniparental disomy.

5.5.3.3 Diagnosis

The diagnosis of uniparental disomy in humans remains challenging as it is not possible by conventional cytogenetic or karyotype analysis. Diagnosis of uniparental disomy is made by using polymorphic microsatellite markers. In general, microsatellite markers are used for specific clinical features for a particular chromosome. Many microsatellite markers are needed to screen for each chromosome segment to diagnose uniparental disomy. Genome-wide screening has proved to be a better option for uniparental disomy diagnosis. Further, SNP arrays have facilitated an increased scope for detecting uniparental disomy due to the convenience of mapping homozygosity and heterozygosity along the length of the chromosomes. SNP array is a subtle method for the detection of isodisomy. SNP array for the detection of heterodisomy is accomplished only after the analysis of parental DNA, which is indispensable for diagnosing uniparental disomy. SNP array also provides a high-resolution method for the diagnosis of genome abnormalities. It simultaneously evaluates copy numbers to detect mosaic gains and losses and copy number changes. Beckwith-Wiedemann syndrome patients with mosaic genome wide paternal uniparental disomy and Silver-Russel syndrome patients with mosaic genome wide maternal uniparental disomy have been reported through a genome wide approach using SNP array. Techniques such as methylation-specific PCR assay and methylation-specific multiplex ligation-dependent probe amplification assay (MS-MLPA) have also enabled the diagnosis of uniparental disomy.

5.5.3.4 Treatment

Molecular analysis of uniparental disomy patients is essential for exact diagnosis, leading to appropriate management of its medical, social, and psychological aspects. It also helps in the accurate assessment of recurrence risk in family members. Among individuals with a normal karyotype, the uniparental disomy occurs due to a de novo mutation, and its risk of recurrence is negligible. If the homozygosity of recessive mutations is responsible for the disorder, the recurrence risk in the patient's sibling is small. The child of a patient is a heterozygous carrier of this mutation. If the imprinted gene is subjected to alteration, leading to UPD, its recurrence risk in the patient's sibling and its child is negligible. In cases with chromosomal translocations, the assessment of recurrence will be more complicated due to the low frequency gametic complementation requirement for generating uniparental disomy. However, its phenotypic outcome can be monitored through deletions. Genetic and epigenetic factors responsible for mutations on the imprinted gene consist of an imprinting control region known as the imprinting center. This center has the property of regulating gene imprinting in the cis region of one parental chromosome. A deletion or a mutation in the imprinting center generates an epigenotype switch. As a result, the maternally inherited chromosome appears like a paternally inherited one in terms of its epigenetic status. defects will be inherited in a Mendelian manner, increasing the risk of recurrence compared with uniparental disomy. Here, the phenotypes of uniparental disomy and deletion appear similar. Investigation of their copy number plays a crucial role in identifying the difference between these two phenotypes, especially in the imprinting center.

5.5.4 Jacobsen Syndrome

Jacobsen syndrome (OMIM147791), also known as 11q terminal deletion disorder, is one of the rare genetic disorders due to the loss of genes located on the long q arm of chromosome 11. One child in 100,000 is born with Jacobsen syndrome. This disorder is more common in girls in comparison to boys. It is also known as terminal deletion disorder as it occurs due to the loss of genes from the terminal region of chromosome 11. This genetic condition was first reported by a Danish researcher Jacobsen in 1973 in multiple members of the same family with 11;21 translocations derived from a carrier patient. Jacobsen syndrome occurs due to the deletion of a DNA fragment from the middle or terminal region of the chromosome 11 long q arm consisting of BSX, NRGN, ETS-1, FL-1, and RICS genes (Teresa et al. 2009). If a small fragment is deleted from the long arm of the 11q consisting of a few genes, the resulting condition is known as partial Jacobsen syndrome. Since the first report of Jacobsen syndrome, more than 200 cases of this disorder have been reported. The clinical features of Jacobsen syndrome have been well known, and symptoms of this disorder are well diagnosed.

5.5.4.1 Symptoms

Pre and postnatal physical growth retardation, psychomotor retardation, facial dysmorphism, and thrombocytopenia are the most common symptoms of Jacobsen syndrome. Malformations of the heart, kidney, gastrointestinal tract, genitalia, central nervous system, and skeleton have been noticed in a sub-set of patients. Ocular, hearing, immunological, and hormonal problems were also present.

More than 60% of patients with Jacobsen syndrome are born at term, premature birth occurs in 30% of the cases, and the delivery is post-term in 10% of the cases. 46% of patients might be subjected to complications during delivery, such as abnormal fetal presentation, premature membrane rupture, and cephalopelvic disproportion. In 65% of the cases, delivery is spontaneous, whereas, in 35%, it is performed through cesarian. 60% of babies born with Jacobsen syndrome have an average birth weight, 40% are underweight and 3% are overweight. The Jacobsen syndrome patients' mean maternal and paternal ages are 27 and 30 years, respectively. 20% of the children die during the first two years of life due to congenital heart disease complications. Life expectancy is unknown for patients who survive the neonatal period and infancy.

5.5.4.2 Etiology

Jacob syndrome occurs due to a deletion in the long q arm of chromosome 11. The size of deletion may vary from 7 - 20 Mb. Breakpoints for the deletion occur within, or distal to sub-band 11q23.3, and the deletions extend to the telomere. Large terminal deletions result in embryonic lethality. Even though the terminal deletion of q11 is responsible for Jacob syndrome in most of the cases (85%), unbalanced translocations resulting from the segregation of a familial balanced translocation are also responsible for the occurrence of the syndrome. A partial deletion of chromosome 11q may also result from an unbalanced translocation occurring de novo or through other chromosomal rearrangements such as ring chromosomes or recombinations of parental pericentric inversions. Here deletion at 11q is further complicated by additional imbalances. In a small subset of patients,11q23.3 breakpoints cluster to rare folate sensitive, fragile site FRA11B due to the extensive expansion of (CGG)n repeats and hypermethylation of adjacent CpG islands. Site A of FRA11B is a (CGG)n repeat in the 5' UTR region of CBL2 protooncogene, mapping between 118,575,702 and 118,775,922 bp. Among 70% of the individuals, 11 copies of this repeat exist, whereas, in the individuals with cytogenetic expression of FRA11B, this repeat is extended to several hundreds of copies. The CCG repeats are expanded to 80 - 100 copies without any fragile site expression. The hypermethylation of the (CCG)n triplets on chromosome 11 could delay the DNA replication of this fragile site resulting in a break or an impaired replication. 10% of cases with Jacobsen syndrome are related to the amplification of the CCG triplet. Deletions 11q might also occur due to recombinations mediated by low copy repeats, palindromic AT-rich repeats, and olfactory receptor gene clusters (Seppanen et al. 2014). Multiple mechanisms are involved in the generation of 11q deletions. In Jacobsen syndrome patients, the origin of the 11q deletion breakpoints occurring proximally to the corresponding band 11q23.3 is maternally inherited, whereas the distal breakpoints are paternally inherited. Genome imprinting is also involved in (CCG)n expansion and methylation.

11q23 region is gene rich, comprising of 342 genes. Among them, 174 have been localized to 11q24.1. The minimal region of expressing Jacobsen syndrome is about 14 Mb. The FL1 gene maps to the deleted 11q24 sub band in Jacobsen syndrome patients and is a protooncogene that interacts with genes involved in the vasculogenesis, hematopoiesis, and intercellular adhesion. The BARX2 gene, located in the minimal deleted 11q region for Jacobsen syndrome is expressed in the development of neural and craniofacial structures. BRX2 is a possible candidate gene for the development of facial dysmorphism and craniosynostosis in Jacobsen syndrome patients. The JAM3 gene, a member of the junction adhesion molecules maps for a 9 Mb region in the 11q distal to D11S1351 and is a candidate gene for hypoplastic left heart trigonocephaly. The B3GAT1 gene located at 133.77 Mb of 11q, expressed in the brain during its early development, might be involved in bipolar disorder observed in some Jacobsen syndrome patients. BSX, a highly conserved homeobox gene expressed during early brain development, is a candidate gene for global cognitive development. It has been suggested that the neurogranin (NRGN) gene, involved in the synapse plasticity and long-term potentiation contributes to auditory attention deficit. FEZ1 and RICS are the hypothetical candidate genes for axonal outgrowth brain development. Genes KCNJ1 and ADAMTS15 are expressed in the liver and kidney of the fetus and might be involved in kidney malformations. TECTA gene codes for the non-collagenous component of the inner ear tectorial membrane might be engaged in neurosensorial deafness. Several genes located in the distal region of 11q, such as EST1, CHK1, BARX2, OPCML, whose role in the progression of Jacobsen syndrome is still unknown.

5.5.4.3 Diagnosis

Patients with the classical Jacobsen syndrome phenotype are diagnosed with mental retardation, facial dysmorphic features, and thrombocytopenia. However, the diagnosis is confirmed by karyotype analysis. Diagnosis of Jacobsen syndrome is difficult in patients with low clinical aspects and borderline mental development. In these cases, the presence of Jacobsen syndrome is confirmed by thrombocytopenia. Children with Jacobsen syndrome are diagnosed through certain features, including short stature, webbed neck, down slanting palpebral fissures, and ptosis. The fetus's prenatal diagnosis of 11q deletion is made through amniocentesis or chorionic villus sampling, cytogenic G banding analysis, and telomeric FISH.

5.5.4.4 Treatment

Children suffering from Jacobsen syndrome require treatment and special attention during the neonatal period. Heart surgery might be inevitable as cardiac malformations can be severe. Feeding children may be difficult, requiring tube feeding. Plate count might reach low average values due to thrombocytopenia and abnormal platelet function, necessitating prophylactic platelets or whole blood transfusion. Patients with Jacobsen syndrome have short stature due to growth hormone IGF-1 deficiency. Growth hormone replacement therapy cannot be performed on these patients as they might have a genetic predisposition to certain malignancies that would be exacerbated by a tumor promoter. Strabismus (a condition where eyes do not line up with one another) can be treated with an eye surgery before the baby attains one year of age. Early intervention by speech, occupational and behavioral therapists will address cognitive and behavioral problems. Recent work has shown that music therapy has been beneficial to some patients with Jacobsen syndrome.

5.5.5 Kleefstra Syndrome

Kleefstra syndrome (OMIM610253), also known as 9q sub-telomeric deletion syndrome, is a rare genetic disorder characterized by intellectual disability. Kleefstra syndrome is inherited in an autosomal dominant manner, either due to deletion of a submicroscopic region in chromosome 9 or a mutation in the gene regulating the euchromatin. This syndrome has been identified in 85 patients so far. Among them 75 with a deletion in chromosome 9 and 10 with a mutation in the euchromatin gene.

5.5.5.1 Symptoms

Kleefstra syndrome is characterized by moderate to severe developmental delay, intellectual disability, childhood hypotonia, distinct facial features such as brachymicrocephaly, sinophrys, unusual eyebrow shape, full everted lower lip, protruding tongue, and prognathism. Most of the individuals with Kleefstra syndrome will have an intellectual disability whose range varies from moderate to severe. They will also have severe expressive speech delay with little or no speech development. General language development will usually be high, making non-verbal communication. Patients with Kleefstra syndrome also display sleep disturbances, stereotypes, and self-injurious behavior. A substantial proportion of the individuals have hypermetropia and hearing impairment from a young age. Congenital heart defects, urogenital defects, epilepsy, and renal disorders have been observed in individuals with Kleefstra syndrome.

5.5.5.2 Etiology

Kleefstra syndrome occurs either due to a microscopic deletion in the long q arm of chromosome 9 (9q34.3) or by the intragenic mutation of the euchromatin histone methyltransferase 1 (EHMT1) gene responsible for haploinsufficiency. Most of the patients with Kleefstra syndrome have 9q34.3 deletion. This deletion is smaller than 3Mb. EHMT1 encodes a histone H3 Lys9 methyltransferase and is involved in chromatin remodeling. Inheritance of Kleefstra syndrome is autosomal dominant as the deletion in one copy of chromosome 9 or mutation in one copy of EHMT1 is sufficient to generate the condition (Siano et al. 2022). Almost all Kleefstra syndrome cases are sporadic. Recently, three familial cases have been reported occurring through a subtelomeric 9q deletion present in a mosaic pattern in their mothers. Under rare extreme conditions, affected patients will inherit segment deleted chromosome 9 from an unaffected parent. Here, the parent carries a chromosomal rearrangement known as balanced translocation, where genetic material is neither gained nor lost. Balanced translocations do not cause any complications. However, they become unbalanced when passed on to the next generation. Children who have inherited unbalanced translocation have a chromosomal rearrangement with a deleted segment of chromosome 9. A few individuals inherit 9q34.3 deletion from the unaffected parent, who is a mosaic for the deletion (deletion only in the sperm and egg cells but not in the somatic cells).

5.5.5.3 Diagnosis

Kleefstra syndrome was diagnosed on a proband with a heterozygous deletion on chromosome 9q34.3. After establishing the phenotypic diagnosis, molecular testing approaches such as chromosomal microarray analysis, single gene testing, and multigene panel testing were made along with karyotype analysis. Chromosomal microarray analysis uses SNP and Oligonucleotide microarrays to detect large deletions that cannot be detected by sequencing analysis. Single gene testing of EHMT1 can identify small intragenic deletions, insertions, missense, nonsense, and splice-site variants, but cannot detect whole gene deletions duplications. 5% of patients with Kleefstra syndrome have an intragenic deletion. This deletion is detected by an MLPA, CMA, qPCR assay that can detect single exon deletions and duplications. Deletions that are not intragenic and too small can be detected by gene-targeted methods such as FISH. Multigene panels for sequence analysis, deletion or duplication, and other non-sequence-based tests are performed to understand the role of EHMTI and other genes in generating intellectual disability. Disruption of EHMTI expression due to balanced chromosome rearrangement can be detected by karyotype and FISH analysis.

5.5.5.4 Treatment

Treatment for patients with Kleefstra syndrome involves a multidisciplinary team with members specializing in handling intellectual deficiencies. Vocational training and speech therapy, physical, sensory integration, and occupational therapy will be required for the patients from an early age. Continuous monitoring and standard treatment are necessary for individuals with renal, cardiac, and urologic issues. This follow-up is a lifelong process. Psychiatric care and behavioral intervention are needed with the possibility of therapy.

5.6 CONCLUSIONS

As evidenced in this chapter, molecular diagnostics will play a pivotal role in improving public health. Molecular diagnostic technologies will be the critical control for medicine, pharmaceuticals, forensic sciences, and devastating biological warfare for the next 100 to 200 years. Genetic testing will provide the basis for the disease diagnosis, characterization, and identification of its modifiers along with its susceptibility. Genetic tests for most common and rare disorders are available and practiced for quick and accurate diagnosis. At present, PCR-based tests are dominating the market, and alternate technologies such as FISH, peptide nucleic acids, electrochemical detection of DNA, biochips, nanotechnology, and proteomics technology are expected to gain momentum in the coming years. However, these tests face a few issues, such as cost-effectiveness, reproducibility, training of laboratory personnel, which can be overcome by developing integrated chip devices that facilitate the identification of changes in genetic makeup, even from single cells. The application of nanotechnology to molecular diagnostics will refine and extend the detection limits further. The utilization of SNPs and other molecular markers will further enhance the diagnostic and prognostic value. These devices will further improve molecular diagnostics, and their implementation will impose additional challenges to public, private research laboratories and the pharmaceutical industry. Soon, molecular diagnostics is expected to provide specific information on each tumor or cDNA sample and the information regarding the aberrant pathways and mutations governing them.

REFERENCES

Amazawa, K., Ogata, T., Ferguson-Smith, A. C. (2010). Uniparental disomy and human disease: An overview. *Am. J. Med. Genet. Part C. Semin. Med. Genet.* 154C:329–334.

Barr, M. L., Plunkett, E. R. (1956). Testicular dysgenesis affecting the seminiferous tubules principally, with chromatin-positive nuclei. *Lancet* 2:853–856.

Bartsch, O., Rasi, S., Hoffmann, K., Blin, N. (2005). FISH of supernumerary marker chromosomes (SMCs) identifies six diagnostically relevant intervals on chromosome 22q and a novel type of bisatellited SMC (22). *Eur. J. Hum. Genet.*13:592–598.

Bélien, V., Gérard-Blanluet, M., Serero, S., Le Dû, N., Baumann, C., Jacquemont, M. L., et al. (2008). Partial trisomy of chromosome 22 resulting from a supernumerary marker chromosome 22 in a child with features of cat eye syndrome. *Am. J. Med. Genet.* 146A: 1871–1874.

Bembi, B., Marchetti, F., Guerci, V. I., Ciana, G., Addobbati, R., Grasso, D., et al. (2006). Substrate reduction therapy in the infantile form of Tay Sachs disease. *Neurology* 66: 278–280.

Berends, M. J., Tan-Sindhunata, G., Leegte, B., Van Essen, A. J. (2000). Phenotypic variability of Cat-Eye syndrome. *Genet. Couns.*12:23–34.

Bertram, L., McQueen, M. B., Mullin, K., Blacker, D., Tanzi, R. E. (2007). Systematic meta-analyses of Alzheimer disease genetic association studies: the Alz Gene database. *Nat Genet.* 39(1):17–23.

Beytía, M. L., Vry, J., Kirschner, J. (2012). Drug treatment of Duchenne muscular dystrophy: available evidence and perspectives. *Acta Myol.* 31: 4–8.

Bezemer, I. D., Doggen, C. J., Vos, H. L., Rosendaal, F. R. (2007). No association between the common MTHFR 677C-T polymorphism and venous thrombosis: results from the MEGA study. *Arch. Intern. Med.* 167: 497–501.

Cardno, A. G., Gottesman, I. I. (2000). Twin studies of schizophrenia: from bow-and-arrow concordances to Star Wars Mx and functional genomics. *Am. J. Med. Genet.* 97: 12–17.

Carsten, G. B., Ching H. W., Susana, Q-R., Nicolas, D., Enrico, B., Ana F., et al. (2014). Diagnostic approach to the congenital muscular dystrophies. *Neuromuscular Disorders* 24: 289–311.

Ceelie, H., Spaargaren-Van, R. C. C., Bertina, R. M., Vos, H. L. (2004). G20210A is a functional mutation in the prothrombin gene; effect on protein levels and 3'-end formation. *J Thromb Haemost.* 2:119–127.

Chin, C. D., Linder, V., Sia, S. K. (2012). Commercialization of microfluidic point-of care diagnostic devices. *Lab Chip* 12(12): 2118–2134.

Choudhary, M. G., Babaji, P., Sharma, N., Dhamankar, D., Naregal, G. R. (2013). Derivative 11;22(Emanuel) syndrome: a case report and a review. *Case. Rep. Pediatr.* 2013:4.

Claudia, C., Laura, F., Donatella, M., Silvia, T., Monica, M., Susanna, E. (2017). Fragile X syndrome: a review of clinical and molecular diagnoses. *Italian Journal of Pediatrics* 43:39.

Clinton, S. M., Meador-Woodruff, J. H. (2004). Abnormalities of the NMDA receptor and associated intracellular molecules in the thalamus in schizophrenia and bipolar disorder. *Neuropsychopharmacology.* 29(7): 1353–1362.

Coon, K. D., Myers, A. J., Craig, D. W., et al. (2007). A high-density whole-genome association study reveals that APOE is the major susceptibility gene for sporadic late-onset Alzheimer's disease. *J. Clin. Psychiatry.* 68(4): 613–618.

Daniel, E. S. (2017). molecular diagnosis of thalassemias and hemoglobinopathies an aclps critical review. *Am. J. Clin. Pathol.* 148:6–15.

David, B., Priya, S., Christine, Lo., Emma, L. B., Robert, W. T., Rita, H., Stephen, W., Patrick, F. C., Marios, H. (2015). Mitochondrial pathology in progressive cerebellar ataxia. *Cerebellum & Ataxias* 2:16–20.

David, J. B., Katharine, B., Carla, M. B., Susan, D. A., Angela, B., et al. (2018). Diagnosis and management of Duchenne muscular dystrophy, part 1: diagnosis, and neuromuscular, rehabilitation, endocrine, and gastrointestinal and nutritional management. *Lancet Neurol.* 17(3):251–267.

Edelmann, L., Pandita, R. K., Spiteri, E., Funke, B., Goldberg, R., Palanisamy, N., et al. (1999). A common molecular basis for rearrangement disorders on chromosome 22q11. *Hum. Mol. Genet.* 8:1157–1167.

Edward, J. W., Sarah, T. (2017). Therapies targeting DNA and RNA in Huntington's disease. *Lancet Neurol.* 16(10): 837–847.

Elizabeth, S. S., Ali, S. C., Karen J. L., Lydia, L .S. (2017). Short stature homeobox-containing haploinsufficiency in seven siblings with short stature. *Case Reports in Endocrinology* Article ID 7287351.

Ellison, J. W., Wardak., Young, M. F., Gehron, R. P., Laig-Webster, M., Chiong, W. (1997). PHOG, a candidate gene for involvement in the short stature of Turner syndrome. *Hum. Mol. Genet.* 6:1341–1347.

Fatemi, S. H., Folsom, T. D. (2009). The neurodevelopmental hypothesis of schizophrenia revisited. *Schizophr. Bull.* 35(3): 528–548.

Francisco, A-Z., Roberto, L (2018). Epigenetics in Turner syndrome. *Clinical Epigenetics* 10: Article No. 45.

Fromer, M. et al. (2014). De novo mutations in schizophrenia implicate synaptic networks. *Nature* 506: 179–184.

Fu, Y. H., Kuhl, D. P., Pizzuti, A., et al. (1991). Variation of the CGG repeat at the fragile X site results in genetic instability: resolution of the Sherman paradox, *Cell* 67:1047–1058.

Gerhard, B. (2011). Short stature due to SHOX deficiency: genotype, phenotype, and therapy. *Horm. Res. Paediatr.* 75:81–89.

Gründer, G., Cumming, P. (2016). The dopamine hypothesis of schizophrenia. *The Neurobiology of Schizophrenia* 2016: 109–124.

Gubala, V. L. F., Ricco, A. J., Tan, M. X., Williams, D. E. (2012). Point of care diagnostics: status and future. *Anal Chem.* 84(2): 487–515.

Guven, M. A., Ceylaner, G., Ceylaner, S., Coskun, A. (2006). Prenatal diagnosis of a case with Emanuel syndrome (supernumemary der (22) syndrome). *Ultrasound Obstet. Gynecol.* 28:512–561.

Hall, S. S., Dougherty, R. F., Reiss, A. L. (2016). Profiles of aberrant white matter microstructure in fragile X syndrome. *Neuroimage Clin.* 11:133–138.

Hamidreza, S. (2018). Critical Review on Thalassemia: Types, Symptoms, and Treatment. *Advancements Bioequiv. Availab.* 1(2):15–18.

Hu, W., MacDonald, M. L., Elswick, D. E., Sweet, R. A. (2015). The glutamate hypothesis of schizophrenia: evidence from human brain tissue studies. *Ann. N. Y. Acad. Sci.* 1338(1): 38–57.

İkbal, A., Gürkan, E., Vatansever, H., Ulusal, U., Tozkir, H. (2015). A case with Emanuel syndrome: extra derivative 22 chromosome inherited from the mother. *Balkan J. Med. Genet.* 18:77–82.

Jacobs, P. A., Strong, J. A. (1959). A case of human intersexuality having a possible XXY sex-determining mechanism. *Nature* 183:302–303.

Jacobsen, P., Hauge, M., Henningsen, K., Hobolth, N., Mikkelsen, M., Philip, J. (1973). An (11;21) translocation in four generations with chromosome 11 abnormalities in the offspring. A clinical, cytogenetical, and gene marker study. *Hum. Hered.* 23:5.

Jentsch, J. D., Roth, R. H. (1999). The neuropsychopharmacology of phencyclidine: from NMDA receptor hypofunction to the dopamine hypothesis of schizophrenia. *Neuropsychopharmacology* 20(3): 201–225.

Jody, L. K. (2011). Factor V Leiden thrombophilia. *Genetics In Medicine* 13(1): 1–16.

Kenny, E. M. et al. (2014). Excess of rare novel loss-of-function variants in synaptic genes in schizophrenia and autism spectrum disorders. *Mol. Psychiatry* 19: 872–879.

Kevin, P., Cohoon, D. O., John, A. H (2014). Inherited and Secondary Thrombophilia. *Inherited and Acquired Thrombophilia* 129(2): 254–257.

Kim, K.W., Jhoo, J. H., Lee, K. U., et al. (1999). Association between apolipoprotein E polymorphism and Alzheimer's disease in Koreans. *Neurosci. Lett.* 277(3):145–148.

Klinefelter, H. F., Reifenstein, E. C., Albright, F. (1942). Syndrome characterized by gynecomastia aspermatogenesis without A-Leydigism and increased excretion of follicle stimulating hormone. *J. Clin. Endocrinol. Metab.* 2:615–627.

Liehr, T., Pfeiffer, R. A., Trautmann, U. (1992). Typical and partial cat eye syndrome: Identification of the marker chromosome by FISH. *Clin. Genet.* 42:91–96.

Lips, E. S. et al. (2012). Functional gene group analysis identifies synaptic gene groups as risk factor for schizophrenia. *Mol. Psychiatry* 17: 996–1006.

Liu, C. C., Kanekiyo, T., Xu, H., Bu, G. (2013). Apolipoprotein E and Alzheimer disease: risk, mechanisms, and therapy. Nat. Rev. Neurol. 9(2): 106–118.

Lubs, H. A. (1969). A marker X chromosome. *Am. J. Hum. Genet.* 21:231–244.

Luo, J., Yang, H., Tan, Z., Tu, M., Luo, H., Yang, Y., et al. (2017). A clinical and molecular analysis of a patient with Emanuel syndrome. *Mol. Med. Rep.* 15:1348–1352.

Mahley, R. W., Rall, S. C Jr. (2000). Apolipoprotein E: far more than a lipid transport protein. *Annu. Rev. Genomics. Hum. Genet.* 1:507–537.

Mahley, R. W., Weisgraber, K. H., Huang, Y. (2006). Apolipoprotein E4: a causative factor and therapeutic target in neuropathology, including Alzheimer's disease. *Proc. Natl. Acad. Sci. U S A.* 103(15):5644–5651.

Malerba, A., Kang, J. K., McClorey, G., et al. (2012). Dual myostatin and dystrophin exon skipping by morpholino nucleic acid oligomers conjugated to a cell-penetrating peptide is a promising therapeutic strategy for the treatment of Duchenne muscular dystrophy. *Mol. Ther. Nucleic Acids* 1: e62.

Marie, B., Christopher J. K., Guy, A. R., Nicolas, D. (2017). Systematic review of autosomal recessive ataxias and proposal for a classification. *Cerebellum & Ataxias* 4:3.

Mark, T. C., John, T. B., Michael, L., John, D. O. (2002). Laboratory Diagnosis of Dysfibrinogenemia. *Arch. Pathol. Lab. Med.* 126: 499–505.

Martin, J. B., Bell, J. (1943). A pedigree of mental defect showing sex-linkage. *J Neurol Psychitr.* 6:154–157.

Maurizio, Z., Francesca, E., Guido, T., Louis, F. (2008). Thrombophilia. *Drug Target Insights* 3: 87–97.

McColgana, P., Tabrizia, S. J. (2018). Huntington's disease: a clinical review. *European Journal of Neurology* 25: 24–34.

McGrath, J., Saha, S., Welham, J., et al. (2004). A systematic review of the incidence of schizophrenia: the distribution of rates and the influence of sex, urbanicity, migrant status, and methodology. *BMC. Med.* 2: 13.

McTaggart, K. E., Budarf, M. L, Driscoll, D. A., Emanuel, B. S., Ferreira, P., McDermid, H. E. (1998). Cat eye syndrome chromosome breakpoint clustering: Identification of two intervals also associated with 22q11 deletion syndrome breakpoints. *Cytogenet. Cell Genet.* 81:222–228.

Mears, A. J., Duncan, A. M., Budarf, M. L., Emanuel, B. S., Sellinger, B., Siegel-Bartelt, J., et al. (1994). Molecular characterization of the marker chromosome associated with cat eye syndrome. *Am. J. Hum. Genet.* 55:134–142.

Mehrez, M. J. (2011). Epidemiology of Prothrombin G20210A mutation in the Mediterranean region. *Mediterr. J. Hematol. Infect. Dis.* 3(1): e2011054.

Michael, A., Stephan, L. (2019). Thalassemias: An Overview. *Int. J. Neonatal Screen.* 5: 16.

Ming, Y. L., Stephan, M. (2015). Thrombophilia. *Vascular Medicine* 20(2): 193–196.

Moh, H. M., Valentinus, B., Mudjiani, B., Shirley, F. L. (2017). Duchenne muscular dystrophy: overview and future challenges. *Aktualn. Neurol.* 17 (3): 144–149.

Mohammad, A. A., Qiurong, Z., Yongping, L. (2015). Factor XII (Hageman Factor) Deficiency: A Very Rare Coagulation Disorder. *Open Journal of Blood Diseases* 5: 39–42.

Nichola, Z. L., Philippa, D. H., Amy, K. R., Victoria, N., Robert, M. F., et al. (2012). Cerebellar ataxia in patients with mitochondrial DNA disease: a molecular clinicopathological study. *J. Neuropathol. Exp. Neurol.* 71(2): 148–161.

Poort, S. R., Rosendaal, F. R., Reitsma, P. H., Bertina, R. M. (1996). A common genetic variation in the 3'-untranslated region of the prothrombin gene is associated with elevated plasma prothrombin levels and an increase in venous thrombosis. *Blood.* 88:3698–3703.

Rao, E., Weiss, B., Fukami, M., Rump, A., Niesler, B., Mertz, A., et al. (1997). Pseudoautosomal deletions encompassing a novel homeobox gene cause growth failure in idiopathic short stature and Turner syndrome. *Nat. Genet.* 16:54–63.

Raymund, A.C.R. (2010). Huntington's disease: a clinical review. *Orphanet Journal of Rare Diseases* 5:40

René, S. K., Iris, E. S., Robin, M. M., Andreas, M-L, Daniel, R. (2015). Schizophrenia. *Disease Primers* 1:1–24.

Renu, S., Monica, S. (2009). Pathogenesis & Laboratory approach to Thrombophilia. *Eastern Journal of Medicine* 14:29–35.

Ripke, S. et al. (2014). Biological insights from 108 schizophrenia-associated genetic loci. *Nature* 511: 421–427.

Roberto, R., Lygia da, V. P. (2001). The frequency of Tay-Sachs disease causing mutations in the Brazilian Jewish population justifies a carrier screening program. *Med. J/Rev. Paul. Med.* 119(4):146–149.

Sandra, M. (2014). Laboratory investigation of thrombophilia. *J. Med. Biochem.* 33: 28–46.

Schinzel, A., Schmid, W., Auf der Maur, P., Moser, H., Degenhardt, K. H., Geisler, M., et al. (1981). Incomplete trisomy 22. I. Familial 11/22 translocation with 3:1 meiotic disjunction. Delineation of a common clinical picture and report of nine new cases from six families. *Hum. Genet.* 56:249–262.

Scott, M. S., Scott, C. W., Kenneth, A. B., Raj, K. et al. (2016). Guidance for the evaluation and treatment of hereditary and acquired thrombophilia. *J. Thromb.* 41:154–164.

Seppänen, M., Koillinen, H., Mustjoki, S., Tomi, M., Sullivan, K. E. (2014). Terminal deletion of 11q with significant late onset combined immune deficiency. *J. Clin. Immunol.* 34: 114–118.

Shahid, R., Muhammad, Z. Z., Zulfiqar, A., Alia, E. (2018). Schizophrenia: An overview. *Clin. Pract.* 15(5): 847–851.

Sharma D, Murki S, Pratap T, et al. (2014). *BMJ Case Rep.* 2014: bcr2014203923.

Siano, M. A., De Maggio, I., Petillo, R., Cocciadiferro, D., Agolini, E., Majolo, M., Novelli, A., Della Monica, M., Piscopo, C. (2022). De novo mutation in kmt2c manifesting as kleefstra syndrome 2: case report and literature review. *Pediatr. Rep.* 14:131–139.

Sienkiewicz, D., Kulak, W., Okurowska-Zawada, B. et al. (2015). Duchenne muscular dystrophy: current cell therapies. *Ther. Adv. Neurol. Disord.* 8: 166–177.

Smaranda, M., Zoltan, B., Anca M., Adina, S., Bianca, Ş., Rodica, B. (2017). Late onset Tay-Sachs disease in a non-Jewish patient: case report. *Acta Medica Marisiensis* 63(4):199–203.

Sokolenko, A. P. Imyanitov, E. N. (2018). Molecular diagnostics in clinical oncology. *Front. Mol. Biosci.* 5:76.

Solovyeva, V. V., Shaimardanova, A. A., Chulpanova, D. S., Kitaeva, K. V., Chakrabarti, L., Rizvanov, A. A. (2018). New approaches to Tay-Sachs disease therapy. *Front. Physiol.* 9:1663.

Strehle, E. M., Straub, V. (2015). Recent advances in the management of Duchenne muscular dystrophy. *Arch. Dis. Child.* 100: 1173–1177.

Tahir, M., Malla, F. A., Dar, Arshad, A. P., Mahrukh, H. Z. (2016). Frequency and pattern of cytogenetic alterations in primary amenorrhea cases of Kashmir, North India. *The Egyptian Journal of Medical Human Genetics* 17(1):25–31.

Tangvarasittichai, S. (2011). Thalassemia Syndrome. In: Kenji Ikehara Eds, Advances in the Study of Genetic Disorders. Intech Open, Croatia. DOI: 10.5772/18051.

Teresa, M., Concetta S. P., Paul, G. (2009). Jacobsen syndrome. *Orphanet Journal of Rare Diseases.* 4:9.

Terry, S. C., Jerman, J. H., Angell, J. B. (1979). A gas chromatographic air analyzer fabricated on a silicon wafer. *IEEE Trans Electron Devices* 26(12): 18801886.

Tormene, D., Grandone, E., De Stefano, V., Tosetto, A., Palareti, G., Margaglione, M., Casta man, G., Rossi, E., Ciminello, A., Valdrè, L., Legnani, C., Tiscia, G. L., Bafunno, V, Carraro, S., Rodeghiero, F., Simioni, P. (2012). Obstetric complications and pregnancy-related venous thromboembolism: the effect of low-molecular-weight heparin on their prevention in carriers of factor V Leiden or prothrombin G20210A mutation. *Thromb Haemost.* 107(3):477–484.

Tropak, M. B., Yonekawa, S., Karumuthil-Melethil, S., Thompson, P., Wakarchuk, W., Gray, S. J., et al. (2016). Construction of a hybrid beta-hexosaminidase subunit capable of forming stable homodimers that hydrolyze GM2 ganglioside in vivo. *Mol. Ther. Methods Clin. Dev.* 3:15057.

Walfisch, A., Mills, K. E., Chodirker, B. N., Berger, H. (2012). Prenatal screening characteristics in Emanuel syndrome: a case series and review of the literature. *Arch. Gynecol. Obstet.* 286:299–302.

Wang, J. (2008). Electrochemical glucose biosensors. *Chem. Rev.* 108(2): 814825.

Waring, S. C., Rosenberg, R. N. (2008). Genome-wide association studies in Alzheimer's disease. *Arch. Neurol.* 65(3):329–334.

Wein, N., Alfano, L., Flanigan, K. M. (2015). Genetics and emerging treatments for Duchenne and Becker muscular dystrophy. *Pediatr. Clin. North. Am.* 62: 723–742.

Weinberger, D. (1999). Cell biology of the hippocampal formation in schizophrenia. *Biol. Psychiatry* 45(4): 395–402.

Zafar, U. K., Elisa M-M., Chris, M. E. (2013). Schizophrenia: Causes and Treatments. *Current Pharmaceutical Design* 19:6451–6461.

6 Gene Therapy

"Many people say they are worried about the changes in our genetic instructions. But these instructions are merely a product of evolution, shaped so we can adapt to certain conditions, which might no longer exist. We all know how imperfect we are. Why not become a little better apt to survive?"

James Watson 1991

6.1 INTRODUCTION

Human beings have been trying to understand the flow of genetic information from parents to offspring. Ancient Greek scholars made the first records of understanding the inheritance of genetic characters, whose concepts were supported and followed for many centuries. Scientific evidence-based genetic studies were provided in 1850 by Gregor Mendel in a series of experiments conducted on pea plants, describing a pattern of inheritance by observations made on separate units, identified later as genes. Until 1953, very little was known about the physical form of the genes. The development of the double-strand DNA model by American biochemist James Watson, and the British biophysicist, Francis Crick, have revolutionized genetics and led to molecular genetics development. In the 1970s, enzymes that can separate the genes and re-insert them into other organisms in pre-determined sites have been discovered, creating the possibility for the development of genetic engineering to produce novel drugs and antibodies. The rapid development of genetic engineering tools has led to the discovery of gene therapy in the 1980s by modifying the genome for inhibiting genetic disorders and other lethal diseases in human beings.

Local modifications in the human genome are the objective of medicine since the role of a gene in inheritance has been established. Gene therapy is one way to improve the expression of a gene by altering its bases or by replacing the mutated nucleotides. Presently, gene therapy is one of the predominant areas of research globally in many laboratories, and most of its applications are at a practical level. Currently, gene therapy trials are being conducted in countries such as the United States of America, Europe, and Australia. The application of gene therapy to treat recessive genetic disorders such as cystic fibrosis, hemophilia, muscular dystrophy, sickle cell anemia, acquired diseases such as cancer, viral infections such as HIV is in progress. This chapter provides details regarding the methods employed for gene therapy, vectors, and ethical issues.

6.2 GENE THERAPY AND GENETIC ENGINEERING

A genetic disorder occurs due to the inheritance of a mutated gene, having harmful effects on an individual. Most genetic disorders are recessive, expressed in a small percentage of the population. Most

DOI: 10.1201/9781003343790-6

of the population are carriers for one or more disorders. Specific genetic disorders are also recorded as higher in an ethnic group. The inheritance of genetic disorders in humans occurs due to improper functioning of a specific gene, change in its sequence, or addition or deletion of nucleotides. The essence of gene therapy is in replacing a defective gene with a healthy one to solve the problem. Patients have received genetically engineered cells as treatments for genetic disorders such as hemophilia and sickle-cell anemia. Currently, genetic engineering of cells from bone marrow for enhancing the immune cells' ability to fight against cancer and resist the infection of viruses such as HIV is practiced. A similar approach may also lead to an effective treatment for hereditary disorders.

6.2.1 Gene Therapy

Gene therapy is a process of gene insertion into the cells and tissues of an individual for the treatment of hereditary disorders. During this process, a deleterious mutated allele is replaced with a functional one. Even though this technology is still in its infancy, its application has achieved some success. Further, scientific progress has been continuing to move gene therapy towards mainstream medicine. The first clinical trial for human gene therapy was carried out in 1989 by Rosenberg and his team to treat severe combined immunodeficiency (SCID) by transferring the adenosine deaminase (ADA) gene into the T-Lymphocytes of a patient. Early success in clinical trials of gene therapy in SCID patients combined with efforts to improve transduction efficiency has provided an arena for the successful transplantation of the ADA gene with increased efficiency. Eventually, this has resulted in low dose conditioning with busulfan, used to engraft gene-modified cells during ADA-SCID setting. After this successful landmark study on SCID, over 1700 clinical trials have been pursued, various diseases and disorders have seen an average of 100 new gene therapy clinical trials since 1999.

For practicing gene therapy, the first logical step is to introduce genes directly into the human cells with disorders due to defects in a single gene (monogenic). This approach has become more challenging to achieve, primarily due to the problems in carrying large DNA fragments and delivering them to the correct location in the large genome. Currently, most gene therapy studies are aimed at controlling the spread of cancer and hereditary disorders that occur due to a single gene defect. Antisense therapy is strictly not a form of gene therapy but a genetically mediated therapy (Yanyu et al. 2018). The human genome is one of the largest, and human gene therapies remain complex with many techniques that need further development. Many genetic disorders and their strict genetic links need to be understood entirely before applying gene therapy. A variety of methods are being practiced for the replacement or repairing of the genes targeted for gene therapy.

1. Insert a normal gene into a non-specific location within the genome to replace a non-functional gene. This approach is the most used and has gained some success.
2. Swap an abnormal gene with a normal gene through homologous recombination.
3. Repair the abnormal gene through selective reverse mutation, which retains its normal function.
4. Alter the regulation of a specific gene.
5. Replace defective mitochondria with normal mitochondria using spindle transfer.

6.2.2 Gene Therapy and Hematopoietic Stem Cells

James Blundell reported the very first published case of blood transfusion in 1824. This obstetrician successfully transferred the cells from one individual to another to carry out their life sustaining activities. After a hundred years, the first successful transplantation of hematopoietic stem cells was performed between identical twins. With the invention of human leukocyte antigen (HLA) typing, this therapy was made applicable to siblings and unrelated donors whose HLAs matched. Most inherited blood cell diseases and stem cell defects are treated by transplanting allogeneic hematopoietic stem cells. Hematopoietic stem cells are considered the ideal targets for gene transfer due to their high potentiality, longevity, and self-renovation capacity. The transplanted genetically normal hematopoietic stem cells can serve as an outgoing source of blood cells for all lineages eliminating these disorders with a single treatment whose benefits last lifelong.

One of the most widespread examples of hematopoietic stem cells and gene therapy is designing vectors for the generation of induced pluripotent stem cells and their differentiation. Patients suffering from chronic liver disease and infection by hepatitis C virus, requiring a liver transplant, may need transplantation of mature hepatocytes or cells differentiated from induced pluripotent stem cells. As the transplanted cells may also be subject to re-infection by the hepatitis C virus, the transfer of the vector encodes a short hairpin RNA, directed against the virus, providing resistance to transfer cells through the re-infection. Resistant cells can restore the normal hepatic function of the liver by repopulating over time.

Even though there are high rates of success with the transplantation of allogeneic hematopoietic stem cells, the outcome of the transplantation keeps decreasing with less well-matched allogeneic donors, such as haploidentical family members, and any unrelated donors. Reduction in the HLA match between donors and recipients will increase the graft rejection and graft versus host disease risk. A hematopoietic stem cell graft rejection will take the patient into a dangerous position leading to anemia, infection, and bleeding. There has been an improvement in methods to minimize graft rejection and graft versus host disease. Autologous hematopoietic stem cell transplantation is a process by which the patient's hematopoietic stem cells are genetically modified. All major immunological complications are avoided with allogeneic hematopoietic stem cell transplantation. Expression of the gene introduced into hematopoietic stem cells is needed in cells such as RBC, neutrophils, or lymphocytes for specific disorders. This approach of an autologous transplant using vectors such as lentivirus have produced clinical benefits against several disorders.

6.2.3 CAR-T Therapy

Chimeric antigen recipient T cell therapy (CAR-T) is immunotherapy involving the manipulation or reprogramming of the patient's T-lymphocytes for minimizing the spread of tumor T cells. CAR-T involves manipulating or reprogramming the patient's T-lymphocytes to minimize the spread of tumor T cells. The utilization of genetically engineered T cells was initiated in the late 1980s for an increased immune response against malignancies. Chimeric antigen receptors are synthetic protein molecules generated by the fusion of a single-chain variable fragment of a monoclonal antibody with the T-cell receptor's signaling and co-stimulatory machinery, which is re-directed to recognize the tumor antigen as CD19 (Park et al. 2018). CART19 comprises of an anti-CD19 single-chain variable fragment linked to a co-stimulatory domain such as CD28 or 4-1BB and CD34 signaling domains through a transmembrane sequence. This construct can recognize a defined tumor surface antigen such as a monoclonal antibody and trigger full T cell activation.

The first generation in the CART was the fusion of a single chain fragment variable to a transmembrane domain and an intracellular signaling unit chain CD3 Zeta by Eshhar and his group members in 1993. This design has successfully combined the active element of a monoclonal antibody with a signaling domain for increased recognition of the tumor specific epitope for the T-cell activation independently without any influence from the major histo compatibility complex (MHCC). Integration of co-stimulating molecules such as CD28 acts as a second activator for enabling a marked proliferation of T-cells and increased expression of cytokines. The second generation of CAR-T has incorporated a co-stimulatory domain by incorporating molecules such as CD134 and CD137, for further enhancement of CAR-T function. The third generation of CAR-T demonstrates an improvement in the T cell activation of the akt route for cell cycle regulation. The latest among these, the fourth generation of CAR-T, has shown the development of T cells for their high persistence compared with other generations. One of the adverse effects of CAR-T therapy is the identification of non-tumor cells that express the target epitope by CAR.

T cells collected from the patient's peripheral blood for the generation of CAR-T cells are activated and transduced with the constructs using viral vectors or with sleeping beauty transposon systems, re-infused into the patient after lymphodepleting chemotherapy for the destruction of lymphocytes and T cells. During this process, former non-tumor specific T cells acquire the ability to recognize CD-19 positive cells, forming activating synapses, including TCR triggering and co-stimulation. CAR-T cells will proliferate and become activated, exerting their functions effectively, traffic around the body, and establishing immunological memory. CAR-T19 therapy has successfully treated patients suffering from relapsing/refractory B-cell acute lymphoblastic leukemia, non-Hodgkin lymphoma, and chronic lymphocytic leukemia (Maude et al. 2014). CAR-T therapy has shown unprecedented success in B-Cell malignancies. Patients of adult B-acute lymphoblastic leukemia have shown survival of more than 7 months after grafting, and more than 80% of relapsed and refractory diseased patients have achieved a complete remission by being alive even after one year. Two licensed products Tisagenlecleucel of Novartis, and Axicabtagene Ciloleucel of Kite Pharma/Gilead, targeting the CD19 antigen, are on the market for B-acute lymphoblastic leukemia and B-cell lymphoma treatment. Many CAR-T cells are in clinical trials for regulating hematological and solid organ malignancies with promising results.

Despite its success, CAR-T treatment has created a list of novel complications and a high range of side effects. The expression of tumor antigens such as CD19 is high in tumor cells. However, they are not expressed in B cells, such as the CD19 antigens of normal and malignant cells. CAR design for targeting the CD19 antigens is not capable of distinguishing them in these two types of cells. Another adverse effect of CART therapy is a tremendous increase in cytokine release syndrome. After CAR-T infusion, activation of the immune system induces a rapid increase in the inflammatory cytokine levels, leading to cytokine release syndrome. CAR-T cell-related encephalopathy syndrome is another major complication developed due to the infusion of CAR-T cells with symptoms ranging from mild confusion to fatal cerebral edema. The design and development of vectors, which can reinforce safety, is in progress to overcome these challenges.

6.2.4 CRISPR/Cas9

Genome editing is a technology that provides the ability to change an organism's DNA by addition, deletion, or alteration of a specific location in the genome. CRISPR-Cas9 is abbreviated for clustered regularly interspaced short palindromic repeats and CRISPR-associated protein 9 is a fast, accurate, economically cheaper genome editing tool, which has created much excitement in the scientific community because of its efficiency. This novel genome-editing tool is an adaptation of a naturally existing genome editing system in *E.coli*. In 1987, Ishino and his team identified a region with a distinctive pattern in the *E.coli* genome. This region consists of a highly variable sequence intercalated by a palindromic repeat with a known function. In 2005, this variable sequence region was identified to be of extra-chromosomal origin, involved in the immune memory against phages and plasmids. Immune memory acts in three phases: The first phase is the adaptation phase during which bacteria (and some archaea) incorporate short foreign sequences known as protospacers into the host genome. These sequences recognize the invading virus DNA, which cleaves into small fragments, and integrates into its own DNA. The second phase is the expression and maturation phase in which a second infection occurs by the same virus. These sequences undergo transcription, resulting in the generation of small RNA fragments known as precursor CRISPR RNAs (crRNAs), which become crRNA on maturation. The third phase is the interference phase in which the mature crRNA will form complexes with a set of associated proteins such as Cas9 to recognize and cut the virus DNA and finally disable it (Gupta and Musunuru, 2014). This system was identified as CRISPR and Cas9 in 2012, becoming a vital gene and genome editing tool.

Three nuclease systems are reported to exist, and these are known as type I zinc finger nucleases (ZFNs), type II CRISPR/Cas nucleases, and type III transcription activator-like effector nucleases (TALENs). The type II system involving Cas9 proteins has been shown to be more promising for genome editing and engineering. Type II CRISPR/Cas9 has originated from *Streptococcus thermophilus*, capable of specifically cleaving the dsDNA. The site-specific disruption of the dsDNA by CRISPR/Cas9 requires a precise complementary base pairing of CRISPR RNA (crRNA) to the target sequence along with a short DNA sequence known as a protospacer-adjacent motif (PAM). Cas9 protein isolated from different organisms recognizes different PAMs. Cas9 protein from *Streptococcus pyogenes* requires a 5'NGG3' sequence that is located adjacent to the target locus. Cas9 protein from *Streptococcus pyogenes* is regulated by both CRISPR RNA (crRNA) and trans-activating RNA (tracrRNA), thus constituting an essential component of the type II CRISPR/cas system. To simplify the programming of Cas9, the dual-tracrRNA:crRNA can be replaced by an engineered single-guide RNA (sgRNA).

CRISPR has enabled the editing of target specific DNA sequences in the genome of any organism on the planet by using three molecules, namely nuclease such as Cas9, which cleaves the ds DNA, the RNA guide involved in guiding the complex towards the target sequence, and he target DNA (Figure 6.1).

The CRISPR/Cas9 belongs to the programmable nuclease-based genome editing technology, where the programmable nucleases can generate mutations at any specific location in an organism's

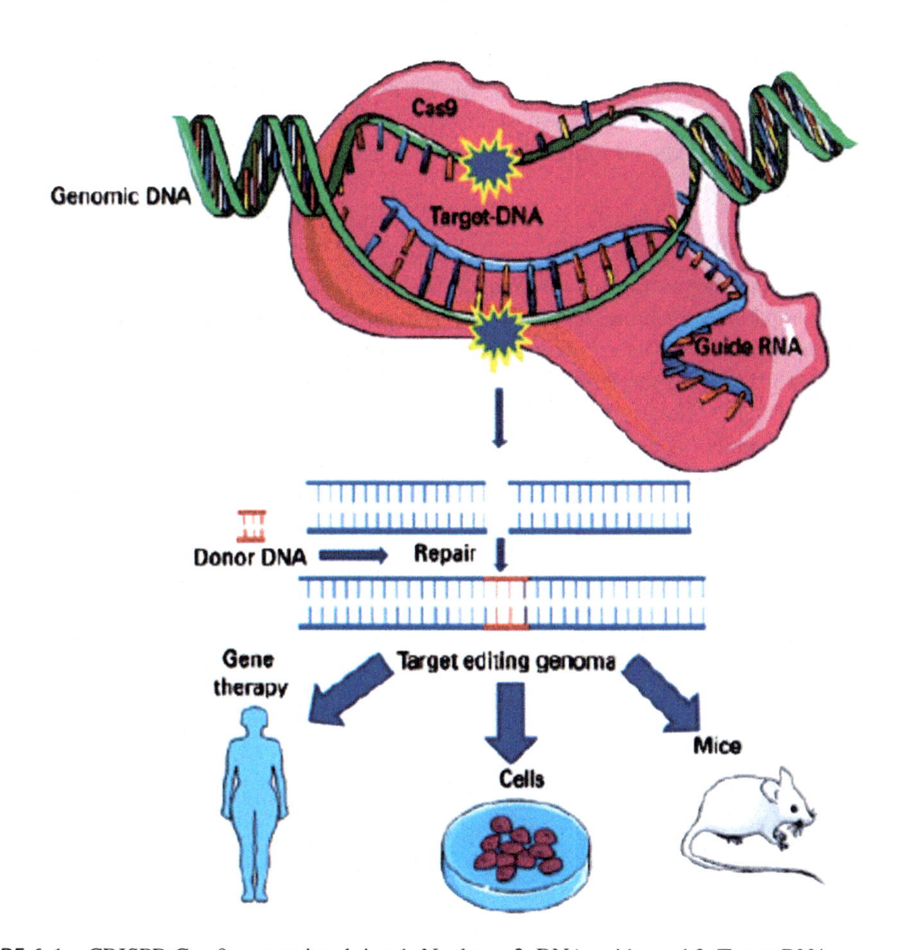

FIGURE 6.1 CRISPR Cas-9 system involving 1. Nuclease 2. RNA guide, and 3. Target DNA.

genome. Further, the endonuclease activity of Cas9 serves as genomic scissors by creating double stranded breaks (DSBs), assuring precise genome editing. A disruption in the DNA sequence will trigger DNA repair mechanisms, including the homology based directed repair and non-homologous end joining, leading to specific knockouts and knockins. Endogenous DNA repair pathways can be exploited for the generation of mutations at desired sites in the genome.

6.2.4.1 CRISPR/Cas9 and Neurodegenerative Disorders

Also known as motor neuron disease or Lou Gehrig's disease is a neurodegenerative disorder that occurs due to the loss of neuromotor function, controlling the voluntary muscles resulting in stiff muscles, twitching of muscles, and rapid decrease in muscle size. There are more than 20 genes involved in the neuropathology of amyotrophic lateral sclerosis. Recent studies have revealed that silencing the SOD1 gene using interfering RNA and artificial micro-RNA (miR-SOD1) has delayed the onset of the disease and prolonged the life span of the SOD1G93A mouse model. Compared with micro-RNA, CRISPR-mediated genome editing has enormous potential for the treatment of amyotrophic lateral sclerosis. CRISPR/Cas9 has been applied for regulating the SOD1 gene in the SOD1G93A mouse model after *in vivo* gene delivery by the adeno associated virus vector. Disruption has significantly reduced the expression of SOD1 mutant protein in the lumbar and thoracic spinal cord, resulting in an enhanced motor function, leading to diminished muscle atrophy. CRISPR/Cas9 treated ALS mice have significantly enhanced neuromotor survivability, delayed onset of the disease, and prolonged their life span compared with controls, confirming the potentiality of CRISPR/Cas9 in the treatment of Amyotrophic Lateral Sclerosis and other neurodegenerative diseases. Researchers from the University of California and Utah successfully restored the mutated hemoglobin gene responsible for sickle cell anemia to its normal function by isolating $CD34^+$ cells from the carriers of sickle cell anemia and editing by CRISPR/Cas9. The results have considerably reduced mutated gene expression and enhanced wild-type gene expression after 16 weeks.

6.2.4.2 Limitations of CRISPR/Cas9

Despite the potentiality for curing several human diseases, CRISPR/Cas9 system has a few issues that make it incompatible and limits its feasibility. The guide RNA of the CRISPR need not wholly match the target sequence for its recognition and degradation. This characteristic feature of Cas9 explores the possibility of off-target effects such as recognition, digestion, and other consequences leading to unpredictable mutations, constituting a significant limitation of this technique. The success of Cas9 depends on the uniqueness of the target sequence as it is capable of tolerating nucleotide mismatches to some extent. Studies conducted on nucleotide mismatches have revealed that the mismatches occurring at the 5′ end are better tolerated than the mismatches occurring at 3′ end as the 3′ end sequence is more significant for target identification. Delivery of sgRNA and Cas9 is another major limitation of this tool. Lentiviral and adenoviral vectors are used for introducing these two components into target cells. Viral vectors also suffer from many drawbacks such as low carrying capacity, tissue tropism of specific organs. Adeno-associated viruses display a high rate of tissue tropism to muscles, liver, brain, and eye. Viral vectors used for therapeutic implications are re-arranged for minimizing virulence. Most undesirable and sometimes harmful consequences might occur even after re-arrangement.

Non-viral approaches such as electroporation, nucleofection, microinjection using glass micro-capillaries, and lipofectamine are the alternatives for safe gene editing. Li and his associates in 2017 discovered an artificial virus for the effective transfer of CRISPR/Cas9 into mice, which is more reliable than lipofectamine (Li et al. 2017). The potential adverse effects of bacteria derived Cas9 need to be completely studied before developing the CRISPR/Cas9 based therapeutics.

6.3 THERAPEUTIC NUCLEIC ACIDS AND THERAPEUTIC POLYNUCLEOTIDES

Initial utilization of nucleic acids as therapeutic agents has started during the application of nucleoside analogs for combating diseases. Rapid development in drug research and modern genetic engineering technologies has led to the discovery of biologically active molecules, which are highly specific to their targets and are involved in the progression of the disease. Numerous gene therapy strategies have been developed and tested. Some of them are based on nucleic acids, which inhibit gene expression at transcriptional or post-transcriptional levels. Development of these strategies has been possible due to a series of pivotal discoveries made during the past 25 years, including, elucidation of the human genome, which enabled the unlocking of several molecular pathways responsible for diseases; discovery of novel RNA molecules such as miRNA and siRNA with complex biological functions; development of RNA analogs such as locked nucleic acids with profound clinical and therapeutic applications; and enhanced bioavailability of nucleic acid-based drugs. Technological advancements made in the de novo synthesis of long DNA constructs, usage of DNA shuffling, bioprospecting, combinatorial nucleotide chemistry, and genetic engineering of viruses have made nucleotide-based gene therapy possible (Geoffrey et al. 2011). Advancements in synthetic biology, systems biology, computational biology, bioinformatics, and nanotechnology have further developed this area and have set a new paradigm for utilizing therapeutic nucleic acids in gene therapy. This nucleic acid-based gene therapy strategy has potential applications including cardiovascular conditions, neurological conditions, inflammatory disorders, cancer, infectious diseases, and even organ transplantation. Nevertheless, the utilization of nucleic acids for the treatment of genetic disorders is still in its infancy. One of the main reasons for this is the involvement of multiple genes in the generation of a genetic disorder. The nature of every gene and its contribution to generating the disorder is still unknown.

DNA inhibition or RNA expression for halting the production of an abnormal protein responsible for a disease by keeping other proteins ineffective is the primary basis for the role of therapeutic nucleic acids in gene therapy. These therapeutic nucleic acids are high molecular weight charged compounds with unique physical and chemical properties different from small drug molecules and are highly unstable in a biological environment. These peculiar properties make them unique, with specific requirements needed for their utilization. Even though various therapeutic nucleic acids are involved in gene therapy, they share a common mechanism of action, mediated by the sequence-specific recognition of endogenous nucleic acids through Watson Crick base pairing. Therapeutic nucleic acids are broadly divided into two categories, DNA-based therapeutics, and RNA-based therapeutics.

6.3.1 DNA-Based Therapeutics

6.3.1.1 Plasmids

Plasmids are high molecular weight, double-stranded DNA capable of encoding specific proteins while carrying the transgene constructs. A plasmid DNA molecule is also considered a prodrug, which employs DNA transcription and translation machinery in a cell immediately after internalization for the biosynthesis of their therapeutic entity, the protein. Accessing the host cell nucleus after their entry into the cytoplasm is essential for plasmid DNA activity. Along with the disease treatment, plasmids are also used as DNA vaccines for genetic immunization. During its infancy, plasmid-based therapy was considered for the treatment of monogenic disorders. The first federally approved human gene therapy protocol was initiated in 1990 to treat adenosine deaminase deficiency. Since then, more than 500 gene therapy processes have been approved and are under implementation. The first successful gene therapy-based cure for severe combined immunodeficiency disorder (SCID) was performed in 2002. In 2003, the Chinese drug regulatory agency approved the first gene therapy product under the trade name Gendicine. It is a recombinant adenovirus, genetically engineered for the expression of tumor suppressor gene p53 to treat head and neck squamous carcinoma. Combining chemotherapy and radiotherapy has improved the efficiency of gendicine three to four-fold with the amelioration of toxicity. Gendicine has also considerably improved the therapeutic effect of radiation in cells affected by pancreatic cancer. Clinical trials are in progress for further improvement. Alipogene tiparvovec is an adeno-associated virus engineered with the lipoprotein lipase gene to treat patients suffering from lipoprotein lipase. The European Union has approved gene therapy using alpogene for patients with lipoprotein lipase deficiency in 2012. As the deficiency of lipoprotein lipase is rare in humans (one in a million), the drug has been tested successfully on only 27 patients.

6.3.1.2 Oligonucleotides for Antisense and Antigene Applications

Oligonucleotides are the short stretches of single-stranded DNA sequences whose cellular internalization can selectively inhibit the expression of a single specific protein through standard Watson-Crick base pairing. The expression of oligonucleotides varies with their applications. For antisense applications, oligonucleotides (antisense oligonucleotides) interact to form a duplex with the mRNA or the pre-mRNA, inhibiting their translation. Further, mRNA processing will be inhibited by degrading the target complex by endogenous cellular RNase H or functional blocking of the mRNA chain by steric hindrance. Both of these processes will consequently result in the inhibition of protein biosynthesis. Initial antisense oligonucleotides were highly sensitive to nuclease enzymes, necessitating modifications. Changes to the sugar backbone and ribose component of the nucleotides have improved their stability, binding strength, and specificity. These oligonucleotides have their phosphodiester or phosphoramidate, or phosphorothioate modified and are considered second generation oligonucleotides. Among these, phosphorothioate modification involving the replacement of one non-bridging oxygen atom by a sulfur atom is the most common second-generation modification of antisense oligonucleotides, which have shown an enhanced uptake by the target cells along with high stability and specificity. Modifying the ribose sugar at 2′ position, either 2′-O-methyl or 2′-O-methoxyethyl will further enhance the resistance of the oligonucleotides with exonucleases. Oligonucleotides with a modified non-sugar furanose ring have shown additional stability to nuclease considered third generation antisense oligonucleotides. Locked nucleic acid (LNA), peptide nucleic acid (PNA), and morpholino phosphoramidates (MF) are some of the common third generation antisense oligonucleotides.

The efficiency and safety of the antisense oligonucleotides belonging to different generations was tested for various neurodegenerative disorders such as Alzheimer's, Parkinson's, Huntington's, and amyotrophic lateral sclerosis. Two antisense oligonucleotides, fomivirisen from the first generation and mipomersen belonging to the second generation have been approved by the Food and Drug Administration of the United States. After successful antisense inhibition of proteins in animal models, the first antisense drug, fomivirsen sodium (Vitravene, Isis Pharmaceuticals, Carlsbad, CA), was approved for cytomegalovirus retinitis treatment in AIDS patients in 1998. Antisense oligonucleotides such as MG98 and ISIS 5132 designed to inhibit the biosynthesis of DNA methyltransferase and c-Raf kinase, respectively, are in human clinical trials for cancer treatment. Synthetic antisense DNA oligonucleotides and oligonucleotide analogs for inhibiting the replication of several infectious agents such as hepatitis C virus, human cytomegalovirus, human immunodeficiency virus, and papillomavirus, have also been designed. For antigene applications, oligonucleotides will have to enter the cell's nucleus to form a triplex with the double-stranded genomic DNA, inhibiting its transcription process and consequently its protein synthesis. For therapeutic applications, oligonucleotides are used to selectively block the expression of proteins implicated in diseases.

6.3.1.3 DNA Aptamers

Aptamers, also known as chemical antibodies, are single stranded synthetic DNA or RNA molecules of 56 - 126 bases length, capable of binding to the nucleotide coding region of the protein with high affinity, serving as decoys. DNA aptamers are short single stranded oligonucleotide sequences with a very high affinity to the target nucleotide sequences. These DNA aptamers are isolated from a large pool of nucleic acids through systemic evolution of ligands by exponential enrichment (SELEX) or aptabid. Aptamers are also considered alternatives to monoclonal antibodies because of their non-toxic and non-immunogenic compositions with a good tissue penetration and can be injected both subcutaneously and intravenously, easy to make and modified under in vitro conditions. DNA aptamers that target the coding regions of lysozyme, thrombin, HIV trans-acting responsive element, hemin, interferon-γ, vascular endothelial growth factor, prostate specific antigen, dopamine, and heat shock factors are under development and are in clinical trials.

6.3.1.4 DNAzymes

DNAzymes are catalytic DNAs isolated by Breaker and Joyce in 1994 through in vitro selection of a random DNA library comprising of 1015 DNA sequences. DNAzymes catalyze biochemical processes such as RNA cleavage, oxidative/hydrolytic DNA cleavage, nucleotide (both DNA and RNA) ligation, and DNA phosphorylation, which are practiced in vivo and under in vitro conditions successfully. To date, no DNAzyme has been found to occur in nature, which might be due to the existence of DNA as double-stranded in most of the cells, and the catalytic core of DNAzymes is single-stranded. Since the first report of DNAzyme, many different types of DNAzymes, especially RNA cleaving DNAzymes, have been extensively used as biosensors and therapeutic agents. RNA cleaving DNAzymes catalyze the cleavage of a single RNA linkage embedded within a DNA strand. A DNAzyme is formed by hybridizing a substrate and enzyme strands, consisting of an active site, enzyme region, and binding sites (Figure 6.2).

A substrate strand containing a single RNA linkage serves as a cleavage site. In the presence of a co-factor, the enzyme forms a defined secondary structure with the catalytic ability to cleave the RNA linkage, resulting in two fragments with a 2′-3′ cyclic phosphate and a 5′OH terminus (Figure 6.3).

This catalytic property is highly specific to the substrate strand, significantly decreasing with every base mismatch. Most DNAzymes exhibit high flexibility for designing the DNA sequences in the binding arms, facilitating the rational design and development of multifunctional DNAzymes using the same co-factor for diverse applications. The versatility of RNA cleaving DNAzymes has resulted in the emergence of RNA cleaving DNAzyme based functional probes, which are further categorized into, biorecognition agents; signal amplification agents; and activatable hosting agents. The biorecognition function of RNA cleaving DNAzymes is achieved by hybridizing RNA cleaving

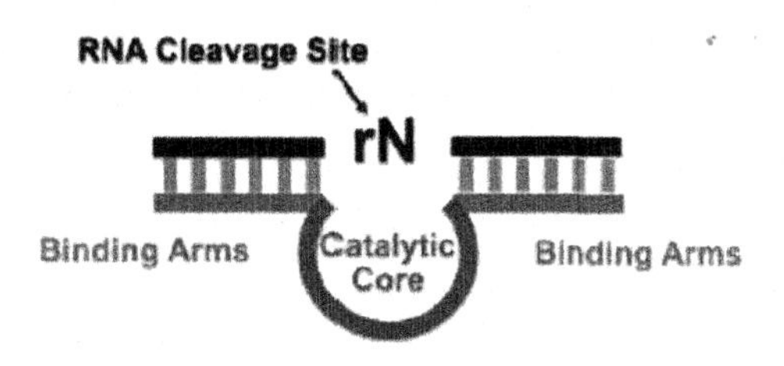

FIGURE 6.2 RNA cleaving DNAzyme.

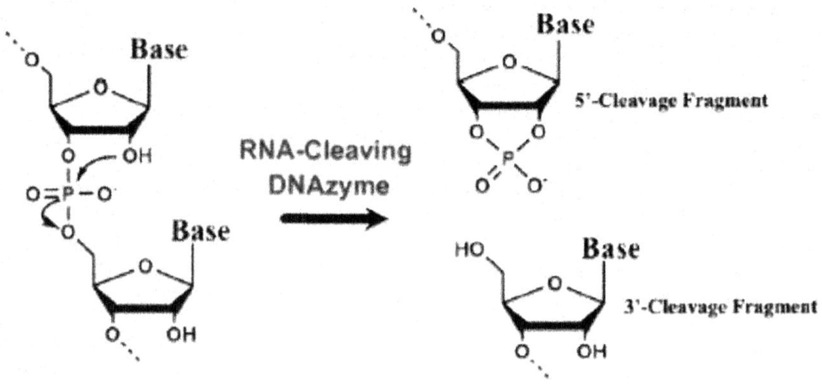

FIGURE 6.3 RNA cleavage catalyzed by DNAzyme giving rise to two fragments.

FIGURE 6.4 Biorecognition function of RNA cleaving DNAzymes.

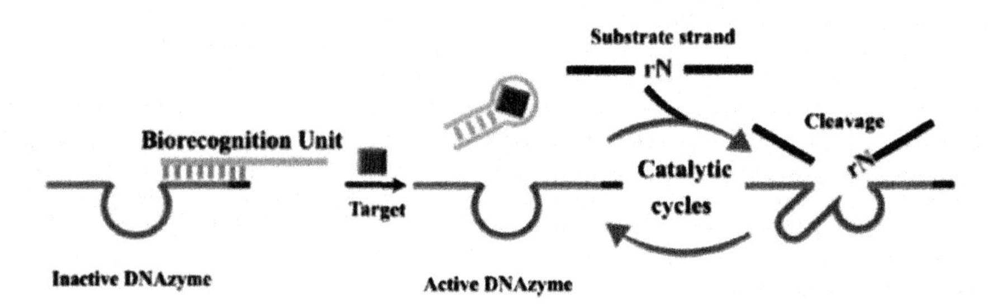

FIGURE 6.5 Activity of RNA cleaving DNAzymes as biocatalysts for signal amplification.

DNAzymes, enzyme strand with substrate strand, forming an RNA duplex. In the presence of the target co-factor, this enzyme strand will form a well-defined 3-dimensional structure and catalyzes the cleavage of the substrate strand into two fragments (Figure 6.4). The biorecognition function is widely used for the detection and imaging of co-factors, especially metal ions.

RNA cleaving DNAzymes also serve as biocatalysts for signal amplification. These enzymes are embedded with an inactivated biorecognition unit. In the presence of the target, the biorecognition unit will be activated and subsequently cleave the substrate strand with the assistance of a co-factor, which would either cause the regeneration of the target or release of the DNAzyme strand to start another new round of catalytic reactions (Figure 6.5).

For hosting, RNA cleaving DNAzymes are activated either by external or internal stimuli, resulting in the release of therapeutic reagents (Figure 6.6). RNA cleaving DNAzymes can assemble into various 3D nanostructures through advanced DNA nanotechnology, providing a biocompatible matrix for intracellular delivery. They can also be incorporated into a variety of nanomaterials, which further improves the efficiency of delivery (Figure 6.6).

Diseases with complex etiologies such as cancer and neurodegenerative disorders such as Alzheimer's disease and Parkinson's disease are targeted using DNA as a therapeutic agent. In addition, DNA vaccines for malaria, AIDS, and many other diseases are under development. DNA

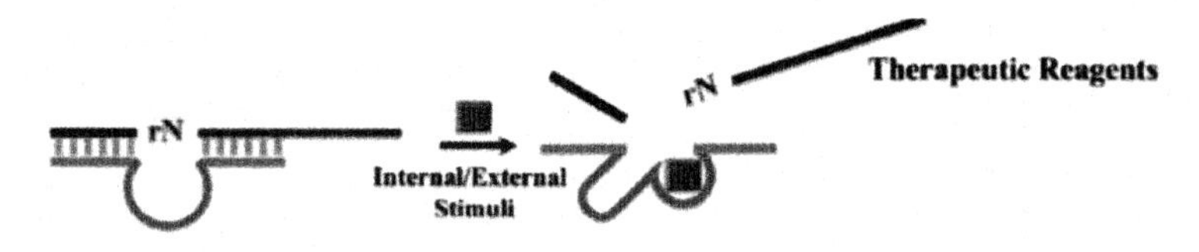

FIGURE 6.6 Hosting activity of RNA cleaving DNAzymes.

vaccines have also been used to prevent allergic responses. Further, the success of gene therapy lies in the successful delivery of the target genes. Modified HIV, lentivirus, adeno, and adeno-associated viruses, and herpes simplex viruses are the most widely used vectors. Non-viral vectors such as liposomes and nanoparticles show some promising results during the non-invasive administration through skin, eyes, and lungs (for details, refer to vectors section).

6.3.2 RNA Based Therapeutics

The biological understanding of RNAs is increasing progressively from an intermediary between DNA and protein to a highly dynamic and versatile molecule that regulates gene function and cells in all living organisms. Breakthroughs in the discovery of catalytic RNAs, transformed into ribozymes, and in RNA interference led to the emergence of RNA-based therapeutics, which further broadened the range of drug targets. Current RNA-based therapeutics are classified based on their mechanism of activity into inhibitors of mRNA translation, RNAi agents, catalytic RNA, and RNA aptamers that bind to proteins and other biomolecules.

6.3.2.1 RNA Aptamers

The ability of RNA to fold into 3D structures and bind to proteins for their inhibition as protein antagonists make them a promising therapeutic agent. RNA aptamers are the single-stranded nucleic acid segments that can directly interact with the proteins to inhibit their functions. Aptamers directly recognize their target proteins based on their complementarity in shape. The binding affinity and ability of the aptamers with the targets are extremely specific, similar to the monoclonal antibodies. RNA aptamers are developed similarly, as DNA aptamers with only change of capping with modified nucleotides such as 2′-O-modified pyrimidines, 2′-amino pyrimidines, or 2′-fluoropyrimidines for preventing their terminal degradation as they are highly susceptible to exonuclease degradation. RNA aptamers have displayed a promising intervention of pathogenic protein synthesis against HIV-1 transcriptase. RNA aptamers capable of binding and inactivating vascular endothelial growth factor (VEGF) have been isolated. Injection of anti-VEGF aptamers into the human eye have either retained eyesight or improved eyesight in 80% of patients without any side effects. Thus, pegaptanib (Macugen, Eyetech Pharmaceutics, and Pfizer) became the first RNA aptamer approved by USFDA for its use in age-related macular degeneration (AMD). RNA aptamers have a unique ability to reach intracellular targets as well as direct binding to extracellular targets. NOX-A-12, 45 nucleotide RNA aptamer specifically antagonizes the CXC chemokine ligand 12/stromal cell-derived factor-1 (CXCL 12/SDF-1), a regulatory chemokine needed for the migration of leukemic stem cells into the bone marrow has been designed. This inhibition of SDF-1 binding to its receptors will lead the leukemia stem cells to re-enter the cell cycle and become vulnerable to chemotherapy. Phase-1 clinical study of NOX-A-12 has proven to be safe in 48 healthy humans. Phase-II study of this aptamer is in progress.

6.3.2.2 RNA Decoys

RNA decoys provide alternate competing binding sites for proteins for acting as translational activators and mRNA stabilizing elements. RNA decoys prevent translation by inducing instability, leading to the destruction of mRNA. Overexpressed short RNA corresponding to CIS regulatory elements are used as decoys of transactivating proteins to prevent transactivators binding to their corresponding CIS-acting elements in viral genomes. RNA decoys are less affected by the infectious agents' variability compared with other nucleic acid agents.

6.3.2.3 Ribozymes

Ribozymes are catalytic RNAs that can act as enzymes in the complete absence of protein. An enzymatic moiety can be included within the antisense oligonucleotide for cleaving the target RNA after RNA–RNA duplex formation. Ribozymes were discovered by Tom Cech and his associates at the University of Colorado in 1982 in *Tetrahymena thermophila*. Later, Sid Altman and his associates at Yale University showed that the RNA component of RNase P, M1 RNA of *E.coli*, could process tRNA precursors without any protein factors. Cech and Yale shared the Nobel prize in 1989 for the discovery of ribozymes.

Due to their high specificity and wide range of target selection, both naturally occurring, and artificially synthesized ribozymes are used to suppress gene function. Ribozymes bind to the target moiety through antisense sequence-specific hybridization, cleave the phosphodiester backbone at the targeted site, resulting in its inactivation. They can selectively bind to the target mRNAs, forming a duplex with a highly distorted conformation of easy hydrolysis. This hydrolysis of mRNA may be used for the targeted suppression of specific genes and for validating the disease-related genes as potential targets for new therapeutic interventions. Among several ribozymes, hammerhead and hairpin ribozymes, whose names derived from their secondary structures, have been prioritized because of their small size and rapid kinetics. Hammerhead ribozymes cleave RNA at the nucleotide sequence U–H (H ¼ A, C, or U) by hydrolysis of a 30–50 phosphodiester bond. Hairpin ribozymes utilize the nucleotide sequence C–U–G as their cleavage target. Ribozymes are used in gene knockout therapy by targeting overexpressed oncogenes such as the human epidermal growth factor receptor type-2 gene implicated in breast cancer and human papillomavirus infection. Angiozyme is the first synthesized ribozyme under evaluation in clinical trials. This ribozyme was designed to combat angiogenesis through the specific cleavage of mRNA, which produces VEGF. The anti-Flt-1 ribozyme was explicitly designed to cleave the mRNAs for the primary VEGF receptors. Healthy individuals well tolerate this ribozyme both after intravenous infusion and subcutaneous bolus administration. Heptazyme (LY466700), a nuclease-resistant hammerhead ribozyme, completed a phase II clinical trial in 28 patients with chronic HCV and was reported safe and well tolerated. One of the significant limitations for the usage of ribozymes in gene therapy is their susceptibility to RNases.

6.3.2.4 Small Interfering RNAs

RNA is known to play a significant role in transferring information between DNA and protein synthesis via translation. RNA is also involved in the structural, enzymatic, and decoding of information during protein synthesis. Walter Gilbert, in 1986 proposed The RNA World. Andrew Fire and his colleagues discovered RNA interference (RNAi) in the nematode *Caenorhabditis elegans* during 1998, where natural RNA interference protected the organism's genome from invasion by mobile genetic elements such as viruses. It is a posttranscriptional mechanism by which a gene is silenced through chromatin remodeling, inhibition of protein translation, or direct mRNA degradation (ubiquitous in eukaryotic cells). In this process, RNA molecules complementary to a gene-specific coding sequence will induce the degradation of its corresponding mRNA and further block its translation. Small interfering RNAs (popularly known as siRNAs) are short double-stranded RNA of 21 - 23 nucleotides in length, complementary to the protein's mRNA sequence whose translation is blocked. Upon administration, SiRNA molecules form an RNA-induced silencing complex (RISC), the binding of which to the target mRNA will induce its degradation mechanisms such as nuclease activity, leading to its silencing. RNAi can downregulate disease-causing genes due to their simplicity and high target specificity. Introducing a foreign double-stranded RNA (dsRNA) into the cytoplasm of the targeted organism can further enhance the sequence-specific degradation of endogenous mRNAs that are homologous to the introduced dsRNA. After entry into the cytoplasm, they are processed by an RNase III enzyme known as Dicer, which cleaves long dsRNAs into short 21 - 23 nucleotide duplexes that have symmetric 2 - 3 nucleotide overhangs with 5′ phosphate and 3′ hydroxyl groups. Initially, it was believed that effective RNAi silencing requires complete sequence homology throughout the target mRNA's length, Later, it was found that seven contiguous complementary base pairs are sufficient to induce direct RNAi-mediated silencing. Attempts are made to utilize the siRNAs for the inhibition of HIV, hepatitis, and influenza infection. Most siRNAs are short double-stranded molecules of only 21 - 23 nucleotides length, which can be chemically synthesized. The foremost advantage of utilizing the siRNAs over DNA oligonucleotides is their delivery into the cytoplasm as more stable duplexes. Suppression of genes other than the target gene leading to off-target effects is the main hurdle that often affects their target delivery into the cells. Immunoreceptors such as toll-like receptors might release cytokines, altering the gene expression.

RNAi technology is also applied to the silencing of oncogenes (mutated), gene amplification, translocations, and viral oncogenes to elucidate their function and enhance their interactions with other genes involved in cellular metabolic pathways. ALN-RSV01 is designed against respiratory syncytial virus infection using siRNA. The N protein of respiratory syncytial virus is the main target for ALN RSV01, tested on 16 different lung transplant recipients under clinical trials and 40 other siRNAs. siRNAs are investigated for the inhibition of HIV, hepatitis, and influenza infection. The siRNA is also involved in allograft rejection. Still, the application of siRNA as a therapeutic agent is very much in its infancy. As siRNAs do not integrate into the genome and offer more safety when compared with plasmid molecules, the possibility of delivering a mixture of siRNAs for targeting multiple disease-causing genes in a single delivery system to control complex diseases is under consideration.

6.3.2.5 MicroRNA (miRNA)

MicroRNAs (miRNAs) are small non-coding RNA molecules of 21 - 25 nucleotides in length and are known to possess regulatory activity. They are partially complementary to mRNA, and their primary function is to downregulate gene expression by translational repression, mRNA cleavage, and deadenylation. They were first discovered in 1993 by Victor Ambrose and Gary Ruvkun in *C. elegans* as negative regulators in the post transcriptional modulation of almost all biological processes. Research on this molecule has identified its association with the pathogenesis of human diseases. Their capability for the regulations of vital biological processes by targeting multiple biomolecules makes them the most efficient therapeutic agents. In the nucleus, miRNAs are transcribed by RNA polymerase II, known as pri-mi RNA consisting of a 5′ cap and a 3′ poly A tail. Pri-mi RNA is processed into pre-miRNA by a microprocessor complex consisting of an RNAse III enzyme Drosha and a DS RNA Pasha, further exported by a karyopherin exportin and a Ran-GTP complex. Ran GTPase binds with an exportin and pre-miRNA to form a nuclear heterotrimer, processed by an RNAse III enzyme Dicer to generate miRNA. The dicer also initiates RNA-induced silencing complex (RISC) formation, responsible for miRNA expression and RNA interference, leading to the gene silencing.

Studies related to the involvement of miRNAs in cancer therapy became evident after their role in chronic lymphocytic leukemia through the downregulation of two human mir genes mir-15a and mir-16-1, in more than two-thirds of patients. Mir-17-92 clusters amplified in B cell lymphomas and mir-155, overexpressed in hematological cancers were the first oncogenic miRNAs to be described followed by several miRNAs identified for lung, ovarian, breast cancers, and glioblastomas. In addition, diseases such as Alzheimer's, Parkinson's, obesity, cardiovascular and autoimmune disorders also show a significant increase in the expression of miRNAs.

Therapeutics involving miRNAs are classified into two types based on their utilization, miRNA antagonists, and miRNA mimics. The utilization of miRNA mimics, also known as miRNA replacement therapy, is achieved in two different approaches. The direct approach involves the utilization of oligonucleotides or virus-based constructs to block the expression, or the introduction of tumor suppressor miRNA, lost because of the ongoing disease. An indirect approach involves utilizing drug therapy to modulate miRNA expression by regulating their transcription and processing. miRNA-122, expressed in hepatocytes, interacts with hepatitis C virus, leading to its proliferation. This discovery has paved the way for developing the first anti-viral agent specifically targeting miRNA, called miravirsen (formerly SPC3649). It is a b-D-oxy-locked nucleic acid-modified phosphorothioate antisense oligonucleotide, targeting liver-specific miRNA-122. It is one of the miRNA antagonists in clinical trials on patients suffering from hepatitis C. Another miRNA antagonist, RG-101, is also under clinical trials for the same process. mi-RNA mimic targeting miRNA-34 is the first to reach the clinical trials stage for hepatocellular carcinoma treatment. miRNAs are also potential targets for malignant plasma cells in multiple myeloma and the modification of insulin release in diabetes mellitus patients.

6.3.2.6 Circular RNA (CircRNA)

RNAs are linear structures with open 3' and 5' ends. Circular RNAs are molecules with a covalent bond between 5' and 3' ends, making them closed and conferring resistance against degradation by exonucleases, thus making them more stable. The quantity of CircRNAs is ten times higher than linear RNA in cytoplasm. Another significant feature of circRNA is its non-coding nature. CircRNAs are non-coding and detected through scrambled exon-enrichment in samples treated with the RNAse, using either TRAP electrophoresis or 2D gel electrophoresis. cricRNAs can bind to RNA-binding proteins and ribonucleoprotein complexes, thus acting as sponges. Because of their binding ability, cricRNAs are also considered as competing endogenous RNAs.

To date, no therapeutic molecules related to the circRNA have been synthesized or produced or approved, but their role in regulating some vital human diseases is implied. Studies are in progress for exploring the potential role of cricRNAs in therapeutics. cANRIL, a circRNA of a non-coding RNA, a risk factor for atherosclerosis. miRNA of sponges Cirs-7, capable of adsorbing and quenching the normal miRNA, is implicated in Parkinson's disease, Alzheimer's disease, and cancer. circRNAs are also considered as potential biomarkers for aging and gastric cancer.

6.4 THERAPEUTIC POLYNUCLEOTIDES

The utilization of nucleic acids comprising either DNA or RNA to treat genetically inherited disorders and diseases such as cancer makes them a potential avenue for their mitigation, and if possible, their elimination. Most nucleic acid therapies are based on the correction of genetic mutations through therapeutic transgene vectors for complementing the mutated or defective genes with wild ones, regulated by heterologous and non-native promoters and polyAsites. Several practical methods accomplish this approach. Vectors delivering the transgene are usually incorporated randomly into genomes, leading to an un-predictable transgene expression, gene silencing, and transforming normal cells into cancerous ones. Even though DNA based therapies were successful in treating monogenic disorders, the prevalence of side effects reduces the therapeutic potential of the treatment. One of the main limitations of classical gene targeting is the inefficiency of the process, with only 106 - 107 transfected cells facilitating homologous recombination between the gene targeting vector and the targeting chromosomes, might be due to the existence of nonnative DNA, drug selection genes and DNA sequences that significantly decrease the stability of the pairing process. Further, drug selection genes present in classical targeting vectors require specific cellular manipulations for their elimination. All these limitations have forced the development of alternate strategies. The development of polynucleotide gene targeting strategies involves utilizing oligonucleotides of

25 - 200 bp in length or polynucleotides of >200 bp in length to induce sequence-specific modifications of target sequences in the genomes has shown promise to a certain extent in terms of its efficiency (Luis, 2008). These strategies are derived by the utilization of oligo or polynucleotides for the regulation of homologous recombination, polynucleotide small DNA fragments (SDFs), triplex-forming oligonucleotides (TFOs), and single-stranded oligodeoxynucleotides (ssODNs/SSOs).

Polynucleotide small DNA fragments carry a single or multiple base alteration to find the sequence homolog and cellular enzymatic pathways that facilitate a homologous exchange between an SDF and target sequence. Small fragment homologous replacement modification utilizes SDF polynucleotides of size varying between 200 and 2000 bp in length. Small fragment homologous replacement has been demonstrated to occur in a variety of cells during both in vivo and invitro conditions, targeting many genes involved in genetic inherited disorders such as sickle cell anemia, β-thalassemia, cystic fibrosis, Duchene's muscular dystrophy, severe combined immune deficiency, α-1 antitrypsin deficiency, and spinal muscular dystrophy. All of these are single-base substitutions, except cystic fibrosis, which occurs due to a 3base deletion in the human CFTR gene. 1% of the immortalized cystic fibrosis airway epithelial cells homozygous for the DF508 mutation were genotypically and functionally corrected by transfection with SDF, which has re-inserted the lost three bases. Studies on DSF-transfected mouse embryonic stem cells have also indicated a drastic reduction of ΔF508 mutated transcript along with cAMP-dependent chloride efflux (Sangiuolo et al. 2008).

Triplex forming oligonucleotides (TFOs) are 10 - 40 bases in length, bind to specific regions in duplex DNA as a third strand to form a triple helix. These triple helices are formed at poly-purine or poly-pyrimidine regions of the DNA, bound to the TFOs by Hoogsteen hydrogen bonds. Due to their high binding efficiency, TFOs manipulate gene sequence and function, such as the induction of sequence-specific mutagenesis in plasmids, mice, and intrachromosomal recombination. TFOs are successfully applied for gene targeting by the tethering of SDF to TFO directed to the proximity of the targeted sequence. Successful TFO mediated correction of adenosine-deaminase-deficient lymphocytes and p53 glioblastoma cells of humans has been demonstrated. Single stranded oligodeoxynucleotides (SSOs) are <200 bases in length, comprising of a single mismatch to the target sequence, mainly at the center of the sequence. Their efficiency was evaluated by plasmids carrying reporter genes. Evaluation of the cellular responses has indicated a more significant number of genes involved in DNA repair, cell cycle arrest, and apoptosis which were upregulated after transfection with SSOs. The frequency of gene correction is five-fold higher when an SSO is co-transfected with a plasmid rather than a lone SSO transfection.

6.4.1 DNA Vaccines

DNA vaccination is the utilization of genetically engineered DNA for producing an immunologic response. DNA vaccination was begun in 1990. A plasmid DNA, upon direct transfection to animal skin and muscle cells under *in vivo* conditions, generates an immune response to viral and non-viral antigens. This method, later termed DNA immunization, has generated much curiosity, leading to a plethora of pre-clinical and clinical trials involving viral and non-viral antigens to induce protective humoral and cellular immune responses in a wide variety of animal models.

DNA vaccination was achieved by selecting DNA plasmids without any eukaryotic origin of replication to ensure that this will not be subjected to replication and its integration into the mammalian genome. Further, plasmids are encoded with antigens. This antigen encoding plasmid DNA upon transfection into the recipient cells through gene gun, electroporation, and injection can induce an immune response against bacteria and viruses, and a powerful mammalian promoter can regulate its expression. Transfected cells express the antigen in the host immune system, without the involvement of any unsafe agents. Mild electroporation into tissue will generate holes in the membrane, eventually increasing the DNA uptake into the cells for its expression. Encapsulating DNA onto the microparticles or utilizing adjuvants such as aluminum has considerably improved the DNA vaccine's potency (Soltani et al. 2018). CpG sequences within the DNA vaccines are the most potent adjuvant, causing the immunostimulation of second-generation DNA vaccine vectors. DNA vaccines can induce immunity even in newborn individuals with high levels of maternal antibodies that bind and neutralize the conventional vaccines. DNA vaccines are also considered to be safe, cheap, and are more stable.

DNA vaccines for malaria, influenza, hepatitis B, HIV, and a few types of cancer are under clinical trial and are shown to be safe and more effective when transfected with gold particles. Despite the enormous expectations and initial success, the efficacy of the DNA vaccines has been low, and it is not sure that the DNA vaccines will be able to perform up to expectations. Second generation DNA vaccines are evaluated for increasing their potency by retaining their safety and immunogenicity. Some of their efficiency depends on techniques that increase their transfection rate or target the DNA to specific cells for increasing their systemic administration, as evidenced by the development of propulsion devices for targeting mucosal cells for the development of HIV vaccine or Langerhans cells.

6.5 THERAPEUTIC GENES AND THEIR ROLE IN CANCER THERAPY

As an outcome of tremendous growth and advancement in molecular biology and genomics, new tools have been designed to treat cancer, known as cancer gene therapy. Information and access to the genes and the establishment of technology that enables the gene transfer and the capability to introduce the therapeutic genes into human cells in vivo and in vitro conditions have made gene therapy one of the possible solutions which can be adapted for curing cancer. Replacement of tumor suppressor genes for correcting a transformed phenotype requires the transduction of every tumour cell in cancer therapy, which the available gene transfer mechanisms cannot achieve. Therapeutic approaches were developed to enhance the transduction and transfection of genes into tumour cells, targeting only a limited population of tumour cells. To increase its reach, gene therapy focused on the delivery of genes encoding the proteins capable of activating the prodrugs for inducing the cytotoxicity of both transduced and neighbor tumor cells. Delivery of the genes encoding the proteins involved in the disruption of cell cycle progression, induction of apoptosis, and genes involved in the modulation of immune response for inducing the anti-tumour immunity represents other significant areas of study. Gene therapy strategies employ the delivery of genes that convert prodrug enzymes into tumour cells, followed by the systemic administration of the non-toxic prodrug. The approach of expressing the drug susceptible genes in tumour cells is known as suicide gene therapy. Cells transduced for prodrug enzyme expression will confer the prodrug conversion into active metabolites, thereby generating cytotoxicity. In contrast, virus directed enzyme or prodrug therapy will improve the percentage of therapy by administering the prodrug in high systemic concentrations for its conversion into cytotoxic metabolite only within the microenvironment of the transduced tumour (Zarogoulidis et al. 2013). Transduction of tumour cells for the expression of non-mammalian or the over-expression of mammalian enzymes has been employed for a selective increase in the conversion of prodrugs into the respective toxic metabolites. A variety of strategies were implemented for the transduction of genes encoding non-mammalian enzymes. Prodrugs have been converted into active antimetabolites to target active cells replicating in the cell cycle. A few therapeutic genes are discussed in this section.

6.5.1 HSV-Thymidine Kinase

Herpes simplex virus-thymidine kinase (HSV-tk) is involved in the specific phosphorylation of antiherpetic nucleoside analogs such as ganciclovir, acyclovir, and bromovinyl-deoxyuridine to monophosphates. In return, cellular kinases will phosphorylate the monophosphorylated nucleotides to triphosphates and be incorporated into the replicating DNA by DNA polymerase-α. Incorporation of these nucleotide analogs into DNA is associated with chain termination and the induction of DNA single-strand breaks. Compared to cells transduced to express HSV- thymidine kinase, mammalian cells are relatively resistant to the toxic effects of these agents in vitro and in vivo. Both retroviral and adenoviral vectors have been tested for the direct transduction of tumor cells with HSV-thymidine kinase (Tenser, 1991). Systemic administration of ganciclovir in the setting of HSV-thymidine kinase-transduced tumor models is associated with substantial regressions. Complete eradication of a few tumors has been achieved even with the transduction of only 10 - 70% of a tumour cell population. These observations contributed to the identification of the "bystander effect." Multiple mechanisms are involved in this effect, such as transferring phosphorylated ganciclovir nucleotides from HSV- thymidine kinase transduced cells to surrounding cells via intracellular gap junctions. In addition, interleukin-6 and interleukin-1 secreted by HSV-thymidine kinase transduced tumor cells treated with ganciclovir may function as paracrine factors in the induction of non-transduced cell death.

6.5.2 Cytosine Deaminase

Cytosine deaminase is expressed in bacteria and fungi but not in mammalian cells. Cytosine deaminase is involved in cytosine deamination to uracil, converting the nontoxic prodrug 5-fluorocytosine to 5-fluorouracil. Metabolism of 5-fluorouracil to the deoxynucleotide monophosphate (FdUMP) inhibits thymidylate synthase activity and blocks the methylation of uridylate to thymidylate. Conversion of 5-FU to fluorouridine triphosphate (FUTP) and incorporation into RNA will also disrupt the function of rRNA and mRNA. 5-FU is an effective, highly toxic anticancer agent. Transduction of tumor cells to express cytosine deaminase and then administration of 5-fluorouracil can obviate normal tissue toxicity by increasing concentrations of 5-fluorouracil selectively in the tumor (Haberkorn et al. 1996). An advantage of the cytosine deaminase/5- fluorouracil combination is that it can generate a potentially substantial bystander effect to compensate for the low transduction efficiencies of the available delivery systems. 5-fluorouracil is capable of readily diffusing in and out of the cells through a non-facilitated diffusion mechanism. Cells transduced to express cytosine deaminase are exposed to a concentration gradient of 5-fluorouracil, which decreases with increasing distance. With as few as 2% of the tumor cells transduced to express cytosine deaminase, significant tumor regressions are induced by the administration of 5- fluorouracil. Also, CD/5-FC therapy was influential in tumor model cells that are not sensitive to 5-fluorouracil. The only disadvantage that prevents the cytosine deaminase/5-fluorouracil combination is the diffusion of 5-fluorouracil into normal surrounding tissues, which is predominantly toxic to dividing when compared with quiescent cells. The cytosine deaminase/5-fluorouracil combination could be more effective against tumors, such as gliomas, hepatomas, and sarcomas, that reside in mitotically quiescent tissues.

6.5.3 Cytochrome P450 2B1

Cyclophosphamide and ifosfamide are relatively nontoxic to cells without their conversion by cytochrome P450 2B1 into active metabolites. Hepatic cytochrome P450 2B1 activates cyclophosphamide to form 4-hydroxycyclophosphamide, leading to phosphoramide mustard, a DNA alkylator, and acrolein, a protein alkylator. These metabolites circulate systemically in both tumor and normal tissues, resulting in toxicity to normal cell populations, such as hematopoietic cells and the gastrointestinal mucosa, with a high proliferative index. Cytochrome P450 2B1 is expressed by hepatocytes and generally not by tumor cells. Thus, transduction of tumor cells to express cytochrome P450 2B1 results in the direct intratumoral metabolism of cyclophosphamide or iphosphamide. This approach has been used to sensitize gliomas expressing cytochrome P450 2B1 to cyclophosphamide. In addition, this approach is associated with a bystander effect induced by diffusion of cyclophosphamide metabolites and is not dependent on cell-to-cell contact.

6.5.4 Suicide Genes

The application of suicide genes for the sensitization of cancer cells through genetic manipulation is achieved by expressing drug activating enzymes for gene-directed enzyme prodrug therapy (GDEPT). Usually, GDEPT is not expressed in cancer cells. Hence, the administration of proper prodrugs should lead to the killing of these cells. However, numerous genes involved in the coding of prodrug converting enzymes from other biological sources such as bacteria, and viruses, serve as promising candidates for the efficient suicide gene therapy of cancer. The *E. coli gpt* gene encodes for xanthine-guanine phosphoribosyl transferase. The conversion of xanthine analog 6-thioxanthine to 6-thioxanthine monophosphate is facilitated by the addition of a ribose phosphate group. Further, 6-thioxanthine monophosphate within the cell is demethylated into 6-thioguanine monophosphate, a potent nucleic acid synthesis inhibitor. The *E. coli DeoD* gene encodes a purine nucleoside phosphorylase for hydrolyzing the nontoxic 6-methylpurine-2′-deoxyribonucleoside into a toxic purine analog 6-methylpurine, in contrast to mammalian purine nucleoside phosphorylase. *E. coli* nitroreductase activates the prodrug CB1954, a weak monofunctional alkylating agent. CB1954 is bioreduced by nitroreductase to a 4-hydroxylamino derivative, converted by thioesters such as coenzyme A into a potent bifunctional alkylating agent. As bioreduction of CB1954 is more efficient with nitroreductase than with cellular DT diaphorase, transduction of tumor cells to express nitroreductase confers sensitivity to the active CB1954 metabolite (Duzgunes, 2019).

6.5.5 Gene Therapy Directed Apoptosis and Cell Control

The discovery of cellular proteins that regulate cell-cycle progression and the induction of apoptosis has provided opportunities to exploit their expression in cancer gene therapy strategies. The following are three such approaches explored by the transduction of tumor cells with cellular genes. When constitutively expressed, it disrupts cell proliferation and induces apoptosis.

6.5.5.1 P53

Induction of the p53 tumor suppressor in the cells is associated with the cellular response to genotoxic agents, oxidative stress, hypoxia, and oncogene expression. As a response to DNA damage, p53 activates a variety of genes involved in cell-cycle arrest (p21), DNA repair (PCNA, GADD45), and the induction of apoptosis (Bax, IGF-bp3). p53 induces apoptosis through a mechanism, which is independent of transcriptional activation. Induction of p53 leads to cell-cycle arrest, DNA repair, or even apoptosis. Induction by p53 is specific to the cell type and depends on the cell's apoptotic threshold. Notably, the induction of tumor-cell apoptosis by cancer chemotherapeutic agents is dependent on normal p53 function. The rate of p53 mutation in human tumors is ~50%, which provides support for the development of gene therapy strategies to confer the expression of normal p53 protein.

Several studies have demonstrated the transduction of p53 into tumor cells of in vitro culture and inducing cell-cycle arrest or apoptosis followed by the inhibition of growth in the tumor xenografts of mice. With the possible limitations in the available gene-delivery systems, a combination of p53 gene therapy and conventional anticancer treatment may achieve a complete tumor cell mortality. The combination of p53 gene therapy with either cisplatin or radiation increases the apoptotic response of tumor cells and the control of tumor xenografts. These findings upon clinical trials have demonstrated the feasibility of delivering p53 by replication-incompetent retroviral and adenoviral vectors into tumors of patients with non-small-cell lung cancer (NSCLC). During the phase I trials, adenovirus-mediated p53 gene transfer was accomplished by computed tomography-guided percutaneous fine-needle injection or by bronchoscopy in 28 NSCLC patients. Evaluation of post treatment tumor biopsy specimens has demonstrated the adenoviral vector in 86% of the patients by PCR and vector-specific p53 expression in 46% of patients by reverse transcriptase (RT)-PCR. These studies demonstrate the feasibility of adenoviral-mediated p53 gene therapy.

6.5.5.2 E2F-1

A member of the E2F transcription factors family regulates the expression of gene products that are concerned with the S phase progression of the cell cycle. Even though E2F-1 shows some oncogenic nature, recent reports support its role as a tumor suppressor. Overexpression of E2F-1 results in apoptosis. The development of spontaneous tumors was observed in E2F-1-deficient mice. Based on these observations, an adenoviral vector has been developed for delivering E2F-1 into glioma cells. These studies have demonstrated that the transduction of E2F-1 into glioma cells has induced apoptosis and inhibited xenograft growth in nude mice. Along with p53, E2F-1 gene therapy has provided an approach to inhibit tumor cell growth and induce apoptosis.

6.5.6 Cytokine Genes

Identification and cloning of the genes responsible for the synthesis of cytokines have enabled the characterization of their pleiotropic effects on the immune system and inflammatory cells. Systemic administration of IL-2 is associated with tumor regressions in subsets of patients suffering from renal cell carcinoma and melanoma. Systemic delivery of cytokines, such as IL-2, which nonspecifically activate the immune system, has been limited by substantial toxicity. By delivering gene encoding for prodrug-converting enzymes, cytokine-based gene therapy can restrict the cytokine expression to the tumor microenvironment. Cytokine genes including IL-1, IL-2, IL-4, IL-6, IL-12, granulocyte-macrophage colony-stimulating factor (GM-CSF), tumor necrosis factor (TNF), IFN-α, and IFN-γ are involved in the induction of antitumor immunity.

Direct in vivo delivery of cytokine genes into tumour cells is an alternative approach under consideration. An IFN-γ encoding plasmid was directly injected into the tumor cells either as a naked DNA or in a cationic liposome and resulted in the production of IFN-γ for 7 days. IFN-γ activates T cells, natural killer (NK) cells, macrophages and induces major histocompatibility complex (MHC) class I and II expression. The gene gun mediated intra-tumoral delivery system has been applied successfully for plasmids expressing IFN-γ, IL-6, TNF, and IL-2. The intramuscular injection of plasmid DNA encoding IFN-α is associated with reductions in tumor growth and the development of metastases by a CD8+ T-cell-dependent mechanism (Santosh et al. 2018). IFN-α is involved in activating the immune system, decreasing tumor cell proliferation, angiogenesis, induction of T-helper pathways, and upregulation of MHC class I expression.

6.5.7 Costimulatory Genes

Two different mechanisms achieve the effective activation of T cells, firstly, through major histocompatibility complex (MHC) molecules, involving antigen-specific interactions with the T-cell receptor (TCR), and secondly, through co-stimulation provided by the interaction of B7 molecules with CD28 or CTLA4 on the T cell surface. Interaction of antigen-TCR will lead to the selection of antigen-specific cytolytic T cells (CTL). However, co-stimulation is essential for an appropriate signaling and clonal expansion of the CTL population. Even though most of the cells express MHC class I molecules for antigen presentation, costimulatory molecules are predominantly reported on professional antigen-presenting cells (APCs), such as dendritic cells, Langerhans cells, B cells, monocytes, and macrophages. Lack of costimulatory molecules on tumor cells results in MHC class I presentation of tumor antigens leading to T cell anergy. Thus, delivering the costimulatory genes into the tumor cells is considered a mechanism for generating antitumor CTL.

A recombinant vaccinia virus is used for transducing weak immunogenic syngeneic murine tumor cells with genes encoding the B7 molecules B7-1 or B7-2 under in vitro conditions. Tumor growth is found to be inhibited following the implantation of transduced tumor cells into immunocompetent mice, indicating that the costimulation of B7 molecules is sufficient to induce antitumor CTL activity. In contrast, mice models whose immune systems are suppressed by irradiation have accepted the B7-transduced tumor cells. Consequently, these mice have also rejected the subsequent challenge with parental non-transduced tumor cells. It was demonstrated that B7 co-stimulation is needed to induce tumor-specific CTL in naïve mice but not required for tumor rejection (Mcnab et al.2007). Transduction of B7 genes into one tumor site may confer rejection of B7-negative tumors at other sites. Transduction of multiple myeloma cells with a recombinant adeno associated virus will induce the expression of B7-1 or B7-2 along with antitumor-specific CTL activity. Activation of T cells involves multiple factors. In this regard, delivering a single immunomodulatory gene might not be sufficient to activate effective antitumor immunity. Cytokines are involved in the expansion of T cell clones. Gene therapy strategies for delivering both a cytokine and costimulatory gene could prove synergistic in inducing antitumor immunity as evidenced by the adenoviral-mediated transduction of IL-2 and B7-1 genes, resulting in a more significant antitumor effect in a breast cancer model.

6.5.8 Tumor-Associated Antigen Genes

Gene therapy using recombinant vectors expressing a tumour-associated antigen was developed as a vaccine for the induction of specific immunotherapy. Genes that express these tumour-associated antigens are classified into three categories:

1. Endogenously expressed non-mutated genes, which are frequently overexpressed in tumors. Non-mutated genes associated with tumors include melanoma/melanocyte differentiation antigens MART-1/MelanA, gp100, tyrosinase, TRP-1, and TRP-2, testicular cancer/testes antigens MAGE, BAGE, GAGE, and NY-ESO-1, carcinoembryonic antigen (CEA), PSA, and DF3/MUC1.
2. Endogenously expressed mutated genes include p53, cdk4, caspase 8, and β-catenin, whose expression leads to an altered protein.
3. Exogenously expressed genes which are antigens that are derived from exogenous sources such as viral transformation, as exemplified in human papillomavirus-positive cervical cancer.

Gene therapy-based tumor vaccination induces immunity against specific tumour- associated antigens, distinguished from nonspecific immune stimulation. These antigens are associated with the transfer of cytokine or co-stimulatory genes. One method of vaccination involves the direct delivery of tumor-associated antigens, by transferring the encoding gene directly into the patient through viral or nonviral systems. In vitro transfection or transduction of cells, generally consisting of tumour-associated antigen genes followed by the re-introduction of these cells into the patient, is another approach. Direct delivery of the tumour-associated antigen gene subcutaneously or intradermally is applied more widely than the ex-vivo approach.

Genes encoding tumour-associated antigens are vaccinated by transfection and viral transduction. During clinical trials, recombinant vaccinia virus expressing carcinoembryonic antigen was successfully administered intradermally. Peripheral blood lymphocytes isolated from vaccinated patients responded to stimulation with a carcinoembryonic antigen in vitro. Genes encoding the melanoma/melanocyte differentiation antigens MART-1 and gp100 have been used to vaccinate patients with metastatic melanoma. A complete long-term response was achieved in a patient vaccinated with an adenovirus expressing melanocyte differentiation antigens, MART-1. Multiple genes that encode for tumor-associated antigens can be constructed in a single plasmid, or the genes can be co-administered using separate plasmids. However, the potential disadvantage of immunization against a single tumor-associated antigen is the possibility of downregulated antigen expression in tumors. Immunization against multiple tumor-associated antigens could potentially decrease the development of immunologic resistance by the tumor cells. Multiple tumor-associated antigens for a specific tumor must be known to overcome the resistance. As of now, only a few well-characterized tumor-associated antigens for most tumors are known (Hubbard et al. 2016). Another powerful approach developed to deliver the tumor-associated antigen genes is the co-delivery of the B7-1 gene with the tumor-associated antigen gene to maximize the antitumor immune response. Recombinant vaccinia viruses encoding for either B7-1 or carcinoembryonic antigen have been administered as a mixture at a ratio of 1:3 to stimulate the anti-CEA response. Similar results have been obtained by vaccination using mixtures of vaccinia viruses expressing B7-1 and MUC1. Further, advanced vaccinia and fowlpox vector systems have been developed to express B7-1, intercellular adhesion molecule (ICAM), and lymphocyte function-associated antigen (LFA)-3 in a single virus for co-administration with vectors expressing tumor-associated antigens.

6.6 GENE KNOCKIN AND GENE KNOCKOUT

The ability to engineer the mouse genome has proved to have much research, medicine, and biotechnology applications. For genome engineering, transgenic mice have played a major role as models for genetic disorders, for the understanding of embryonic development, and for evaluating therapeutics. Transgenic mice and their cell lines have further accelerated research by allowing scientists to attribute gene functions, explore genetic pathways, and manipulate proteins' cellular, physiological, and biochemical properties. The latter half of the 1980s has witnessed a predominant development in the methodology, which has enabled the generation of transgenic mouse models with gene-specific null mutations and modifications at a specific chromosomal position of a given gene, which was an apparent need right from the inception of mouse model studies for human disease. Several clinical studies of patients have revealed that the genetic basis of their disease is the outcome of either loss of gene function such as cystic fibrosis, CFTR, or genetic changes leading to the modification of expression levels such as leukemogenic diseases, or the appearance of unique activities not occurring in healthy individuals such as Marfan syndrome (Mathew, 2016). The development of mouse models for these diseases required novel strategies to generate animal lines with identical genetic changes. The gene-specific character of these proposed changes requires a shift in the experimental approach facilitating the ability to modify them. Identification of and the subsequent ability to manipulate pluripotent stem cells from mouse embryos have provided an opportunity. Landmark studies by a series of independent investigators, including Evans and Kaufman, Oliver Smithies, and Mario Capecchi, awarded the 2007 Nobel Prize in Medicine have led to a unique strategy for the generation of mouse models with gene-specific changes for the development of null alleles with Mendelian-inheritable traits. This strategy, summarily known as gene targeting or generation of knockout mice, exploited the accessibility and plasticity of embryonic stem cells and experimental strategies to mediate gene-specific changes in these cells.

6.6.1 GENE TARGETING

Homologous recombination in embryonic stem cells is now a routine method for modifying the mouse genome at a specific locus. This technique was initially developed for site-directed mutagenesis in yeast, which was successfully adapted for mammalian cells, resulting in the possibility of deletion, point mutation, inversion, or translocation in mice. The homologous recombination step is accomplished by generating a piece of DNA identical to the locus of interest except for the alteration and a drug resistance marker. This engineered piece is swapped to replace the original DNA fragment. The DNA construct to be introduced into the genome of the embryonic stem cells should contain the mutation with several kilobases of DNA, homologous to the mouse genome, flanking the mutation. These flanking sections are possible locations of the recombination occurrence. Homologous recombination in embryonic stem cells is an infrequent event with a probability of less than 0.01%. The vector must contain genes conferring drug resistance or sensitivity so that the researchers can enrich their population for cells that have taken up the DNA. Even with this selection, most surviving embryonic stem cells have randomly integrated a new DNA fragment rather than recombining at the exact locus of interest.

6.6.2 Gene Knockin

Gene knockin is a process of target-specific insertion of an exogenous gene at the desired locus in the genome to modify an endogenous gene function. Gene knockin is applied for studying the fate of mutations responsible for genetic disorders in humans. Compared with random integration, which inserts a gene at multiple sites causing gene disruption, gene knockin inserts the desired gene only in the specific locus, making it safe and more reliable. The general procedure for gene knockin is similar to that of conditional gene knockout through homologous recombination. However, in gene knockin, the endogenous sequence is replaced with a mutated gene without any disturbance. The following steps achieve the procedure for the gene knockin:

1. Selection of a DNA construct containing a specific gene such as mutated or reporter oncogene.
2. These genes with flanking sequence known as loxP undergo inverse recombination when exposed to the sites carrying Cre-recombinase enzyme, resulting in the omission of intervening DNA.
3. Introduction of the modified gene into the endogenous locus.
4. Embryonic stem cells carrying the modified gene are inoculated into a cavity of an early mouse embryo and implanted into a surrogate female mouse, where the mouse embryo matures into a chimeric mouse with a germline carrying the transgenic gene.
5. Resulting offspring of the chimeric mouse carry the knockin gene.

This technique allows tissue-specific expression of the mutant gene from the endogenous promoter, avoiding random integration. Complete control of the genetic environment surrounding the overexpression, and likely, the DNA does not incorporate itself into multiple genome locations. Site-specific gene knockin provides a high consistency in the transgene expression among the generations as the expressed gene is present as a single copy. As the targeted transgene is not interfering with the critical locus, the resulting phenotype is undoubtedly due to the exogenous protein expression.

Gene knockin has a wide range of applications in human health, specifically in the modeling of human diseases and skin diseases. Knockin mouse models are used for the study of skin diseases such as epidermolytic hyperkeratosis. Knockin mice are also used to model Huntington's disease with the advantage of regulating the CAG repeat number using a suitable protein expressed under an endogenous Hdh promoter. Deviations in the interaction of therapeutic agents with a specific molecular target restrict the usage of wild-type mice as pre-clinical tools in obtaining efficiency and performing safety testing. For evading such problems, mouse models with comparable characteristics were found in humans for testing a specific drug target. MF-63 and a selective PGES-1 inhibitor were introduced in mPGES-1 knockin mice models for examining the occurrence of inflammation. MF63 has shown a solid hindrance to human mPGES-1 in human cells under in vivo conditions.

Additionally, the substitution of the human taconic gene with the human ptges minigene was made. Treatment of knockin mice with MF63 has prevented the synthesis of PGE2 and exhibited an inflammatory response in the affected mice. In the animal model, gene knockin replaces genes of interest with mutated alleles to mimic human disease. Gene knockin is more compatible with an inducible system, which permits the study of the effects of mutations that closely resemble the causes of human disease. Gene knockin can be applied to study the effect of altered genes and their functions, enabling a functional protein study to be carried out in-depth. Despite having many applications, knock in still suffers from a few disadvantages, including multiple side effects due to the combination of transgenes and the artifacts leading to unspecific protein-protein interaction due to the forced overexpression of genes. Gene knock in has also proved to be expensive, and its mouse models do not entirely resemble the pathology of human diseases. Despite these drawbacks, many researchers rely on knockin mice to study the exogenous expression of a protein. Although the generation of knockin mouse models has avoided many problems of the traditional transgenic mouse, this technique needs much development concerning time for assembling the vector and identifying embryonic stem cells that have undergone homologous recombination.

6.6.2.1 Knockin Editing

For logistical and regulatory reasons, the genome-editing approach is not feasible for the correction of the mutations responsible for diseases among individuals. Gene therapy clinical trials using next-generation lentiviral and self-inactivating retroviral techniques have shown promising safety and efficacy results, however long-term risks remain unknown. Genes such as *CD40L* cause X-linked hyper-immunoglobulin M (X-HIGM) syndrome, requiring regulatory expression. Unregulated expression of the CD40 ligand has rectified the deficiency, but its persistent expression has led to malignant lymphoproliferation after its delivery by the lentivirus vector, making itself an ideal candidate for therapeutic genome editing. A double stranded break will be generated in the DNA at a specific site in the genome using either regular or chimeric restriction endonuclease to perform the genome editing. Breaks on the DNA will trigger the cell repairing machinery, which will repair the DNA probably through two mechanisms, repair by the non-homologous joining of the ends, leading to the creation of small insertions or deletions, inactivating the genetic element, and repair by homologous recombination using a donor molecule with long homology arms, creating defined nucleotide changes in the genome. Hubbard and his colleagues in 2016 combined the utilization of transcription activator like effector nucleases (TALENs) with adeno-associated virus 6 (AAV6) to generate homologous recombination-mediated editing of the CD40L gene in the T cells derived from both wild type donors and patients suffering from CD40L deficiency. AAV6 has been designed to knockin a wildtype CD40L cDNA so that the cDNA can be expressed using endogenous regulatory elements and correct downstream disease-causing mutations. A vital aspect of the knockin vector is utilizing the redundant codon message system to create the cDNA, which codes for the same protein, but is diverged at the nucleotide level. This unique feature prevents the homologous recombination machinery from pre-mature recombination before integrating the entire gene (Figure 6.7).

Using the knockin strategy for genome editing, the efficiency of the targeted knockin in both wild type and patient derived T cells has been increased to 50%. It was also possible to engineer cells that can regulate the transgene, such as the unedited *CD40L* gene. Despite the success, several issues are yet to be solved, such as high multiplicities if infection arrives along with high recombination frequencies, with high translational implications. Multiplicities of infection can be minimized by designing a high-quality vector. With slight modifications, this technique has tremendous implications for patients with illnesses due to genetic disorders

6.6.3 Gene Knockout

Gene targeting is a process applied for altering a specific gene to discern its biological function better. An engineered mutation can be re-directed back to its genetic locus through homologous recombination. Thus, a genomic clone can be directly utilized to create a mutation in a selected gene, and homologous recombination can be employed to create a loss of function mutation. One of the most common applications of gene targeting is to produce knockout organisms. Gene knockout is a

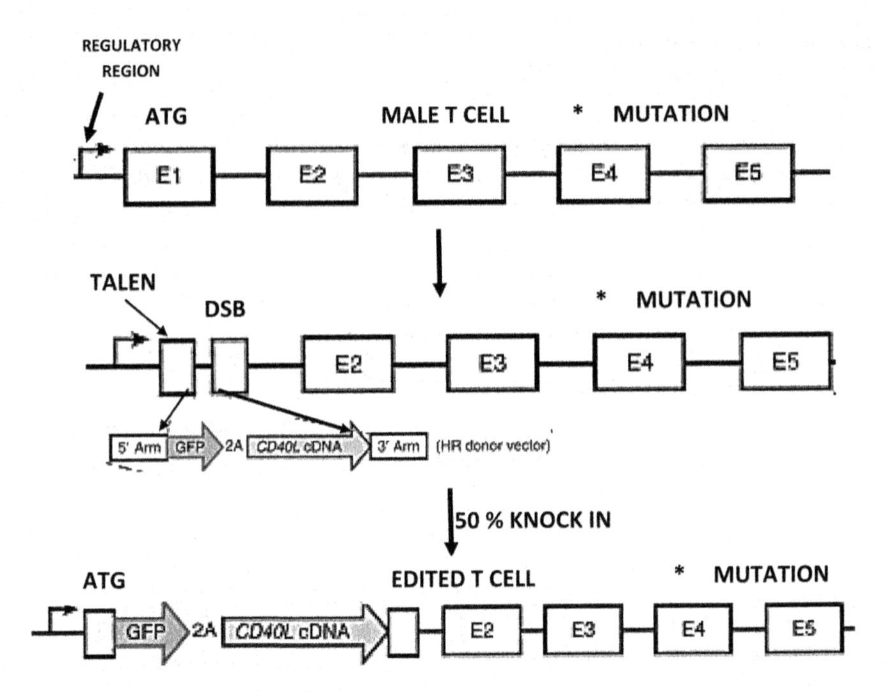

FIGURE 6.7 Strategy for using Genome editing to perform cDNA knockin as a functional gene correction strategy for correcting downstream mutations. The TALEN mRNA is delivered into the T cells. The homologous recombination between the endogenous gene containing a nuclease induced DSB and the donor vector. ATG, endogenous start codon; E1-E5, exons 1-5; TCR, T-cell receptor.

genetic technique by which an organism's genetic makeup is altered by engineering specific genes and making them inoperative. Organisms carrying the inoperative genes are known as knockout organisms or simple knockouts, used for assigning a specific function to an unknown gene whose sequence is available. Knocking out two, three, and four genes are double, triple, and quadruple knockouts. In most cases, a gene's significance and function cannot be wholly determined either by recognizing the amino acid motifs in its protein sequence or by examining its closely related family members. One of the ideal ways to identify a protein's biological function is gene inactivation through gene targeting, which disrupts the targeted ORF and blocks its expression. Gene knockout is the opposite of the gene knockin technique, applied to silence thousands of genes.

The gene knockout mouse is a valuable tool for geneticists to delineate the gene function in its physiological homeostasis. Mice are the ideal human analogs for most biological processes as both genomes share 99% of the similarity. Mice are helpful experimental animals due to their small size, short life span and are highly prolific. Targeted deletion of a gene from the mouse genome provides an essential means for geneticists to determine the biological role of a genetic allele. Besides studying gene function, animal models have also been used as invaluable models for studying genetic disorders. A mutation in the human gene will disable its protein. Its corresponding knockout mouse can be used to study its pathophysiology and in the development of therapies. Some genetic diseases in humans are also involved in the expression of a mutated protein. Point mutations, insertions, deletions (both micro and macro) responsible for most genetic disorders in humans can be mimicked using gene targeting. So far, 11,000 genes in the mouse genome have been successfully silenced using knockout, and a global effort is underway to make a knockout mouse for all its 25,000 genes. Homologous recombination can also be employed to disrupt a gene, which swaps the open reading frame of the original gene with a mutated one. The mutated mouse protein can be targeted for studying the genetic effects of its alteration protein.

6.6.3.1 Conditional Gene Knockout

In some knockout mouse models, the severity of the phenotype precludes the role of a gene in tissue organogenesis. As evidenced in gene knockout mice, 15% of them have mutations that lead to developmental lethalities. The application of Cre/loxP technology for creating conditional knockout mice has solved this problem. Cre/lox technology has been derived from bacteriophage P1. The CRE recombinase will excise any specific region located between two loxP sites. The loxP site is 34 bp in length, which can be genetically targeted around an exon in a gene. Mice generated through this process contain a floxed allele, flanked by loxP sites in all their tissue but which is phenotypically wild. These floxed mice are bred with Cre expressing transgenic animals, in which the promoter used to drive Cre expression will determine the site of gene deletion. For the spatial inactivation of a gene in the mouse genome, a cell type-specific promoter limits Cre expression in a particular tissue. Even though less preferred, flippase (Flp) recombinase isolated from *Saccharomyces cerevisiae* provides a similar means of locus re-arrangement. Like Cre, flippase will also excise the DNA flanking between 34 bp length sequence, known as FRT sites. By using either Cre/loxP or the Flp/FRT, gene expression can be disrupted in both spatial and temporal manner, and the lethality of the knockout mouse phenotype can also be defeated.

6.6.3.2 Other Applications of Gene Knockout

Gene knockout technology has exciting applications in biomedical research. Mouse models are used to study the progress of genetic disorders at the molecular level for a better understanding of a specific gene contribution to disease and design the drugs that can control its protein. One example is the knockout strategy to control sickle cell anemia by switching off a single gene to keep the blood young forever. Targeting the BCL11A gene, whose expression will switch the body's hemoglobin from fetal to adult. Stuart Orkin and his colleagues of Harvard Medical School, Boston, successfully knocked out the BCL11A gene from mice. As a result, the knocked-out mice have produced fatal hemoglobin 20 times more than expected in their blood without any sickle-shaped cells while keeping their vital organs healthy. Gene knockout has wide applications in drug development to discover next-generation therapies for curing diseases based on the human genome.

6.6.3.3 Limitations

Despite the promising applications that gene knockout technology brings for biomedical research, drug discovery, and development, it suffers from severe limitations. One of the significant drawbacks is the death of many gene knockout mice due to developmental defects. Even if the mouse survives, they will have a different phenotypic and physiological trait than their human counterparts, as identified in p53 knockout mice, which develop an entirely different range of tumors than humans do. p53 knockout mice also developed lymphomas and sarcomas, whereas humans developed cell-derived cancers. Because of these differences, the role of the same gene responsible for a specific function in humans and mice cannot be predicted. This aspect adversely limits the utilization of mouse models for human disorders. Still, their embryos can be used as models for experimentation. Another major disadvantage is its cost. Production of custom knockout mice is expensive with cost ranging from $3,000 to $30,000. Still, gene knockout mice are commercially available and can be purchased from organizations such as The Jackson Laboratory (Jax) in Bar Harbor, Maine.

6.7 METHODS OF GENE DELIVERY

Gene therapy is applied for the correction of defective genes responsible for disorders in human beings. Over the past 20 years, this technology has undergone remarkable progress, particularly in gene transfer and expression technology. Currently, two different approaches are being implemented for the transfer of genetic material into human chromosomes. Firstly, ex vivo technique involves the surgical removal of the cells from the patients, injection of the target DNA into the cells to correct the disease, and their division in cultures. New tissues formed are transferred into the targeted location of the patient. Secondly, in vivo technique is the direct injection of therapeutic DNA into human cells. One common feature in these two mechanisms is administering the exogenous DNA into the target cell across the membrane, which should be integrated into the host genome and performs its function and undergoes replication during cell division for its long-lasting expression. Currently, three methods of gene delivery are available: physical methods; viral vector-based methods, and non-viral vector-based methods.

6.7.1 Physical Methods

Gene delivery using physical methods is one of the most targeted gene transfer systems adapted mainly due to its simplicity. This methodology uses either mechanical or physical force to overcome the membrane barrier and facilitates intracellular gene transfer. Physical methods involve both electrical and mechanical-based gene transfer systems to directly deliver the target DNA fragment or a plasmid containing the transgene into cells without any involvement of the cytotoxic or immunogenic substances commonly seen in viral or nonviral vectors.

6.7.1.1 Microinjection

Microinjection refers to the process of using a glass micropipette for injecting a liquid or a jelly substance at a microscopic or macroscopic level. Here, the target is always a living cell with some intercellular space. Microinjection is a mechanical process that uses an inverted microscope with a magnification power of 200x. Alternatively, a dissecting stereo microscope with a magnification of 40 - 50x or a traditional compound upright microscope with a magnification of 200X is also used (Park et al. 2001). For cellular or injection, the target cell is positioned under the microscope using two micromanipulators, one holding the pipette and the other holding a microcapillary needle of diameter ranging from 0.5 to 5 μm. The diameter of the needle will be more prominent while injecting stem cells into an embryo. These micromanipulators penetrate the cell membrane and or the nuclear envelope for introducing either the naked DNA or vector consisting of DNA into the target cell. Microinjection was applied for the first time by Wolf and his colleagues in 1990 for the successful gene transfer to muscle cells in mice, which were later applied to the skin, cardiac muscle, liver, and solid tumour. The main application of microinjection is DNA vaccination. However, this technique suffers from low efficiency in gene delivery and the limitation of its transfection to the needle track

6.7.1.2 Particle Bombardment

Also known as ballistic injection, microprojectile gene transfer or gene gun is another mechanical method of cell transformation that uses high density, sub-cellular sized particles, accelerated to a high velocity for carrying either DNA or RNA into the living cells. This technique was initially developed for gene transfer into plant cells in 1987. Due to its level of success, it has been applied for gene transfer into mammalian and human cells both under in vivo and ex vivo conditions. Particle bombardment is a mechanical method readily applicable to various biological systems and effectively overcomes the physical barriers to gene transfer, such as the cell wall of plants. It is a versatile technique that can be used for transient expression studies and for creating stable transformants. In this method, DNA-coated gold particles are propelled against the cells using a high voltage electronic discharge or spark discharge for their forced intracellular DNA transfer. The particle bombardment method of gene transfer was successfully applied to transfer genes into liver, skin, and muscle cells and is highly appropriate for gene delivery into skin cells for vaccination and immune therapy. This method can also deliver DNA in precise doses and was utilized during the vaccination against influenza virus and in gene therapy for ovarian cancer treatment. Genes delivered by this method are transiently expressed, with considerable cell damage occurring at the center of the discharge site.

6.7.1.3 Electroporation

Electroporation, also known as electro-permeabilization, is a significant increase in the electrical conductivity and permeability of the cell plasma membrane caused by an externally applied electric field. It is used in molecular biology to introduce a genetic material of other organisms, such as loading it with a molecular probe, a drug that can change the cell's function, or a piece of coding DNA. Neumann and his colleagues in 1982 have used electroporation for the first time for transferring the exogenous DNA into the mammalian cells by generating pores on the cell membrane through electric pulses. Electroporation can be best applied on cells that are suspended in solution when compared with solid tissue. This procedure is also highly efficient for introducing foreign genes into mammalian cells for the production of knockout mice during tumor treatment, gene therapy, and cell-based therapy. Electroporation has been clinically tested for DNA based vaccination and cancer treatment. The only limitation of this process is the low transfection rate.

6.7.1.4 Sonoporation

Kim and his colleagues developed this gene delivery method in 1996, which applies ultrasound waves for the temporary permeabilization of cellular DNA by the cell membrane. Sonoporation combines the gene delivery with the possibility of restricting the permeability to the area where the ultrasound is applied. Gene delivery efficiency is regulated by the intensity of the pulse, frequency of waves, and duration of the exposure. Sonoporation is successful in gene transfer to the cornea cells, brain, central nervous system cells, bone, peritoneal cavity, kidney, pancreas, liver, dental pulp, muscle, and heart under in vitro conditions. However, this technique is still in its infancy and is improvised for in vivo conditions.

6.7.1.5 Magnetofection

Magnetofection is a physical method that utilizes a magnetic field for performing transfection. Magnetic iron oxide nanoparticles coated with cationic lipids or polymers are employed to form a complex with DNA through electrostatic interaction. These magnetic particles are concentrated on the targeted cells by applying a magnetic field, forcing the cellular uptake of the DNA. Targeted DNA is released into the cytoplasm depending on the composition of the magnetic nanoparticles. Magnetofection has been successfully applied on a wide range of hard-to-transfect cells by other non-viral methods. This method has been well accepted for invitro gene delivery and may also be applicable for in vivo delivery.

6.7.1.6 Hydroporation

This hydrodynamic-based gene delivery system has been reported independently in two groups, led by Liu and Zhang in 1999. This method is one of the most efficient, applied for intrahepatocyte gene delivery in rodents. Hydroporation is performed by rapid injection of a large volume of DNA solution through the tail vein of rodents. This results in a transient enlargement of its fenestrae, which generates a transient membrane defect on their plasma membrane leading to gene transfer. DNA injection volume equal to 8 - 10% of their body weight and an injection time of 5 - 7 seconds is the optimal condition for gene delivery in mice. A similar approach has increased lobe-specific gene delivery into the liver and skeletal muscle cells of pigs. The hydrodynamic method has also ensured that there is no significant tissue damage while delivering the gene into the liver and muscle cells of small and also large animals. This image-guided and lobe-specific hydrodynamic gene delivery system has great potential to be a part of future human gene therapy.

6.7.1.7 Photoporation

Zeira and his associates in 2003 first reported this laser-assisted gene delivery system. This technique uses a single laser pulse to generate the transient pores on the cell membrane to enter the DNA.

Here, the gene delivery efficiency is controlled by the laser's focal point size and pulse frequency like electroporation. Recently, the gene delivery efficiency has been improved by using carbon black nanoparticles for generating photoacoustic force upon laser stimulation. This technique still needs to be more sophisticated for application under both in vivo and in vitro conditions.

6.7.2 Viral Vector-based Gene Delivery Systems

Gene delivery through viral vectors is accomplished by using both DNA and RNA viruses after deleting the disease-causing genes from their genomes. The purpose of utilizing the RNA-based viral vector system is to provide a long-term transgene expression through gene integration. DNA-based vectors will induce transgene expression in episomal form without integration. Characteristic features of virus-based vectors are displayed in the Table 6.1.

Expression of foreign genes through gene therapy has been achieved in the kidney, heart muscle, eye, and ovary tissues. Gene therapy through viral vectors has been successful in treating cancer, AIDS, and muscular dystrophy. Viruses that are used as vectors in gene therapy are categorized into four types: retroviruses, adenoviruses, adeno-associated viruses, and herpes simplex viruses. Viral-based vectors are used in more than 75% of global gene therapy clinical trials on humans. However, many of them have not been approved to date. One of the highly practiced and approved gene therapy treatments, gendicine delivers the transgene using a recombinant adenoviral vector.

6.7.2.1 Retroviral Vectors

Most first-generation retroviral vectors are derived from oncoretroviruses such as Moloney Murine leukemia virus (MMuLv), which could not transfer the genes into non-dividing cells. This drawback has been overcome by the utilization of the lentivirus family of retroviruses such as human immunodeficiency virus type 1 (HIV-1), bovine immunodeficiency virus (BIV), feline immunodeficiency virus (FIV), and Simian immunodeficiency virus (SIV), which could successfully transfer the genes into non-dividing cells. Retroviruses used in gene therapy cannot replicate inside their host cells and are replication deficient. Even though this characteristic feature is needed to ensure the safety of the vectors during gene therapy, it might impose restrictions on the amount of the virus administered into the human body. Retrovirus mediated gene therapy has been widely used in the treatment of melanoma and ovarian cancer, adenosine deaminase deficiency, severe combined immune deficiency, and Gaucher's disease (a rare genetically inherited disorder that occurs due to the deficiency of the enzyme glucocerebrosidase resulting in the accumulation of lipids in toxic quantities in bone marrow, liver, and spleen). Retroviral vectors are also used to transfect primary human endothelial and smooth muscle cells, a class of rigid cells to transfer.

TABLE 6.1
Characteristic Features of Viral Vectors Used for Gene Therapy

Name	Genome	Carrying Capacity	Activity
Adeno Virus	ds DNA	10 Kb	Dividing and Non-dividing cells
Helper Dependent Adeno Virus	dsDNA	35 Kb	No vector integration High DNA load
Adeno Associated Virus	ss DNA	5 Kb	Sustained gene expression Low immune response
Herpes Simplex Virus	dsDNA	30 Kb	Effective for CNS cells
Onco retrovirus	ss RNA	8 kb	High transduction efficiency
Lenti virus	ss RNA	8 Kb	Sustained gene expression
Foamy virus	ss RNA	9.2 Kb	Forms stable transduction

6.7.2.2 Adenoviral Vectors

Adenoviruses are composed of double stranded DNA in their genomes used in cancer gene therapy to deliver therapeutic DNA to patients suffering from metastatic breast, ovarian, and melanoma cancers. The severe immune response of the host contributes to the limited survival of the adenovirus in the target cells resulting in transient expression of the therapeutic gene as the viral DNA will be lost with time. The first-generation adenoviral vectors carried the therapeutic genes of size 7 Kb into the targeted cells. The generation of gutless adenoviral vectors lacking the viral genes has facilitated the delivery of therapeutic genes up to 30 Kb with decreased toxicity. Adenoviral gene transfer to the COS-7 cells was higher than the one achieved by liposomal gene delivery systems. Oncolytic adenoviral vectors have been developed for replication in tumour cells, amplifying the viral inoculum, leading to the destruction of infected tumour cells. H101 and E1b55K-deleted oncolytic adenoviruses were approved for cancer treatment by the Republic of China due to a response rate of 79% in patients undergoing chemotherapy. Oncolytic adenovirus ONYX-015 of Onyx Pharmaceuticals, Emeryville, CA, USA, is under phase III clinical trials.

6.7.2.3 Adeno-Associated Viral Vectors

Derived from a non-pathogenic parvovirus, adeno-associated viral vector has provided an alternative means of long-term gene expression with reduced risk and adverse reactions. These viruses are not associated with any human infections. The site of the adeno-associated viral DNA integration into the human genome is located on chromosome 19. During the engineering of the vector, most of the virus genome has been deleted to reduce the host's adverse response to the infection. The viral REP and CAP genes were replaced with transgene sequences, and a non-rescuable adeno-associated virus helper plasmid encodes its capsid protein. The size of the therapeutic gene insert is a significant limitation for adeno-associated viruses. The maximum length of the gene that can be incorporated is 5 Kb. Phase I studies are in progress for the gene therapy of hemophilia B using second generation adeno-associated 2 vector, through intramuscular injection in the initial stages and intraportal injection into the liver cells during later stages.

6.7.2.4 Herpes Simplex Viral Vectors

These vectors are the derivatives of a human neurotropic virus with dsDNA of 150 Kb as the genome, which encodes 80 proteins. This virus is transmitted through direct contact and replicates on the skin and mucosal membranes before infecting CNS cells. This virus exhibits high lytic and latent activity, leading to cell death. The high latent activity of this virus makes it stay in its host cell for an extended time. Vectors were generated from herpes simplex virus by deleting immediate early protein genes, making them incapable of replicating inside their host cell. Due to their large genome size, herpes simplex viral vectors can carry a size of 30 Kb. As herpes simplex virus is neurotropic, studies are focused on its utilization in gene therapy for brain tumors. Recent studies have revealed that this vector-mediated gene transfer has blocked pain transmission or reversing chronic pain under in vivo conditions. Many clinical trials are in progress for their utilization in the gene therapy of CNS disorders.

6.7.2.5 Other Viral Vectors

In addition to the four types of viral vectors, many other viruses are considered for their gene therapy utilization. Even though most of them are still in the pre-clinical trial stages, their development has tremendous scope. Several members of alphavirus such as Semliki Forest virus, Sindbis virus, and Venezuelan equine encephalitis are being developed for gene delivery. Despite their toxicity, their prominent neurotropism has made their utilization as vectors for CNS cells valuable. Another possibility in this area is blending the available vectors for to obtain new and more desirable ones. One among them is the combination of adenoviruses and adeno-associated viruses. Inserting a double stranded adeno-associated viral genome into adenoviral capsid produces a combination of adeno-associated and adenoviral ITR. Another approach involves the insertion of adeno-associated viruses into gutless adenovirus and their engineering for transient expression of its Rep protein for site-specific integration and elimination of insertional mutagenesis. Herpes simplex virus/adeno-associated virus and herpes simplex virus/Epstein-Barr Virus are the other hybrid viral vectors considered for gene therapy. Attempts have also been made to coat the viral vectors with lipids or polymers to hide the virus from the host immune system upon in vivo administration, increase the vector circulation duration, and prevent degradation.

6.7.3 Non-Viral Based Gene Delivery Systems

Gene transfer through non-viral delivery systems comprising natural or synthetic compounds is another primary strategy for gene therapy. Here, the DNA, proteins, polymers, or lipids are formed as complexes and can transfer genes with high efficiency compared to viral vectors. The non-viral approach of gene delivery rolls back to the work of Avery, MacLeod, and McCarthy, who demonstrated the changes that occur in the cellular phenotype following exogenous DNA exposure. Calcium phosphate-mediated gene transfection is the first non-viral gene transfer technique, which has gained wide acceptance. Since then, the identification of many non-viral gene transfer techniques has gained momentum and many of them are in clinical trials. Non-viral-based gene delivery systems have an advantage over viral delivery systems mainly due to the lack of immune response and ease of formulation and assembly. Non-viral vectors are classified into three major categories, naked DNA delivery systems: polymeric delivery systems, and liposomal delivery systems.

6.7.3.1 Naked DNA Delivery Systems

The administration of pure DNA in its isolated form into cells through either ex vivo delivery or in vivo delivery comprises naked DNA delivery systems. The ex vivo method of naked DNA delivery has been successfully employed for gene delivery into endothelial and smooth muscle cells. In vivo delivery of naked DNA was first described in 1990, and its efficiency can be further improved upon administration in a pressure-mediated fashion using particle bombardment technology or electroporation. Applying these two techniques has resulted in the localized delivery of naked DNA readily into skin or muscle cells.

6.7.3.2 Polymeric Gene Delivery Systems

Cationic polymers constitute a significant set of non-viral gene delivery systems as they can readily form a complex with the negatively charged anionic DNA molecule. These action mechanisms of these polycomplexes are based on the generation of a positively charged complex due to the electrostatic interaction of these cationic polymers with anionic DNA. Polymers such as polyethyleneimine, poly-L-chitosans, polyamidoamine, polyallylamine, cationic proteins, chitosan, peptides, and dendrimers are successfully applied for in vivo and in vitro gene transfer mechanisms by condensing DNA into small particles and preventing its degradation. Like the cationic polymers, synthetic polymers such as protective interactive noncondensing polymers (PINC), poly-L-lysine, cationic polymers, and dendrimers offer an alternative to cationic lipids as a vehicle for gene delivery into the target cells. Polymer lipids are easy to make and are flexible for any further modifications. Combining the polyethylene glycol for prolonged half-life in blood with a cationic polymer for DNA condensation has resulted in attractive properties of DNA-polymer complex for their delivery into tumor cells. Phase I and phase II clinical trials are in progress to treat cystic fibrosis and ocular degenerative diseases using DNA-polymer complexes (also known as polyplexes). Lipid hybrids combined with polymers have been designed further to improve the gene delivery in the target cells. The DNA will be pre-condensed with poly-L-lysine, protamine, histone, and other synthetic polypeptides and wrapped using cationic liposomes, anionic liposomes and amphiphilic polymers. Encapsulating DNA with a biodegradable polymer will induce a controlled DNA release into the targeted cell over weeks or months. A few polymers such as chitosan have hypercholesterolemia, which makes them highly unfavorable for use.

6.7.3.3 Liposomal Delivery Systems

Liposomes are one of the most effective and versatile non-viral vectors developed for DNA therapeutics. They were initially described in 1965 as a cell membrane model and were applied to deliver nucleic acids into the cells. Liposomes are microscopic particles consisting of concentric lipid bilayers enclosing an aqueous compartment and are formed spontaneously when a film of lipids is hydrated in an aqueous solution. Among all of the liposomes developed to date, liposomes with cationic lipids are the most active during gene delivery. Cationic liposomal formulations such as lipofectamine (Invitrogen, Carlsbad, CA), effectene (Qiagen, Valencia, CA), and tranfectam (Promega, Madison, WI) are commercially available.

Liposomes offer several advantages as gene transfer systems such as:

1. They are economical to produce and will not cause disease.
2. They offer a degree of protection to the DNA from nuclease-based degradation.
3. They are capable of transferring DNA fragments equal to the size of a chromosome.
4. Liposomes can be targeted to specific cells or tissues.
5. Liposomes can overcome problems such as immunogenicity, and contamination that are common in the case of viral vectors. Successful liposomal delivery of DNA and RNA has been reported in various cell types, including tumor epithelial cells, endothelial cells, hepatocytes, and muscle cells through intra tissue or intravenous injection methods.

6.8 CONCLUSIONS

With the increased insight into human genetics, gene therapy has been progressing steadily for decades, supported by the development of vectors (both viral and non-viral) with increased efficiency and the development of new technology such as CRISPR-Cas9 editing. The success rate in treating patients suffering from congenital diseases and monogenic disorders is very high, whereas surgical and pharmacological interventions did not yield good results. However, the knowledge and the wisdom acquired in gene therapy suggests a careful assessment of the situation to overcome the toxicity of the processes and to undertake extensive pre-clinical and clinical studies. The promise of successful gene therapy application in several fields of medicine is bright, as evidenced by a high percentage of clinical trials on this technology.

REFERENCES

Bao, S., Thrall, B. D., Gies, R. A., et al. (1998). In vivo transfection of melanoma cells by lithotripter shock waves. *Cancer Res.* 58:219–221.

Bodles-Brakhop, A. M., Heller, R., Draghia-Akli, R. (2009). Electroporation for the delivery of DNA-based vaccines and immunotherapeutics: current clinical developments. *Mol. Ther.* 17:585–92.

Charlotte, G., Rebecca, H., Antonio, P., Reuben, B. (2018). Cancer immunotherapy with CAR-T cells – behold the future. *Clinical Medicine* 18: 4: 324–328.

Choate, K. A., Khavari, P. A. (1997). Direct cutaneous gene delivery in human genetic skin disease. *Hum Gene Ther.* 8:1659–1665.

Choi, J. G., Dang, Y., Abraham, S., Ma, H., Zhang, J., Guo, H., Cai, Y., Mikkelsen, J. G., Wu, H., Shankar, P., et al. (2016). Lentivirus pre-packed with Cas9 protein for safer gene editing. *Gene Ther.* 23: 627–633.

Colosimo, A., Guida, V., Antonucci, I., Bonfini, T., Stuppia, L., and Dallapiccola, B. (2007). Sequence specific modification of a beta-thalassemia locus by small DNA fragments in human erythroid progenitor cells. *Haematologica* 92: 129–130.

Cong, L., Ran, F. A., Cox, D., Lin, S., Barretto, R., Habib, N., Hsu, P. D., Wu, X., Jiang, W., Marraffini, L. A., et al. (2013). Multiplex genome engineering using CRISPR/Cas systems. *Science* 339: 819–823.

Duzgunes, N. (2019). Origins of suicide gene therapy: methods and protocols. 10.1007/978-1-4939-8922-5_1.

Ewa, K-K. I. D., Judyta, J., Wojciech, M., Joanna, W. (2018) CRISPR/Cas9 Technology as an Emerging Tool for Targeting Amyotrophic Lateral Sclerosis (ALS). *Int. J. Mol. Sci.* 19: 906.

Fu, Y., Foden, J. A., Khayter, C., Maeder, M. L., Reyon, D., Joung, J. K., Sander, J. D. (2013). High-frequency off-target mutagenesis induced by CRISPR-Cas nucleases in human cells. *Nat. Biotechnol.* 31: 822–826.

Giulliana, A. R. G. (2017). Gene therapy: advances, challenges and perspectives. *Einstein.* 15(3):369–375.

Goncz, K. K., Prokopishyn, N. L., Abdolmohammadi, A., Bedayat, B., Maurisse, R., Davis, B. R., And Gruenert, D. C. (2006). Small fragment homologous replacement mediated modification of genomic beta-globin sequences in human hematopoietic stem/progenitor cells. *Oligonucleotides* 16: 213–224.

Guo, H., Leung, J. C. K., Chan, L. Y. Y., et al. (2007). Ultrasound-contrast agent mediated naked gene delivery in the peritoneal cavity of adult rat. *Gene Ther.* 14:1712–1420.

Gupta, R. M., Musunuru, K. (2014). Expanding the genetic editing tool kit: ZFNs, TALENs, and CRISPR-Cas9. *J. Clin. Invest.* 124(10):4154–4161.

Gwiazda, K. S., Grier, A. E., Sahni, J., Burleigh, S. M., Martin, U., Yang, J. G., Popp, N. A., Krutein, M. C., Khan, I. F., Jacoby, K., et al. (2016). High efficiency CRISPR/Cas9-mediated gene editing in primary human T-cells using mutant Adenoviral E4orf6/E1b55k "Helper" proteins. *Mol. Ther.* 24: 1570–1580.

Haberkorn, U., Oberdorfer, F., Gebert, J., Morr, I., Haack, K., Weber, K., Lindauer, M., Kaick, G., Schackert, H. (1996). Monitoring gene therapy with cytosine deaminase: In vitro studies using tritiated 5-Fluorocytosine. Journal of Nuclear Medicine 37: 87–94.

Habib, N. A., Ding, S. F., el-Masry, R., et al. (1996). Preliminary report: the short-term effects of direct p53 DNA injection in primary hepatocellular carcinomas. *Cancer Detect Prev.* 20:103–107.

Hartikka J, Sawdey M, Cornefert Jensen F, et al. (1996). An improved plasmid DNA expression vector for direct injection into skeletal muscle. *Hum Gene Ther.* 7:1205–1217.

Heller, R., Jaroszeski, M. J., Glass, L. F., et al. (1996). Phase I/II trial for the treatment of cutaneous and subcutaneous tumors using electrochemotherapy. *Cancer.* 77:964–971.

Hickman, M. A., Malone, R. W., Lehmann-Bruinsma, K., et al. (1994). Gene expression following direct injection of DNA into liver. *Hum Gene Ther.* 5:1477–83.

Hsu, P. D., Lander, E. S., Zhang, F. (2014). Development and applications of CRISPR-Cas9 for genome engineering. *Cell* 157(6):1262–78.

Hubbard, N., Hagin, D., Sommer, K, et al. (2016). Targeted gene editing restores regulated CD40L function in X-linked hyper-IgM syndrome. *Blood* 127(21):2513–2522.

Hynynen, K., McDannold, N., Martin, H., et al. (2003). The threshold for brain damage in rabbits induced by bursts of ultrasound in the presence of an ultrasound contrast agent (Optison). *Ultrasound Med. Biol.* 29:473–81.

Ishino, Y., Shinagawa, H., Makino, K., Amemura, M., Nakata, A. (1987). Nucleotide sequence of the iap gene, responsible for alkaline phosphatase isozyme conversion in Escherichia coli, and identification of the gene product. *J. Bacteriol.* 169: 5429–5433.

John, C. B., John, J. R. (2012). RNA-based therapeutics- current progress and prospects. *Chem. Biol.* 19(1): 60–71.

Kim, H. J., Greenleaf, J. F., Kinnick, R. R., et al. (1996). Ultrasound-mediated transfection of mammalian cells. *Hum. Gene Ther.* 7:1339–46.

Korashon, L. W., Jennifer, A., Hans-Peter, K. (2011). Hematopoietic Stem Cell Expansion and Gene Therapy. *Cytotherapy* 13(10): 1164–1171.

Li, L., Song, L., Liu, X., Yang, X., Li, X., He, T., Wang, N., Yang, S., Yu, C., Yin, T., et al. (2017). Artificial virus delivers CRISPR-Cas9 system for genome editing of cells in mice. *ACS Nano* 11: 95–111.

Liu, F., Song, Y., Liu, D. (1999). Hydrodynamics-based transfection in animals by systemic administration of plasmid DNA. *Gene Ther.* 6:1258–1266.

Luis, A-S. (2008). Nucleic acids as therapeutic agents. *Current topics in medicinal chemistry* 8: 1379–1404.

Maggio, I., Liu, J., Janssen, J. M., Chen, X., Gonçalves, M. A. (2016). Adenoviral vectors encoding CRISPR/Cas9 multiplexes rescue dystrophin synthesis in unselected populations of DMD muscle cells. *Sci. Rep.* 6: 37051.

Manome, Y., Nakayama, N., Nakayama, K., et al. (2005). Insonation facilitates plasmid DNA transfection into the central nervous system and microbubbles enhance the effect. *Ultrasound Med Biol.* 31:693–702.

Marco, R., Saad, S. K. (2017). Next generation chimeric antigen receptor T cell therapy: going off the shelf. *Bio Drugs.* 31(6): 473–481.

Marraffini L. A., Sontheimer, E. J. (2010). CRISPR interference: RNA-directed adaptive immunity in bacteria and archaea. *Nat. Rev. Genet.* 11(3):181–90.

Matthew, H. P. (2016). Knock-in editing: it functionally corrects! *Blood* 127(21): 2507–2509.

Maude, S., Frey, N., Shaw, P. A. et al. (2014). Chimeric antigen receptor T cells for sustained remissions in leukemia. *N. Engl. J. Med.* 371: 1507–1517.

Maude, S. L., Laetsch, T. W., Buechner, J. et al. (2018). Tisagenlecleucel in children and young adults with B-cell lymphoblastic leukaemia. *N. Engl. J. Med.* 378: 439–48.

Mcnab, G. L., Ahmad, A., Mistry, D., And Stockley, R. A. (2007). Modification of gene expression and increase in alpha1-antitrypsin (alpha1-AT) secretion after homologous recombination in alpha1-AT-deficient monocytes. *Hum. Gene Ther.* 18: 1171–1177.

Neumann, E., Schaefer-Ridder, M., Wang, Y. et al. (1982). Gene transfer into mouse glioma cells by electroporation in high electric fields. *EMBO J.* 1982; 1:841–845.

Ormond, K. E., Mortlock, D. P., Scholes, D. T., Bombard, Y., et al. (2017). Human germline genome editing. *Am. J. Hum. Genet.* 101(2):167–176.

Park, J. H., Riviere, I., Gonen, M., et al. (2018). Long-term follow-up of CD19 CAR therapy in acute lymphoblastic leukemia. *N. Engl. J. Med.* 378: 449–459.

Park, S. W., Gwon, H. C., Jeong, J. O., et al. (2001). Intracardiac echocardiographic guidance and monitoring during percutaneous endomyocardial gene injection in porcine heart. *Hum. Gene Ther.* 12:893–903.

Prausnitz, M. R., Mikszta, J. A., Cormier, M., et al. (2009). Microneedle-based vaccines. *Curr. Top Microbiol Immunol.* 333:369–93.

Geoffrey, R. S., Soya, K., Dieter, C. G. (2011). Oligo/polynucleotide-based gene modification: strategies and therapeutic potential. *Oligonucleotides* 21(2): 55–75.

Richard, A. M., David, G., Anastasia, L., Donald, B. K. (2017). Hematopoietic stem cell gene therapy – progress and lessons learned. *Stem Cell* 21(5): 574–590.

Sangiuolo, F., Scaldaferri, M. L., Filareto, A., Spitalieri, P., Guerra, L., Favia, M., Caroppo, R., Mango, R., Bruscia, E., Gruenert, D. C., et al. (2008). CFTR gene targeting in mouse embryonic stem cells mediated by Small Fragment Homologous Replacement (SFHR). *Front Biosci.* 13: 2989–2999.

Santosh, R. P., Ibrahim, A. A-Z., Raghuram, P. H., Neeta, M., Nidhi, Y., Mohammad, K. A. (2018). *International Medical Journal* 25 (6):361–364.

Sheyn, D., Kimelman-Bleich, N., Pelled, G., et al. (2007). Ultrasound-based nonviral gene delivery induces bone formation in vivo. *Gene Ther.*15:257–66.

Shimamura, M., Sato, N., Taniyama, Y., et al. (2004). Development of efficient plasmid DNA transfer into adult rat central nervous system using microbubble-enhanced ultrasound. *Gene Ther.* 11:1532–1539.

Shimamura, M., Sato, N., Taniyama, Y., et al. (2005). Gene transfer into adult rat spinal cord using naked plasmid DNA and ultrasound microbubbles. *J. Gene Med.* 7:1468–1474.

Soltani, S., Farahani, A., Dastranj, M., Momenifar, N., Mohajeri, P., Emamie, A. D. (2018). DNA vaccine: Methods and mechanisms. *Advances in Human Biology* 8: 132–139.

Sonoda, S., Tachibana, K., Uchino, E., et al. (2006). Gene transfer to corneal epithelium and keratocytes mediated by ultrasound with microbubbles. *Invest. Ophthalmol. Vis. Sci.* 47:558–564.

Tata, D. B., Dunn, F., Tindall, D. J. (1997). Selective clinical ultrasound signals mediate differential gene transfer and expression in two human prostate cancer cell lines: LnCap and PC-3. *Biochem. Biophys. Res. Commun.* 234:64–67.

Tenser, R. B. (1991). Role of herpes simplex virus thymidine kinase expression in viral pathogenesis and latency. Intervirology 32:76–92.

Todaro, M., Quigley, A., Kita, M., Chin, J., Lowes, K., Kornberg, A. J., Cook, M. J., and Kapsa, R. (2007). Effective detection of corrected dystrophin loci in mdx mouse myogenic precursors. *Hum. Mutat.* 28: 816–823.

Van D. L. H. S., Hannaman, D. (2010). Electroporation for DNA immunization: clinical application. *Expert Rev. Vaccines.* 9:503–517.

Wells, D. J. (2004). Gene therapy progress and prospects: electroporation and other physical methods. *Gene Ther.* 11:1363–1369.

Wolff, J. A., Malone, R. W., Williams, P., et al. (1990). Direct gene transfer into mouse muscle in vivo. *Science.* 247:1465–1468.

Yanyu, P., Xiaoyang, H., Chunsheng, Y., Yanqun L., Guan, J (2018). Advances on chimeric antigen receptor modified T-cell therapy for oncotherapy. *Molecular Cancer* 17:91.

Zarogoulidis, P., Darwiche, K., Sakkas, A., Yarmus, L., Huang, H., Li, Q., Freitag, L., Zarogoulidis, K., Malecki, M. (2013). Suicide gene therapy for cancer-current strategies. J Genet. Syndr. Gene. Ther. 9(4):16849.

Zeira, E., Manevitch, A., Khatchatouriants, A., et al. (2003). Femtosecond infrared laser - an efficient and safe in vivo gene delivery system for prolonged expression. *Mol. Ther.* 8:342–350.

Zhang, G., Budker, V., Wolff, J. A. (1999). High levels of foreign gene expression in hepatocytes after tail vein injections of naked plasmid DNA. *Hum. Gene. Ther.* 10:1735–1737.

7 Applications of Gene Therapy

"We might anticipate the interchange of chromosomes and segments. The ultimate application of molecular biology would be the direct control of nucleotide sequences in human chromosomes, coupled with recognition, selection, and integration of the desired genes. It will only be a matter of time before polynucleotide sequences can be grafted by chemical procedures onto a virus DNA."

Joshua Lederberg 1963

7.1 INTRODUCTION

Gene therapy involves inserting a functional gene into a set of malfunctioning cells to correct their dysfunction or to generate a novel function. This technology is currently being applied to treat hereditary diseases such as cystic fibrosis, combined immunodeficiency syndromes, muscular dystrophy, and hemophilia by replacing a deleterious mutant with a functional allele. Scientific breakthroughs in gene therapy are continuing to move therapy toward mainstream medicine, where scientists are focusing on the logical step of introducing the genes directly into the specific organ to replace the defective ones responsible for the disorder. The biology of human gene therapy remains complex, needing much evaluation. The genetic link in many diseases needs to be understood before using the appropriate gene therapy. The public policy of using genetically engineered material on human beings is becoming more widely debated, with most of the participants representing the fields of biology, government, law, medicine, philosophy, politics, and religion, bringing out different views to the discussion. Applications of gene therapy in the control of human diseases are discussed in this chapter.

7.2 APPLICATIONS OF GENE THERAPY

7.2.1 Prevention of Irradiation Damage to Salivary Glands

Salivary glands are postmitotic, well-differentiated epithelial cells with a slow turnover rate and are expected to be relatively resistant to radio waves. Salivary glands are susceptible to infrared radiation. The mechanism of their damage by infrared radiation is yet to be studied. Radiotherapy is applied for the treatment of most head and neck cancers, with most of the patients receiving 50 - 70 Gy irradiation for 5 to 7 weeks. As a result of the radiation, salivary gland tissue is damaged, and the patients suffer from considerable morbidity from the IR-induced salivary hypofunction. Infra-red radiation generates breaks in the DNA also causing oxidative stress by generating toxic, reactive oxygen species (ROS). Superoxide radicals generated by the exposure of cells to radiation are dismutated by

DOI: 10.1201/9781003343790-7

superoxide dismutase to form H_2O_2, which will be further metabolized by catalase and glutathione peroxide to water and oxygen. Administration of MnSOD isoform-plasmid liposomes has protected the lungs, buccal cavity, esophagus, urinary bladder, and intestine (Greenberger and Epperly, 2007).

To restore the salivary gland's function in patients treated with infrared radiation, cDNA of the water channel protein aquaporin 1(AQP1), is added to the acinar cells using Ad5 vector. Administration of AQP1 has led to a dramatic increase in fluid secretion in rats and miniature pigs without significant toxicological effects. This construct is in the phase-1 level of clinical trials for testing in patients receiving infra radiation for head and neck cancer.

7.2.2 Autoimmune Disorders

An autoimmune disorder is a condition when a human immune system fails to detect its body parts and attacks them by releasing proteins known as autoantibodies. Autoimmune disease targets either one organ or the whole body and constitutes a significant, unmet clinical challenge. Even though no single autoimmune disorder is highly prevalent, around 80% of them have been identified to date, affecting 20% of the global population, mostly constituting females (up to 75%). Most of these diseases are incurable and are challenging to manage. With an increase in the number of autoimmune disorders targeting the human population, there is a pressing need for novel approaches to their treatment. Traditional pharmacological approaches for treating these disorders entail the synthesis of small, diffusible compounds either orally or by injection. These approaches have yet to provide ideal agents for use in autoimmune diseases. Recent advancements in research have identified some proteins having the potential for improvement, but they are challenging in long-term administration.

7.2.2.1 Gene Therapy for Autoimmune Disorders

Gene therapy provides an opportunity to deliver protein products, and therapeutic nucleic acids, such as antisense RNA, with more efficiency than traditional methods. Further, in situ production of the gene products would eliminate the need for their frequent re-administration. Delivery of the therapeutic genes and their subsequent expression has the potential for high localization. The greatest strength of gene therapy is its ability to produce high, sustained concentrations of therapeutic macromolecules within a defined anatomical location. For the gene therapy of autoimmune diseases, there is a need for the vectors to have long term gene expression or to enable themselves for frequent re-administration (Toskos et al. 2000). Integration of vectors, such as retrovirus and adeno associated virus, provides the best prospects for long-term gene expression into host cell chromosomes. For non-integrating vectors such as adenovirus, which are only transiently expressed, repeated administration will probably require the vectors to be rendered non-immunogenic. For most applications, therapeutic genes that code for human native proteins are likely to be used. Here, the gene products should not be antigenic. Viral coat proteins are highly antigenic, whose repeated administration requires the patient to be immunosuppressed or tolerant to the virus. Making patients tolerant to potentially pathogenic agents may not be possible, but utilization of immunosuppressive drugs, already used during the treatment of autoimmune diseases, could be of good help. Comparatively, the application of non-viral vector systems has the edge of being non-antigenic.

7.2.2.2 Gene Therapy for Grave's Disease and Lupus Disease

Transferring the desired gene into the target tissues can be achieved by applying both in vivo and ex vivo strategies. For in vivo, vectors are injected directly into the body, and for ex vivo, cells are isolated, genetically altered outside the body, and then injected into the individual. Comparatively, ex vivo gene delivery is more tedious, expensive, and cumbersome. Nevertheless, this system permits the selection and scrutiny of genetically altered cells before re-implantation. Moloney-based retroviral vectors will infect the replicating cells that readily divide in vitro but rarely in vivo. For these reasons, ex vivo gene delivery has made the most significant initial progress, and most human gene therapy trials apply this strategy. Delivery of the genes locally into discrete areas of disease activity will produce fewer side effects, which is more suitable for autoimmune diseases such as Graves' Disease, which primarily affects a specific anatomical location. Systemic gene delivery is better suited for the treatment of disseminated autoimmune disorders such as lupus disease. It is crucial to eliminate the immune system response to an inciting autoantigen to cure an autoimmune disorder. Candidate genes include antigen presentation, T lymphocyte activation, and cytokine function. A vital distinction between curing and treatment of diseases is concerned with a required period of gene expression.

For the elimination of autoimmune disorder, the concerned gene needs to be expressed only for a short period, whereas the gene that treats the disease needs to be expressed throughout the patient's life. At present, the best way to eliminate an autoimmune disorder and bring a permanent cure is by eliminating the autoreactive T cells and induce selective tolerance to the inciting antigen. T cells are eliminated by inducing the genes that encode for a Fas ligand. Selective tolerance is accomplished by downregulating the interactions between the co-stimulatory pairs such as CD40/CD40L and B7/CD28. This strategy does not require any knowledge of the autoantigen. The pathophysiology of an autoimmune disorder identifies an imbalance in Th-1 and Th-2 lymphocyte activities and suggests gene therapy for addressing this imbalance. Utilization of interleukin-12, the product of Th-1 lymphocytes for the treatment of lupus disease, which occurs due to excessive Th-2 activity, and interleukin-10 for the treatment of inflammatory bowel disease, occurs due to excess Th-1 activity.

7.2.2.3 Sjogren's Syndrome

Sjogren's Syndrome is the second most common autoimmune disorder in the United States affecting 1-4 million individuals with 90% of the affected individuals being females. The etiology of this disorder is not clear, and the current methodology for treatment is analgesic. The focal lymphoid cell infiltration mainly characterizes Sjogren's syndrome in the salivary and lachrymal glands, along with the involvement of other glands and organs. In the absence of conventional treatment for this disorder, gene therapy is beneficial to the patients. The transfer of immunomodulatory genes such as anti-inflammatory interleukin-10 and vasoactive intestinal peptides into the salivary glands might reduce the autoimmune condition by increasing salivation and providing relief. cDNAs of human interleukin-10 and vasoactive intestinal peptide were transfected into the submandibular glands of a female nonobese diabetic mouse using AAV2 vectors for providing a stable transgene expression with a bit of immune reaction. Administration of these two genes has resulted in the preservation of salivary flow rate and a reduction in focal autoimmune sialadenitis (Lodde et al. 2006). Once the pathogenesis and the etiology of Sjogren's syndrome are entirely understood, the gene transfer mechanism can be applied to control this autoimmune disorder.

7.2.2.4 Pre-Clinical and Clinical Studies

Pre-clinical studies performed on various animal models provide a hopeful optimism on the role of gene therapy in autoimmune disorders. The immune responses of mice have been selectively modulated by the intramuscular injection of DNA encoding interleukin-12. Injection of genes encoding IL1Ra, IL1sR, and TNFsR has provided an anti-arthritic effect. Injection of decoy oligonucleotides containing NF-kB and AP-11 has also shown a similar effect. Anti-arthritic genes have also been transferred into the mice's hematopoietic stem cells, which produced a high, lifelong circulating gene product.

In 1989, Dr. James A. Wyngaarden, Director of the National Institute of Health, approved the first protocol for inserting a foreign gene into the immune cells of humans who had cancer. In 1990, French Anderson and his colleagues performed the first gene therapy on a four-year-old girl born with severe combined immune deficiency (SCID). This process was highly successful. Over 300 clinical gene therapy trials were performed on more than 3000 individuals in the next few years. The first human gene therapy trial for the autoimmune disorder was started in 1996 with 9 patients who were suffering from rheumatoid arthritis were treated by the retroviral delivery of human

IL1Ra. Gene therapy offers a promising avenue in treating autoimmune disorders with encouraging pre-clinical data leading to human trials, providing optimism about the future development of this approach for handling a broad spectrum of autoimmune disorders. After identifying the appropriate genes and establishing a primary platform, a future process will be to identify an appropriate candidate gene and develop a gene delivery system for long-term gene expression.

7.2.3 Systemic Protein Deficiencies

Salivary glands and endocrine glands display several standard features, particularly in producing high protein levels for their export into the bloodstream. These features can be considered for the treatment of systemic single protein deficiency disorders. Presently, systemic single protein deficiency disorder treatment involves administering a recombinant protein by bolus injection, like insulin injection for diabetes mellitus and erythropoietin for anemias. Expression of the Epo gene in the kidney epithelial cells secretes Epo protein into the bloodstream. Gene therapy with the Epo gene was achieved using adeno associated virus 2 vectors to transfer Epo cDNA into the salivary glands of male mice and rhesus macaques. After administration, Epo serum levels were stable for 54 weeks in male mice and 6 months in rhesus macaques. This study has demonstrated that targeting the salivary glands as depot organs for gene transfer during the treatment of systemic single-protein deficiency disorders has future potential.

7.2.4 Spinal Disorders

The problem of spinal disorders remains more formidable in orthopedics, with its scope of diagnosis ranging from acute traumatic injuries such as fractures and subluxations to highly chronic degenerative conditions such as spondylosis. To date, most of the specific conditions underlying spinal disorders are unknown. Biochemical studies oriented to disc degeneration are significant in many studies, thus making the intervertebral disc the primary focus for research and investigations. However, only a few approaches to prevent disc degeneration are available, demanding novel approaches, including gene therapy. Even though disc degeneration occurs naturally with aging, clinical observations have demonstrated an early disc degeneration adjacent to spinal fusion along with annular disruptions in the nucleus pulposus with a progressive decrease in the proteoglycan content leading to dehydration. Loss of proteoglycans may also alter the loading of facet joints and other structures, leading to degenerative changes. High concentrations of proteoglycans in the nucleus pulposus will generate a swelling pressure and helps to maintain the height and load-bearing ability. Treatment or prevention of disc degeneration using genetically modified proteoglycan content is one of the potential strategies. Thompson and his colleagues in 1991 demonstrated the transfection of hTGF- β1 cDNA into a canine disc tissue has stimulated in vitro proteoglycan synthesis, suggesting the value of this growth factor in the treatment of disc degeneration. Transfection of genes into intervertebral discs and production of therapeutic growth factor proteins in human cells is in clinical trials using efficient adeno virus mediated gene transfer into nucleus pulposus cells with a small possibility of immune reaction and long-term gene expression (Nishida et al. 2008). There is still a long way to go in gene therapy for the treatment of intervertebral disorders, mainly due to the lack of understanding of the effect of growth factors in the biological process and mechanical function of the spinal cord. Once these obstacles are overwhelmed, gene therapy can be the most valuable treatment of spinal cord disorders.

7.2.5 Gene Therapy for Cancer

Cancer is caused due to changes that occur in human genes inherited from parents and are responsible for the traits and characteristics. These changes that occur in the human genome are mainly due to mutations affecting protein synthesis. Over the past five years, survival rates for pancreatic cancer of 4%, for lung cancer 15%, for liver cancer 7%, and for glioblastoma, a common form of brain cancer 5%, remain abysmally low. Prostate and breast cancer are highly amenable to treatment, and survival rates cross 80%. Still, they respond poorly to treatment at later stages of the disease and are responsible for more than 60,000 deaths per year. Current cancer treatment involves a lot of side effects, which is a negative aspect. Systemic toxicity of the chemotherapy will lead to acute and delayed nausea, mouth ulcerations, and mild cognitive impairments. Gene therapy is designed to modify cancer cells by replacing the mutated genes with the normal ones using either a vector or a liposome. The development of gene therapy for cancer treatment poses multiple options, with numerous in vitro and pre-clinical animal models having shown a wide range of testing with appropriate agents with remarkable efficacy. Gene therapy for cancer treatment is divided into three categories, immunotherapy, oncolytic virotherapy, and gene transfer.

7.2.5.1 Immunotherapy

Boosting the patient immune system to target and destroy cancer cells has been tried intensively over the past 100 years, and minimum success has been achieved because cancer cells tend to evolve mechanisms that elude immune detection. A wide range of techniques have been applied to overcome this problem. Recombinant cancer vaccines are injected into patients through gene therapy to cure or contain the host immune system for recognition of the cancer cells by presenting them with high antigenic and immunostimulatory cellular debris. Cancer cells harvested from the patient and cultured under in vitro conditions are engineered to be more recognizable to the host immune system by adding one or more cytokine genes to produce pro-inflammatory immune system molecules or high levels of antigenic proteins. These genetically altered cells are cultured under in vitro conditions, killed, and their cellular contents are injected as a vaccine.

In vivo immunotherapy is also being attempted by delivering immunostimulatory genes, mainly cytokines, into the tumor. These genes will produce proteins that will unmask cells from immune elusion in the cancer cell and trigger antitumor antibodies. Another immunotherapy attempt supported by gene therapy is to directly alter the host immune system to sensitize it to the cancer cells by the utilization of mononuclear blood cells or bone marrow followed by the addition of a tumor antigen or a stimulatory gene to the selected cells for an immune reaction leading to the annihilation of cancer. Gene therapy of cytokine MDA-7 (IL-24), an inducer of cancer cell death, is under clinical trial for its ability to cause a systemic immune reaction in malignant melanoma patients (Menzes et al. 2014). Many attempts have been made to bolster the immune response over the years. After cloning MDA-7 cDNA into a replication-incompetent adenovirus, a vector was injected intratumorally to the patient, which induced apoptosis. Further, this treatment led to a complete and partial response in 2 of 28 patients. A systemic immune activation was also observed in 22 patients along with local apoptosis.

7.2.5.2 Oncolytic Virotherapy

This gene therapy involves utilizing genetically engineered viral vectors to destroy the cancer cells, with the rest of the body remaining unaffected. Oncolytic vectors are designed to infect the cancer cells and induce their death through viral propagation and expression of cytotoxic proteins leading to cell lysis. Viral vectors such as vaccinia, adenovirus, type-1 herpes simplex virus, reovirus, and Newcastle disease virus are utilized for oncolytic therapy due to their natural ability to target cancer cells and the ease with which they can be genetically manipulated. Preliminary trials on mammalian models involving the oncolytic vectors were successful. Oncolytic viral vectors have increased the survival benefits and reduced the metastasis of colon and urinary bladder cancer cells in murine models. An oncolytic virus designed to destroy the osteosarcoma in the canine model has prolonged the survival rate in immunocompetent dogs suffering from syngeneic osteosarcoma. However, oncolytic virotherapy has sustained several stumbling blocks while addressing humans. Most human beings have antibodies to common viruses leading to an immune response, which clears the virus even before it infects the cells. The utilization of replication competent viral particles calls for increased safety precautions, making clinical trials more expensive and cumbersome. During a clinical trial, after the application of the modified vaccinia virus to treat breast and prostate cancer, patients were isolated in a specialized hospital facility for a week to ensure that the virus was completely cleared before allowing them back into the general population.

Due to these limitations, not many clinical trials were made with oncolytic therapy. ONYX-015 is one of the most notable oncolytic adenoviral therapies (Anwar et al. 2022). ONYX-015 is a genetically engineered adenovirus that lacks viral E1B protein. ONYX-015 will not replicate in cells with a normal p53 pathway due to the lack of E1B protein, essential for RNA export during viral replication. Due to mutations, cancer cells have a deficiency in the p53 pathway, allowing ONNYX-015 to replicate and lyse them. Cancer also exhibits altered RNA export mechanisms that permit the export of viral RNA even in the absence of E1B protein. ONYX-015 has completed phase-I and phase-II trials on head and neck squamous cell carcinoma, resulting in tumor regression correlated with the tumor p53 status.

7.2.5.3 Gene Transfer

Therapy either by gene transfer or gene insertion is one of the most exciting treatments that have emerged from gene therapy involving the introduction of a foreign gene into a cancer cell or surrounding tissue. Several genes with various functions, such as suicide genes, anti-angiogenesis, and cellular stasis genes, have been proposed for gene transfer therapy along with several different viral vectors for the delivery of these genes. Among these vectors, replication-incompetent adenovirus is one of the most used during clinical trials. The methodology employed for the gene transfer also includes non-viral methods such as naked DNA transfer, glycodendrimer DNA coating, and electroporation. Initial efforts for implementing gene transfer therapy have faced difficulties. Clinical trials involving gene transfer have suffered from gene silencing resulting in the transferred gene either not being expressed or expressed only for a limited period. Delivery of the therapeutic gene into the target cells must elicit a response, and additional precautions should be taken to prevent the integration of the therapeutic gene into unwanted cells. Despite these difficulties, prostate, lung, and pancreatic tumors were successfully treated in animal models using various genes and transfer methods. Treatment using TNFerade is currently in stage II of clinical trials. TNFerade consists of a cytokine TNF-α gene expressed under transcriptional control of a radiation inducible promoter. TNF-α gene has potent anticancer properties with high systemic toxicity, and gene therapy using the TNF-α gene provides targeting of cancer cells using intratumoral injections and a promoter activated during radiation therapy. Upon injection of TNFerade, the patient will receive radiation therapy to the tumor, which will activate the gene for producing the TNF-α molecule, promoting cell death in the targeted cancer cells and their surrounding cells with the combination of radiation therapy.

7.2.6 Gene Therapy for Eye Diseases

> **Blindness is one of the most life altering conditions a person could experience and the difficulties it causes for mobility and finding employment. Visual impairment is associated with a host of other issues, including insomnia, anxiety, depression, and even the risk of suicide. Restoring proper vision would make an almost unimaginable improvement in quality of life**
>
> *William Hauswirth, Ophthalmologist at the University of Florida, Gainesville*

The retina is the light-sensitive tissue located at the back of the eye, involved in converting light into electrical impulses, sent to the brain through the optic nerve, which allows humans to see. The retina is composed of specialized cells that react to light and process visual signals crucial to vision. Photoreceptor cells are the nerve cells commonly known as rods and cones that convert the light that strikes the retina into electrochemical signals, filtered through a complex network of bipolar cells, amacrine cells, and horizontal cells, before reaching retinal ganglion cells. Axons, the long projections of those cells that form the optic nerve that carries the signals from the retina to the brain's visual cortex are interpreted as images. Degeneration of the retina leads to blindness. Most retinal disorders occur due to the loss of photoreceptor cells, depleting the eye's sensitivity to light. In age related macular degeneration, loss of photoreceptor cells occurs due to the failure of epithelial cells forming a layer at the back of the retina known as retinal pigment epithelium (Lyndon et al.2018). This layer maintains the healthy status of the photoreceptor cells by cleaning the toxic products formed due to photochemical reactions in the retina and provides nutrients. Degeneration of retinal ganglion cells is a principal reason for blindness.

Blindness also occurs due to heredity. Inherited disorders such as Leber congenital amaurosis (LCA) cause the degeneration of the retina, leading to blindness (Allikmets, 2004). Gene therapy is also applied to treat eye diseases, with several clinical trials under progress. Current gene therapy requires the insertion of a needle through the retina to inject the engineered virus behind it. This procedure might disrupt the fragile retina cells and limit the genes' delivery to a limited, targeted tissue region. Drs. John G. Flannery and David V. Schaffer at the University of California, Berkeley have improvised this approach by using a harmless adeno-associated virus, which cannot pass through the retinal layers to reach the targeted photoreceptor cells. The location of the virus injection has also changed from retinal cells to the vitreous humor, a more accessible gel-like fluid at the center of the eye. Adeno associated virus has been genetically engineered to develop variants that can pass through the retinal layers. Millions of this adeno associated virus variations have been injected into the vitreous humor of transgenic mice and adult macaques for selecting the viruses that can reach photoreceptor cells.

The most effective modified adeno associated virus is 7m8, used to deliver genes throughout the retina and optic nerve when injected into the vitreous humor. 7m8 was tested in the models of two genetically inherited disorders, X-linked retinoschisis, and Type-II Leber congenital amaurosis (Brainbridge et al. 2008). X-linked retinoschisis occurs due to mutations in the gene encoding for the protein retinoschisin, leading to the splitting of the retina. Type-II Leber congenital amaurosis occurs due to mutations in the *RPE65* gene located on chromosome 1. Injection of a 7m8 vector carrying the functional copies of the genes into the vitreous humor has improved vision in mouse models. Upon injection of the vector into an ordinary monkey's eye, the virus has penetrated dotted into the cells across the retina but entirely into the delicate vision area known as the fovea. Preparation of a 7m8 vector for its penetration into the human eye cells has led to clinical trials with a maximum time needed for the entire process being 15 minutes. Attempts to knock out the genes or halt the process leading to the death of retinal cells during age-related macular degeneration are also in the process.

7.2.7 Gene Therapy for Cardiovascular Disorders

The development of gene therapy for cardiovascular diseases is challenging due to difficulties in gene delivery into most heart cells. The potential of gene therapy in cardiovascular disorders has become evident during the 1908s, after successful intra-arterial gene transfer using catheter techniques (Deng, 2020). Very soon, conditions such as hyperlipidemias, in-stent restenosis, vein graft stenosis, heart failure, arrhythmias, refractory angina, and peripheral vascular disease have become targets for gene therapy after the pioneering work of Isner and his colleagues in 1996 involving the application of gene transfer for the treatment of peripheral vascular disease. Viral vectors are more extensively used for gene delivery into cardiovascular cells due to their high efficiency than non-viral methods. Vectors of retroviruses, adenoviruses, and adeno associated viruses are used. Despite their application in several non-cardiac gene delivery systems, retroviruses cannot efficiently transduce cardiomyocytes as they require the targeted cells to have inactive cell division for their integration and function. As lentiviruses do not require the cell to have inactive cell division, they have been used for cardiac cell gene therapy. The use of adenoviral vectors for local endovascular catheter-mediated gene therapy in humans is also under trial. Non-viral methods exist such as intramyocardial injection, coronary perfusion, and pericardial delivery.

7.2.7.1 Gene Therapy for Coronary Artery Disease

Coronary heart disease occurs because of insufficient blood flow into the myocardium due to atherosclerosis (deposition of cholesterol and fats inside arteries). Coronary heart disease is transforming into the primary cause of death globally, which is minimized by advances in pharmacology and revascularization using bypass surgery and angioplasty. These approaches cannot be tested on patients with severe angina pectoris and refractory angina. Therapeutic angiogenesis has been developed as an alternative, involving the administration of genes for angiogenesis growth factors including VEGF, FGF, hepatocyte growth factor, platelet-derived growth factor, and hypoxia-inducible factor (Grossman et al. 1995). Among all these, clinical trials were mainly focused on VEGF and FGF to augment collateral vessel development. During the last 10 years, several clinical trials and random control trials such as Euroinject One, KAT, REVASC, NOTHERN, NOVA, VEGF-Neupogen, and GENASIS have been tested for the delivery of VEGF into myocardial cells. Gene transfer has also been done during bypass surgery using epicardial injections. Currently, five angiogenic gene therapy trials have reported final results.

7.2.7.2 Gene Therapy for Heart Failure

Heart failure is a problem, increasing alarmingly in the elderly population, who are being treated through the use of device-based treatments and pharmacotherapies and these have improved patient survival. The 5-year survival for individuals with heart failure is about 50%. Advances in the etiology of the disease have developed few gene therapy approaches, which are under clinical trials. The CUPID2 trial involving adeno-associated virus (AAV)1-sarcoplasmic reticulum calcium ATPase 2a (SERCA2a) is on a clinical trial for patients with chronic systolic heart failure or non-ischemic cardiomyopathy. Adenoviral adenylyl cyclase 6 is being tested in congestive heart failure patients using a dose escalation format, with short-expression duration and immunological reactions as primary concerns. Adeno associated viruses have shown some long-term efficiency in clinical trials. Clinical trials involving sarcoendoplasmic reticulum calcium-ATPase 2a (SERCA2a), SDF-1, and adenylate cyclase-6 (AC6) are also in progress for gene therapy against heart failure (Stewart et al. 2006). The expression and function of SERCA2a have decreased the rate of heart failure by reducing transient calcium levels responsible for systolic heart failure. The *C*alcium *U*pregulation by *P*ercutaneous administration of gene therapy *I*n Cardiac *D*isease (CUPID) clinical trial examines the safety and efficacy of SERCA2a gene therapy in heart failure.

7.2.7.3 Gene Therapy for Vein Graft Stenosis

Vein graft stenosis is a common problem that occurs after undergoing bypass surgery. Most bypass surgeries use autologous saphenous vein as a conduit for bypass grafting, which threatens failure under long-term conditions. Gene therapy is an approach to prevent pathological vein graft remodeling immediately after implantation during its adaptation to higher blood pressure. Vein grafts are excellent sources for the intraluminal or perivascular gene transfer during the preparation of grafts. Metalloproteinase tissue inhibitors are candidates for potential gene therapy, and these are under pre-clinical trials. Vascular grafts and dialysis anastomosis stenosis are also being treated using gene transfer with VEGF reducing the stenosis by enhancing nitric oxide production and prostacyclin in the graft tissue. This study is still under phase I and II of clinical trials.

7.2.8 Gene Therapy for Neuro Degenerative Disorders

Gene manipulation of the central nervous system has witnessed rapid progress with a broad field of experiments allowing the expression of novel genes and the overexpression of endogenous proteins, leading to clinical trials. Repairing the nervous system with its complex structures is exceptionally challenging, and it is achieved through two approaches, ex vivo and in vivo. In the ex vivo approach, gene transfer is performed in cell cultures under in vitro conditions, and the cells are transplanted into the organism. Most of the gene delivery trials into somatic cells have used the ex vivo approach. The gene is directly transplanted into the organism in the in vivo approach. Human neurological disorders are complex, and certain dysfunctional areas in the human brain are challenging to access. Despite these difficulties, intracerebral grafts of fetal and adult-derived cells are the sources for the ex vivo gene delivery due to their diverse origins (Ulrike et al. 1996). It has been identified that cells with diverse origins can survive transplantation into the brain and can reinforce the deficiency. Cell transplantation has successfully reversed behavioral abnormalities in animal model systems concerning central nervous system damage and neurological disorders. Genetically modified cells have provided promising results for intracerebral transplantation. These cells have very few chances of rejection as these cells are autologous. Modern molecular biology techniques can be applied to genetically modified cells to produce a highly controlled and broad range of the desired factor. In neurodegenerative disorders, cells lose trophic factors such as nerve growth factors (NGF), brain-derived neurotrophic factor (BDNF), Neurotrophins NT-3, NT-4, and NT-5. The transplantation of modified cells can restore the production of these factors. The delivery of neurotransmitters and neuromodulators in neuro degenerated mammalian models has restored neuronal function in the brain. Immortalized and non-neuronal cell lines C6, neuroblastoma, and AT20 have been used for gene therapy.

7.2.8.1 Gene Therapy for Neurodegenerative Disorders in Model Organisms

Identification of novel genes whose mutation leads to neurological disorders provides an opportunity for their treatment. Further, identifying their gene products, de-lineation of their cellular functions, and cell death may provide new options for therapy.

7.2.8.1.1 Parkinson's Disease

Parkinson's disease is a progressive nervous system disorder that generates movement disorders. It is a widespread disorder with a prevalence of up to 1%. This disorder is characterized by the loss of dopamine-producing neurons known as dopaminergic neurons. The current treatment features oral L-Dopa therapy, which will be less effective with the progression of the disorder and increased side effects. All of these features make Parkinson's disease ideal for gene therapy. During the last decade, the discovery of genes linked to familial forms of Parkinson's disease has dramatically improved our understanding of the possible molecular basis of the disorder. Based on these findings, novel approaches have been proposed for protecting neuronal functions and to halt disease progression. Enzyme tyrosine hydroxylase is responsible for the biosynthesis of L-Dopa. Introduction of a gene whose expression synthesizes tyrosine hydroxylase into cells can increase the concentrations of L-Dopa. Various methods have been developed for the delivery of genes into animal models, including viral and non-viral systems. The direct transfer of the tyrosine hydroxylase gene into degenerated striatum has been achieved using herpes simplex virus and adeno associated virus vector systems. Another twenty clinical trials are in progress for the long-term expression of tyrosine hydroxylase.

In another attempt, genetically modified cells that produce tyrosine hydroxylase are utilized. Fibroblast cells transfected with tyrosine hydroxylase genes are implanted into the striatum could reduce the rotational behavior in 6-hydroxy dopamine lesioned rats. However, the number of expressing cells were decreased with increasing duration. Methods for extension of transgene expression are in progress. Cell death during Parkinson's disease is being connected to oxidative stress, lack of neurotrophic support, and exposure to toxins. Oxidative stress is responsible for the death of implanted cells in transgenic mice. Overexpression of Cu/Zn superoxide dismutase in transgenic mice has shown them to be more resistant to oxidative stress-induced neuronal damage (Turunen et al. 2006). Transfer of neurons from Cu/Zn SOD overexpressed transgenic mice into immune-suppressed mammalian models has shown a four times higher survival of genetically modified neurons.

7.2.8.1.2 Huntington's Disease

Huntington's disease is an inherited neurological disorder that brings uncontrollable body movements, and generates emotional instability, damaging all intellectual faculties. This disease occurs due to an increase in a trinucleotide CAG repeat in the IT15 gene located on chromosome 4 which encodes for glutamine in the Huntington protein. The severity of the disease depends on the number of repeats, and phosphorylation of Huntington's protein is neuroprotective. Increased phosphorylation of Huntington's protein is either due to the chemical inhibition of calcineurin with FK506 or by the activation of serine/threonine kinase Akt. It has been demonstrated that the RCAN1 gene encodes calcineurin and controlled overexpression of RCAN1 is a viable avenue for the gene therapy of Huntington's disease. Davies and his colleagues in 2007 have shown that the isoform of RCAN1, known as RCAN1-1L, decreases in the human brain when affected by Huntington's disease. They have also demonstrated that increased RCAN1-1L rescues from the toxicity of Huntington's disease. Other isoforms of RCAN1, RCAN1–1S, RCAN1–4, and RCAN1-1L are expressed in nerve cells and inhibit calcineurin. All three RCAN1 proteins are being considered for therapeutic studies for the prevention of Huntington's disease.

7.2.8.1.3 Alzheimer's Disease

Alzheimer's disease is a common dementia and degenerative brain disorder that can be either inherited or acquired. The etiology of this disease is characterized by progressive dementia caused by critical atrophy, neuronal loss, neurofibrillary tangles, senile plaques, and vascular deposits of β-amyloid in the cerebral cortex and hippocampus. Currently, no disease modifying therapies are in existence for Alzheimer's disease due to a minimal understanding of its neuropathology. Existing approved treatments target neurotransmitter acetylcholine for strengthening the parasympathetic nervous system. Cholinergic drugs show efficacy in 50% of patients forcing gene therapy to improve neuro function. CERE-110 is a gene therapy product developed by Ceregene, Inc., San Deigo, CA, the USA, to improve brain nerve cells in patients with Alzheimer's disease. CERE-110 is an engineered nerve growth factor (NGF) gene whose protein controls the development and maintenance of sympathetic neurons. Engineered NGF gene is injected into Alzheimer's patients using an adeno associated vector system at the nucleus basalis of the Meynert location in the brain, where neuron death occurs rapidly. The administered NGF gene is intended to maintain the survival rate of nerve cells. CERE-110 has been injected in doses of 2.0 x 10.011vg (vector genomes)/mL during clinical trials. CERE-110 is also under phase II clinical trials for the treatment of Parkinson's disease.

7.2.8.1.4 Brain Tumors

Brain tumors are one of the major areas of gene therapy, especially in the case of inheritance. The current therapeutic strategy comprises of the direct killing of tumor cells, the production of new tumor antigens on the cell surface for the induction of tumor rejection, and finally, the transfer of drug sensitivity genes into the tumor cells. Thymidine kinase is the most feasible target for this strategy applied on a wide range of animal models during clinical trials. Transfection of tumor cells with the thymidine kinase gene will enable them to metabolize the antiviral drug ganciclovir into ganciclovir triphosphate, a cytotoxic drug that causes cell death. Tumor cells transduced with the thymidine kinase gene are sensitive to ganciclovir treatment. The efficiency of the viral vectors was improved by the implantation of producer cells, which can supply the vector continuously. Retroviral vectors, which can infect rapidly dividing tumor cells and kill them, are used for this approach. Further, to stimulate the immune response for increasing tumor rejection, the possibility of delivering interleukins and granulocyte-macrophage stimulating factors is being investigated. The immune response against tumor cells was successfully increased after vaccination with irradiated tumor cells. Patients with primary or metastatic brain tumors have been treated with cells that produce thymidine kinase through the herpes simplex virus and ganciclovir during clinical trials. The results of this trial are yet to be announced.

7.2.8.1.5 Amyotrophic Lateral Sclerosis

Also referred to as Lou Gehrig's Disease, amyotrophic lateral sclerosis is a progressive degenerative disease that affects the brain and spinal cord neurons and eventually leads to their death. As a result, the brain loses the ability to initiate and control muscle movement, causing total paralysis to the patient in the later stages of the disease. Over the past 10 years, potential drugs and hormones have been discovered for the treatment of amyotrophic lateral sclerosis. The major limitation is in getting these drugs into the brain and spinal cord of the patients by passing through the blood-brain barrier. This barrier keeps many drugs and hormones away from the brain and spinal cord. Gene therapy can be applied to bypass the barrier and deliver the therapeutic agent to the targeted location. Researchers at the Salk Institute and Johns Hopkins University have demonstrated that gene therapy can be applied for motor neurons, increasing animal survival in amyotrophic lateral sclerosis mice. Further, two therapeutic proteins, insulin-like growth factor 1 (IGF-1) and glial cell-derived neurotrophic factor (GDNF) have been compared. IGF-1 was more effective than GDNF, which can be given in the later stages of disease in the mouse model for delaying disease progression. For this, a genetically modified virus carrying the instructions for IGF-1 is injected into muscle cells, where it can infect motor neurons, and enable them to make IGF-1 protein. In this process, a one-time virus injection into the muscle cells will enable the nerve cells to pick up the viral vector and transport it to the spinal cord for its work. Upon entry to the motor neuron or spinal cord cell, the virus will make the cell a factory for the rapid production of the IGF-1 protein. In this manner, a large amount of protein can be continuously synthesized within the spinal cord and constantly provide therapy for the disease.

7.2.9 Gene Therapy and Dermatological Disorders

Skin is an external and highly versatile organ, best suitable for gene therapy. Transduction of the dermis with transferrin, erythropoietin, growth hormone, apolipoprotein E, and the like, will produce a high level of secretion of the gene product, and its physiological effect can be high. The efficacy of the secretion can be further enhanced by modified progesterone receptor driven transcription. Keratinocyte is considered to be the target for dermatological gene therapy due to its easy accessibility for isolation and in vitro tissue engineering. Keratinocyte synthesizes 70 proteins and secretes them into surroundings. Fibroblast cells, melanocytes, macrophages, and endothelial cells are also considered to be target cells for gene therapy. Due to the ease of gene transfer, the dermatological disorder is a favorite target for gene therapy (Tanusree et al.2020). Other than this, skin cancer, chronic wounds, and intractable inflammatory disease are also the focus of attempts for gene therapy. To date, gene therapy to the skin is in the experimental stage and is not in regular use. The main reasons for this are the complexity of the human skin and the cost of the clinical trials. Despite these constraints, one gene therapy drug has been recently approved in Europe to treat lipoprotein lipase deficiency and clinical studies are in progress for chronic and genetic dermatological disorders.

7.3 DIFFICULTIES WITH GENE THERAPY

Acquiring the ability to modify the human genome has been the objective of medicine since the identification of DNA as genetic material and the basic heredity unit. Gene therapy can improve genetic composition by correcting the mutated or altered genes through treatment. For performing gene therapy, various strategies have been described that are under trial. Most gene therapy is currently limited to the laboratory, where its applications are under experimental conditions. Most clinical trials are being conducted in the USA, Europe, and Australia, mainly on recessive gene disorders such as cystic fibrosis, muscular dystrophy, sickle cell anemia, acquired diseases such as cancer, and pathological infections such as HIV (Tebas et al. 2014). Despite the initial success, many techniques need development to achieve success.

A normally functioning gene is inserted into the genome during the gene therapy process to replace the abnormal gene responsible for causing the disorder. One of the toughest challenges faced in this process, achieved by using a vector, is in releasing the gene into target cells. The design of the vector should be more specific with high efficiency of gene release and not be recognized by the immune system. It should also be purified in large quantities for its large-scale availability. The vector should restore normal cell functions upon injection into the patient by transferring the targeted gene without inducing any allergic reactions or inflammatory processes. The vector also should be safe for the environment and to the professionals who work with it. Even though the utilization of viruses as vectors has produced success for gene therapy, their utilization has presented several limitations. The presence of viral genetic material in the vector is a potential aggravating factor, as it can induce an acute immune response and has the possibility of oncogenic transformation.

Since the conduction of the first clinical gene-therapy trial, much attention with considerable promise has been given to this field. There has been substantial investment from the public and private sectors, resulting in an increase in the level of research activity. Numerous preclinical and clinical studies conducted on animal models have provided proof for potential clinical applications. However, their clinical progress has been a little slow. A major setback for gene therapy occurred in September 1999, when a death resulting from a gene-therapy trial occurred. Jesse Gelsinger, an 18-year-old man, died during a clinical trial at the University of Pennsylvania while being treated with ornithine decarboxylase, a deficient hepatic enzyme, using a modified Ad5 vector. As per the investigation, Gelsinger died due to a massive immune reaction to the modified Ad5 vector. This case has led to congressional and Food and Drug Administration (FDA) hearings on the conduct of clinical gene-therapy trials as well as a temporary hold, subsequently lifted, on all adenoviral-vector clinical trials. An investigation by the FDA has also reported numerous potential violations in how this clinical trial was conducted and monitored. After 5 years of investigations, in February 2005, the case was finally settled. Due to the death of Gelsinger, gene therapy has experienced an intense phase of criticism and skepticism. As an appropriate outcome, all gene-therapy trials are subject to stringent regulations by the National Institute of Health (NIH) and the FDA.

Gene therapy is currently subdivided into two categories: somatic cell gene therapy and germline gene therapy. Genetic manipulation of germlines is a subject of debate as it involves a lot of ethical and moral issues. In 2015, a group of Chinese researchers announced the genetic modification of embryonic cells using CRISPR-Cas9. Another Chinese group also reported the insertion of CCR5 gene mutation into the embryonic stem cells for conferring HIV resistance (Callaway 2016). Genetic analysis has confirmed the successful modification of 4 embryos among 26, which reveals the need to improvise the technique and conductance of pre-clinical and clinical trials testing in animal models. In the UK, the first project on editing healthy human embryos has been approved, whereas American research groups remained conservative by not supporting this type of experiment (Giulliana and Raquel de, 2017).

7.4 CONCLUSION

Gene therapy remains a solid and active field of research aimed to develop novel and practical techniques for the treatment of various genetic disorders. Since the optimization of human genetics, gene therapy has advanced through time by optimizing the vectors, introducing novel techniques, and the induction of pluripotent stem cells in combination with genome editing. Initial success has solidified the purpose of gene therapy treatment for congenital and monogenic disorders, where pharmacological and surgical interventions do not show any result. The design of novel vectors with increased efficiency and specificity will further enhance an understanding of inflammatory response, thus improving safety. Gene therapy displays a promising role in future medicine with an increased percentage of preclinical and clinical trials.

REFERENCES

Alexander, J. H., Hafley, G., Harrington, R. A., Peterson, E. D., Ferguson, T. B., Jr., Lorenz, T. J., Goyal, A., Gibson, M., Mack, M. J., Gennevois, D., et al. (2005). Efficacy and safety of edifoligide, an E2F transcription factor decoy, for prevention of vein graft failure following coronary artery bypass graft surgery: PREVENT IV: a randomized controlled trial. *The Journal of the American Medical Association* 294: 2446–2454.

Allikmets, R. (2004). Leber congenital amaurosis: a genetic paradigm. *Ophthalmic Genet.* 25(2):67–79.

Anwar, M., Arendt, M. L., Ramachandran, M. et al. (2022). Ixovex-1, a novel oncolytic E1B-mutated adenovirus. *Cancer Gene Ther.* https://doi.org/10.1038/s41417-022-00480-3.

Bainbridge, J. W., Smith, A. J., Barker, S. S., Robbie, S., Henderson, R., et al (2008). Effect of gene therapy on visual function in Leber's congenital amaurosis. *N. Engl. J. Med.* 358(21):2231–2239.

Boyd, R. F., Boye, S. L., Conlon, T. J., Erger, K. E., Sledge, D. G., Langohr, I. M., Hauswirth, W. W., Komáromy, A. M., Boye, S. E., Petersen-Jones, S. M., Bartoe, J. T. (2016). Reduced retinal transduction and enhanced transgene-directed immunogenicity with intravitreal delivery of rAAV following posterior vitrectomy in dogs. *Gene Ther.* 23(6):548–56.

Callaway, E. (2016). Second Chinese team reports gene editing in human embryos. *Nature.* 08 April 2016:19718.

Conte, M. S., Bandyk, D. F., Clowes, A. W., Moneta, G. L., Seely, L., Lorenz, T. J., Namini, H., Hamdan, A. D., Roddy, S. P., Belkin, M., et al. (2006). PREVENT III Investigators (2006). Results of PREVENT III: a multicenter, randomized trial of edifoligide for the prevention of vein graft failure in lower extremity bypass surgery. *J. Vasc. Surg.* 43: 742–751.

Davies, K. J., Ermak G., Rothermel B. A., Pritchard M., Heitman J., Ahnn J., Henrique-Silva F., Crawford D., Canaider S., Strippoli P., Carinci P., Min K. T., Fox D. S., Cunningham K. W., Bassel-Duby R., Olson E. N., Zhang Z., Williams R. S., Gerber H. P., Perez-Riba M., Seo H., Cao X., Klee C. B., Redondo J. M., Maltais L. J., Bruford E. A., Povey S., Molkentin J. D., McKeon F. D., Duh E. J., Crabtree G. R., Cyert M. S., de la Luna S., Estivill X. (2007). Renaming the DSCR1/Adapt78 gene family as RCAN: regulators of calcineurin. *Federation of American Societies for Experimental Biology Journal* 21: 3023–3028.

Deng, J., Guo, M., Li, G. et al (2020). Gene therapy for cardiovascular diseases in China: basic research. Gene Ther. 27: 360–369.

Deniz, D., Leah, C. B., Ryan, R. K., Meike, V., Lu, Y. (2013). In vivo-directed evolution of a new adeno-associated virus for therapeutic outer retinal gene delivery from the vitreous. *Sci. Transl. Med.* 5(189):189ra76.

Fisher, L. J., Gage, F. H. (1993). Grafting in the mammalian central nervous system. *Physiol. Rev.* 73: 583.

Freed, W. J. (1992). Direct interaction with target-derived glia enhances survival but not differentiation of human fetal mesencephalic dopaminergic neurons. *Rest. Neurol. Neurosci.* 56(1): 55–60.

Gamlin, P. D., Alexander, J. J., Boye, S. L., Witherspoon, C. D., Boye, S. E. (2019). SubILM Injection of AAV for Gene Delivery to the Retina. *Methods Mol. Biol.* 1950:249–262.

Gennady, E., Karl, J. H., Kevin, T. C., Sean, S., Kelvin, J. A. D. (2009). Regulator of calcineurin (RCAN1-1L) is deficient in Huntington disease and protective against mutant huntingtin toxicity in vitro. *The Journal of Biological Chemistry* 284 18: 11845–11853.

George, S. J., Wan, S., Hu, J., MacDonald, R., Johnson, J. L., and Baker, A. H. (2011). Sustained reduction of vein graft neointima formation by ex vivo TIMP-3 gene therapy. *Circulation* 124 (11, Suppl): S135–S142.

Giulliana, A. R. G., Raquel de, M. A. P. (2017). Gene therapy: advances, challenges and perspectives. *Einstein.* 15(3):369–375.

Greenberger, J. S., Epperly, M. W (2007). Antioxidant gene therapeutic approaches to normal tissue radioprotection and tumor radio sensitization. *In Vivo* 21: 141–146.

Grossman, M., Rader, D. J., Muller, D. W., Kolansky, D. M., Kozarsky, K., Clark, B. J., 3rd, Stein, E. A., Lupien, P. J., Brewer, H. B., Jr., Raper, S. E., et al. (1995). A pilot study of ex vivo gene therapy for homozygous familial hypercholesterolemia. *Nat. Med.* 1: 1148–1154.

Hauswirth, W. W., Aleman, T. S., Kaushal, S., Cideciyan, A. V., Schwartz, S. B., Wang, L., Conlon, T. J., Boye, S. L., Flotte, T. R., Byrne, B. J., et al. (2008). Treatment of Leber congenital amaurosis due to RPE65 mutations by ocular subretinal injection of adeno-associated virus gene vector: short-term results of a phase I trial. *Hum Gene Ther.* 19(10):979–90.

Hedman, M., Hartikainen, J., Syvänne, M., Stjernvall, J., Hedman, A., Kivelä, A., Vanninen, E., Mussalo, H., Kauppila, E., Simula, S., et al. (2003). Safety and feasibility of catheter-based local intracoronary vascular endothelial growth factor gene transfer in the prevention of post angioplasty and in-stent restenosis and in the treatment of chronic myocardial ischemia: phase II results of the Kuopio Angiogenesis Trial (KAT). *Circulation* 107: 2677–2683.

Isner, J. M., Pieczek, A., Schainfeld, R., Blair, R., Haley, L., Asahara, T., Rosenfield, K., Razvi, S., Walsh, K., and Symes, J. F. (1996). Clinical evidence of angiogenesis after arterial gene transfer of phVEGF165 in patient with ischaemic limb. *Lancet* 348: 370–374.

Jacobson, S. G., Cideciyan, A. V., Ratnakaram, R., Heon, E., Schwartz, S. B., Roman, A. J., Peden, M. C., Aleman, T. S., Boye, S. L., Sumaroka, A., et al. (2012). Gene therapy for Leber congenital amaurosis caused by RPE65 mutations: safety and efficacy in 15 children and adults followed up to 3 years. *Arch. Ophthalmol.* 130(1): 9–24.

Kastrup, J., Jørgensen, E., Fuchs, S., Nikol, S., Bøtker, H. E., Gyöngyösi, M., Glogar, D., and Kornowski, R. (2011). A randomised, double-blind, placebo-controlled, multicentre study of the safety and efficacy of BIOBYPASS (AdGVVEGF121.10NH) gene therapy in patients with refractory advanced coronary artery disease: the NOVA trial. *Euro Intervention* 6: 813–818.

Kastrup, J., Jørgensen, E., Rück, A., Tägil, K., Glogar, D., Ruzyllo, W., Bøtker, H. E., Dudek, D., Drvota, V., Hesse, B., et al.; Euroinject One Group (2005). Direct intramyocardial plasmid vascular endothelial growth factor-A165 gene therapy in patients with stable severe angina pectoris: a randomized double-blind placebo-controlled study: the Euro inject One trial. *J. Am. Coll. Cardiol.* 45: 982–988.

Kate, L., Dishart, Lorraine, M. W., Laura, D., Andrew, H. B. (2003). Gene Therapy for Cardiovascular Disease. *Journal of Biomedicine and Biotechnology* 2: 138–148.

Katz, A. B., Taichman, L. B. (1999). A partial catalogue of proteins secreted by epidermal keratinocytes in culture. *J. Invest. Dermatol.* 112:818–821.

Laitinen, M., Mäkinen, K., Manninen, H., Matsi, P., Kossila, M., Agrawal, R. S., Pakkanen, T., Luoma, J. S., Viita, H., Hartikainen, J., et al. (1998). Adenovirus-mediated gene transfer to lower limb artery of patients with chronic critical leg ischemia. *Hum. Gene Ther.* 9: 1481–1486.

Le Meur, G., Lebranchu, P., Billaud, F., Adjali, O., Schmitt, S., Bezieau, S., Pereon, Y., Valabregue, R., Ivan, C., Darmon C, et al. (2018). Safety and long-term efficacy of AAV4 gene therapy in patients with RPE65 Leber congenital Amaurosis. *Mol. Ther.* 26(1):256–268.

Lodde, B. M., Mineshiba, F., Wang, J., Cotrim, A. P., Afione, S., Tak, P. P., Baum, B. J. (2006). Effect of human vasoactive intestinal peptide gene transfer in a murine model of Sjogren's syndrome. *Ann. Rheum. Dis.* 65: 195–200.

Lorenz, B., Gyuru, S. P., Preising, M., Bremser, D., Gu, S., Andrassi, M., Gerth, C., Gal, A. (2000). Early-onset severe rod-cone dystrophy in young children with RPE65 mutations. *Invest. Ophthalmol. Vis. Sci.* 41(9):2735–2742.

Lyndon, D. C., Kate, F., Odysseas, G., Julie, K., Yvonne, H. L., et al. (2018). Clinical study of an embryonic stem cell-derived retinal pigment epithelium patch in age-related macular degeneration. *Nat. Biotechnol.* 36(4):328–337.

Menezes, M. E., Bhatia, S., Bhoopathi, P., Das, S. K., Emdad, L., Dasgupta, S., Dent, P., Wang, X. Y., Sarkar, D., Fisher, P. B. (2014). MDA-7/IL-24: multifunctional cancer killing cytokine. *Adv Exp Med Biol.* 818:127–153.

Nabel, E. G., Plautz, G., and Nabel, G. J. (1990). Site-specific gene expression in vivo by direct gene transfer into the arterial wall. *Science* 249: 1285–1288.

Nishida, K., Suzuki, T., Kakutani, K., Yurube, T., Maeno, K., Kurosaka, M., Doita, M. (2008). Gene therapy approach for disc degeneration and associated spinal disorders. *Eur Spine J.* 17 Suppl 4:459–66.

Ripa, R. S., Wang, Y., Jørgensen, E., Johnsen, H. E., Hesse, B., and Kastrup, J. (2006). Intramyocardial injection of vascular endothelial growth factor-A165 plasmid followed by granulocyte-colony stimulating factor to induce angiogenesis in patients with severe chronic ischaemic heart disease. *Eur. Heart J.* 27: 1785–1792.

Seppo, Y-H., Andrew, H. B. (2017). Cardiovascular gene therapy: past, present, and future. *Molecular Therapy* 25(5): 1095–1106.

Stewart, D. J., Hilton, J. D., Arnold, J. M., Gregoire, J., Rivard, A., Archer, S. L., Charbonneau, F., Cohen, E., Curtis, M., Buller, C. E., et al. (2006). Angiogenic gene therapy in patients with nonrevascularizable ischemic heart disease: a phase 2 randomized, controlled trial of AdVEGF(121) (AdVEGF121) versus maximum medical treatment. *Gene Ther.* 13: 1503–1511.

Stewart, D. J., Kutryk, M. J. B., Fitchett, D., Freeman, M., Camack, N., Su, Y., Siega, A. D., Bilodeau, L., Burton, J. R., Proulx, G., et al. (2009). VEGF gene therapy fails to improve perfusion of ischemic myocardium in patients with advanced coronary disease: results of the NORTHERN trial. *Mol. Ther.* 17: 1109–1115.

Tanusree, S., Somenath, S., Dwijendra, N. G. (2020). Gene Therapy and its application in Dermatology. *Indian J. Dermatol.* 65(5): 341–350.

Tebas, P., Stein, D., Tang, W. W., Frank, I, Wang, S. Q., Lee, G., et al. (2014). Gene editing of CCR5 in autologous CD4 T cells of persons infected with HIV. *N. Engl. J. Med.* 370(10):901–10.

Thompson, J. P., Oegema, T. R., Bradford, D. S. (1991). Stimulation of mature canine intervertebral disc by growth factors. *Spine* 16:253–260.

Toskos, G. C., Nepom, G. T. (2000). Gene therapy in the treatment of autoimmune diseases. *J. Clin. Invest.* 106(2):181–183.

Turunen, P., Puhakka, H. L., Heikura, T., Romppanen, E., Inkala, M., Leppänen, O., and Ylä-Herttuala, S. (2006). Extracellular superoxide dismutase with vaccinia virus anti-inflammatory protein 35K or tissue inhibitor of metalloproteinase-1: Combination gene therapy in the treatment of vein graft stenosis in rabbits. *Hum. Gene Ther.* 17: 405–414.

Ulrike, B., Luigi, N., Inder, M. V., Didier, T., Fred, H. G. (1996). Applications of gene therapy to the CNS. *Human Molecular Genetics* 5: 1397–1404.

8 Applications of Molecular Diagnostics

8.1 INTRODUCTION

Molecular diagnostics involve the detection of pathogens and mutations in DNA and RNA for diagnosis, sub-classification, prognosis, and inheritance using the standard and latest molecular biology techniques, well supported by genomics information. Localization, identification, and characterization of genes is needed for an understanding of their deficiencies and of the changes that are responsible for a disorder or disease. Nucleic acid forms the basis for the diagnosis of a genetic disorder. Novel diagnostic tools will determine the interaction of genes with proteins and focus on the patterns of interaction in different types of cells. The capturing of this information as expression patterns is also known as molecular signatures for diagnosing diseases. Molecular diagnostics translate novel discoveries and technologies into functional clinical tests for their predictive response to specific therapies. These technologies offer the potential to move into prognostics, with more than 80% of tests performed today being for the detection of infectious diseases and management. The other 20% are performed in forensic medicine, paternity testing, tissue typing, and oncology. The advantage of molecular diagnostic testing is that one can perform it on a tiny tissue obtained from biopsies or fine-needle aspirates and archival material.

The molecular diagnostics industry is emerging as a powerful player in health care and is one of the fastest-growing areas of clinical testing. The molecular diagnostics market is growing at 15 - 17% annually. Such steady growth in the short term is due to the maturation of technology and the development of diagnostic applications. Expansion of the market into developing countries will lead to increased acceptance and market penetration. Traditionally, diagnostic testing was confined to infectious diseases such as HIV and HPV, genetic analysis for disorders such as cystic fibrosis, and oncology (to a lesser extent) such as leukemia and lymphoma. New molecular test-based diagnosis and the management of several diseases other than cancer, genetic disorders, and infectious diseases are emerging, including testing of immunity, metabolism, nervous system, and cardiovascular function.

Additionally, technologies that can analyze multiple genes (up to 10,000), single cells, or new metabolites are available to provide rapid information for diagnosing, classifying, and managing disease. The rapid development of diagnostics in medicine makes it possible to have a genetic readout to assist in diagnosing or treating a variety of diseases. Further, genomics will fuel diagnostics as every gene-based therapy will need a diagnostic test. With more than 30,000 genes in the human genome, the commercialization of at least 15,000 gene-based diagnostic tests is expected to fuel the diagnostic marketplace, followed by pharmacogenetic and predisposition diagnostics. Rapid progress in molecular diagnostics might change health care. With its broad applications in agriculture, bioterrorism, and environmental science, diagnostics is a primary focus. As a result, the

DOI: 10.1201/9781003343790-8

development of new therapeutics based on molecular testing approaches is in progress. Molecular diagnostics is changing the practice of medicine and will continue to do so in the foreseeable future. The broad applications of molecular diagnostics and testing, its role in rare genetic disorders, pharmacogenetics, pharmacogenomics, personalized medicine, genetic risk assessment, genetic counseling, transplants, epigenetics, and forensics are discussed in this chapter.

8.2 APPLICATIONS OF DIAGNOSTICS IN TESTING FOR RARE GENETIC DISORDERS

Rare disorders are progressive genetic disorders that affect a limited number of individuals in chronically debilitating clinical conditions that are sometimes life threatening. The recognition of globally prevalent rare disorders is one of the significant health concerns. The majority of rare disorders are monogenic and occur due to a variation in a single gene responsible for altering its function. The base prevalence rate of rare disorders as per the World Health Organization is 1 in every 2000 individuals. However, this number changes from country to country based on the prevalence in their particular population. More than 7000 rare disorders have been identified, and this number is steadily increasing with the addition of the members of the list. Around 80% of rare disorders have a genetic origin, whereas the distinct cause for the occurrence of the other 20% of disorders is not well understood (Angural et al. 2020). They might occur by the environment (as in the case of Jamaican vomiting sickness and mesothelioma), infections (maternofetal measles), or immunological (juvenile chronic arthritis). There is a considerable variation in the severity and expression of these rare disorders. Many of them are congenital in onset and continue to exist with a poor disease prognosis over an individual's lifetime. In contrast, some individuals show symptoms (for the same disorder) in the latter half of their life, making it difficult to diagnose them. An estimated 50% of rare disorders occur in children, with 30% dying during infancy before attaining five years and 12% of the children dying between 5 - 15 years of age. Even though these disorders are of low prevalence, their cumulative burden on public health is high (Chelsea et al. 2020). Around 350 million individuals across the globe suffer from rare disorders, which amounts to 10% of its population. These figures are alarming, indicating that the number of individuals suffering from rare disorders is more than those suffering from common diseases such as diabetes (20.8 million individuals).

The diagnosis of rare disorders has been a significant challenge. The very first diagnosis of rare diseases begin in the early 20th century, totally based on biochemical parameters. Cytogenetic methods developed during the 1960s were improvised as molecular genetics and molecular cytogenetics in the last decades of the 20th century. Recently, diagnostic procedures based on genome sequencing started emerging, giving a helping hand to medicine for addressing rare disorders. There is a moderate increase in the efficiency in diagnosing rare genetic disorders with the Sanger sequencing approach for single genes to identify human genetic variations. Applying the gene-by-gene process for testing genetic heterogeny was time-consuming and expensive. Hence, health systems do not take this approach. Moreover, an appropriate diagnostic hypothesis has been essential for a successful genetic diagnosis, provided by next-generation sequencing technology (NGS). NGS has revolutionized the diagnostics of genetic testing by identifying the gene responsible for Mendelian disorders. Until 1986, only 40 genes responsible for genetic disorders were identified. The application of the positional cloning approach for mapping the gene, cloning the region, and sequencing has led to the identification of more than 1000 genes, including those for cystic fibrosis and Huntington's disease. The application of NGS has led to an exponential increase in the number of genes associated with Mendelian disorders. The use of NGS has also significantly contributed to the diagnosis of rare genetic disorders. NGS can rapidly and cost-effectively analyze from just a few to hundreds of genes simultaneously, whole-genome and even whole exome. Application of NGS has ended the diagnostic odyssey. It has led to clinical outcomes such as a change in disease management, enhanced surveillance, prognosis expectations, medication or dietary changes, prenatal genetic diagnosis, preconception carrier screening, and genetic testing of asymptomatic relatives. Three different and critical sequential approaches for detecting rare genetic disorders are, targeted sequencing panel, whole exome sequencing, and whole-genome sequencing.

8.2.1 Targeted Sequencing Panel

These panels have a specific number of genes or coding regions within genes known to harbor mutations responsible for a disorder. These targeted panels are sequenced at a greater depth than the whole exome or whole genome at a low cost. Detected variants are limited to selected genes, producing a low volume of data. Therefore, the workload required for the interpretation is lower. As gene discovery is rapid and the knowledge regarding their functions is expanding, these panels need frequent updates. However, the sequencing panels are limited to detecting structural variants, repetitive elements, and mitochondrial genome variants.

8.2.2 Whole Exome Sequencing

Whole exome sequencing (WES) covers the protein-coding region, up to 1 - 2% of the genome, harboring about 95% of disease responsive mutations. WES enables the identification of gene variants that are not associated with humans. WES can be performed in two ways, a preselected panel comprising of a selected set of genes related to the patient's phenotype, the examination of rare and potentially damaging variants and comparison of the patient's phenotype with these genes. The second approach enables the identification of gene variants that we are yet to discover. Incomplete coverage of the regions, limited ability to detect the variations, variants responsible for the somatic mosaicism, and structural and deep intronic variants are some of the limitations of the whole-exome sequencing approach

8.2.3 Whole Genome Sequencing

Whole-genome sequencing (WGS) can unveil a significant portion of the human genome and discover novel genes and gene regulators at a low cost. The data obtained through WGS can address the complications in inheritance models. It is a strength in that it can identify the genetic reason with a single test. WGS can discover a broad range of genetic variations, including single nucleotide variants (SNVs), indels, and structural variants such as copy number variants and translocations. The whole-exome and whole-genome have great potential in the diagnosis of rare disorders, which enables the analysis of many genes in a single test, resulting in incidental findings and rare variants of unknown significance (Kym et al. 2019). Equipment unavailability, the complexity of the genome, and current bioinformatics algorithms are some of the limitations of the whole-genome sequencing approach to diagnosing rare disorders.

Despite technological advancements for diagnosing rare disorders, more than half of the patients with suspected genetic disorders remain without a definite diagnosis. An estimate of more than 200 million people live with unresolved genetic disorders globally, and there might well be more than this because of the non-recognition of genetic disorders. Patients with undiagnosed genetic disorders are particularly vulnerable to specific unmet needs. Without appropriate diagnosis, the patient might not have proper treatment, which could have irreversible consequences on their health. International projects such as The Voice of Rare Disease Patients in Europe, Global Commission to End the Diagnostic Odyssey for Children with a Rare Disease, and Undiagnosed Diseased Network for Children have been created to support those who remain undiagnosed with genetic disorders.

8.3 APPLICATIONS OF DIAGNOSTICS IN PHARMACOGENETICS AND PHARMACOGENOMICS

Pharmacogenetics refers to the study of inherited differences, such as variations in drug metabolism. Due to the existence of variability among individuals before a response to a drug or therapy, it is a difficult task to predict the degree of effectiveness of a medication for a particular patient. It is well known that clinical factors such as body size, age, sex, hepatic and renal functions, and pharmacological factors such as differences in metabolism, drug distribution, and drug-directed proteins are well known to affect drug response. Pharmacogenetics studies an individual's reaction to a drug at the genetic level. It is a primary clinically proven application that has the potential to bring a revolution in drug therapy. Due to this, diseases are treated and controlled to enhance the quality of a patient's life. Mostly, the one medication fits all medicines cannot effectively work for all individuals (Jan and Cornelius, 2009). It is difficult to envisage the individual benefiting from the one with adverse side effects. Pharmacogenetics studies the molecular mechanisms that account for variation in response to the drug due to inherited characteristics. Pharmacogenetics involves the utilization of limited and specific genetic markers, and its application in the clinical management of an individual is known as personalized medicine. The goal of personalized medicine is to use an individual's genetic data to prevent genetic diseases, diagnose a disease at its early stages or initiate treatment to cure diseases.

After elucidating the human genome, advancements in technology and bioinformatics have led to the exponential expansion of molecular genetic research. As a result, there has been a shift in the research paradigm from monogenic disorders to genetically complex and more prevalent cancer, diabetes, cardiovascular and psychiatric disorders, adversely affecting public health to a greater extent. Pharmacogenomics is defined as the study of variations in DNA and RNA characteristics concerning a drug response. Specifically, pharmacogenomics evaluates the molecular determinants at genome, proteome, and transcriptome levels.

Due to the rapid advancement in genomic medicine, the genetic biomarkers can move away from empirical and population-based approaches into a stratified one, putting an end to the trial-and-error approach to therapy. Only 50% of patients were estimated to respond positively to medication. The other 50% might experience therapeutic delays or might not be medicated precisely as per the symptoms. Few individuals might also face adverse drug reactions, leading to death under extreme conditions. Identifying genetic factors might predispose a patient to adverse drug reactions, thereby preventing such reactions. Genetic testing might help predict drug interactions among individuals. CYP2D6 metabolizes 20 - 25% of the drugs in the human body. Patients with multiple CYP2D6 copies are rapid drug metabolizers and may not achieve therapeutic plasma levels at a regular dose. Conversely, individuals with few functional CYP2D6 genes are slow drug metabolizers, potentially causing drug levels to exceed the therapeutic range (Jeffrey and Geoffrey, 2003). Subjecting an individual to molecular screening tests will help identify many diseases at earlier stages, where the conditions might be favorable for the treatment and curing of a disease. A DNA test will be critical to disease prevention as the genetic information is accessible long before beginning a disease process. Genetic tests are more cost-effective than phenotypic tests, as delayed diagnosis might lead to morbidity and increased medical procedures. Genome monitoring assays will also help to detect the recurrence of a disease and completion of the treatment. Genome-based prognosis tests can assess the risk of disease progression and will help in the process related to the treatment of a disease. After initial reluctance, pharmaceutical companies have started applying pharmacogenomics and biomarkers for drug development, with this becoming a pervasive strategy. The application of pharmacogenetics and genomics in phase II clinical trials is for developing new molecules with reduced risk and cost (Juliette et al. 2016). For the development of therapeutics that target an individual's genetic makeup, it is imperative to create predictive diagnostic genetic tests known as companion diagnostics, along with novel drug candidates. Companion diagnostics will enable stratified patient subsets for their correlation with therapeutic outcomes, forming a critical and necessary complement to targeted drug therapies. Pharmacogenetic and genomic tests will help identify patients who are safely responding to the approved but not prescribed medications due to toxicity and inefficacy issues.

8.3.1 Commercially Available Pharmacogenetic and Pharmacogenomics Tests

Hundreds of pharmacogenetic and genomic tests are commercially available or are performed in laboratories and are waiting for their final clearance. The FDA approved many DNA and protein-based tests for in-vitro diagnostic analysis. Earlier, these tests were intended to target a single gene mutation to predict diseases and risk for the patient. Novel tests can evaluate thousands of genes and more than 100 genetic variations. The Hercep test (Dako) was one of the first test kits approved in 2001 by the Center for Devices and Radiological Health to detect HER2 (Human Epidermal Growth Factor 2) overexpression in breast cancer tissue. The HER2 gene copy number measurement is through FISH. Complex multigene kits for diagnosing breast cancer, such as FDA approved 70 gene based MammaPrint of Agendia, are available. This kit helps stratify early-stage breast cancer patients into low and high-risk groups for long term disease management. The Amplichip CYP450 test (Roche) was the first pharmacogenetic test approved by the FDA in 2005 for genotyping 27 CYP2D6 and 3 CYP2C19 alleles that are associated with different drug metabolizing phenotypes. DMET Plus Panel (Affymetrix) covers a wide range of genetic variations, including SNPs, insertions, deletions, trialleles, and copy number variations influencing drug metabolism. This panel identifies 1936 drug metabolizing biomarkers located in 235 genes. The PhyzioType (Genomas) system employs a biostatistical algorithm and 384 SNPs from 222 genes. This product is waiting for FDA approval and patent. PGxPredict: The CLOZAPINE test (PGxHealth) can detect a nucleotide polymorphism in the HLA-DQB1 gene for predicting the likelihood of clozapine-induced agranulocytosis. Pharmacogenetic tests for detecting the HLA-B*1502 allele of carbamazepine-induced Stevens-Johnson syndrome are also performed. This test is recommended for the Asian descendants as the reaction is 10 times higher in this population group. A list of other pharmacogenetic tests is provided in Table 8.1.

8.3.2 Clinical Significance of Pharmacogenetics and Pharmacogenomics

8.3.2.1 Cancer

Phenotypic and genotypic variations are found in cancer patients with identical types and stages. Cancer therapies mainly target cell surface receptors or downstream effector molecules, facilitating

TABLE 8.1
List of Commercially Available Pharmacogenetic and Pharmacogenomic Tests for the Detection of Diseases

S.No.	Disease	Drug Name	Manufacturing Company/Test
1.	HIV infection	Maraviroc	Trofile
2.	Fungal infection	Voriconazole	Roche
3.	Attention-deficit hyperactivity disorder, depression, obsessive compulsive disorder	Atomoxetine fluoxetine	Roche
4.	Colorectal cancer	Capecitabine	Thera Guide
5.	Breast cancer	Trastuzumab	HercepTest
6.	Acute lymphocytic leukaemia	Azathioprine, 6-MP, Thioguanine	Prometheus TPMT Genetics
7.	Chronic myelogenous leukaemia	Nilotinib	Invader UGT1A1 Assay
8.	Colorectal cancer	Irinotecan	Invader UGT1A1 Assay

mutations in the signaling pathway to influence drug sensitivity and resistance. One of the significant outcomes of pharmacogenomics research in oncology is enabling mutations to inform the treatment decisions and predict patient outcomes. There is a considerable improvement in the analysis of cancer tissues and the stratification of patients. Most of this progress has been made in identifying molecular features that determine tumor classification, prognosis, treatment, and response. Pharmacogenetic tests for diagnosing different cancers have been developed and are commercially available. One of the best among these is the HER2 receptor amplification test for breast cancer and treatment with trastuzumab (Genentech). HER2 is overexpressed in one quarter of the patients with breast cancer. Overexpression of oncogene HER2 is correlated with enhanced tumor formation, metastasis, poor prognosis, and resistance to chemotherapy. Pharmacogenetic testing for breast cancer and its treatment with trastuzumab has become integral as the degree of variability in HER2 gene expression helps determine the patient's response to the treatment. Pharmacogenomics is also responsible for the significant advancement in lung cancer treatment. Erlotinib (genentech) and gefitinib (astrazeneca) are tyrosine kinase inhibitors designed to target epidermal growth factor receptors (EGFR) shown to influence predisposition to lung cancer.

8.3.2.2 Cardiology

The role of pharmacogenomics is rapidly increasing in recent years, making promising discoveries. The discovery of two anti-thrombotic drugs, warfarin and clopidogrel, has further strengthened their role in human health. Novel anticoagulant agents such as dabigatran and etaxilate mesylate entered the market after approval by the FDA in 2010. Apart from this, oral coumarin anticoagulants, including warfarin, acenocoumarol, and phenprocoumon, have been the standard coagulants for the treatment of thromboembolic disorders for the past 65 years. Despite their high efficacy, these drugs have a narrow therapeutic window and pose an increased bleeding risk during initial treatment. A substantial variation existed among individuals concerning oral coumarin anticoagulants depending on the patient's age, sex, body mass index, vitamin K intake, smoking, requiring frequent monitoring and dosage adjustment. Inter-individual differences in oral coumarin anticoagulants dose response are also significantly influenced by genetic variations in two enzymes, CYP2C9 and VKORC1. CYP2C9 metabolizes oral coumarin anticoagulants and VKORC1is the target for these two drugs. Variations in CYP2C9*2 and CYP2C9*3 alleles have decreased the CYP2C9 enzymatic activity, inhibiting the metabolism of oral coumarin anticoagulants. Polymorphisms in VKORC1-1639 G>A influence pharmacodynamic response to coumarins. Polymorphisms in these two enzymes have enforced the revision of warfarin by including pharmacogenetic information in 2007 by the FDA. However, medical societies such as the American College of Chest Physicians have not changed their guidelines due to inadequate data from prospective studies. Clinical trials are underway to determine the influence of pharmacogenetic oral coumarin anticoagulants dosage on treatment.

8.3.2.3 Psychiatry

Genome-wide association studies have identified genetic variants that provide new molecular targets for antipsychotic and antidepressant agents. Antipsychotic medications affect the dopamine pathway and its components, with dopamine receptor genes DRD2 and DRD3 displaying polymorphisms. An association between serotonin receptor genes HTR2A and HTR2C, and their outcomes was identified. Depression studies have identified treatment outcome association of genes in the serotonergic and noradrenergic systems. Specific associations have been reported for polymorphisms in the 5-HTTLPR serotonin transporter (SLC6A4), HTR2A, and HTR1A serotonin receptor genes. Genome wide association studies have identified the genetic elements that respond to lithium, an ion with anti-suicidal and mood-stabilizing effects. In fact, its biochemical mechanism of response is unclear. Studies have implied that the genes responsible for the inositol pathway might be involved in lithium action. CYP450 isoenzymes metabolize most psychiatric drugs. Specifically, CYP2D6, CYP1A1, CYP3A4, and CYP2C19 isoenzymes metabolize antidepressants and antipsychotic agents. Studies have reported polymorphisms in CYP2D6 associated with the side effects of the drug, risperidone. Genotyping of the CYP2D6 gene may identify patients who need to be monitored for risperidone serum levels. Other findings also demonstrated a correlation between the genetic variants for CYP2D6 and serum levels of the anti-depressants venlafaxine, nortriptyline, and paroxetine. Depression patients with CYP2D6 duplicated genes are ultra-metabolizers of nortriptyline and fail to respond to treatment. Patients with two non-functional copies of the CYP2D6 gene were poor metabolizers of tricyclic anti-depressants with elevated plasma levels of the drugs.

8.4 APPLICATIONS OF DIAGNOSTICS IN PERSONALIZED MEDICINE

The market value of molecular diagnostics is more than $3 billion and grows at 25% per year. Molecular diagnostics is predicted to revolutionize the drug discovery, development process, selection customization, dosage, routing existing and new therapeutics, and personalize medical care. Many drugs prescribed for various diseases show limited efficacy in 70% of patients. These phenotypically derived drugs treat only the symptoms of a disease but do not cure them. These drugs are not acceptable in the long run. Adverse drug reactions caused by the failure to predict their toxicity in individuals and toxic drug-drug interactions are responsible for more than 100,000 deaths and more than 2 million hospitalizations, costing billions of dollars for health care (Mateja et al. 2021). It has been predicted that 20 - 40% of patients receive the wrong drugs during treatment. Elucidation of the human genome and expansion of proteomics research combined with emerging technologies such as functional imaging, biosensors, and bioinformatics have produced many unprecedented changes in the health care system by introducing the genetically prescribed medication approach.

Personalized medicine prescribes specific therapeutics that are best suited for an individual. For a given disease, the rate of disease progression for each person is unique, and every individual uniquely responds to a drug. Personalized medicine detects disease predisposition, provides early disease diagnosis, assesses its prognosis provides pharmacogenomic measurements of drug efficacy and risk of toxicity, and monitors the status of the illness until the outcome. Personalized medicine takes advantage of molecular understanding of disease for optimizing drug development. Personalized medicine will influence the population's preventive resources and therapeutic agents before they succumb to a disease. Selection of drug targets and optimal drug dosage, monitoring of the patients for short and advanced clinical trials at low cost, predicting the response of individuals to drugs and their toxic effects, reducing the overall cost of drug development, increasing the drug value, provision for improvement of effective health for all individuals in a population are the goals of personalized medicine in drug development. Personalized medicine can substantially enhance the productivity of drug discovery. By identifying the correct gene, proper pathway, and exact target, one can develop an appropriate drug for the treatment of a disease. This approach reduced the number of new compounds under phases II and III of human clinical trials by 20%. The number of patients in phase III trials was reduced by 10%, and the length of the phase III trials was reduced by 20%. The development of biomarker-based drugs has considerably reduced the compound failures during clinical trials and an estimated savings of up to $500 million per drug launched.

8.4.1 SNP Genotyping and Personalized Medicine

The EU council has considered personalized medicine as a medical model system that uses characteristics of an individual's phenotypes and genotypes such as molecular profile, medical imaging for designing the right therapeutic strategy for an individual, determining the predisposition to a disease, and targeted prevention of a disease. Personalized medicine has the potential for optimization, timely delivery, and treatment closing, enabling patients to receive maximum benefit with the least risk associated with toxicity and trial and error process. Molecular diagnostics play a pivotal role in personalized medicine by embedding at every stage of the process, from providing reliable information for the treatment of the patient to precise monitoring of its effects. Molecular diagnostics identify specific biological traits in an individual's genome pinning the diagnosis, prognosis, and predictions of disease recurrence. Further advancements in the technology can enable personalized medicine to classify disease subtypes and identify patients based on their response to a therapy or a modality. An individual's response to a drug is a complex interaction of genetic and non-genetic factors comprising genetic variants in the drug target, genes involved in the disease pathway, and drug-metabolizing enzymes (Stuart et al. 2011). All of these enzymes can predict a drug's efficacy or toxicity. More than one million SNPs are available as genetic markers for genotyping and phenotype studies. Various technologies, along with bioinformatics support, have led to the booming mining of high-quality sequence variations, revealing numerous loci that can generate resourceful data for patient management. Genotyping uses a variety of techniques such as oligonucleotide genomic arrays, gel and flow cytometry, sequencing, mass spectrometry, microarrays, and gene chips for increasing the rate of data generation and analysis. As per the Wellcome trust, more than 2.1 million SNPs have been identified and have been deposited into the SNP consortium public database (URL: www.snp.cshl.org). A high-resolution SNP map will expedite the identification of complex diseases such as asthma, diabetes, atherosclerosis, and psychiatric disorders. SNP technology is focused on detecting a predisposition to cancer, predicting toxic responses to drugs, and selecting the ideal combination and individual anticancer drugs.

Despite its widely recognized applications, the clinical practice of molecular diagnostics is still limited. Present-day health care systems are not designed to support personalized medicine and diagnostic technologies. Even though molecular diagnostic tests can influence therapy and its selection, reimbursement of these tests is very limited. New pricing and reimbursement models are supported by rewarding innovation and value to increase the utilization of these tests.

8.5 APPLICATIONS OF DIAGNOSTICS IN FAMILY RISK ASSESSMENT

Risk assessment is a vital part of genetic testing. Risk percentage calculation should be as accurate as possible for the counselor to make decisions for the patient and his family members. Genetic risk refers to the probability of an individual carrying a specific disease associated with a mutation or being affected by a genetic disorder. Results of genetic testing such as mutations, polymorphism markers, presence of risk factors, genetic tests of parents and siblings carrying a specific mutation, the ethnic background of a parent and overall mutation rate for the ethnicity, frequency of the mutations in the family, and population are factors considered for calculating the genetic risk. Genetic risk data obtained from population studies is the most common starting point for risk assessment studies. A good example is the genetic risk variation between ethnic groups described for the disease cystic fibrosis occurs due to the mutation in the CFTR (cystic fibrosis transmembrane conductance regulator) gene. The distribution of the diseased allele for the CFTR gene varies between different ethnic groups, such as Caucasians, providing the inputs for calculating the risk of the disease. The possible misinterpretation of the data for allelic bias introduced by migrating populations, environmental factors, and misclassification of the outcome are the other factors considered for assessing the risk of a determined population. Elucidation of the human genome and the development of molecular biology techniques have led to close integration of genetic and epidemiological studies, considerably decreasing the misinterpretation. As a result, a flood of information reflecting the association between genetic variation and disease-related outcomes has been generated. The number of possible associations and polymorphisms is completely calculated by characterizing millions of SNPs in the human genome (Nader et al. 2019). The application of common genetic variations associated with studies to generate risk profiling using multiple vulnerability genes might play a pivotal role in the early diagnosis of high-risk individuals and patients.

Genetic counseling has been developed to address the clinical and social consequences of Mendelian disorders. Consequently, this has become an integral part of genetic testing. The focus of genetic counseling is on assessing genetic risk, educating at-risk individuals and their family members about disease management and manifestation, education on reproductive options, and providing the psychological and emotional support to cope with untreatable disorders. Baye's theorem states that the likelihood of a disease or a condition after undergoing a test depends on the test's specificity, sensitivity, and probability of disease before conducting the test. Baye's theorem states that the probability of a disease or a condition after undergoing a test depends on the specificity and the sensitivity of the test and the probability of disease before conducting the test. The first part of the theorem depends on the technology, and the second part is more complex, comprising factors influencing probability assessment.

Genetic testing is a complex approach made from the inputs provided by genetic counselors, clinicians, nurses, and researchers. For single-gene disorders, the risk assessment is calculated on:

1. The outcome of the results emerging from genetic testing of symptomatic individuals or patients,
2. Assessing the relatives of the patients with a disorder,
3. Prenatal diagnosis, through which the amount of risk is calculated,
4. Predictive evaluation for a mutation causing the gene to develop symptoms.

Genetic testing provides information to health care providers for assigning the information to healthy patients to reduce the risk of their developing a disorder. This reduction should be more accurate, requiring continuous efforts to identify small changes in the background of many risk factors involved.

8.6 APPLICATIONS OF DIAGNOSTICS IN TRANSPLANTATION

Accurate diagnosis forms the main prerequisite for precision medicine during organ transplantation. The rejection of a transplant by the recipient immune system has been classified into T cell-mediated rejection and antibody-mediated rejection. Advanced diagnostic tools such as C4D staining and solid phase HLA antibody testing have provided a more accurate diagnosis of these two rejection entities. However, the unmet need in transplantation is the utilization of more precise diagnostics, which was dominated by the application of omics technologies and solid phase HLA antibody testing platforms. HLA testing has evolved over 40 years with comprehensive molecular HLA typing and sensitive solid phase HLA antibody screening methodologies facilitating the donor-recipient humoral risk assessment. The sensitivity of donor-specific cross match has been considerably increased by AHG-augmentation and flow cytometry-based methods. The introduction of solid-phase antibody testing has improved the sensitivity and specificity of HLA and non-HLA antibody identification. Single HLA antigen bead assays were further modified to distinguish complement-fixing antibodies from non-complement fixing ones. Even though the complement-fixing HLA antibodies are associated with higher rejection rates and allograft loss, non-complement binding HLA antibodies have also been documented. The development of rapid, high throughput sequencing and PCR technology for HLA typing has allowed more accurate recipient and donor HLA genotype definition and classification. High resolution HLA genotyping can differentiate certain unique epitopes for which a recipient may make specific antibodies, determining the humoral risk profile of the recipient.

Elucidation and understanding of the human genome have improved knowledge of the donor and recipient genotypes for transplantation. Utilization of genome-wide association studies using high throughout genotyping gives tools for discovering the genes related to pathogenesis. Whole-genome sequencing using next-generation sequencing technology serves as a biomarker for graft injury, providing the donor-derived cell-free DNA circulated for organ transplantation. As a result, molecular diagnostics uses circulating cell-free DNA as an indicator of cell damage and subsequent lysis, resulting from cytotoxic immune responses. Donor-derived cell-free DNA is measurable in the blood and urine of transplant recipients, serving as a direct marker indicating the allograft injury. In female recipients with organ transplantation from a male donor, the Y chromosome was considered a marker to identify the donor-derived cell-free DNA. Sigdel and his colleagues (2015) analyzed donor-derived cell-free DNA in urinary samples as an indicator of the injury incurred during a kidney transplant. Urine samples of female renal transplant and male donor kidneys were analyzed for the Y chromosome after biopsy matching. The average donor derived cell-free DNA was significantly higher when compared with stable recipients. De Vlaminck and his colleagues (2014) have developed genome transplant dynamics for identifying the donor-derived cell-free DNA regardless of the gender, using SNPs distributed across the genome to discriminate between donor and recipient DNA molecules. With this model, De Vlaminck correlated increased donor-derived cell-free DNA levels and acute rejection in patients with heart transplantation. Circulating cell-free DNA was isolated from plasma samples, purified, and sequenced. Genome transplant genomics has been shown to display better characteristics than the commercially available gene expression assay allo map test for monitoring rejection after heart transplantation. In the recipients of lung transplantation, Donor derived cell-free DNA can be used as a marker for graft injury to improve the performance of transbronchial biopsy. The differences in the transplanted tissue mass between bilateral and single lung transplant recipients were studied by checking the donor-derived cell-free DNA levels in the plasma of lung transplant recipients diagnosed with acute or chronic rejection. Donor derived cell-free DNA levels were elevated in samples of patients with rejection (Michel et al. 2016). The SRY gene on donor derived cell-free DNA is now considered a means for quantifying the level of rejection and assessing the health status in liver transplanted patients.

Along with the DNA, RNA is also used as a biomarker for assessing the level of transplantation. Identifying mRNA transcripts as biomarkers for the response to allograft injury is the method applied for the early identification of an allograft injury. Identifying the pathogen-related transcripts will also provide an opportunity for diagnostic analysis and allows targeting of pathophysiologic processes. Antibody-mediated rejection of a graft leads to an upregulation of a non-coding RNA cluster in endothelial, monocyte, and natural killer cells. Based on their size, these non-coding RNAs are classified into long and small ones. Long non-coding RNAs (lncRNAs) are the intracellular noncoding ribonucleotides that assist in the regulation of genome and proteome. These lncRNAs can be used as markers for acute rejection of the renal transplant, as they can be detected in the patient's blood. Compared with the controls, RP-11-395P13.3-001 and RP11-354P17.15-001 were upregulated in the urine samples of patients suffering from acute rejection. Urinary lncRNA RP11-354P17.15-001 might serve as a biomarker for acute renal kidney transplant recipients. T-cell immunoglobulin mucin domain 3 (TIM3) gene plays a vital role in the acute rejection of renal transplants. TIM3 serves as the identifier for proinflammatory TH1, TH17, and other cytotoxic T-lymphocyte, natural killers, and primary T-effector cells. Antirejection therapy has decreased TIM-3 mRNA expression in acute renal kidney transplant recipients, which has reduced their utility for predicting disease prognosis after treatment. There is mRNA expression of 10 genes DUSP1, CFLAR, ITGAX, NAMPT, MAPK9, RNF130, IFNGR1, PSEN1, RYBP, and NKTR in the blood or biopsy of a graft in patients subject to rejection. MicroRNAs (miRNAs) are also regulated in organ transplantation recipients. miR-122 and miR-148a were upregulated in recipients experiencing hepatic acute allograft rejection (Mengela et al. 2013). Using a TaqMan low-density array human microRNA panel containing 365 mature human miRNAs, overexpression of miR-142-5p, miR-155, and miR-223 can be noticed in acute rejection patients. RNAs used as the biomarkers for the graft and for acute rejection is given in Table 8.2.

8.6.1 Role of Short-Term Repeats in Organ Transplantation

Short tandem repeat (STR) analysis is being used for understanding the variations between the donor and recipient DNA before organ transplantations. STR analysis and the identification of amelogenin

TABLE 8.2
List of RNAs Used as Biomarkers for Diagnosing the Graft Rejection

Organ	Sample	RNA
Kidney	Urine	lncRNAs RP11-354P17.15-001 and RP-11-395P13.3-001
Heart	Blood	mRNA CD-154
Lung	Blood	mRNA ICOS
Kidney	peripheral blood	mRNA TIM-3
Kidney	Blood	mRNA CFLAR, DUSP1, IFNGR1, ITGAX, MAPK9, NAMPT, NKTR, PSEN1, RNF130, RYBP, CEACAM4, EPOR, GZMK, RARA, RHEB, RXRA, SLC25A37
Liver	Plasma	miR-122 and miR-148a
Kidney	Blood	miR-142-5P, miR-155 and miR-223

gene on the X and Y chromosomes is performed on leukemia patients for allogeneic bone marrow transplantation to reconstitute the recipient bone marrow with donor stem cells for the treatment of a variety of hematological malignancies by establishing complete engraftment. STR profiles of donor and recipient help in assessing the engraftment status after transplantation. The presence of only the donor allele indicates complete engraftment, and mixed chimerism shows the presence of both donor and recipient markers in different loci. STR assay for chimerism is helpful for the follow-up of post-liver transplantation. 4% of the donor lymphocytes are present in the recipient up to 3 weeks after transplantation. After four weeks, the persistence of donor lymphocytes can cause graft-versus-host disease (GVHD) in the recipient, which remains undiagnosed until six weeks after transplantation.

8.7 APPLICATIONS OF DIAGNOSTICS IN EPIGENETICS

Epigenetics studies heritable and non-encoded genetic changes in an organism by triggering the genes on and off. It brings specific modifications to the genes and activates them by histone acetylation and DNA methylation. Epigenetic changes modulate gene expression in humans and alter the cellular pathways making an individual susceptible to diseases such as pulmonary fibrosis and liver disease. Most epigenetic changes are stable and heritable. Chromatin modifications are independent of the traits embedded in the DNA and are responsible for the phenotype. The nucleosome is the core unit in the chromatin, consisting of 147bp DNA folded around histone octamers composed of two histone proteins, H2A, H2B, H3, and H4. H2A and H2B proteins exist in multiple forms differing in their primary amino acid sequence and histone sites (Shruti et al. 2022). Changes in the chromatin structure will permit the DNA to be accessed by transcriptional and regulatory complexes.

8.7.1 Mechanisms of Epigenetic Changes

Three types of mechanisms are responsible for changes to the chromatin structure.

8.7.1.1 DNA Methylation

DNA methylation occurs by the addition of the methyl group to the genomic DNA, which leads to the modification of the gene function and affects expression. Covalent addition of the methyl group to the 5th carbon of cytosine ring resulting in 5-methyl cytosine is a highly characterized DNA methylation process. 1.5% of the DNA in the human genome is methylated to 5-methyl cytosine. These methyl groups bind to the major DNA groove, inhibiting its transcription. Transmission of gene expression from parental cells to daughter cells occurs through DNA methylation. This mechanism of epigenetic regulation is economical, involving epigenetic markers in a 5-C position in the cytosine ring. Differences in DNA methylation among different genomes are associated with epigenetic memory formation. The level of DNA methylation is sharply decreased during early embryogenesis among mammals, increasing during later stages through de novo methylation. The existence of CpG islands and GC enriched regions is one of the most interesting regions and patterns of DNA methylation. Various genes in the human and other mammalian genomes contain many unmethylated CpG dinucleotides in their 5' regions. DNA methylation is also responsible for the inactivation of the second X chromosome in females. DNA methylation is closely associated with gene silencing. Silent genes in the inactivated X chromosome are heavily methylated after gene silencing. DNA methylation is a mechanism for recognizing the genes to be silenced. Their promoter regions are also methylated for their irreversible inactivation in somatic cells right from the embryonic stage. Protein complexes comprising DNA methyl transferases recognize the modified nucleosomal histones H3-K9. Methylation prevents the binding of transcription factors to the promoter regions (García-Giménez et al. 2017). Alternatively, it signals to the binding proteins for more reliable gene expression. The regions of unmethylated DNA are associated with actively expressed genes, with their promoters not undergoing DNA methylation due to their non-association with chromatin structures involved in methylation.

An elegant technique, bisulfite PCR, can identify methylation sites in the genome. For performing this PCR, a small amount of the DNA is treated with bisulfite to convert unmethylated cytosines into uracils. Primers for the PCR should have 200 - 300 bases of CpG island between them containing enough CpG dinucleotides. Upon amplification, the product is pyrosequenced, or its symmetrical methylation positions in DNA chains are analyzed by ligating a linker to its 3' terminus, which links the complementary DNA chains of CpG islands in the amplified product.

8.7.1.2 Histone Modification

A posttranslational modification occurs to the histone proteins, such as methylation, phosphorylation, acetylation, sumoylation, and ubiquitylation. Post translational changes to the histone proteins will alter the chromatin structure or recruit the histone modifiers, affecting gene expression. Histone modifications are also involved in diverse biological processes such as transcriptional activation or inactivation, chromosome packaging, DNA damage, and or repair.

8.7.1.3 Non-coding RNAs

Non-coding RNAs are transcribed from DNA but are not translated into proteins. These RNA forms regulate gene expression at the transcriptional and post-transcriptional level through histone modification, DNA methylation, gene silencing, and heterochromatin formation. Non-coding RNAs are also known to modulate drug effects. Non-coding RNAs are divided into two groups, long non-coding RNAs of length up to 200 nucleotides, and short non-coding RNAs of the length 30 nucleotides which are further divided into three major classes: microRNAs (mi RNAs), short interfering RNAs (si-RNAs), and piwi interacting RNAs (piRNAs). The role of miRNAs in frequent dysregulation of malignancies and tumor cell drug resistance have been elucidated. miRNAs act as onco or suppressor mi-RNAs (Vickers et al.2011). Modulation of miRNA deficiencies either by antagonists or mimics is a novel area of drug research for the restoration of the gene regulation network and its pathways, leading to drug resistance.

Epigenetic mechanisms will determine the organization of genes in the nucleus of a cell. They will influence their expression by changing the conformation of chromatin and the accessibility of the DNA to transcriptional factors, and its mechanics. Epigenetic changes such as posttranslational histone modifications and DNA methylation will result in the activation, silencing, or position of genes, regulating their expression.

8.7.2 Epigenetic Biomarkers

Epigenetic changes that occur in an organism across the genome represent an orchestrated phenomenon that can change the transcriptional outcome of a genetic code. Identifying epigenetic changes associated with a disease and factors promoting such changes serve as biomarkers for identifying these changes while making clinical decisions. An epigenetic marker is stable and reproducible during sample processing, can be measured in the body fluids or in the tissues with primary preparations such as fresh, frozen, formalin fixed paraffin embedded for predicting the risk of future disease development, defines or detects a disease, can say something about the natural history of a disease, predicts a disease outcome, responds to a treatment or a therapy.

Epigenetic markers are more advantageous than genetic ones. They fill the clinical gaps by showing the extent to which a genetic trait is controlled. They incorporate information from the environment and lifestyle of an individual and provide information regarding the natural history of a disease, considered to be bio-archives. Many epigenetic biomarkers such as miRNAs and PTMs are highly stable in plasma, serum, urine, saliva, semen, and vaginal secretion fluids. They are also highly stable and are detected in primary tissue preparations.

8.7.3 Epigenetics Based Tests Performed in Clinical Laboratories

Epigenetic based clinical tests have been developed and commercialized. So far, two methylation-based diagnostic assays have been commercialized for diagnosing neurodevelopmental disorders due to Prader-Willi syndrome and Angelman syndrome. Non-invasive prenatal testing (NIPT), also known as prenatal cell free DNA, was developed to screen for genetic abnormalities in the fetus. During this process, DNA from both mother and fetus is extracted from a maternal blood sample and screened for chromosomal abnormalities such as Downs, Edwards, and Patau syndrome, sex chromosomal abnormalities, and Rh blood type. Differential methylated regions are used for differentiating the maternal and cfDNA. Methylation of the maspin gene in the maternal plasma of 66 pregnant women, 55 normal fetuses, and 11 carrying trisomy18 fetuses was analyzed using methylation-based restriction endonucleases and chemical processing of DNA. Median unmethylated maspin concentrations were significantly increased in a woman carrying trisomy 18 fetuses compared with the normal ones. Unmethylated and methylated maspin concentrations are useful as potential biomarkers for diagnosing trisomy 18 in the cfDNA during the first trimester of the pregnancy, irrespective of the sex and genetic variations of the fetus. The thyramir miRNA classifier and ThyGenX oncogene panel consist of markers for the prediction of thyroid cancer. miRNA expression in samples is measured by RT-qPCR using a custom-designed miRNA PCR panel based on the expression of 10 miRNAs miR-29 b-1–5p, miR-31–5p, miR-138–1–3p, miR-139–5p, miR-146 b-5p, miR-155, miR-204–5p, miR-222–3p, miR-375, and miR-551 b-3p. Rosetta GX MI-lung is used to diagnose lung cancer patients. The combination of thyramir and ThyGenX has produced a sensitivity of 89% and specificity of 85%. The histone ELISA kit diagnoses auto-immune disorders. This kit permits the semi-quantification of autoantibodies against histones in human serum (Stuart, 2011). The presence of histones can be used in the differential diagnosis of systemic lupus erythematosus and other rare autoimmune disorders such as scleroderma that show nuclear proteins in the serum.

8.8 APPLICATIONS OF DIAGNOSTICS IN FORENSIC SCIENCES

Forensics is a novel methodology for establishing the truth based on evidence for the identification or by being reduced to an available motif. Medico-legal and criminal identification are the integrated parts of forensic identification with a probative value. This value resides in the ability of an expert to compare the evidence at a crime scene with reference evidence. Forensics can compare traces of blood, saliva, or any biological sample at the crime scene with those found on the suspect's clothes or with samples from the victim. Scientific progress and technological advancement have made medico-legal identification more accurate. Their role is becoming more significant in cases related to civil, family, criminal law, and catastrophes with many victims. Sir Alec Jeffreys used DNA fingerprinting as evidence in forensic science by proving that variations in the DNA sequence among two individuals can be used as a tool for the identification of the suspect in a crime scene. Sir Jeffreys pioneered genetic fingerprinting for forensic personal identification cases, including crimes, filiation, consanguinity, sexual abuse, and immigration. The discovery of the PCR technique by Kary Mullis in 1983 has opened new doors to DNA analysis in forensics. DNA is extracted from a biological human micro trace containing nucleated cells and is subjected to amplification using PCR. Since the European Council issued Recommendation No.92 for the acceptance of DNA analysis as evidence in the court, many countries have accepted DNA analysis in criminal justice.

8.8.1 Methods in Forensic Sciences for Identification of Humans

8.8.1.1 Autosomal STR Profiling

Since their discovery in the 1990s, microsatellite or short tandem repeats are a highly genotyped genome component for distinguishing individuals, identifying culprits, and exonerating the innocent. ST repeats are the gold standard for the identification of humans in forensic investigations. The nomenclature for the STR and their allelic variants was made in 1993 by the DNA commission of the International Society of Forensic Genetics (ISFG). They comprise mono, di, tri, tetra, penta, and hexanucleotide repeats. In most cases, tetranucleotide repeats are considered for genotypic analysis in the forensic sciences. An individual can be either homozygous with the same number of repeats or heterozygous with a different number of repeats. STR profiling is also used for testing the paternity or maternity of an individual, disaster victim identification, rape perpetrators identification, and kinship testing. STR markers are advantageous as they can be analyzed rapidly at a low cost (Bukyya et al. 2021). In 1997 the Federal Bureau of Investigation (FBI) launched the Combined DNA Index System (CODIS) database, which includes 13 - 16 autosomal highly polymorphic loci located on the non-coding regions of different chromosomes and amelogenin gene located on the X and Y chromosomes for determining the sex. CODIS combines DNA analysis with computational technology at local, state, and national levels to compare DNA profiles in digital mode. CODIS comprises of two indices, the forensic index, which contains the DNA from a crime location, and the offender index consisting of profiles from convicted offenses and crimes.

Apart from their utilization in forensics, STRs also have applications in medical genetic research due to the association of trinucleotide STRs with genetic disorders. STR kits in forensic labs are manufactured by Life Technologies, Promega, and Qiagen companies. Using these kits, around 15 - 16 STRs can be analyzed through multiplex PCR amplification, separated on capillary gel electrophoresis, followed by automated DNA sequencing.

8.8.1.2 Y-Chromosome Analysis

Due to its existence only in males, the Y-chromosome is useful in the crime scene, specifically while studying the male-female ratio in body fluid mixtures such as rape. By analyzing the STR content of the Y-chromosome, the male component information can be obtained (Shewale et al. 2003). Y-STR profiling is specifically helpful in tracing the identity of vasectomized or azoospermic rapists who do not leave any sperm traces for identification.

8.8.1.3 Mitochondrial DNA

As mitochondria are maternally inherited, all matrilineal share the same haplotype. The number of mitochondria copies per cell varies from 200 to 1700, increasing the probability of survival compared to nuclear DNA. Forensic analysis of the mitochondria includes the biological samples of old, severely degraded DNA and analysis of samples with very low DNA concentrations, such as hair shafts. Mitochondrial DNA has been used to identify Tsar Nicholas II and his brother Georgij Romanov (Ivanov et al. 1996).

8.8.1.4 Autosomal SNPs

SNPs have low heterozygosis when compared to repeats. SNP typing carries an advantage with a template size of 50 bp compared with STRs, which need a DNA template of 300 bp for obtaining good profiling. Due to this, SNPs have become vital in analyzing degraded DNA samples. SNP typing has been applied to identify the victims of the world trade center disaster in 2001 (Marchi et al. 2004).

8.8.1.5 Biomarkers in Forensics

Technology advancements have permitted the identification of biomarkers in forensic identification and will continue to improve. Studies have demonstrated the role of mRNA in forensics. Some mRNA is stable in body fluids, such as blood, semen, and saliva, and can be used as a tool for their identification in forensic cases. miRNA is also gaining popularity as a forensic marker for its identification in body fluids. The advantage of miRNA markers is their small size of around 18 - 22 bp and more stable when compared with mRNA. Variations in CYP206, a drug-metabolizing enzyme, are also considered to be a biomarker, as this leads to adverse drug effects followed by death (Levo et al. 2003).

8.9 CONCLUSION

The molecular basis for the diagnosis of human disorders refers to the detection, diagnosis, and prognosis of mutations in DNA. The combination of molecular diagnosis and clinical laboratory has rapidly developed over recent decades, with much of the support obtained from molecular biology. The identification and nuanced characterization of disease are vital for its accurate diagnosis. Discovery of the gene responsible for a disorder will provide enormous value to the insights into the disorder, allowing physicians to assess its pre-disposition and also to design or implement diagnostic methods. Molecular diagnostics in its current form is a clinical reality with its roots deep in gene expression and its function.

Molecular diagnostics will continue to be a significant source of global public health, facilitating the detection, characterization, and prognosis of a disease, monitoring the drug response, and assisting in the identification of genetic modifications and disease susceptibility. A wide range of detection tests is available to study the variations in the DNA and to study the gene expression. To implement these tests, hurdles such as which test to employ, choice of the technology, equipment needed, cost-effective issues, accuracy, reproducibility, and training of the personnel should be crossed. Presently, PCR-based detection systems are dominating diagnostics. Alternative technologies that aim to explore genome complexity without PCR involvement are expected to gain momentum in the market. The development of integrative chip devices for genetic readouts of single cells and biomolecules, along with proteome-based testing, will further improve molecular diagnostics and makes it accessible to society.

REFERENCES

Angural, A., Spolia, A., Mahajan, A., Verma, V., Sharma, A., Kumar, P., Dhar, M. K., Pandita K.K., Rai, E., Sharma, S. (2020). Review: Understanding Rare Genetic Diseases in Low Resource Regions Like Jammu and Kashmir–India. *Front. Genet.* 11:415.

Bukyya, J. L., Tejasvi, M. L. A., Avinash, A, P. C. H., Talwade, P., Afroz, M. M., Pokala, A., Neela, P. K., Shyamilee, T. K., Srisha, V. (2021). DNA Profiling in forensic science: A review. *Glob. Med. Genet.*8(4):135–143.

Chelsea, E. L., Kaela, S. S., Melissa, W., Victor, F. (2020). Rare genetic diseases: nature's experiments on human development. *Science.* 23: 101123.

De V. I., Valantine, H., Snyder, T. M., et al. (2014). Circulating cell-free DNA enables noninvasive diagnosis of heart transplant rejection. *Sci. Transl. Med.* 6:241ra77.

García-Giménez, J. L., Seco-Cervera, M., Tollefsbol, T. O., Romá-Mateo, C., Peiró-Chova, L., Lapunzina, P., Pallardó, F. V. (2017). Epigenetic biomarkers: Current strategies and future challenges for their use in the clinical laboratory. *Crit. Rev. Clin. Lab. Sci.* 54(7–8):529–550.

Ivanov, P.L., Wadhams, M.J., Roby, R.K., Holland, M.M., Weedn, V.W., Parsons, T.J. (1996). Mitochondrial DNA sequence heteroplasmy in the Grand Duke of Russia Georgij Romanov establishes the authenticity of the remains of Tsar Nicholas II. *Nat. Genet.* 12:417–420.

Jan, W., Cornelius K. (2009). Current applications and future trends of molecular diagnostics in clinical bacteriology. *Anal. Bioanal. Chem.* 394:731–742.

Jeffrey, S. R., Geoffrey S. G. (2003). The integration of molecular diagnostics with therapeutics implications for drug development and pathology practice. *Am. J. Clin. Pathol.* 119:26–36.

Juliette, P-F., Kaisa, I-C., Lotte, S., Anja, S., Roman, R., Denis, H., Mark, L. (2016). Enabling equal access to molecular diagnostics: what are the implications for policy and health technology assessment. *Public Health Genomics* 19:144–152.

Kym, M. B., Taila, H., Leslie, G. B., Richard, A. G. A., Micheil, I., Olaf, R., John, B., Sally, L. D., Nebojsa, J., Timo, L., Deborah, M. I., Karen, T., Axel, V., Gareth, B. (2019). A diagnosis for all rare genetic diseases: The Horizon and the Next frontiers. *Cell* 177: 32–37.

Lee C. V. (2011). Pharmacogenomics in clinical practice reality and expectations. *Pharmacy and Therapeutics* 36: 412–450.

Levo, A., Koski, A., Ojanprena, I., Vuori, E., Sajantila, A. (2003). Post-mortem SNP analysis of CYP2D6 gene reveals correlation between genotype and opioid drug(tramadol) metabolite ratio in blood. *Forensic Sci. Int.* 135:9–15.

Marchi, E. (2004). Methods developed to identify victims of the World Trade Centre disaster. *Am. Labor.* 36:30–36.

Marwaha, S., Knowles, J. W., Ashley, E. A. (2022). A guide for the diagnosis of rare and undiagnosed disease: beyond the exome. *Genome Med.* 14**:** 23.

Mateja, V., Karin, W., Aleš, M., Borut, P. (2021). Improving diagnostics of rare genetic diseases with NGS approaches. *Journal of Community Genetics.*12:247–256.

Mengela, M., Campbella, P., Gebela, H., Randhawaa, P., Rodrigueza, E. R., Colvin, R, Conwayg, J., Hachemh, R., Halloranb, P. F., Keshavjeeg, S., Nickersoni, P., Murpheyj, C. et al. (2013). Precision diagnostics in transplantation: from bench to bedside. *American Journal of Transplantation* 13: 562–568

Michael, N., Tara, S., Minnie, S. (2016). Advances in diagnostics for transplant rejection. *Expert Rev. Mol. Diagn.*16: 1121–1132.

Nader, I., Al, D. M, Walid, Q. (2019). Genomics and precision medicine: molecular diagnostics innovations shaping the future of healthcare in Qatar. *Advances in Public Health.* 2019:3807032.

Pollard, S., Weymann, D., Dunne, J. et al. (2021). Toward the diagnosis of rare childhood genetic diseases: what do parents value most? *Eur. J. Hum. Genet.* 29: 1491–1501.

Ralph, K. I., Laurence, M. D. (2004). Pharmacogenomics and pharmacogenetics: future role of molecular diagnostics in the clinical diagnostic laboratory. *Clinical Chemistry* 50:1526–1527.

Salah, A. M. A. (2017). Application of Molecular diagnostics in Microbiology: a review. *Sud. Med. Lab J.* 5: 75–84.

Samantha, P., Deirdre, W., Jessica, D., Fatemeh, M., John, B., James B.S.W., Jan, M. F., Sylvia S-I., Nick, D., Alison, M. E., Mark, H., Larry, D. L., Dean, A. R. (2021). Toward the diagnosis of rare childhood genetic diseases: what do parents value most? *European Journal of Human Genetics* 29:1491–1501.

Shewale, J. G., Sikka, S. C., Schneida, E., Sinha, S. K. (2003). DNA profiling of azoospermic semen samples from vasectomized males by using Y-PLEX 6 amplification kit. *J. Forensic Sci.* 48:127–129.

Shruti, M., Joshua, W. K., Euan, A. A. (2022). A guide for the diagnosis of rare and undiagnosed disease: beyond the exome. *Genome Medicine* 14:23

Sigdel, T. K., Vitalone, M. J., Tran, T. Q., et al. (2015). A rapid noninvasive assay for the detection of renal transplant injury. *Transplantation* 2:97–101.

Stuart, A. S. (2011). Personalizing medicine with clinical pharmacogenetics. *Genetics In Medicine* 13: 987–995.

Vickers, K. C., Palmisano, B. T., Shoucri, B. M., et al. (2011). MicroRNAs are transported in plasma and delivered to recipient cells by high-density lipoproteins. *Nat. Cell Biol.* 13:423–433.

Vinkšel, M., Writzl, K., Maver, A., Peterlin, B. (2021). Improving diagnostics of rare genetic diseases with NGS approaches. *J. Community. Genet.* 12(2):247–256.

9 Case Studies

9.1 INTRODUCTION

A genetic disorder occurs due to one or more abnormalities in a person's DNA. Most genetic disorders are inherited from one generation to another. Genetic conditions such as hypercholesterolemia (a genetic predisposition to high cholesterol levels) and cystic fibrosis are specific to a particular ethnic group. Genetic analysis of patients with symptoms will confirm the occurrence or the chance of inheriting a genetic disorder. Advancement in technology has further increased understanding of the genetic mechanisms responsible for genetic disorders, which have facilitated the development of early diagnosis, intervention, and treatment to prevent diseases or minimize their severity. This chapter offers case studies involving genetic tests for the diagnosis of genetic disorders.

9.2 FACTORS RESPONSIBLE FOR THE OCCURRENCE OF GENETIC DISORDERS

Genetic disorders occur due to a mutation inherited or generated as a body's response to mutagens and biological agents such as viruses. Diagnosis of a genetic disorder comprises three major components, physical examination, detailed family history, and clinical laboratory testing. Physical examination and primary care may not be able to diagnose a genetic disease but can be used along with detailed family history and differential diagnosis using available laboratory tests. Another major factor responsible for the occurrence of genetic disorders is an occurrence of a condition within family members, disclosed on pedigree and detailed examination of the family history. The occurrence of multiple miscarriages, stillbirths, and childhood deaths are mostly due to genetic disorders. The occurrence of health conditions in adults such as heart disease, dementia, and cancer in relatives at a young age might also be due to genetic predisposition. Other clinical symptoms such as developmental delay, mental retardation, congenital abnormalities, dysmorphologies involving the heart, facies, and growth problems also indicate the involvement of a genetic disorder or disorders. Some rare physical features such as wide-set or droopy eyes, flat face, short fingers, and tall stature may not involve genetic disorders. However, their occurrence should be confirmed by a genetics specialist. While genetic conditions appear predominantly during childhood, genetic disorders might also appear during adolescence or adulthood. In most cases, a genetic disorder remains undetected for several years until some factor or a metabolite, or toxin triggers the onset of the disorder. In these cases, detailed family history and physical examination will help in the early diagnosis

DOI: 10.1201/9781003343790-9

9.3 GENETIC TESTING

Depending on the type of abnormality, three main types of genetic testing are performed in laboratories, categorized into cytogenetic, biochemical, and molecular for the detection of abnormalities in chromosome structure, protein function, and DNA sequence, respectively. Cytogenetic tests involve the examination of chromosomes in dividing human cells under a microscope. T lymphocytes are the most readily accessible for cytogenetic analysis due to their easy collection from blood. Cells from bone marrow, amniotic fluid, and other tissue biopsies are also cultured for cytogenetic studies. After cell culture, chromosomes are fixed, spread on a microscope slide, and stained. Distinct banding patterns of the chromosomes will provide the analysis of chromosome structure. Biochemical tests involve the analysis of proteins and hormones performing multiple functions. Mutation in any of these proteins will lead to a failure in their function. Depending on the type of function, biochemical tests will measure protein activity and metabolite concentrations. These tests require a tissue sample consisting of the protein. The molecular analysis involves the testing of the DNA for mutations. For small mutations, direct DNA testing is the most effective method. A DNA test can be performed on any type of tissue, and it is required in small concentrations. Few genetic disorders such as cystic fibrosis occur due to multiple mutations on a single gene (CFTR), making molecular testing challenging.

9.4 CASE STUDIES

9.4.1 Huntington's Disease

A 62-year-old woman visited a clinic for neurologic and psychologic examination and genetic testing. As mentioned in Figure 9.1, her cousin suffering from mild chorea was recently diagnosed with Huntington's disease (HD). Her maternal uncle is reported to have similar symptoms but has never undergone any genetic testing. The patient's brother, aged 65, also recently tested positive for HD. Her mother was diagnosed as a paranoid schizophrenic but never with this specific disorder. The patient feels that she has become clumsier, having mild trouble while walking. She has denied any other symptoms.

Physical and neurological examinations are regular except for mild problems with tandem gait. The patient proceeded with genetic testing after appropriate counseling and signed informed consent.

In the DNA analysis by using PCR, an expanded trinucleotide CAG repeat in the first exon of the Huntington gene was detected. The expanded CAG repeat allele produces a longer fragment than the normal allele. The PCR product was sequenced on the ABI capillary sequencer. The sequencing of a

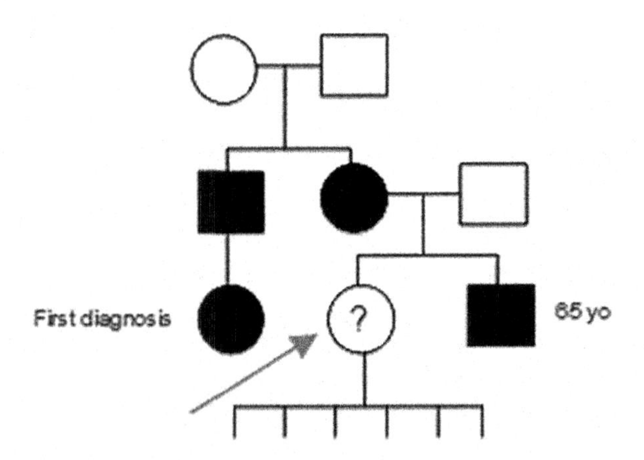

FIGURE 9.1 Pedigree of a woman suffering from neurologic and psychologic issues.

normal allele has provided two peaks, with the first one representing 19 repeats and the second one representing 21 repeats. The CAG repeat lengths are the same for the replicate sample. The capillary sequence result for the patient has also shown two peaks, one peak in the normal range and the other one in the expanded range. The number of CAG repeats in the patient's DNA was calculated using the formula [(size of band in bp - 48 bp)/3] + 2. The normal allele has a CAG repetition in the range of 9 - 26 and an expanded allele of >35 repeats. Also, an intermediate range consisting of 27 - 35 repeats was detected. Based on the DNA analysis, the patient's diagnosis was positive for Huntington's disease with a high end of reduced penetrance range.

This patient has inherited one normal allele, and one expanded allele from either of her patients with 40 CAG repeats in the tested DNA. Complete penetrance begins with 40 repeats; sometimes, patients with 41 - 42 repeats may not even manifest symptoms. The number of differences between the number of CAG repeats between the replicates shows that it was 39.43 rounded off to 39, and 39.6 rounded off to 40 repetitions. The patient has six children whose medical records are unavailable.

9.4.2 Down Syndrome

Down syndrome is one of the most frequent chromosomal disorders in the USA, occurring once every 800 births. This chromosomal disorder occurs due to a non-disjunction during egg development and failure in the segregation of the 21st chromosome pair during meiosis. In 3 - 4% of cases, this disorder occurs due to Robertsonian translocation, a process by which a chromosome attaches to another. As a result, the number of chromosomes will be 45 instead of 46. Here, the same amount of genetic material is shared by 45 chromosomes, whose offspring have a 25% Down syndrome karyotype. In 1 - 2% of cases, Down syndrome occurs due to non-disjunction after conception, leading to a mosaic pattern of inheritance. As a result, some cells are trisomic, and some are normal. A karyotype of mosaicism can explain Down syndrome in a child. It can also be diagnosed during the prenatal period through definite amniocentesis and chorionic villus sampling for Robertsonian translocation and premature birth of a child with Down syndrome. Women younger than 30 years bear a 1% risk of recurrent Down syndrome. Women older than 30 years will have the same risk percentage, but the recurrence of Robertsonian translocation is high. The risk of Down syndrome and other chromosomal disorders will increase with increasing maternal age. Prenatal diagnosis of women older than 35 years, forms the leading target group for births with Down syndrome. A fetus with ONTDs carries a high risk of having a child with Down syndrome due to low fetoprotein levels. This is 20% of the fetuses tested for Down syndrome using fetoprotein levels in the serum.

9.4.3 Fragile X Syndrome

Also known as Martin-Bell Syndrome, it is one of the most common genetically inherited disorders responsible for mental retardation affecting 1 in every 6000 males and 10,000 females. This disorder occurs due to an abnormality in the fragile site of the FMR 1 gene at locus Xq27.3 on the X-chromosome, leading to an increase in the number of CGG trinucleotide repetitions. As this mutation is X-linked, males are more severely affected than females, exhibiting mental retardation with characteristic physical features and behavioral alterations. An elongated narrow face with a large forehead with frontal bossing, long prominent chin with large anteverted ears, hypoplasia of the middle third of the face, mandibular protrusion, and macro orchadism (large, bilateral testes) are the predominant clinical manifestations of fragile X syndrome. ogival palate, cleft palate, the presence of mesiodens, dental hypo mineralization, and abrasion of the occlusal surfaces and incisal edges increase the dimensions of the dental crowns in the mesiodistal, and cervicoocclusal orientation, producing severe bone-dental discrepancies are the frequent intraoral anomalies of fragile-X-Syndrome. The presence of multiple supernumerary teeth, orodental findings in a maxillary and mandibular arch, and other physical features were seen in a 14-year-old Fragile X Syndrome boy.

A 14-year-old boy with multiple extra teeth in both upper and lower jaw was referred by a general dental practitioner to the Department of Pedodontics and Preventive Dentistry, SCB Dental College and Hospital, Cuttack, Odisha, India. On clinical examination, the child exhibited behavioral disorders with autistic features. He has an elongated narrow face with a large forehead, a prominent chin, frontal bone, hypotonia, hyperlaxity of the ligaments, and cognitive deficiency. Testes were bilaterally enlarged, and the testicular volume was >30 cc using Prader orchidometer beads (Figure 9.2). The intraoral examination has revealed multiple supernumerary teeth on the upper as well as lower dental arch, crowding, high-arch palate, macroglossia, cleft palate, and multiple carious teeth. Panoramic radiographic evaluation has identified the presence of multiple teeth in the anterior region of both maxillary and mandibular dental arches with congenital absence of teeth in 18, 28, 38, and 48. The family history shows some mental retardation in his sibling. Due to the presence of mental retardation, physical characteristics, and macrochordism, the patient was diagnosed with fragile X Syndrome.

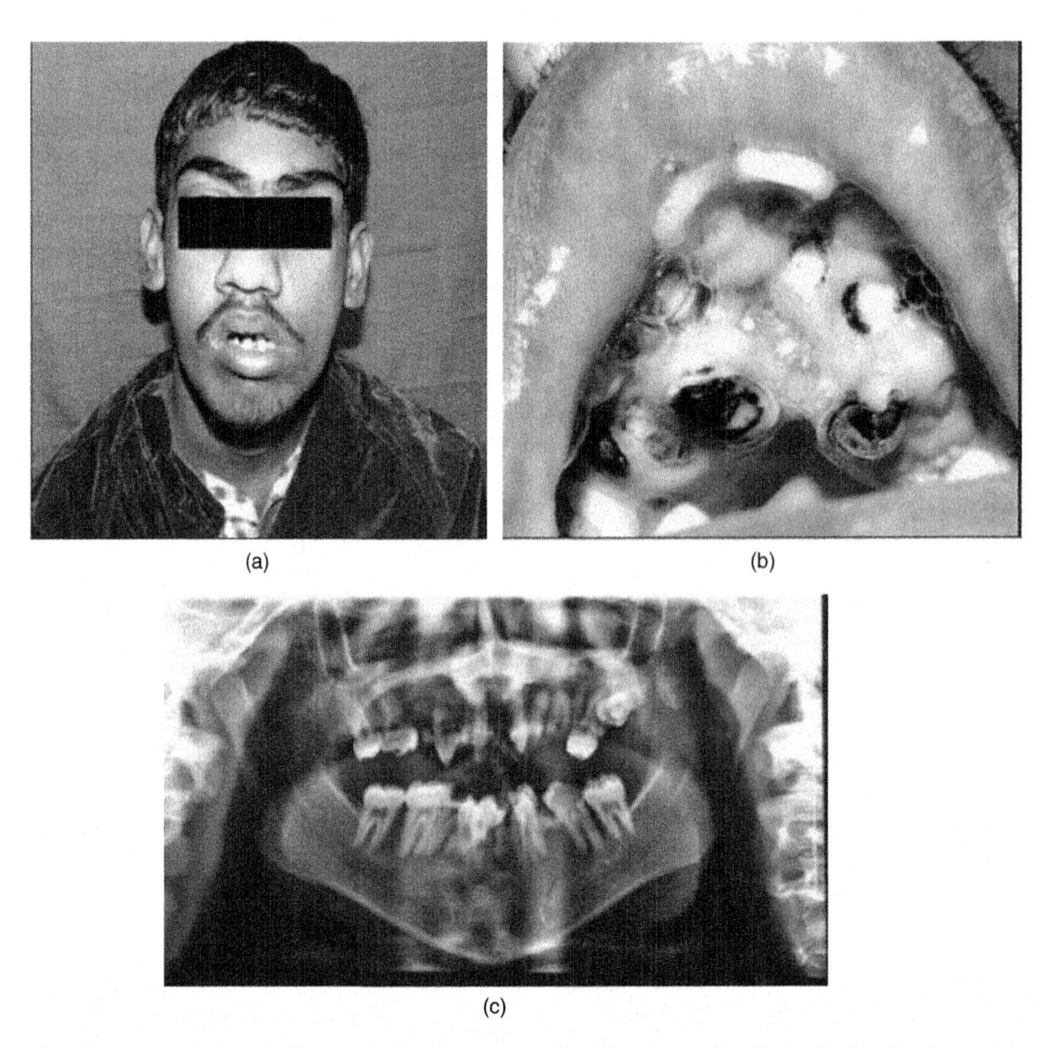

FIGURE 9.2 A. Extraoral view revealed an elongated and narrow face with a large forehead and prominent chin. B. Intraoral view revealed multiple supernumerary and carious teeth in relation to both upper and lower dental arches. C. Panoramic radiograph revealing the presence of multiple teeth in the anterior region of both maxillary and mandibular dental arches with congenital absence of teeth in relation to 18, 28, 38, and 48.

Delayed developmental milestones and delayed mental development are the first diagnostic sign of Fragile X syndrome as described by Purdon Martin and Julia Bel in 1943. There were no symptoms until the child attained 8 - 9 years due to a remarkable attenuation of clinical manifestations during childhood. The facial and body features will be more evident during puberty. In this case, the patient was of pubertal age and displayed attenuated facial features, mental retardation, autistic behavior, macroorchidism, mandibular prominence, ogival, and cleft palate. The presence of supernumerary teeth is rare, and their association with fragile X syndrome is infrequent. DNA analysis and PCR test would be highly reliable in this case but were not performed due to low socioeconomic status and their significant prognostic value in further management. The presence of orofacial alterations in mentally disabled male children provides an initial diagnostic clue for establishing fragile X syndrome in dentistry.

9.4.4 Duplication of Chromosomes and Trisomy

A female infant is born after 36 weeks of gestation by a spontaneous vaginal delivery with micrognathia, clenched fists with overlapping fingers, mild hypotonia, and rocker bottom feet. The infant also has structural heart defects such as large VSD, small ASD, PDA, and left superior vena cava. Echocardiography identified thickened redundant, dysplastic atrioventricular and semilunar valves, a tricommisural bicuspid, thickened aortic valve with right atrial and ventricular dilation. Head ultrasound had identified a mega cisterna magna variation in the central nervous system. At birth, the baby had a weak cry and poor respiratory effort with a decreased air entry. After having been shown to parents, the infant was shifted to an infant special care unit for further care. On admission, IV fluids were administered at 80 ml Kg/day. Babygrams revealed hazy lungs, enlarged heart, stomach distended with air, and subcutaneous emphysema. The patient's respiratory condition worsened after 40 hours of the birth, and she was pronounced dead. Mother was 37 years old and had undergone a ruptured ectopic pregnancy 14 years before. She had declined amniocentesis and aneuploidy screening during pregnancy (Boluda-Navarro et al. 2021).

A combination of genetic studies has diagnosed the possible reason. Due to insufficient cell quantity, karyotype and metaphase FISH studies could not be completed. I-FISH analysis was performed on the patient's peripheral blood with specific probes for chromosomes 13, 18, 21, X, and Y, respectively, displaying three hybridization signals for chromosome 18. The genetic factor was identified as trisomy of chromosome 18. Chromosomal microarray was performed using the patient's peripheral blood. It consists of 2,696,500 oligonucleotide probes and 7,43,304 SNP probes. The microarray was examined through Affymetrix cytoscan HD, and the results were analyzed using the chromosome analysis suite (Figure 9.3).

Karyoview of the patient has indicated the presence of three copies of chromosome 18 and a 2.40 Mb gain in chromosome 22q11.21 (Figure 9.4). The duplicated region contains 33 genes described in Online Mendelian Inheritance in Man (OMIM).

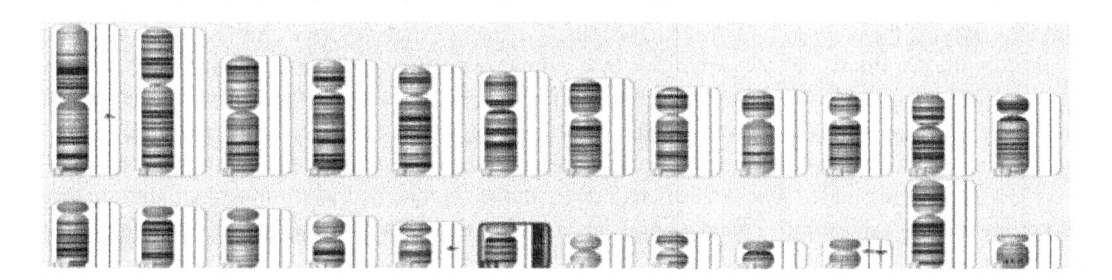

FIGURE 9.3 Chromosome Analysis Suite (ChAS) of a patient. Arrowed regions indicate a gain of chromosome 18.

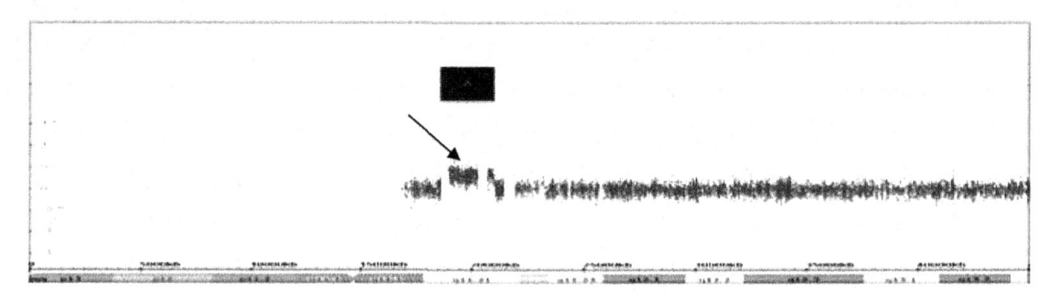

FIGURE 9.4 2.40 Mb gain in chromosome 22q11.21.

Among them, nine genes, PRODH (606810), SLC25A1 (190315), CDC45L (603465), GP1BB (138720), TBX1 (602054), COMT (116790), TANGO2 (616830), RTN4R (605566), and SCARF2 (613619) can cause disease. Despite the 22q11.2 duplication in the patient's genome, her phenotypic features such as rocker bottom feet, clenched fists, overriding fingers, heart malformations, and microcephaly are most consistent with trisomy18. Heart defects including VSD and ostium secundum ASD, micrognathia, muscular hypotonia, and microcephaly occur due to 22q11.2 duplication. Overlapping features of these two genotypic imbalances make it difficult to determine the genotype-phenotype correlations in this patient. Delineation of increased microdeletion and microduplication syndromes with high variability of unpredictable phenotypes will be difficult to analyze as in the present case with trisomy 18 interfered by small CNVs.

9.4.5 Chromosomal Abnormalities and Heteromorphism in Couples Leading to Recurrent Abortions

Loss of pregnancy without intervention before the 20th week is known as spontaneous abortion. The incidence of spontaneous abortions in diagnosed pregnancies is 15 - 20%. In couples with two or more abortions, the prevalence of both structural and numerical chromosomal abnormalities was identified. Couples with these abnormalities upon having pregnancy might lead to miscarriage, a stillborn child, or an infant born with severe congenital and mental disabilities. A descriptive study was conducted to investigate chromosomal abnormalities and heteromorphism in 75 couples with three or more abortions and 140 couples with two abortions before the third-month pregnancy, along with 40 control individuals. 5 ml of blood was cultured, and these couples' cultured peripheral blood lymphocytes were used for chromosomal analysis. After the planting and harvesting stages, the chromosomes at metaphase were stained with giemsa.

Among 65 patients with two consecutive abortions, four had chromosomal heteromorphism, and two had chromosomal abnormalities. Among 75 couples with three abortions, seven were found to have chromosomal heteromorphism, and four patients had chromosomal abnormalities (Chen, 2012). Among the couples with two consecutive abortions, six had abnormal children, and 38 had no children. Among the couples with three abortions, 7 had abnormal children, and 48 had no children. Regarding the chromosomal abnormalities in a woman with two abortions, one had chromosomal abnormalities, four had chromosomal heteromorphism, and 60 were normal. Among the women with three abortions, four had chromosomal abnormalities, and five were heteromorphic. Among 65 men whose wives had two abortions, one had abnormalities in his karyotype, and two were heteromorphic. Among the males whose wives had three abortions, one had a chromosomal abnormality, and two were heteromorphic. This data presents a higher chromosomal abnormality rate in women than in men.

Translocations in 22 couples with a history of recurrent abortion or birth of two or three stillborn children or malformed children were investigated. Translocation carriers in women with two

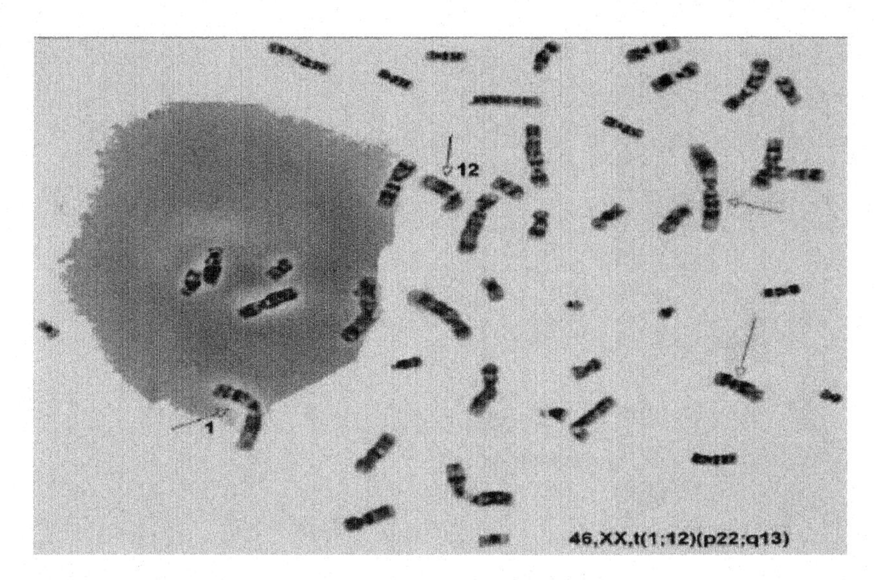

FIGURE 9.5 Translocation between chromosomes 12 and 1 in one woman with consecutive abortion.

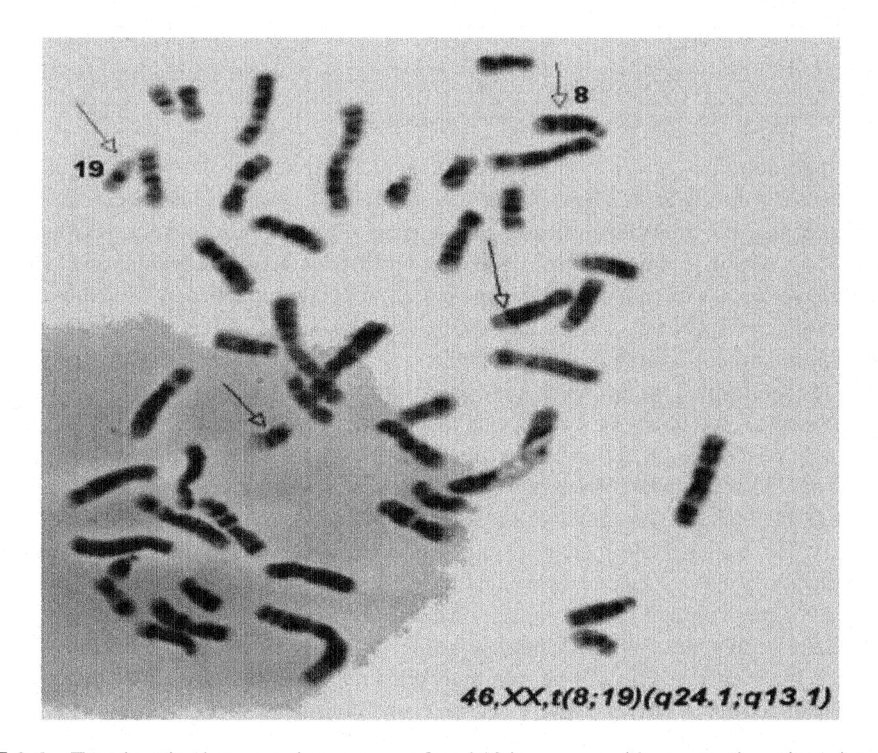

FIGURE 9.6 Translocation between chromosomes 8 and 19 in women with consecutive miscarriage.

abortions were 3.7% and with three abortions was 5.3%. Translocations of chromosomes 1,7,12, and 22 led to abortions, while translocation of chromosomes 5, 9, 14, and 21 led to the miscarriage or the birth of disabled children (Figures 9.5, 9.6 and 9.7). The real risk depends on the specific chromosomes, size of the segment involved in the rearrangement, sex of the transmitting parent, and mode of the assessment.

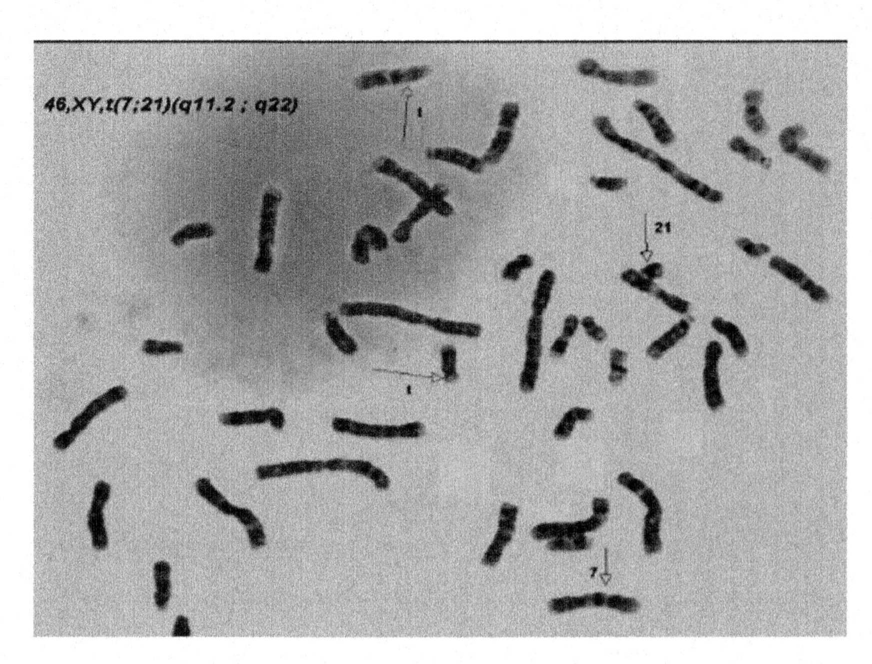

FIGURE 9.7 Translocation between chromosomes 7 and 21 in a man where his wife has recurrent abortion.

9.4.6 Congenital Anomalies in Newborn Infants

Congenital malformations are disabilities that occur due to structural, functional, behavioral, and metabolic disorders during birth. These malformations lead to a permanent change in the body structure during prenatal life. Mechanisms leading to congenital malformations are largely unknown but may be due to errors in embryonic cell proliferation, differentiation, migration, and programmed cell death. It may also occur due to differences in cell-to-cell communication, leading to a variety of effects with a large spectrum of severity. Genetic changes, including chromosomal abnormalities and mutant genes, account for 15% of the anomalies, and the environmental factors are responsible for 10% of the malformations, whereas a combination of these factors resulting in multifactorial inheritance produces 20 - 25% of the malformations.

A study was conducted at Assiut University Hospital. Upper Egypt for seven months from 1st March 2007 to 1st October 2007. Neonates with congenital anomalies were selected from 5000 newborn babies (3052 males, 1936 females, and 12 with a disorder of sex development) born in the department of Obstetrics and Gynecology. Complete maternal history regarding age, residence, parity, previous births, drug intake, acute and chronic illnesses during pregnancy, pregnancy complications such as diabetes mellitus and pre-eclampsia, obstetric complications such as multiple pregnancy, oligohydramnios, polyhydramnios, radiation exposure, and bleeding were taken. Family history had been recorded in the cases with similarity, chromosomal abnormalities, and stillbirths. Family pedigree was considered for all cases. Family pedigree was considered for all cases. Newborn babies were thoroughly examined for gestational age, birth weight, jaundice, pallor, cyanosis, and minor or gross external anomalies. Karyotype was performed for all malformed babies (103 cases). Statistical analysis was performed using a t-test and chi-squared test to identify the significance of the variables.

Among the 5000 live-born infants, 103 were found with congenital abnormalities with a prevalence of 65 males and 38 females. In the 103 babies, anomalies in the skeletal system had the highest frequency, followed by chromosomal abnormalities, circulatory system anomalies, CNS anomalies, genital organs anomalies, gastrointestinal tract anomalies, eye and ear anomalies, and urinary system anomalies. Normal karyotypes were detected in 75 cases, and chromosomal abnormalities were

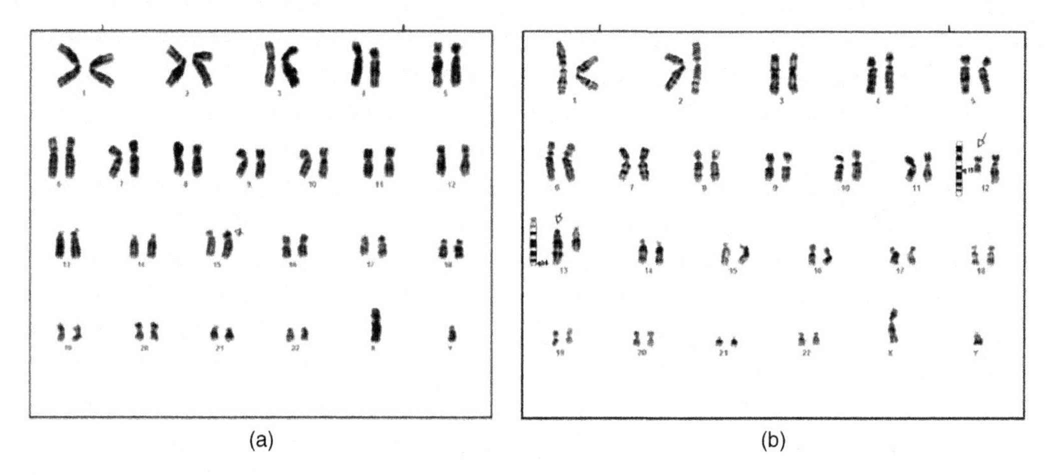

FIGURE 9.8 Karyotypes with chromosomal abnormalities f.

A: 46 X, Y, der (15;21 Q10;q10) +21.
B: 46 X, Y, t(12;13) (q15;q34).

found in 28 cases classified into numerical and structural abnormalities. Numerical abnormalities were present in 22 cases with Down syndrome in 16, Turner syndrome in 3, Edward syndrome in 2, and Patau syndrome in 1. Structural abnormalities were found in 6 cases with Down syndrome in 3, and iso Turner syndrome in 3 cases each. A balanced translocation of [(12;13) (q15; q14)] existing as dysmorphic features and undescended testis was observed in one case (Figure 9.8). A deletion in chromosome 9 (q11; q31) existing as disorder of sex development (DSD) was present in one case. Patients with normal karyotypes might be due to genetic syndromes, teratogenic symptoms, or the cause could be detected (Donald et al. 2017).

9.4.7 Tree Man Syndrome

The 'Tree Man' phenomenon made it into mainstream media when a man was reported with excessive growth of warts forming treelike bark mainly on hands and feet and less sporadically on the neck and face. The disease was apparent from childhood when the patient suffered a minor knee cut while playing. Small warts emerged from the close vicinity of the wound and later progressed rapidly to the whole body. The development of warts from a wound has attracted a professor from the University of Maryland to investigate him in greater detail. He was diagnosed with epidermodysplasia verruciformis (EV) and commenced treatment regimens in collaboration with Indonesian doctors to formulate the best therapeutic strategies for the disease. Epidermodysplasia verruciformis is a mouthful to say but saying the name is much easier than having it. EV is a rare autosomal recessive hereditary skin disorder. Lewandowsky and Lutz first described this disorder in 1922 as an epidermal nevus. In 1939, Sullivan and Ellis described a close relationship between EV and a high risk of skin cancer. Mutations in the EVER1 or EVER2 genes on chromosome 17 lead to EV. The disorder develops due to a defect of cell-mediated immunity, leading to an abnormal susceptibility of patients to a specific group of HPV genotypes known as EV HPV. Most of these are HPV 5 and 8; more rarely, HPV 14, 17, 20 and 47, and all have oncogenic potential. The condition usually has an onset between the ages of 1 and 20 but can occasionally present in middle age. In recent years, two susceptibility loci for EV mapped to chromosome 17 have identified two novel genes, EVER1, and EVER2. Mutations play a crucial role in the development of EV. Studies conducted to determine the function of EVER proteins have shown their ability to participate in the maintenance of zinc balance in cells. Biallelic null mutations of either TMC6 or TMC8 encoding EVER1 and EVER2 account for about half of the patients and families displaying EV. These genes are widely expressed throughout the body, including in leukocytes, but patients with null mutations display no consistent abnormalities of development or function of any subset of leukocytes.

A total of 501 patients have been described worldwide, with most cases sporadic. A few reports also indicate familial cases. If a person contracts EV, they will probably develop wart-like lesions all over the body. Warts sometimes progress into long bark-like tumors. Because of this, epidermodysplasia verruciformis is also called 'tree man syndrome'. There are only 600 cases a year and only a handful of the extreme tree-like cases exist. Although the inheritance of epidermodysplasia verruciformis is mostly autosomal recessive, sporadic, sex-linked, and autosomal dominant inheritance have also been described. In those cases of autosomal recessive inherence, there may be a history of consanguinity in the parents of the afflicted individual. Cases of atypical inheritance may also be associated with chronic lymphopenias. Regardless of the mode of inheritance, chronic infection with human papilloma virus (HPV) is the characteristic feature of this phenotype.

Abul Bajandar of Bangladesh (popular as tree man) has been suffering from epidermodysplasia verruciformis (EV) for the last two decades, which has led to a bark-like outgrowth on his body (Figure 9.9). So far, he has undergone 25 surgeries to remove these growths from his hands and feet. Nevertheless, the condition relapsed with some outgrowths becoming several inches long, giving unbearable pain. This condition is due to epidermodysplasia verruciformis linked with a high risk of skin cancer, leading to wart-like lesions on cover parts of the body. In a few cases, such as Bajandar, warts progress into a long bark-like tumor, giving the name 'Tree Man Syndrome'. The inherited conditions make the person highly susceptible to infections caused by human papilloma virus (HPV), one of the most common sexually transmitted infections.

Tree syndrome is a lifelong malady. EV is treated by alleviating the symptoms through medication and surgical procedures as there is no definite cure for the condition. Treatment includes the application of liquid nitrogen, topical ointments containing salicylic acid, and cryotherapy, to destroy the warts. Limiting the patient from exposure to the sun's rays is another vital aspect of preserving the patient's skin from developing cancer.

In another case, a 10-year-old girl from Bangladesh, Sahana Khatun, is being treated for the tree man syndrome. Ms. Khatum was admitted to Dhaka Medical College and Hospital with unusual

FIGURE 9.9 Abul Bajandar from Bangladesh, dubbed 'Tree Man' suffers from epidermodysplasia verruciformis (EV). *Courtesy: Times of India.*

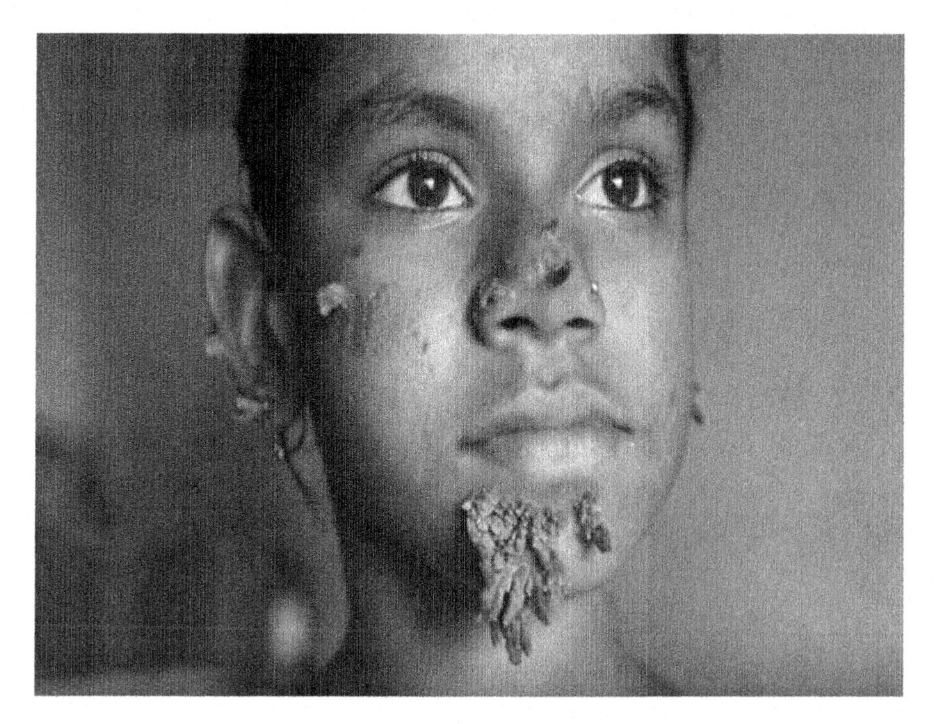

FIGURE 9.10 Sahana Khatun, a 10-year-old girl from Bangladesh suffering from Tree Man syndrome.

marks and growths on her face (Figure 9.10). Bark-like warts were growing on her chin, nose, and ear. It all began as rashes on her face eight years ago and these transformed into tree-like warts last year. The symptoms appear to be a variety of epidermodysplasia verruciformis, responsible for an unusual skin disorder.

9.4.8 Emanuel Syndrome

Emanuel syndrome is an inherited disorder that occurs due to an imbalance in chromosomes leading to the formation of supernumerary marker chromosomes. 9% of these supernumerary chromosomes are the derivatives of chromosome 22. [der (22)] is a supernumerary chromosome with the following karyotype:

47, XX+der (22)t(11;22) (q23;q11) in females, and
47,XY+der(22)t(11;22)(q23;q11) in males (rare).

This rare disorder was named Emanuel syndrome in 2004 (OMIM no. 609029) and 100 cases have been reported to date. Balanced male and female carriers have 0.7% and 3.7% chances of having der(22). Patients of Emanuel syndrome have distinct facial dysmorphism with a prominent forehead. They are also severely mentally disabled (Madan Gopal et al. 2013).

A young mother of 22 years of age was reported with a male neonate whose parents were consanguineous. Most of the anti-natal period of the infant was uneventful except for less-marked abdominal enlargement and low fetal movements. After the full-term vaginal delivery, the infant was tiny, with a 2.2 Kg weight and a head circumference of 32 cm. The infant had a remarkable facial appearance with a prominent forehead, dilated veins, widely separated eyes with down slanting palpable fissure, broad nasal bridge, prominent philtrum, bilateral large and low set ears with pre-auricular pit. The infant has a small penis of 1.5 cm with completely descended testes. Echocardiography revealed a

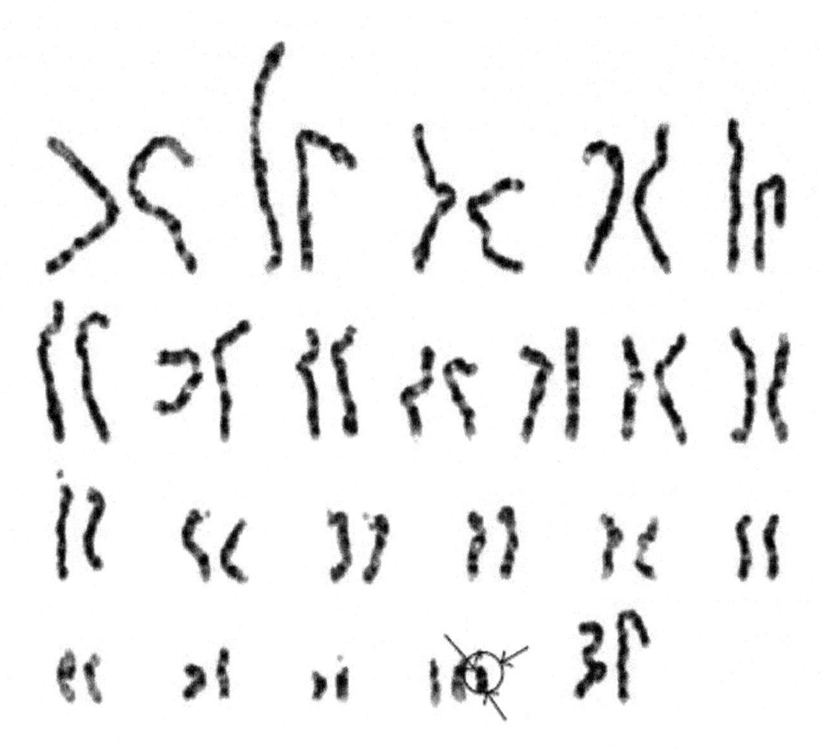

FIGURE 9.11 Infant's karyotype showing an extra supernumerary marker chromosome.

moderately large, subaortic ventricular septal defect, and his right kidney was missing. He had a mild hearing loss with an un-remarkable ophthalmological assessment. Karyotype analysis using G-banding has shown an extra supernumerary marker chromosome with supernumerary derivative (22)t(11;22) at the 550th band (Figure 9.11).

Karyotyping analysis of his parents to identify the origin of this supernumerary marker chromosome has shown that the mother was a balanced carrier 46, XX,t(11;22)(q23.3;q11.2) (Figure 9.12).

During the first three years, the infant had central hypotonia and developmental delay, with all of the parameters remaining well below the average growth. He showed developmental delay with all parameters that remained well below the third percentile.

9.4.9 Jacobsen Syndrome

It is a contiguous disorder due to a deletion in the long q arm of chromosome 11. The size of the deletion varies from 9 - 20 Mb. Jacobsen reported this disorder in 1973 in a family that inherited unbalanced translocation from a carrier. In most reported cases, the deletions occur in the q arm towards the centromere, giving a distinct phenotype. The clinical feature of Jacobsen syndrome is retardation in growth, psychomotor, facial dysmorphism, thrombocytopenia, and pancytopenia (Figure 13). There are malformations in the heart, kidney, gastrointestinal tract, genitalia, skeletal and nervous systems. The severity of the symptoms will increase in patients with increased deletion size. 85% of the cases occur due to deletions, and the other 15% are due to imbalanced segregations of the chromosomes (Rodríguez-López et al. 2021). Two phenotypes of this syndrome were identified, complete Jacobsen syndrome when BSX, NRGN, ETS-1, FLI-1, and RICS genes are deleted, and partial Jacobsen syndrome when the patient is haploinsufficient of some of these genes. To date, 200 cases of the disorder have been reported, with an estimated occurrence of 1 in every 1,00,000 births. The ratio of the female/male occurrence is 2:1.

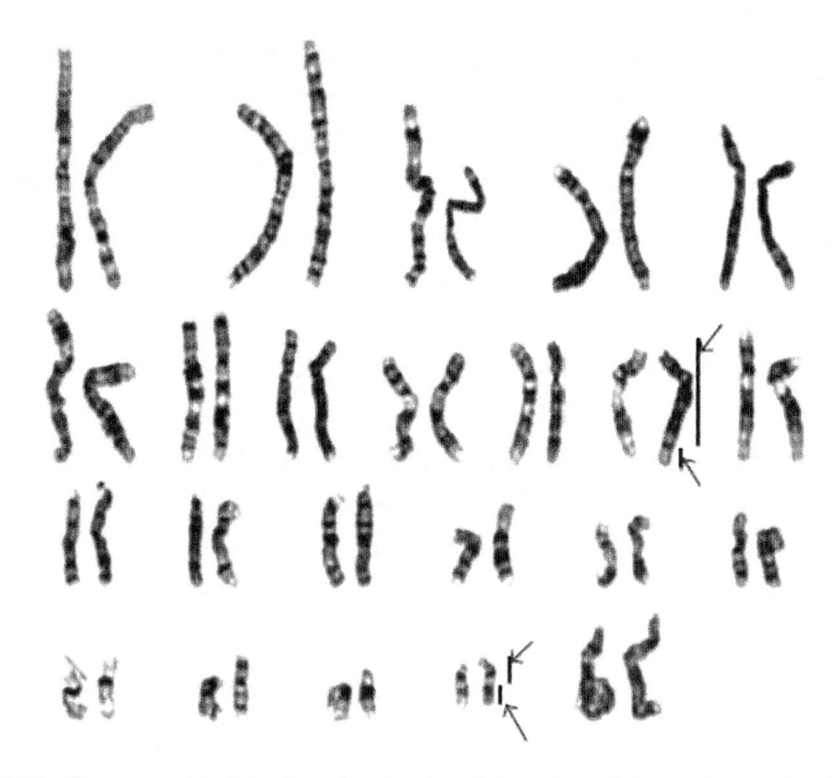

FIGURE 9.12 Karyotype of the infant's mother showing a balanced non-Robertsonian translocation between chromosomes 11 and 12.

9.4.9.1 Case Study

A six-year-old boy from Bulgaria was referred to The Laboratory of Molecular Genetics, Clinical Analysis Service, Consorcio Hospital, Valencia, Spain, for a genetic consultation due to facial abnormalities, loss of consciousness, and clonic movements. This boy was born following a normal pregnancy in a healthy, non-consanguineous Bulgarian family with a weight of 2.28 Kg, 45 cm length, and 31 cm head circumference. The phenotypic evaluation of the boy was performed using the HPO database and Face2Gene software. Abnormal facies with a small chin, low-set ears with posterior positioning, hypoplasia of the earlobes and lips, short neck, and short stature. Blood analysis has detected thrombocytopenia with platelet levels between 90-100x109 per L. He also suffered from bronchitis, otitis infections, and constipation and he underwent diaphragmatic hernia surgery at nine months and congenital cryptorchidism at three years. The boy also suffers from bronchitis, otitis infections, and constipation. MRI of the brain and diagnostic ultrasounds of the abdomen, thorax, heart, and encephalogram have revealed cryptorchidism with undescended testes of different sizes present in the inguinal canal and congenital heart malformations.

A high-resolution whole-genome SNP/CNV hybridization was performed with a Cytoscan HD array, with 3 million markers covering the whole genome. There was a heterozygous loss of a 6.8 Mb region in the chromosome 11 long arm. A partial Jacobsen syndrome was diagnosed in this boy due to the haploinsufficiency of three genes, ET-1, FL-1, and RICS, essential for the development of the disorder (Teresa et al. 2009). The deleted 6.8 Mb region encompasses 71 genes. Among them, 22 are OMIM genes (Figure 9.13). This deletion is de novo as both the parents had a normal karyotype confirmed by a high-resolution G-band analysis. CNVs were expected due to their existence in reference to population. Loss of heterozygosity (LOH) regions also did not reveal consanguinity or uniparental disomy.

FIGURE 9.13 11q24.3-25 deleted region in the chromosome 11 long arm of a Bulgarian boy.

9.4.10 Uniparental Disomy and Chediak-Higashi Syndrome

Uniparental disomy is a genetic condition in which both the homologs of a chromosome are inherited from one parent. Uniparental disomy might range from a single fragment to an entire chromosome. The rate of estimation of uniparental disomy is 1:3500 births. Uniparental disomy of some chromosomes does not adversely affect the phenotype, whereas a few can lead to abnormalities that occur through genomic imprinting. Uniparental disomy occurs as hetero disomy or isodisomy, depending on the transmission of the chromosomes. Uniparental disomy can also be partial if a single chromosome fragment is affected. (Scriver et al. 2001)

Chediak-Higashi syndrome is a rare immune disorder characterized by immunodeficiency and severe hemophagocytic lymphohistiocytosis due to altered cytotoxic lymphocytes, mis-sorting proteins in neutrophils, and loss of natural killer cell function. Patients with Chediak-Higashi syndrome have oculocutaneous albinism and thin silvery-white skin with a predisposition to bleeding, infections, and neurological dysfunction. This disorder is characterized by large pathognomonic cytoplasmic lysosomal vesicles in granulocytic cells resulting in lysosomal dysregulation (Yamazaki et al. 2003). Chediak-Higashi syndrome occurs due to mutations in the lysosome trafficking regulator (LYST) gene (also known as CHS1) localized on the long arm 1q42.1-1q42.2 of chromosome 1. The LYST gene consists of 53 exons, and highly conserved domains such as BEACH, PH, and WD40 at its C-terminal end. These domains are involved in vesicular transport, regulation of lysosome size, fission, and secretion (Cullinane et al. 2013). Different types of mutations in the LYST gene are responsible for the intensity of the disorder. Loss of function mutations is associated with severe forms, leading to death. Missense mutations are milder forms with increased chances of survival.

9.4.10.1 Case Study

A 3-year-old girl born to a non-consanguineous parent after a normal pregnancy and cesarean delivery was hospitalized with a high fever, cough, nasal secretions, and no appetite. Earlier, this girl suffered from a non-febrile respiratory infection and was treated with antibiotics. She has incomplete ocular albinism without pigmentation in the peripheral retina. She has no record of any serious infections or excessive bleeding. Even there is no family history of albinism or hypopigmentation or even deaths during the early stages of life. She was diagnosed to have pancytopenia and oculocutaneous albinism. In her peripheral blood and bone marrow samples, purple-colored inclusions were detected in lymphocytes and granulocytes. Intense vacuolization with purple inclusion was also noticed in other cells (Boluda et al. 2021). Cytogenetic analysis by C-banding and FISH has revealed a normal female phenotype (46, XX). Upon PCR amplification of the LYST gene cDNA and sequencing, the genotype revealed a homozygous frameshift c.8380dupT in exon 32, leading to a premature stop codon (NP_000072.2: p. Tyr2794Leufs*8) (Figure 9.14).

This variant might alter the functions of BEACH, PH, and WD40 domains, removing protein function (Figure 9.15). This phenotype is considered pathogenic by the American College of Genetics and Genomics.

The inheritance pattern of the disorder was determined through segregation analysis, revealing her mother as the carrier of the variant and her father as the wild type. Paternity tests and family

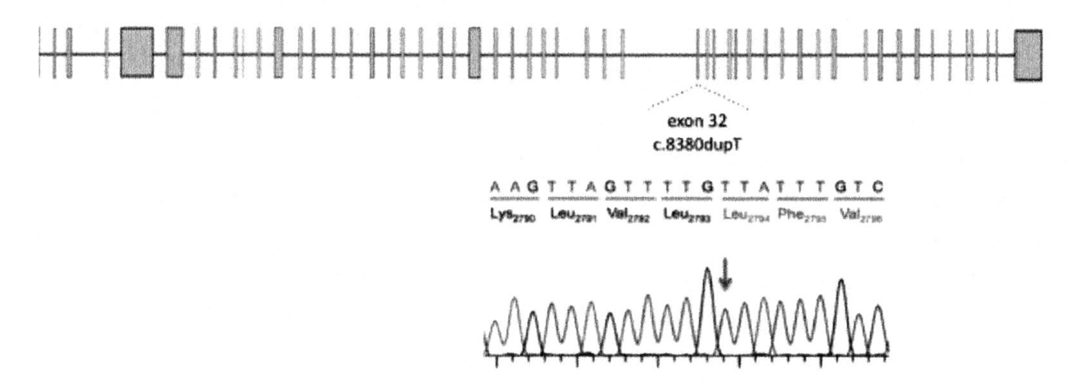

FIGURE 9.14 Structure of a LYST gene and frame shift identified in exon 32, giving rise to a premature stop codon.

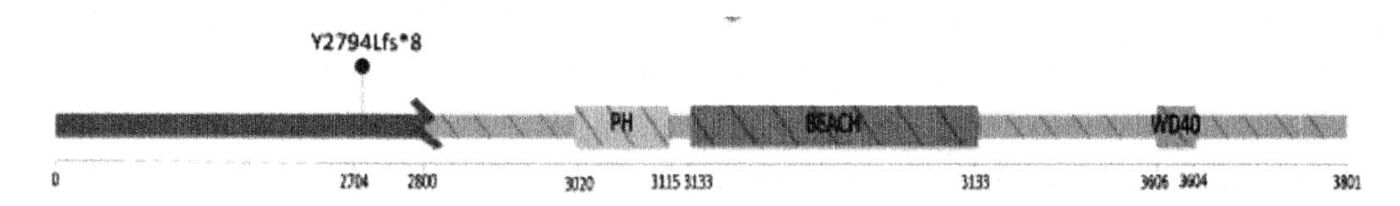

FIGURE 9.15 Mutation in exon 32 of a LYST Gene giving a truncated protein and altering the domain functions.

relationships were carried out using the SNP array. The chromosomal regions containing the LYST gene were assessed for copy number variations and loss of heterozygosity. No copy number variations were observed in the parent DNA. A loss of heterozygosity was found on the telomere of chromosome 1 long arm encompassing 25 Mb (1q4144) in the patient's DNA, accounting for a variation in the homozygosity of the LYST gene (Nagle et al. 1996). The patient was admitted to hospital repeatedly with frequent high fever, impairment, and splenomegaly of 5 - 15 cm. She was treated with dexamethasone and etoposide, to which she responded positively. After the remission, the patient had a stem cell transplant from an unrelated donor with HLA compatibility 10/10. After two years of the transplant, the patient maintained complete donor chimerism with normal physiological and neurological development without needing hospitalization.

9.5 CONCLUSION

Genetic disorders occur due to one or several abnormalities in an individual's genome. Most genetic disorders are inherited from one generation to another, affecting individuals' health. To date, more than 10,000 genetic disorders have been identified globally. Fueled by the technological advancement after the elucidation of the Human Genome Project, molecular diagnostics has offered tests to predict genetically inherited disorders. Further technological advancements will enable laboratories to offer genetic tests with increased efficiency at more affordable prices. A newborn genetic test is one of the most common modes of genetic testing, where every newborn is screened for genetic disorders with the hope of early detection to prevent the onset of symptoms and minimize the severity of the disease. Carrier testing tests the parents for carrying the alleles for genetic disorders and passing them to their children. This test is particularly for individuals with a family history of genetic disorders. Prenatal testing is to detect the changes in genes in the fetus. This test is offered to couples with an increased risk of having a baby with a genetic disorder. Pre-dispositional genetic testing identifies individuals with a risk of developing a genetic disorder before the onset of symptoms. This test can identify the mutations which increase the chance of developing a disorder in an individual. By providing these insights, medicine can identify the solutions for the individuals and their families struggling to resolve the genetic disorders and make life more healthy, easy, and simple.

REFERENCES

Boluda-Navarro, M., Iba´ ñez, M., Liquori, A., Franco-Jarava, C., Mart´ınez-Gallo M., Rodr´ıguez-Vega, H., Teresa, J., Carreras, C., Such, E., Zu´ ñiga, A´., Colobran, R., Cervera, J. V. (2021). Case report: partial uniparental disomy unmasks a novel recessive mutation in the lyst gene in a patient with a severe phenotype of che´diak-higashi syndrome. *Front. Immunol.* 12:625591.

Chen, H. (2012). Atlas of genetic diagnosis and counseling. Springer Science Business Media, LLC, New York.

Cullinane, A. R., Schäffer, A. A., Huizing, M. (2013). The BEACH is hot: A LYST of emerging roles for BEACH-Domain containing proteins in human disease. *Traffic* 14:749–766.

Donald, E. T., Hai, W., Jianli, D. (2017). Detection of an underlying 22q11.2 duplication in a female neonate with trisomy 18. *Lab Medicine* 48:372–375.

Favier, R., Akshoomoff, N., Mattson, S., Grossfeld, P. (2015). Jacobsen syndrome: advances in our knowledge of phenotype and genotype. *Am. J. Med. Genet. Part. C. Semin. Med. Genet.* 169: 239–250.

Gelehrter, T. D., Collins, F. S, Ginsburg, D. (1998). Principles of medical genetics. 2nd edition. Williams & Wilkins, Baltimore.

Grossfeld, P. (2017). Brain hemorrhages in Jacobsen syndrome: a retrospective review of six cases and clinical recommendations. *Am. J. Med. Genet.* 173: 667–670.

Lachiewicz, A. M., Dawson, D. V., Spiridigliozzi, G. A. (2000). Physical characteristics of young boys with fragile X syndrome: Reasons for difficulties in making a diagnosis in young males. *Am. J. Med. Genet.* 92:229–236.

Madan Gopal, C., Prashant B., Nitin, S., Dilip, D., Gururaj, N., Vijay, S. R. (2013). Derivative 11;22 (Emanuel) syndrome: a case report and a review. *Case Reports in Pediatrics.* 2013: 237935.

Mahowald, M. B., McKusick, V. A., Scheuerle, A. S., Aspinwall, T. J. (2001). Genetics in the clinic: clinical, ethical, and social implications for primary care. St. Louis: Mosby, Inc.

Nagle, D. L., Karim, M. A., Woolf, E. A., Holmgren, L., Bork, P., Misumi, D. J., et al. (1996). Identification and mutation analysis of the complete gene for Chediak–Higashi syndrome. *Nat. Genet.* 14:307–311.

Prayas, R., Arpanna, S., Jayanta Kumar, D., Prasanna Kumar, S., Jitendra Kumar, D. (2017). Fragile X syndrome: a rare case report with unusual oral features. *Contemporary Clinical Dentistry* 8:650–652.

Ridaura-Ruiz, L., Quinteros-Borgarello, M., Berini-Aytés, L., Gay-Escoda, C. (2009). Fragile X-syndrome: literature review and report of two cases. *Med. Oral. Patol. Oral Cir. Bucal.* 14: e434–439.

Rodríguez-López, R., Gimeno-Ferrer, F., Montesinos, E., Ferrer-Bolufer, I., Luján, C. G., Albuquerque, D., Cataluña, C. M., Ballesteros, V., Pérez-Gramunt, M. A. (2021). Immune deficiency in Jacobsen syndrome: molecular and phenotypic characterization. *Genes* 12: 1197.

Scriver, C. R, Beaudet, A. L., Sly, W. S., Valle, D. (2001). The molecular and metabolic basis of inherited disease. McGraw-Hill, New York.

Sherman, S., Pletcher, B. A., Driscoll, D. A. (2005) Fragile X syndrome: diagnostic and carrier testing. *Genet. Med.* 7:584–587.

Teresa, M., Concetta, S. P., Paul, G. (2009). Jacobsen syndrome. *Orphanet Journal of Rare Diseases* 4:9–18.

Thompson, M. W., McInnes, R. R., Willard, H. F. (1991). Genetics in Medicine, 5th Edition. W.B. Saunders Company, Philadelphia.

Yamazaki, S., Takahashi, H., Fujii, H., et al. (2003). Split chimerism after allogenic bone marrow transplantation in Chediak-Higashi syndrome. *Bone Marrow Transplant.* 31:137–140.

Glossary

N- Acetyltransferase polymorphism: The polymorphism of a drug metabolizing enzyme N-Acetyltransferase, a phase-II conjugating liver enzyme, that catalyzes the *N*-acetylation (deactivation) and *O*-acetylation (activation) of arylamine carcinogens and heterocyclic amines.

Achondroplasia: A bone growth disorder leads to short-limbed and disproportionate dwarfism, due to a mutation in the fibroblast growth factor receptor 3 gene (FGFR3), inherited in an autosomal dominant fashion.

Adeno-Associated Viral Vectors: The adeno-associated viral vector has provided an alternative means of long-term gene expression with reduced risk and adverse reactions. These are derived from a non-pathogenic parvovirus.

Alzheimer's Disease: A complex heterogeneous neurodegenerative disorder that occurs due to the interaction of genetic and epigenetic factors. It is an aggressive form of dementia also manifesting memory, language, and behavioral deficits.

Amyotrophic Lateral Sclerosis: Also referred to as Lou Gehrig's Disease, amyotrophic lateral sclerosis is a progressive degenerative disease that affects the brain and spinal cord neurons and eventually leads to their death.

Angelman Syndrome: Also known as happy puppet syndrome, caused by a deletion in the maternal chromosome 15 or uniparental disomy. This disorder is associated with extrapyramidal symptoms and signs such as unsteady gait with stiff upper arms, hyperactivity, seizures, severe mental retardation with the absence of mental speech, or paroxysms of inappropriate laughter.

Ataxia: An impaired coordination condition due to poor voluntary muscle movement.

Ataxia-Telangiectasia: A chromosome instability syndrome caused by defects in the ATM gene involved in a multitude of pathways regulating genome stability maintenance, cell cycle, and programmed cell death.

Autoimmune Disorders: Conditions in which a human immune system fails to detect its body parts and attacks them by releasing proteins known as autoantibodies.

Autosomal Dominant: Autosomal dominance is a pattern of inheritance that occurs if the gene in question is located on one of the 1-22 or non-sex chromosomes. 'Dominant' means a single copy of the disease-associated mutation is sufficient to cause the disease.

Autosomal Recessive: An autosomal recessive inheritance gene is located on one of the 1 - 22 autosomes or non-sex chromosomes. Autosomes do not affect an offspring's gender. 'Recessive' means that two copies of the gene are necessary to have the trait or disorder, with one copy inherited from the mother and the other from the father.

Autosomal Recessive Ataxias: These are neurological disorders characterized by degeneration or abnormal development of the cerebellum and spinal cord.

Cat Eye Syndrome: This is a rare genetic disorder in humans leading to the development of vertical colobomas in the eyes of the patients due to chromosome 22 partial tetrasomy.

Chediak-Higashi Syndrome: This is a rare immune disorder characterized by immunodeficiency and severe hemophagocytic lymphohistiocytosis due to altered cytotoxic lymphocytes, mis-sorting proteins in neutrophils, and loss of natural killer cell function.

Chimeric Antigen Recipient T Cell Therapy (CAR-T): This is immunotherapy, involving the manipulation or reprogramming of the patient's T-lymphocytes, to minimize the spreading of tumor T cells.

Circular RNA: These are molecules with a covalent bond between the 5' and 3' ends, making them closed and conferring resistance against degradation by exonucleases, thus making them more stable.

Classical Cytogenetic Techniques: Traditionally chromosomal aberrations were studied in human cells by banding techniques such as Q-banding, G-banding, R-banding, C-banding, T-banding, banding of the nucleolar organizing region, and high resolution banding, which allows precise identification of rearrangements in chromosomes.

Comparative Genomic Hybridization (CGH): A technique that allows the detection of changes to the copy number without cell culture.

Congenital Anomalies: Congenital malformations are the disabilities that occur due to structural, functional, behavioral, and metabolic disorders during birth. These malformations lead to a permanent change in the body structure during prenatal life.

CRISPR-Cas9: Is a fast, accurate, economically cheaper genome editing tool, which has created much excitement in the scientific community because of its efficiency. This novel genome editing tool was adapted from a naturally existing genome editing system in *E.coli*.

Cystic Fibrosis: One of the most severe life-shortening multisystem monogenetic disorders, caused by the dysfunction of chloride channels in exocrine glands.

Denaturing High-Performance Liquid Chromatography (dHPLC): A novel technology that can completely replace the gel electrophoresis step for the analysis of PCR fragments. This technology can be applied for genotyping, loss of heterozygosity (LOH) determinations, detection of DNA mutations, and polymorphisms.

DNAzymes: Catalytic DNA, involved in RNA cleavage, oxidative/hydrolytic DNA cleavage, nucleotide (both DNA and RNA) ligation, and DNA phosphorylation under in vivo as well as under in vitro conditions successfully.

DNA vaccination: The utilization of genetically engineered DNA for producing an immunologic response.

Down syndrome: Occurs due to the trisomy of chromosome 21 (each cell has three copies of chromosome 21 instead of two). In a few cases, Down syndrome is reported to occur when a fragment or a portion of chromosome 21gets translocated to another chromosome during spermatogenesis or oogenesis of the parents.

Emanuel Syndrome: This is a rare genetic disorder that occurs due to the inheritance of additional genetic material on chromosomes 11 and 22.

Factor V Leiden Mutation: This mutation is one of the most common and frequently inherited prothrombic conditions. The occurrence of this disorder is due to a mutation in factor V. It is a genetic disorder characterized by an inadequate anticoagulant response to Activated Protein C (APC) is a natural anticoagulant protein, which cleaves and inactivates procoagulant factors such as Va and VIIIa, leading to the downregulation of thrombin.

Friedreich Ataxia: This is a form of progressive autosomal recessive ataxia, sets during childhood and leads to ambulation between 12 - 15 years. It considerably reduces life span as it often affects the heart. Friedreich ataxia occurs due to an abnormal GAA trinucleotide repeat expansion in the frataxin gene located on mitochondria.

Fluorescent In Situ Hybridization (FISH): This is one of the most sensitive assays for the localization of specific nucleic acid sequences and detection of numerical chromosome abnormalities, structural chromosomal rearrangements, and cryptic abnormalities.

Fragile X Syndrome: A clinical and cytogenetic entity, known to be responsible for several intellectual disabilities due to the X chromosomal inheritance.

Genetic Analysis: The process of analyzing a heredity process either through pedigree or by applying techniques such as FISH.

Gene Expression Analysis: Also known as gene expression profiling, is a powerful technique applied in molecular biology, genetic engineering, and molecular diagnostics for checking the expression of thousands of genes simultaneously. In the context of cancer, gene expression profiling is used for diagnosing and classify tumors more accurately.

Gene Therapy: The process of inserting genes into the cells and tissues of an individual for the treatment of a disorder, such as hereditary during which a deleterious mutated allele is replaced with a functional one.

Gene Knockin: A process of target specific insertion of an exogenous gene at the desired locus in the genome to modify an endogenous gene's function. Gene knockin is applied for studying the fate of mutations that are responsible for genetic disorders in humans.

Gene knockout: A genetic technique by which an organism's genetic makeup is altered by engineering certain genes and making them inoperative.

Haploinsufficiency: Occurs when only 50% of the required protein is synthesized in a cell. Haploinsufficiency results from either loss of function mutation or one copy of the gene getting deleted.

Haplotype: This is a set of genes of an organism inherited from a single parent, namely, mother or father. The word haplotype is derived from haploid, describing the cells having a single set of chromosomes, and from the genotype, which provides the genetic makeup of an organism.

Hematopoiesis: This is the process of formation, development of blood cells in humans, other animals, and their maturity. In humans, blood and its components are formed in highly specialized cells with a short life cycle of about 120 days for human erythrocytes, around 5 days for leukocytes, 4 days for thrombocytes, and from several days to several months for lymphocytes.

Heritability: This is the proportion of a phenotypic variance attributed to its genetic variance and the extent to which individual genetic differences contribute to individual phenotypic differences.

Herpes Simplex Viral Vectors: These vectors are the derivatives of a human neurotropic virus with dsDNA of 150 Kb as the genome, which encodes 80 proteins. This virus is transmitted through direct contact and replicates on the skin and mucosal membranes before infecting CNS cells.

Hip Dysplasia: This is nine times more common in females than in males. One of the foremost environmental factors contributing to hip dysplasia is the baby's response to the mother's hormones during pregnancy.

Huntington Disease: A monogenic disorder, autosomal dominantly inherited due to increased CAG repeats in exon 1 of Huntingtin gene located in the 4p16.3 locus on the short arm of chromosome 4, producing Huntingtin protein.

Human Leukocyte Antigen (HLA) Typing: This technique is used to match recipients and donors for bone marrow and hematopoietic stem cell transplantation.

Hypervariable Minisatellite Repeats: Are the short tandemly repeated sequences of 6 - 50 bp, present throughout the human genome. The number of repeats at any given locus is highly polymorphic between individuals. Hypervariable minisatellite repeats are beneficial resources for DNA fingerprinting.

Hypophosphatemia: A type of vitamin D resistant rickets which is genetically transmitted through an X-linked dominant.

Hypervariable Region Polymorphism of mt DNA: In human mt DNA, the HV region is 11.1 Kb long, divided into hypervariable region I, II, and III regions of 400 bp each. These regions are the focal sources of polymorphisms. Sequence variations among individuals are largely found in these two HV1 and HV2 regions.

Jacob's Syndrome: XYY male syndrome is a genetic condition that occurs due to an extra copy of the Y chromosome in their cells, resulting in 47 chromosomes.

Jacobsen syndrome: Occurs due to deletions in the q arm of chromosome 11. Delayed development such as speech and neuromotor skills are the outcomes of the deletion.

Kleefstra Syndrome: Kleefstra syndrome (OMIM610253), also known as 9q sub-telomeric deletion syndrome, is a rare genetic disorder characterized by intellectual disability. Kleefstra syndrome is inherited in an autosomal dominant manner, either due to deletion of a submicroscopic region in chromosome 9 or a mutation in the gene regulating the euchromatin.

Klinefelter Syndrome: Most common genetic disorders occur in male humans due to changes in sex chromosomes. This disorder is a clinical manifestation that occurs due to an extra X chromosome in males.

Leber's Disease: Characterized by the loss of vision during adult life, due to mutations in the mitochondrial genome.

Linkage Analysis: This is a method applied to establish the carrier status of female 'at-risk' carriers and prenatal diagnosis. In many cases, linkage analysis has been replaced by mutational analysis but in a small number of families in whom the mutation cannot be identified, linkage analysis remains the only way for the genetic diagnosis of carriers.

Lynch Syndrome: This occurs due to the accumulation of microsatellite instability mutations in the coding regions of the *MSH2* and *MLH1* genes. People born with an inherited defect in these genes have a great risk of developing hereditary non-polyposis colorectal cancer.

Marfan Syndrome: It is an autosomal dominant disorder. This rare hereditary connective tissue disorder affects multiple parts of the human body.

Microsatellite Repeats: Also known as short tandem repeats, are among the most variable loci in the human genome. These repetitive sequences are usually di, tri, or tetranucleotide repeats with rare penta and hexanucleotide repeats.

MicroRNAs (miRNAs): A group of small non-coding RNA molecules of 21 - 25 nucleotides in length and are known to possess regulatory activity. They are partially complementary to mRNA, and their main function is to downregulate gene expression by translational repression, mRNA cleavage, and deadenylation.

Multifactorial Inheritance: The condition where multiple factors comprising of both genetic and environmental factors are involved in the inheritance of a trait. Here, a combination of genes from both of the parents and unknown environmental factors produces a trait or a condition.

Multiplex PCR: A widespread molecular biology technique for amplification of multiple targets in a single PCR experiment. In this multiplexing assay, more than one target sequence can be amplified by using multiple primers.

Multiplex Ligation-Dependent Probe Amplification (MLPA): A multiplex PCR technology that utilizes up to 45 probes, with each probe specific for an exon of a specific gene of interest for evaluating the relative copy number of each gene.

Muscular Dystrophy: This is where progressive motor paralysis occurs predominantly in males due to the decrease or absence of dystrophin protein.

Myotonic Dystrophy: A muscular dystrophy occurs due to a group of long term genetic disorders resulting in muscular impairment and prolonged muscle tensing.

Open Neural Tube Defects (ONTDs): During pregnancy, every human spinal cord begins its formation as a flat plate of cells, which further rolls into a tube called a neural tube. If an entire neural tube or a part of the neural tube fails to close, leaving an opening, it is known as ONTD. Spina bifida (Open spinal cord), anencephaly (Open skull), and encephaloceles (protrusion of the brain or its coverings through the skull) are the general open neural tube defects.

Phenylketonuria: A disorder caused by mutations in the PAH gene, encoding phenylalanine hydroxylase catalyzing a reaction of hydroxylation of phenylalanine to tyrosine. This metabolic disorder leads to intellectual disability and behavior problems.

Prader-Willi Syndrome: Occurs due to the loss of function of genes on chromosome 15, characterized by developmental delay, neonatal hypotonia, and postnatal development of hyperphagia with obesity, short stature, hypogonadism, and mild to moderate mental retardation.

Realtime PCR: This technique is based on the utilization of fluorescent labeled probes for detection, confirmation, and quantification of the PCR products, which are being generated in real time.

Retroviral Vectors: Retroviruses used in gene therapy cannot replicate inside their host cell and are replication deficient. First-generation retroviral vectors are Moloney Murine Leukemia Virus (MMuLv). Second generation retroviruses are human immunodeficiency virus type 1 (HIV-1), bovine immunodeficiency virus (BIV), feline immunodeficiency virus (FIV), and simian immunodeficiency virus (SIV).

Rett Syndrome: A genetic disorder occurring due to genetic defects in the MECP2 gene involved in several pathways that regulating genome activity.

Reverse Transcriptase PCR: This is a technique of cDNA synthesis from mRNA, mediated by the enzyme reverse transcriptase through a process known as reverse transcription, followed by the amplification of a specific cDNA by PCR. One of the most valuable and sensitive techniques for mRNA detection and its quantitation that is currently available.

Ribozymes: These are catalytic RNAs that can act as enzymes in the complete absence of protein. An enzymatic moiety can be included in the antisense oligonucleotide for cleaving the target RNA RNA–RNA duplex formation.

Robertsonian Translocations: Breakpoints occur in the short arms of two acrocentric chromosomes (both homologous and non-homologous) followed by their fusion to form dicentric chromosomes.

Schizophrenia: Is a chronic psychiatric disorder that occurs due to the heterogeneous genetic and neurobiological combination, heavily influencing the early brain development of the fetus, reflected in a variety of psychotic symptoms, including hallucinations, delusions, disorganization, motivational and cognitive dysfunctions.

Sickle Cell Disease: Also known as sickle cell anemia or depanocytosis, is a blood disorder characterized by the RBC that has acquired a rigid, abnormal sickle shape during deoxygenation, due to abnormal hemoglobin, leading to hemolysis and anemia.

SNP: One of the most common sources of variations in the human genome is the single base mutation substitutions of one nucleotide by another, known as single nucleotide polymorphisms (SNPs).

Single-Strand Conformational Polymorphism (SSCP): This is one of the simplest techniques applied for the detection of mutations responsible for human genetic disorders.

Sjogren's Syndrome: This is an autoimmune disorder characterized by salivary and lachrymal gland focal lymphoid cell infiltration and the involvement of other organs.

Spectral Karyotyping (SKY): Multicolour FISH (M-FISH), allows the staining of all chromosomes with different colours, utilizing the concentrations and combinations of different fluorescent dyes.

Spinal and Bulbar Muscular Atrophy: An X-linked neurological disorder occurs due to the expansion of microsatellite repeats in the ORF of the androgen receptor gene, leading to its repression.

Spinal Muscular Atrophy (SMA): This is a neuromuscular disorder, inherited as an autosomal recessive trait, characterized by symmetric proximal muscle weakness due to degeneration of the anterior horn cells of the spinal cord.

Suicide genes: These are genes responsible for cell death through apoptosis. They are used in gene therapy.

Tay-Sachs Disease: This is a severe autosomal recessive disorder that leads to the destruction of the brain and spinal cord.

Temporal Temperature Gradient Gel Electrophoresis (TGGE): This is an effective and sensitive technique applied to identify and characterize genetic polymorphisms of nuclear and mitochondrial genomes.

Testicular Feminization Syndrome: Also known as androgen insensitivity syndrome, it occurs at a frequency of about 1 in 65,000 males. People afflicted with this syndrome are chromosomally males (44A+XY), with physical development as females. They have female external genitalia, a blind vagina without a uterus. Testes may be present either in the labia or in the abdomen. These people are sexually sterile, and the condition cannot be reversed even by injecting the androgen (male hormone).

Thalassemia: Also known as Mediterranean anemia is a genetically inherited disorder that affects hemoglobin synthesis in humans.

Thrombophilia: Is a hypercoagulable state, characterized by the increased tendency to develop thrombosis, obstructive clot formation either in an artery (arterial thrombosis) or veins (venous thrombosis).

Trans-Species Polymorphism: Trans-species polymorphism (TSP) is the mechanism of identical allele occurrence in related species. Similarity developed due to either convergence or introgression is excluded from trans-species polymorphism.

Tree Man Syndrome: Epidermodysplasia verruciformis (EV), popularly known as 'Tree Man' syndrome is a phenomenon with excessive growth of warts forming treelike bark mainly on hands and feet and on the neck and face.

Turner Syndrome: This occurs due to either partial or complete loss of one of the sex chromosomes (monosomy). In most cases, the second X chromosome in females will be missing, leading to a variety of symptoms.

Uniparental Disomy: This is when both chromosomes or homologous regions or chromosomal segments are inherited from a single parent, namely, mother or father.

Waardenburg Syndrome: This is a set of genetic conditions responsible for hearing loss, hair, skin, and eye pigmentation, inherited in an autosomal dominant pattern.

Williams Syndrome: This is a chromosomal disorder caused due to microdeletions at 7q11.23 leading to an imbalance of 20 - 30 genes, adversely affecting several pathways.

Wolf-Hirschhorn Syndrome: This occurs not due to inheritance but due to deletions in the p arm of chromosome 4 known as 4p16, leading to intellectual disability and seizures.

X-Chromosome Inactivation: This is an imbalance in the human X chromosomes generated when the inheritance of two X chromosomes from both father and mother to daughter is corrected and controlled by a unique dosage compensation mechanism, known as X-chromosome inactivation, also known as Lyonization. It is the inactivation process of one X chromosome copy during the early stages of fetus development.

X-linked Ataxia: This is an emerging group of ataxia disorders characterized by a defect in the cerebellum due to mutations and imbalances in the genes located on the X chromosome.

X-linked Dominant: X-linked dominant is a rare way to pass down a trait or disorder through families. A single abnormal gene on the X chromosome can cause a dominant sex-linked disease.

X-linked Recessive: This is a mode of inheritance. Here, a mutation in a gene on the X-chromosome causes the phenotype to be expressed in males who are hemizygous for the gene mutation (with only one X chromosome) and in females who are homozygous for the gene mutation (they have a copy of the gene mutation on each of their two X chromosomes).

Y-Linked Disorders: The traits associated with the Y chromosome are known as holandric or Y-linked traits. At least 14 such traits have been identified because of their transmission from father to son. One such holandric trait is hypertrichosis pinnae auris responsible for abnormal long hair on the outer pinna.

Index

D

E

F

U

V

W

X

Y